THE BIOLOGY AND ECOLOGY OF TINTINNID CILIATES

THE BIOLOGY AND ECOLOGY OF TINTINNID CILIATES

MODELS FOR MARINE PLANKTON

Edited by

John R. Dolan
Marine Microbial Ecology
Laboratoire d'Océanographie de Villefranche-sur-Mer

David J.S. Montagnes
Institute of Integrative Biology
University of Liverpool

Sabine Agatha
Department of Organismic Biology
University of Salzburg

D. Wayne Coats
Smithsonian Environmental Research Center

Diane K. Stoecker
University of Maryland Center for Environmental Science

WILEY-BLACKWELL

A John Wiley & Sons, Ltd., Publication

Wiley-Blackwell is an imprint of John Wiley & Sons, formed by the merger of Wiley's global Scientific, Technical and Medical business with Blackwell Publishing.

Registered office: John Wiley & Sons, Ltd, The Atrium, Southern Gate, Chichester, West Sussex, PO19 8SQ, UK

Editorial offices: 9600 Garsington Road, Oxford, OX4 2DQ, UK
The Atrium, Southern Gate, Chichester, West Sussex, PO19 8SQ, UK
111 River Street, Hoboken, NJ 07030-5774, USA

For details of our global editorial offices, for customer services and for information about how to apply for permission to reuse the copyright material in this book please see our website at www.wiley.com/wiley-blackwell.

Library of Congress Cataloging-in-Publication Data
The biology and ecology of tintinnid ciliates: models for marine plankton / edited by John R. Dolan ... [et al.].
p. cm.
Includes bibliographical references and index.
ISBN 978-0-470-67151-1 (cloth)
1. Tintinnida. I. Dolan, John R. (John Richard), 1955–
QL368.T55B56 2012
579.4'9–dc23

2012010713

A catalogue record for this book is available from the British Library.

Wiley also publishes its books in a variety of electronic formats. Some content that appears in print may not be available in electronic books.

Cover image: Morphological diversity in the plankton. A view of a single microscope field (20× objective) of plankton net tow material from the Bay of Villefranche in January 2011. Image by J.R. Dolan.
Cover design by: Simon Levy Associates

Set in 9/11 pt PhotinaMT by Toppan Best-set Premedia Limited
Printed and bound in Malaysia by Vivar Printing Sdn Bhd

1 2013

CONTENTS

The color plate section can be found between pages 136 and 137

CONTRIBUTORS

Sabine Agatha
Department of Organismic Biology, University of Salzburg
Hellbrunnerstrasse 34, 5020 Salzburg, Austria
Sabine.Agatha@sbg.ac.at

Tsvetan R. Bachvaroff
Smithsonian Environmental Research Center
P.O. Box 28, 647 Contees Wharf Rd.
Edgewater, MD 21037, USA
bachvarofft@gmail.com

D. Wayne Coats
Smithsonian Environmental Research Center
P.O. Box 28, 647 Contees Wharf Rd.
Edgewater, MD 21037, USA
coatsw1@gmail.com

John R. Dolan
Marine Microbial Ecology
Laboratoire d'Océanographie de Villefranche-sur-Mer
CNRS Université Paris VI UMR 7093
Station Zoologique, B.P. 28
06230 Villefranche-sur-Mer, France
dolan@obs-vlfr.fr

Micah Dunthorn
University of Kaiserslautern
Department of Ecology
Erwin-Schroedinger Street
67653 Kaiserslautern, Germany
dunthorn@rhrk.uni-kl.de

Takashi Kamiyama
Tohoku National Fisheries Research Institute
Fisheries Research Agency
3-27-5, Shinhama, Shiogama, Miyagi 985-0001, Japan
kamiyama@affrc.go.jp

Jere H. Lipps
John D. Cooper Archaeology and Paleontology Center
Santa Ana, CA, USA
jlipps@fullerton.edu
and
University of California Museum of Paleontology
Berkeley, CA 94720, USA
jlipps@berkeley.edu

George B. McManus
Department of Marine Sciences
University of Connecticut
1080 Shennecossett Rd.
Groton CT 06340, USA
george.mcmanus@uconn.edu

David J.S. Montagnes
Institute of Integrative Biology
University of Liverpool
BioSciences Building
Crown Street
Liverpool, L69 7ZB, UK
dmontag@liv.ac.uk

Michèle Laval-Peuto
Chemin des Campons 144
06480 La Colle-sur-Loup, France
mpeuto.moreau@orange.fr

Richard W. Pierce
P.O. Box 132
North Attleboro
MA, USA
rw_pierce@verizon.net

Luciana F. Santoferrara
Department of Marine Sciences
University of Connecticut
1080 Shennecossett Rd.
Groton CT 06340, USA
luciana.santoferrara@uconn.edu

Paul Simon
Max Planck Institute for Chemical Physics of Solids
Nöthnitzer Strasse 40, 01187 Dresden, Germany
paul.simon@cpfs.mpg.de

Thorsten Stoeck
University of Kaiserslautern
Department of Ecology
Erwin-Schroedinger Street
67653 Kaiserslautern, Germany
stoeck@rhrk.uni-kl.de

Diane K. Stoecker
University of Maryland Center for Environmental Science
Horn Point Laboratory
2020 Horns Point Road
P.O. Box 775
Cambridge MD 21613, USA
stoecker@umces.edu

Michaela C. Strüder-Kypke
Department of Molecular and Cell Biology
University of Guelph
Guelph, ON N1G 2W1, Canada
michaela.strueder.kypke@uoguelph.ca

Chapter 1

INTRODUCTION TO TINTINNIDS

John R. Dolan[1]

[1]Laboratoire d'Océanographie de Villefranche-sur-Mer, CNRS Université Paris VI, Villefranche-sur-Mer, France

The Biology and Ecology of Tintinnid Ciliates: Models for Marine Plankton, First Edition. Edited by John R. Dolan, David J.S. Montagnes, Sabine Agatha, D. Wayne Coats, and Diane K. Stoecker.

1.1 WHY A BOOK ON TINTINNID CILIATES?

Tintinnids are "curious" because their morphology is unusual for planktonic organisms: it is a shelled cell. There are other groups of shelled organisms in the plankton, such as foraminifera and pteropods, but very few are as diverse as the tintinnids. They show an astounding variety of forms in their shells (loricae), and dozens of different types can be found in just a few liters of seawater (Dolan & Stoeck 2011). Nonetheless, all have a similar fundamental trophic role in the pelagic ecosystem: that of grazers at the base of the food web feeding on nanoplankton, for the most part. They are, therefore, ecologically united as planktonic ciliates, mainly consumers of nanoplankton, and morphologically and phylogenetically united as shelled, choreotrich ciliates. The biology and ecology of this group merit investigation if we are to understand how species-rich planktonic food webs work.

Tintinnids attracted the attention of some of the first microscopists, who were fascinated by such "shelled infusoria". From the early oceanographic expeditions, the variety of forms of their loricae catalogued from the plankton net-tow material served as an example of biodiversity in the plankton. As they are among the few protist organisms sampled using plankton nets, they were for a time considered a major component of the microzooplankton (zooplankton ranging in size from 20 to 200 μm), grazers of the smallest planktonic algae in the ocean (Beers 1982). Beginning in the 19th century and still today, tintinnids have been the focus of detailed studies that together form a considerable body of literature consisting of hundreds of reports. In recent years, monographic treatment of tintinnids has seen something of a renaissance with the appearance of comprehensive works on tintinnids of the South Atlantic (Alder 1999), Brazilian territorial waters (Gomes et al. 2006), Lebanese coastal waters (Abboud-Abi Saab 2008), and the Adriatic Sea (Krsinic 2010). The literature on tintinnids constitutes a wealth of information covering a surprisingly wide range of topics and approaches. This information has, however, remained scattered across hundreds of journal articles and monographs published over nearly 150 years.

This volume is an attempt to bring together material covering distinct aspects and topics to allow a comprehensive view of the biology and ecology of tintinnids. The ultimate goal in assembling this material is to provide a digest (a complete compendium would required a much longer book) showing the complexity of the different facets of a well-circumscribed group of single-celled eukaryotes, organisms of the marine plankton.

Although tintinnid ciliates are characteristic of the marine plankton, there are a few species found in freshwaters. Freshwater tintinnids often appear to be ephemeral in lake plankton, blooming and then virtually disappearing. In this book, we focus on marine plankton; freshwater forms will not be treated, but mentioned only in passing.

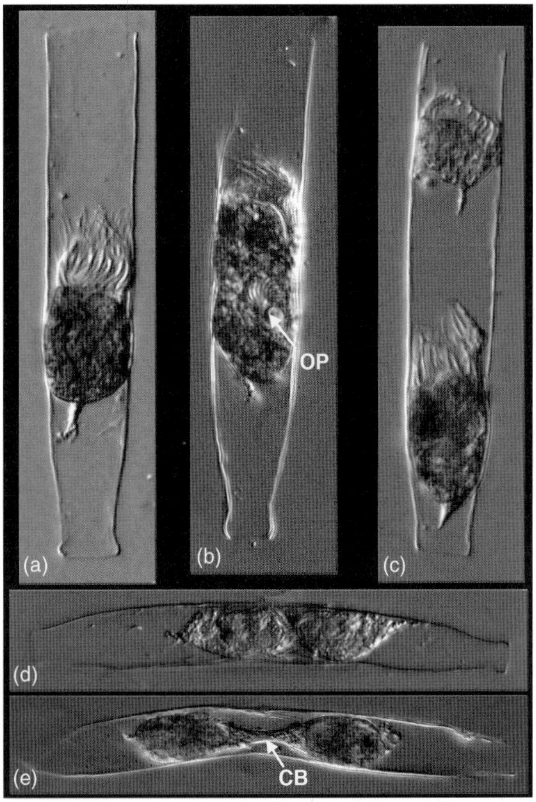

Fig. 1.1 Life-cycle stages in *Eutintinnus inquilinus*. (a) A trophont (feeding stage) cell. (b) An early stage of binary fission showing the oral primordium (OP), which will develop into a new mouth. (c) The final stage of cell division. (d) An early stage of conjugation. (e) A conjugating pair showing a well-developed cytoplasmic bridge (CB) through which nuclear material will be exchanged. Loricae are about 100 μm long. Images are of Lugol's-fixed cells; sample from a mixed culture provided by F. Lombard. For color version, see Plate 1.1.

1.2 WHAT IS A TINTINNID CILIATE?

Formally, tintinnids are ciliate protists of the class Spirotrichea, subclass Choreotrichida, order Tintinnida. Among the unicellular eukaryotes, the common characteristics of ciliates are possession of cilia during all or some part of the life cycle, nuclear dualism (macro- and micronuclei), and sexual recombination through conjugation, separate from reproduction which in tintinnids and most other ciliates is through binary fission (Fig. and Plate 1.1). Ciliates in the marine plankton are largely dominated by members of the class Spirotrichea, all of which have mouth cilia in the form of polykinetids (trichia), which are compound cilia that resemble bristles more than fine hairs arranged in a rough circle, more or less a spiral, around the oral cavity. Most marine planktonic ciliates are members of either the subclasses Choreotrichia or Oligotrichia. In both groups the oral polykinetids are well developed, serving both for locomotion and feeding. The subclass Oligotrichia (meaning "few trichia"), comprises forms with the oral polykinetids arranged in an open circle around the oral cavity and few cilia on the cell, such as in the familiar (often cone-shaped) oligotrichs of the genera *Strombidium* and *Laboea*. In contrast, species in the Choreotrichia have oral polykinetids arranged in closed circle around the oral cavity and usually relatively conspicuous rows of cilia on the cell surface. Choreotrich ciliates include species of the genera *Strobilidium* (often sphere-shaped), *Strombidinopsis* (often carrot-shaped), as well as species in which the ciliate cell is always surrounded by a shell (lorica), the tintinnid ciliates.

Besides the lorica, there are cytological characteristics that distinguish tintinnids from other choreotrichs (Fig. 1.2). The oral ciliature includes structures called tentaculoids, which appear to be related to prey capture (see Chapter 4). The posterior end of the cell is a contractile, stalk-like, peduncle which connects the cell to its lorica. The tintinnid cell surface has a distinct field of rows of short cilia, the somatic ciliature (see Chapter 3). These very brief characterizations are based on the detailed descriptions and analyses found in Lynn (2008), Lynn & Small (2000), and Corliss (1979).

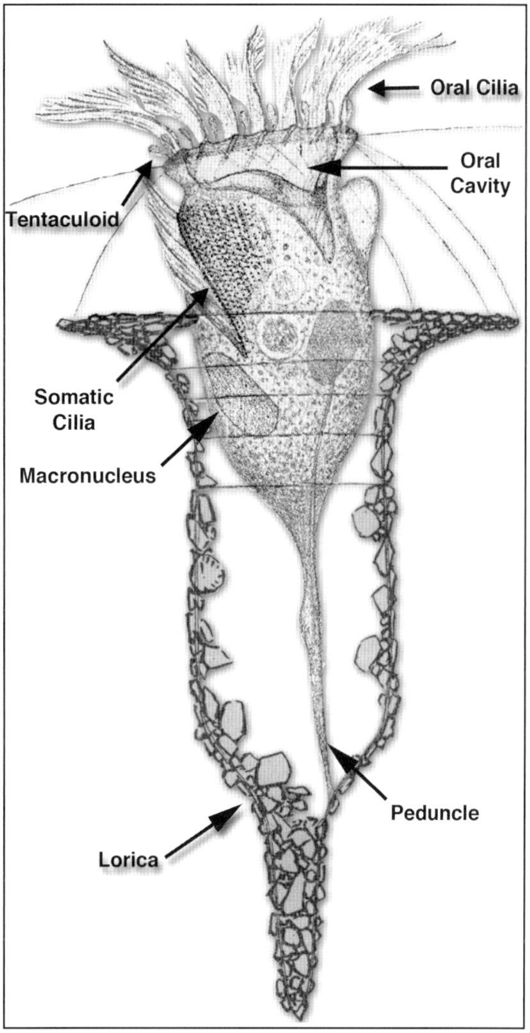

Fig. 1.2 Basic morphology of a tintinnid in a cut-away drawing of *Tintinnopsis campanula*, adapted from Fauré-Fremiet (1924). The oral cilia are comprised of oral membranelles arranged in closed circle around the funnel-shaped oral cavity. Interspersed between the oral membranelles are tentaculoids, structures presumed to be related to prey capture. The cell surface is covered in part by rows of short cilia, the somatic cilia. The macronucleus is often evident. The posterior end of the ciliate cell is attached to the lorica by a contractile peduncle.

1.3 THE LORICA AS THE DEFINING CHARACTERISTIC OF TINTINNID CILIATES

In addition to being the only choreotrich ciliates with loricae, tintinnids are also the only ciliates of the plankton with loricae. Although there are cytological characteristics that distinguish tintinnids, the defining one is its lorica or shell (although other cytological characters distinguish tintinnids from other choreotrichs or "naked oligotrichs": see Chapter 3). The lorica, although of various shapes ranging from tubular to vase- or bowl-shaped, is always open at one end, the oral end, and closed or tapered at the opposite, or bottom end. Within the lorica the tintinnid cell is attached near the bottom end by a contractile portion of the ciliate cell, a pedicle or peduncle (see Fig. 1.1). While feeding, the cell extends out of the lorica and the tintinnid is propelled mouth-end forward. If disturbed, the cell retracts into the lorica.

The contractile behavior and the lorica as the defining character were apparent in the text of the first description of a tintinnid by O.F. Müller published in 1778 (Box 1.1). The "creature" was described as *Trichoda inquilinus*; all ciliates were placed in the genus *Trichoda* by Müller; *inquilinus* – "lodger" or "occupant" – refers to the ciliate as the inhabitant of a structure. Remarkably, not only was the contractile behavior noted but also that reproduction was by cell division. Müller noted that the "mother cell" (the proter that retains the original mouth) abandons the lorica to the newly formed daughter cell. The form he described might be the species now called *Eutintinnus inquilinus* (shown in Fig. 1.1), known to attach to particles (Fauré-Fremiet 1908; Jonsson et al. 2004). The appellation *Tintinnus inquilinus* was given to the species by Schrank (1803) who removed it from *Trichoda*, giving it its own genus, without unfortunately noting why he chose the term "*Tintinnus*", which most commonly refers to the ringing of a bell.

Very early on it was noted that distinct types of lorica exist among tintinnids and that these may represent natural groups (Claparède & Lachmann 1858). A completely lorica-based taxonomy and phylogeny was introduced by Kofoid & Campbell (1929, 1939) and identifications are today still based on lorica morphology. However, culture studies and recent molecular phylogenies have shown that lorica characteristics do not always reflect phylogenetic relationships (see Chapters 2 and 3). Nonetheless, for practical reasons, tintinnids traditionally have been divided into two groups corresponding to easily distinguished lorica types: one with agglutinated (or agglomerated) loricae composed of particles and another, those with hyaline, generally transparent, loricae. The two lorica types very approximately correspond to different habitats in which tintinnid species with the different loricae are the dominant forms: coastal and open waters (see Chapter 10).

Mineral particles are used in the "agglutinated" loricae of species in the genera *Tintinnopsis*, *Stenosemella*, *Tintinnidium*, and *Leprotintinnus*. These forms are generally restricted to coastal waters. In some genera there are species that incorporate the hard parts of some microalgae. The coccoliths from coccolithophorid algae are used by some species (*Codonella*, *Dictyocysta*, *Codonellopsis*, and *Acanthostomella*) or the remains of diatoms (*Laackmanniella*). The arrangement of "agglutined particles" can be quite precise. For example, in *Codonella elongata*, the coccolithophores are all oriented in the same "face-out" direction (Fig. 1.3). However, not all species show precision in particle use. Examples of the variety of forms with agglutinated lorica are shown in Fig. and Plate 1.4. These species generally dominate tintinnid assemblages in coastal waters but some can be found in open waters.

"Hyaline loricae" groups tintinnids of a very large variety of shapes, sizes, and architectures whose sole common characteristic is a lack of extraneous particles in the lorica. The morphologies range from the transparent small (50 µm diameter) bowl-shaped *Ascampbellia tortulata* to the large, conical *Cyttarocylis cassis* which shows a sculpted surface. Although some hyaline lorica species are found in coastal waters (e.g. species of *Helicostomella*, *Favella*) typically they are the dominant forms in open-water tintinnid assemblages. A sampling of the diversity of hyaline-lorica forms is shown in Fig. and Plate 1.5.

The term "lorica" refers to the armor or protective clothing worn by roman soldiers. Dujardin (1841) credits Ehrenberg (1832) with using the word in his Latin text to describe the protective structures, *Panzer* in the German text, of tintinnids as well as other protists and rotifers. Indeed, the hard structures of many different protists appear early in the fossil record and likely evolved as protection against predation by other protists (Porter 2011). Among planktonic organisms, the seemingly odd morphologies of present-day forms are thought to be the result of a "watery arms race" between prey and predator (Smetacek 2001; Hamm & Smetacek 2007). In contrast, Kofoid & Campbell (1939) stated that the lorica probably did not evolve as protective device because it affords little protection against large metazoan predators such as copepods. However, tintinnids are subject to predation from a very wide range of predators (Chapter 5) as well as

BOX 1.1 Translation of the 1779 Latin text of O.F. Müller

Trichoda inquilinus (T. lodger): a sheathed trichoda, with a hyaline, cylindrical, shell and a little footstalk

This is among the smallest creatures, and with the slow work of a simple lens, the eye observes a very small point with great mobility. The shell, whether egg-shaped, oblong, smooth, is so glassy transparent that otherwise invisible things might be seen. For this creature is a swimmer, which you might say has a pivoted footstalk . . .

I have been able to observe these things for a long time with my eyes unaided, being apparent in great numbers. The most apt name should account for the quick motion and the rotation of the hairs. Often the organ of the hairs is spread out resembling a comb, then the more distinct, outermost, blackish teeth, in the hyaline jelly of the true center, which are show various movements, are seen.

The clear footstalk runs through the middle of the sheath, fixed though at the wide base. Often it is contracted, the hairs are together, the little head is submerged, and the side strings of the sheath stand forth like a pregnant creature stretching. Slowly the head retreats back, with the hairs, whether thin tentacles or spread about, extend, barely occupying the entrance of the sheath, intently moving for nutrition. I saw it withdraw within the sheath and be spread out at the top of the head.

Having been disturbed by dust (and I don't know where the center is), it becomes very bare, it adheres to very many things and moves about slowly; it swims around vehemently with the head clinging on to the outside of the sheath with the work of the hairs; if it is clinging on to the base of the sheath, fixed only a little, whether extended in a perpendicular or oblique fashion, an alternate movement of the top of the head follows, moving in water continuously or less so.

I have found two swimming together lengthwise. Whether it was at the base of the sheath, or lying higher up, or lower than the top, and the head was pulled about in every sense: at length it tore itself apart, front from back.

I saw a certain one, the heads of which were more than threefold, and they fully occupied the whole sheath even with the footstalk inconspicuously or for a short time; also two by two the little creatures lived in the sheath with their own footstalks.

An extension developed, while I watched the little creature growing. The head is divided in the middle and the bottom at the base of the footstalk, while what happens must be due to the workings of the hairs. By what means the middle of the sheath is truly divided I know not even though keeping them in clean and pure sea water, even in entirely non-renewed water, through the whole of winter 1776, 1777 and 1778.

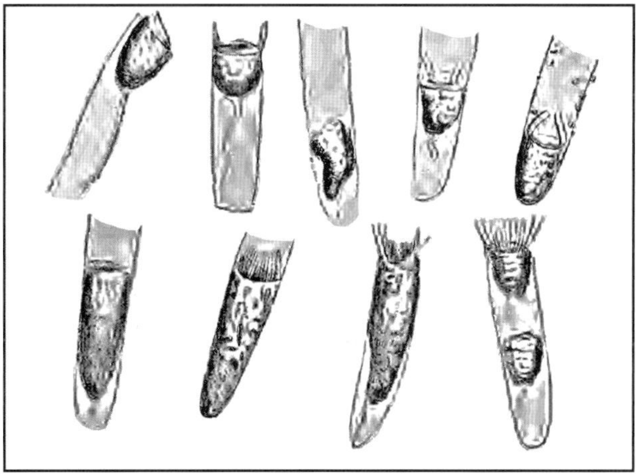

Fig. B1.1 The lodger occupies various places; small, scaly sheaths called Monads are found together and sticking together in the fifth image. A series of fertile mothers having taken positions to divide, laden and mature, and at the last image a mother is now swimming away. *O.F. Müller (1779) pp. 8–9 (Latin translation by B. Scott, University of Liverpool).*

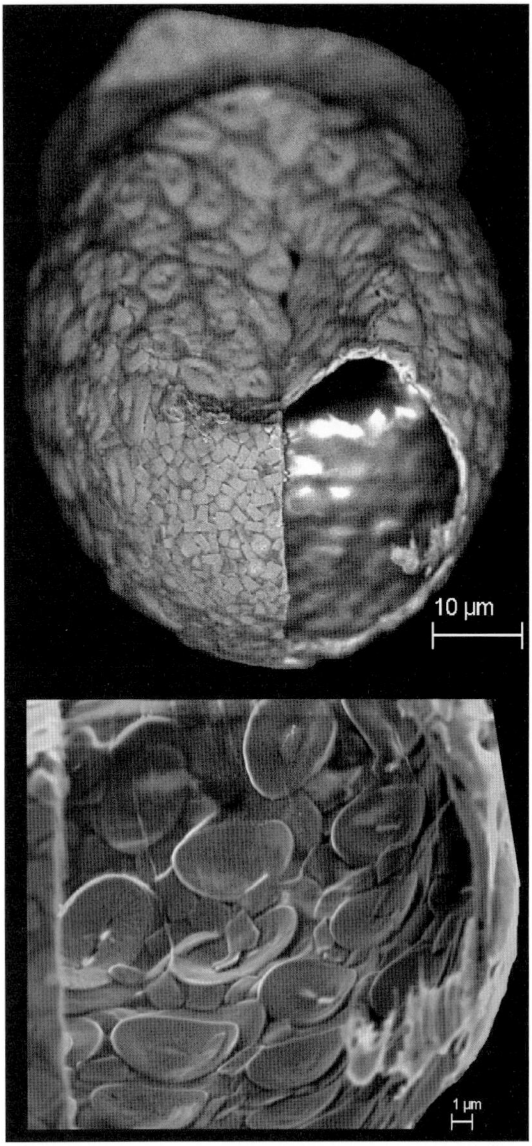

Fig. 1.3 Scanning electron micrograph of a specimen of *Codonella elongata* from the Bay of Villefranche. The upper panel shows the lorica with a bottom portion cut away with a high-energy electron beam. The lower panel shows the interior surface of the lorica. Note that the lorica is agglomerated with the coccolithes of *Helicosphaera carteri* oriented as they are on the living algae. Images are courtesy of François Guyot and Charles Bachy. An image of a live specimen of *C. elongata* is shown in Fig. 1.4.

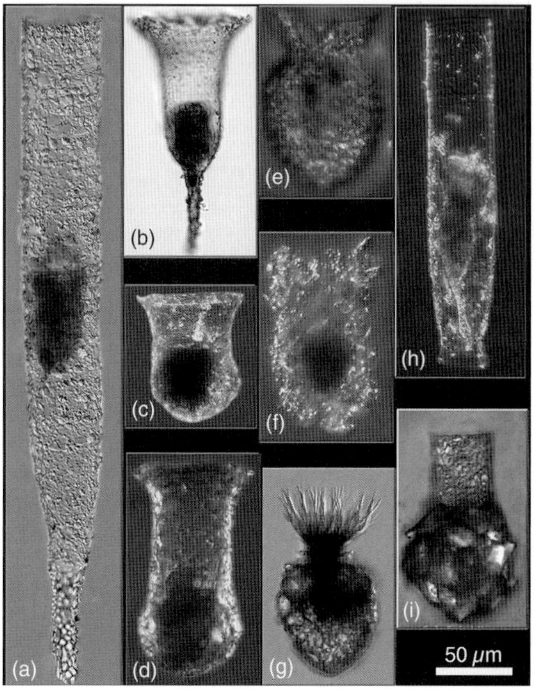

Fig. 1.4 Some tintinnid species with agglutinated loricae: *Tintinnopsis radix* (a), *T. campanula* (b), *T. dadayi* (c), *T. spiralis* (d), *Codonella elongata* covered with coccoliths (e), *Tintinnidium* sp. (f), *Stenosemella ventricosa* (g), *Leprotintinnus pellucidus* (h), and *Codonellopsis schabi* (i). Species names are attributed based on lorica morphology. All the specimens are Lugol's-fixed cells except for *Codonella elongata* which was a live specimen. For color version, see Plate 1.4.

parasitic infections (Chapter 6). In reality, the identity of the major source of mortality for tintinnids is unknown so it is difficult to assess the possible use of the lorica as a protective device.

Many explanations other than "armor" for the lorica have been suggested. These include that of a flotation aid in the case of the hyaline lorica (Kofoid 1930; Campbell 1926), alternatively as aid in quickly sinking away from predators in the case of the agglutinated lorica of *Stenosemella* (Capriulo et al. 1982), and maintaining directionality in swimming in the case of the elongate nail or spike-shaped lorica of oceanic forms (Kofoid & Campbell 1939). A lorica, like projecting spines or trailing structures in other small zooplankton, may reduce swimming speed while enhancing the capture of food particles by increasing fluid motion

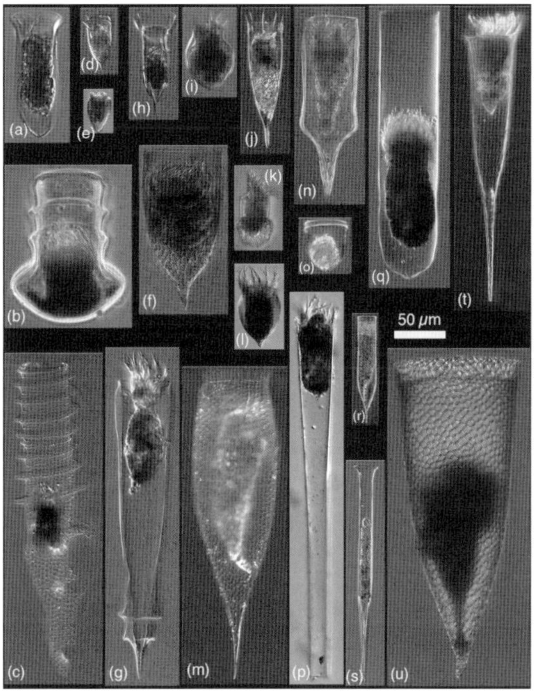

Fig. 1.5 Examples of tintinnid species with hyaline loricae: *Amphorides quadrilineata* (a), *Amplectella collaria* (b), *Climacocylis scalaria* (c), *Acanthostomella conicoides* (d), *Protorhabdonella simplex* (e), *Epiplocylis blanda* (f), *Xystonellopsis paradoxa* (g), *Ormosella trachelium* (h), *Proplectella elipisoida* (i), *Dadayiella ganymedes* (j), *Dictyocysta lepida* (k), *Metacylis mediterranea* (l), *Parafavella parumdentata* (m), *Parundella messinensis* (n), *Ascampbellia tortulata* (o), *Eutintinnus stramentus* (p), *Undella hyalina* (q), *Helicostomella subulata* (r), *Salpingella acuminata* (s), *Rhabdonella spiralis* (t), and *Cyttarocylis cassis* (u). All the specimens are Lugol's-fixed cells. For color version, see Plate 1.5.

around the oral cilia (Emlet & Strathmann 1985). The lorica can also serve as an attachment device. Jonsson et al. (2004) found that in some species of *Eutintinnus*, the tintinnid can reversibly attach its lorica to detrital particles thus increasing its feeding rate owing to the "tethering effect" (Strathmann & Grünbaum 2006) and reducing its susceptibility to copepod predation. Lastly, the lorica may furnish protection against ultraviolet radiation (Armstrong & Brasier 2005), allowing tintinnids to exploit the near-surface waters. There appears no reason not accept the possibility that the variety of loricae found among tintinnids may serve a variety of functions.

1.4 HISTORY OF TINTINNID STUDIES

Early mentions of tintinnids are mostly in taxonomic treatises and consist of little more than simple listings in species catalogues of "infusoria" (i.e., Müller 1776, Schrank 1803; Ehrenberg 1832; Dujardin 1841). The first detailed consideration of tintinnids was by Claparède & Lachmann (1858–1860, part 1, pp. 192–221) who provided notes on the morphology and ecology of most of the 17 known species, and placed them all in a single genus, *Tintinnus*. They remarked on the abundance of tintinnids in marine waters compared with their rarity in freshwaters, and the diversity of lorica types. The appearance of more marine biological studies of tintinnids marked the last quarter of the 19th century.

In 1873 Ernst Haeckel published the first paper devoted exclusively to tintinnids. He not only described new species from waters off Messina (Italy) and Lanzarote (Canary Islands), but also the development and release of what he thought were reproductive spores. We now know that he was the first to describe parasites in tintinnids (see Chapter 6). A few years later appeared the observations of Herman Fol (1881, 1883, 1884) on tintinnids from the Bay of Villefranche-sur-Mer, describing investigations into the chemical nature of the lorica (see Chapter 2) and some new species. These reports were followed by a series papers from the Zoological Station in Naples first by Géza Enzt, senior (1884, 1885) and then Eugene von Daday (1886, 1887) who published the first monographic review of tintinnids (Daday 1887). The number of known species rose to 70 with the monograph; he also included a key to the species listed (many of which were new) and reviewed reports of anatomy and what little was known of physiology as well as biogeography.

Detailed observations on feeding and fine structure did not begin until the early 20th century, perhaps because maintaining tintinnids in the laboratory proved difficult (as Müller had remarked back in 1778). Emmanuel Fauré-Fremiet published in 1908 the first extensive report on a tintinnid, now known as *Eutintinnus*, possibly the same species Müller observed, based on both living and fixed specimens. This was followed by Géza Entz junior (1909b) who notably included the study of fine-structure based on thin sections of what is now known as *Favella*. These studies of tintinnids, based on specimens from near-shore waters, were soon eclipsed by the development of biological oceanography. In terms of tintinnid studies, oceanography turned the focus toward the open sea, first with regard to taxonomy and new species, and later to the role of tintinnids in marine ecosystems.

Distinct forces prompted the development of biological oceanography in the late 19th and early 20th centuries. These ranged from the primary one of the need to understand the dynamics underlying exploited marine populations, especially herring and cod, to nationalistic desires to show mastery of the seas and science (Mills 1989). Regardless of the precise reasons behind the great oceanographic expeditions of the early 20th century, the study of tintinnids profited greatly. Although singularly neglected by the English Challenger Expedition, in the subsequent oceanographic expeditions of many countries tintinnids were granted the same attention as that given to other planktonic groups such as diatoms, dinoflagellates, and copepods. The major publications include Karl Brandt's monographs from the German 1899 Plankton Expedition (Brandt 1906, 1907), the Hans Laackmann reports from the German South Polar Expedition of 1901–1903 (Laackmann 1907, 1910), Alphonse Meunier's reports from the Duke of Orleans Arctic expedition of 1907 (Meunier 1910), and Eugen Jörgensen's monograph from the Danish Thor Expedition to the Mediterranean in 1908–1910 (Jörgensen 1924). These largely taxonomic treatises, cataloguing species occurrences and describing new forms, reached their apex with the reports of Charles Kofoid and Arthur Campbell on material from the Agassiz Expedition (Fig. 1.6) of the USA to the Eastern Tropical Pacific in 1904–1905 (Kofoid & Campbell 1929, 1939).

In describing the species found in material from the Agassiz expedition, Kofoid and Campbell produced two monographs. The first, "A conspectus" (Kofoid & Campbell 1929), was a compendium of previously described forms as well as new species from the expedition material and other samples they collected. At that time the literature contained descriptions over 1750 different forms, variously described as species, subspecies, and varieties of tintinnids. The "conspectus" ascribed species status to 705 forms in 51 genera, all based uniquely on features of the lorica (Kofoid & Campbell 1929). Some years later, Kofoid and Campbell published a second report, which concerned exclusively the tintinnids from the Agassiz Expedition. In this second monograph they re-named and re-arranged many tintinnid genera and described some new species yielding a new total of 726 species (Kofoid & Campbell 1939). Other notable studies include those of Hada working in Japanese waters and the Western Tropical Pacific (1937, 1938), and Campbell's report on material from the last voyage of the Carnegie throughout the Pacific (1942). Although such efforts in descriptive zoology continued in the latter half of the 20th century, their pace slowed considerably with a shift in focus toward the role of tintinnids in marine food webs.

A turn toward ecological studies was belated because tintinnids and other small protist grazers had long been suspected to be of importance. Naive views on marine food webs can be understandably summarized by the saying "Big fish eat little fish" (Fig. 1.7). However, by the late 1800s it was recognized that, rather than plant and organic matter supplied by rivers feeding small fish, plankton formed the base of marine food webs. Furthermore, plankton production was likely in some manner linked to exploited fish populations. This view provided the scientific justification of Victor Hensen's Plankton Expedition in 1889, which is generally considered as the first campaign of biological oceanography (Mills 1989). The studies of Kofoid (1897) and Lohmann (1901) had shown that most of the biomass in the plankton was missed using plankton nets, even the "fine silk" used to sample for tintinnids and other small plankton.

By the 1920s careful studies of coastal fish populations and food webs, for example the herring food web, revealed the complexity of predator–prey relationships in the plankton (Fig. 1.8 and see Chapter 5, Fig. 5.4). Thus, by the 1930s, the planktonic community was known to be dominated, at least in terms of biomass, by very small organisms and the community as a whole was characterized by complex predator–prey relationships. Nonetheless, studies on the ecology of tintinnids and other small grazers of the plankton did not begin in earnest until the 1960s.

The period of relative inactivity from the 1930s to the 1950s was attributed by John Beers (1982) to the development of pigment methods in phytoplankton studies which replaced the use of microscopic analysis of water samples. Meanwhile, zooplankton studies continued to rely on net sampling. The effect was then that microzooplankters, such as tintinnids, were overlooked or ignored because few people were examining

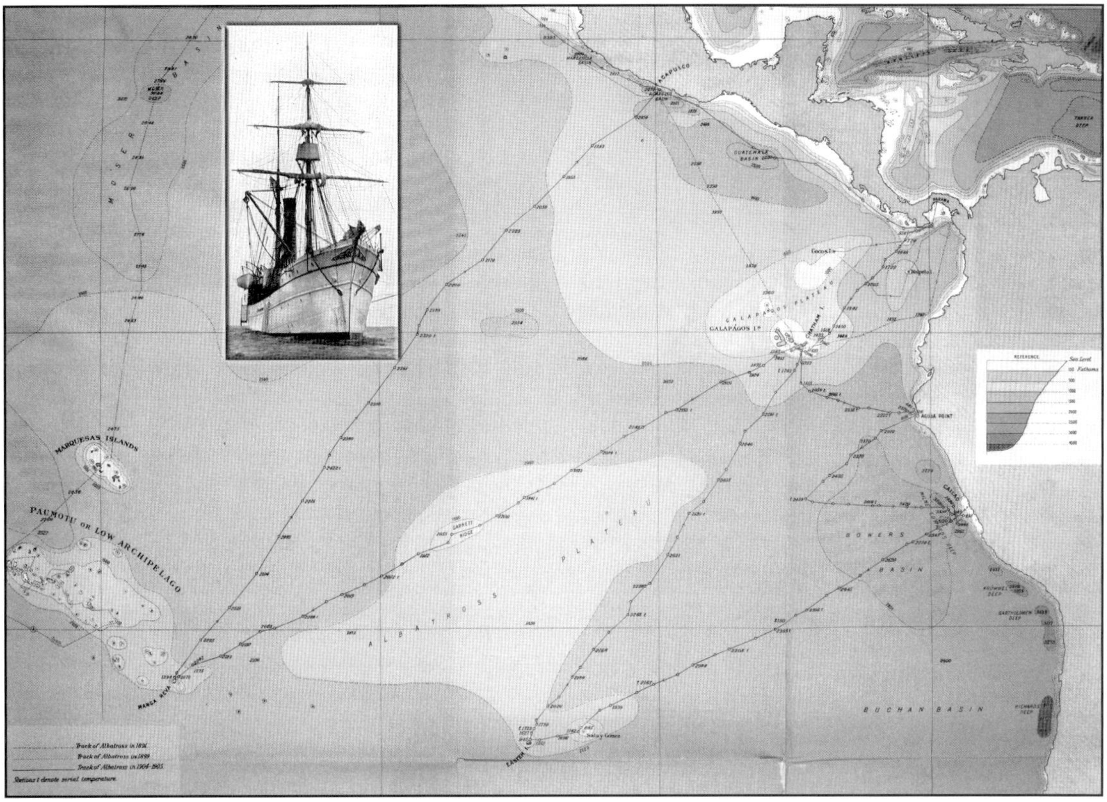

Fig. 1.6 Station map of the Agassiz expedition (from Agassiz 1906). Inset image, courtesy of the National Oceanic and Atmospheric Administration (NOAA) image library, shows the *Albatross* on station. Louis Agassiz organized the oceanographic expedition which explored the Eastern Tropical Pacific in 1904–1905 with the assistance of the US navy using the US Fisheries Commission steamer *Albatross*. The scientific staff included two notable "assistants": C.A. Kofoid and H.B. Bigelow. The material gathered during the expedition was the basis for the two widely used monographs of Kofoid & Campbell (1929, 1939).

water samples with a compound microscope. What occurred in the early 1960s was firstly that zooplankton researchers were pointedly reminded that plankton nets were not sampling all the grazers, that is all of the zooplankton (Banse 1962; Hansen & Andersen 1962). Furthermore, the use of size-fractionation reminded phytoplankton researchers that small phytoplankton often accounted for most of the standing stock (see, for example, McAllister et al. 1960) and these small cells were not usually thought of as major food items of the zooplankton sampled with nets.

Field studies focusing on microzooplankton, in particular those of Beers & Stewart (1967, 1969), showed unequivocally that microzooplankton represented a considerable biomass. Microzooplankton were then apparently "re-discovered" in the 1960s so that by the end of the decade, John Ryther (1969) in "Photosynthesis and fish production in the sea" could simply state:

> "Intermediate between the nannoplankton and the carnivorous zooplankton are a group of herbivores, the microzooplankton, whose ecological significance is a subject of considerable current interest. Representatives of this group

Fig. 1.7 The naive view of marine food webs as "Big fish eat little fish" was illustrated by Pieter Bruegel the elder in 1557, as an allegory of human greed. Until the late 19th century, the base of the marine food chain (nourishing the littlest fish) was thought to be matter supplied by river input and shallow water plants. Image of a lithograph version, © Trustees of the British Museum.

include protozoans such as Radiolaria, Foraminifera, and Tintinnidae, and larval nauplii of microcrustaceans. These organisms, which may occur in concentrations of tens of thousands per cubic meter, are the primary herbivores of the open sea."

Microzooplankton were placed, for a time, at the center of a new view of marine food webs (Fig. 1.9). However, it was recognized that protozoan ecology, as a whole and including planktonic protists, was sorely under-developed (Corliss 1973). Perhaps not surprisingly, in the ensuing years, the ecology and physiology of tintinnids received considerable attention.

An extensive series of studies was published by Ken Gold covering culture (Gold 1968, 1970, 1971, 1973), coastal communities (1975, 1977), as well as anatomy and physiology (Gold 1969a, b, 1979; Gold & Morales 1975a, b, 1976a, b, c; Gold & Pollinger 1971). Gold's

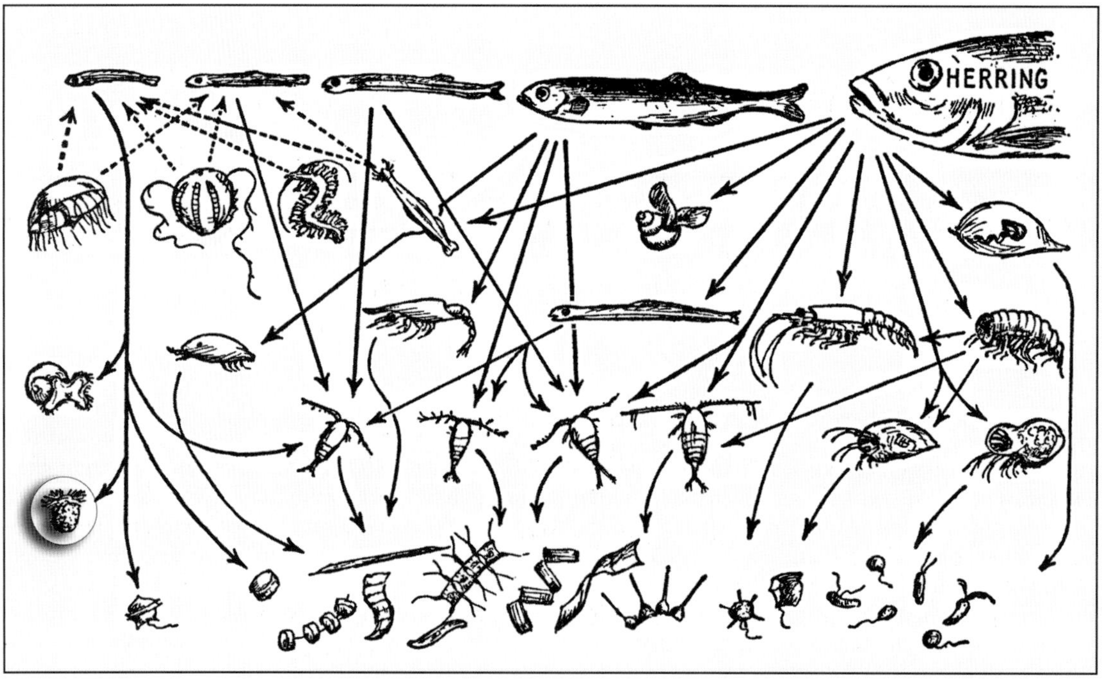

Fig. 1.8 By the end of the 19th century, planktonic organisms were recognized as the base of a complex food web supporting exploited fish populations. The drawing by Hardy (1965) depicts the herring food web based on the relationships described in his 1924 herring monograph. Note in the lower left a tintinnid. Figure adapted from Hardy (1965).

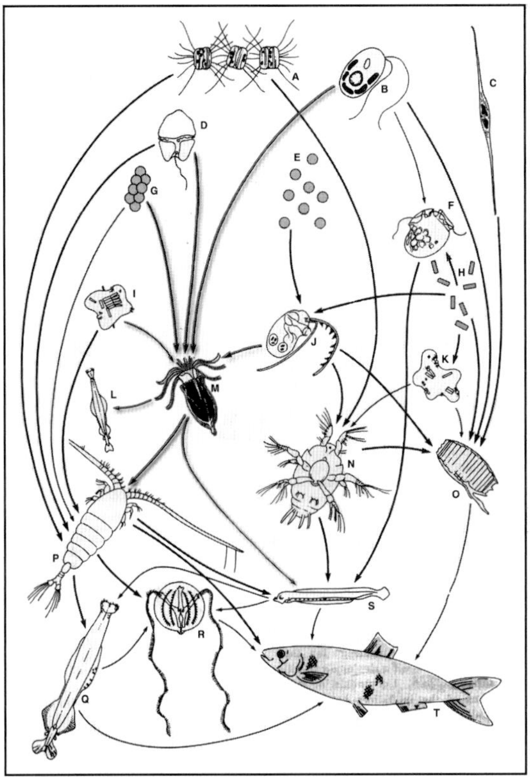

Fig. 1.9 Trophic relations of microzooplankton after Conover (1982). Heavy arrows denote direct relations. (a) A diatom. (b) An autotrophic nano-flagellate. (c) A *Ceratium* (dinoflagellate). (d) A gymnodinid dinoflagellate. (e) Single cells of autotrophic prokaryotes. (f) A dinophysid dinoflagellate. (g) Clumped cells of autotrophic prokaryotes. (h) Single cells of heterotrophic prokaryotes. (i) Detrital particle with attached bacteria. (j) Heterotrophic nanoflagellate. (k) Detrital particle. (l) Small chaetognath. (m) Tintinnid ciliate. (n) Copepod nauplius. (o) Appendicularian. (p) Copepod. (r) Ctenophore. (s) Large chaetognath. (t) Planktivorous fish. © Collection of Musée Océanographique de Monaco, Fondation Albert Ier, Prince de Monaco.

laboratory investigations along with the further field studies of Beers & Stewart (1970, 1971) reiterated the probable importance of microzooplankton, of which tintinnids were an obvious component. Perhaps consequently, between 1974 and 1984, worldwide at least ten PhD theses specifically addressed the ecology of tintinnids in planktonic ecosystems: Blackbourn (1974) in northwest Canada, Johansen (1976) in eastern Canadian, Hedin (1976) in Sweden, Heinbokel (1977) in California, Rassoulzadegan (1975) in France, Souto (1979) in Argentina, Damodara (1981) in Southern India, Brownlee (1982), Capriulo (1982), and Verity (1984) in the northeast USA. During this time extensive cytological investigations of tintinnids were carried out (Hedin 1975a; Laval 1971, 1972; Laval-Peuto 1975, 1976a,b, 1977; 1981; 1983).

The period of the mid-1970s to the mid-1980s also saw the development of our current view of the structure of planktonic ecosystems, a view that recognizes the importance of picoplankton, both autrophic and heterotrophic. This period then included the proposition of the "microbial loop" and the beginnings of the modern field of aquatic microbial ecology. The reassessment of the structure of planktonic communities, as distinct both qualitatively and quantitatively from a linear view of primary producer to herbivore to fish, was prompted by several discoveries (or re-discoveries), as follows.

1. Both primary production and phytoplankton biomass are dominated by small (<20 μm) cells in most systems (as reviewed in Malone (1980)).

2. Phytoplankton are a considerable direct source of dissolved organic carbon, excreting a significant portion of the carbon fixed through photosynthesis (e.g., Sharp 1977; Sellner 1981).

3. The abundance of bacteria is high, much higher than previously thought (revealed through the development of direct-counting techniques using fluorescent stains) and relatively invariant (i.e., Hobbie et al. 1977).

4. Bacterioplankton are active with bacterial secondary production of the same order of magnitude as primary production in many systems (Fuhrman & Azam 1980).

The conclusion that a substantial and active microbial community exists in the plankton was inescapable. However, its structure was unclear. Field and laboratory experiments with tintinnids, taken as typical microzooplankton, had shown high grazing rates on nanoplankton size prey (Heinbokel 1978a, b; Heinbokel & Beers 1979; Rassoulzadegan 1978, Rassoulzadegan & Etienne 1981; Stoecker et al. 1981). However, there was little evidence that microzooplankton were major consumers of bacteria-sized organisms. In fact, work with freshwater ciliates suggested that ciliates typical of the marine microzooplankton were poor candidates for the role of bacteriovores (Fenchel

1980a, b, c, d). In contrast, studies of the feeding and abundances of marine heterotrophic nanoflagellates strongly supported their proposition as bacteriovores (Fenchel 1982a, b, c, d). There remained the need to explain high bacterial abundances and growth rates.

Most of the pieces of the puzzle, perhaps first noted by Pomeroy (1974), were put together with the proposition of the microbial loop by Azam et al. in 1983. The "loop" was one of dissolved organic matter. It is excreted by both autotrophic and heterotrophic microbes as well as metazoan zooplankton and it supports bacterial production. Heterotrophic nanoflagellates consume the bacterial production and grazing by microzooplankton accounts for the production of heterotrophic nanoflagellates. A simplified contemporary view of the microbial loop in the plankton is shown in Fig. 1.10. Not shown in the figure are complexities such as mixotrophy in the phytoplankton, parasitic organisms (see Chapter 6), and the large variety of predators known to feed on microzooplankton such as tintinnids (see Chapter 5).

Not long after the formal proposition of the microbial loop, the true complexities of the microbial component of marine plankton became apparent. It was found that across all microbial size-classes, taxonomic affinity was a very poor predictor of the basic trophic attribute of autotroph, heterotroph, or mixotroph. The existence of a complex microbial food web characterized by many connections among microbes of blurred trophic roles (prokaryotic and eukaryotic alike), as well with higher trophic levels, was recognized (e.g., Sherr & Sherr 1988). Attention was diverted from tintinnids in part because there is, in reality, no "typical" microzooplankter, and furthermore tintinnids are usually but a minority component of the assemblage "protist microzooplankton" (see Chapter 9). In addition, research activity in plankton food-web dynamics slowed somewhat throughout the 1990s. In the oceanographic community there was a clear shift in focus to prokaryote communities. The development of molecular tools made prokaryotes much more amenable to study (e.g., Sherr & Sherr 2008a) and dissolved organic matter, the substrate supporting heterotrophic prokaryotes, was found to be a pool of carbon much more dynamic than previously thought (Hedges 2002). In recent years most tintinnid studies have focused on questions of diversity and distributions (e.g. Fonda Umani et al. 2011; Sitran et al. 2009; Thompson & Alder 2005) or taxonomy and phylogeny, notably using a combiniation of both molecular and classic

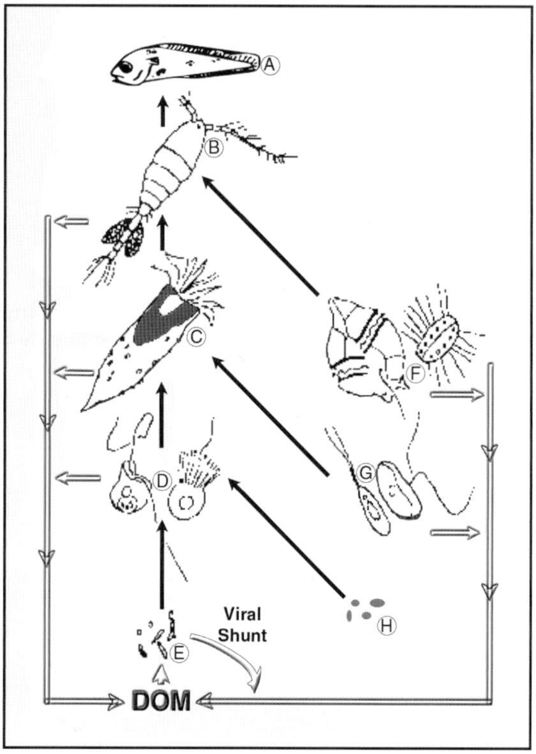

Fig. 1.10 Simplified contemporary view of the microbial loop in the marine plankton and the microbial food web based on the relationships described in Fenchel (1988). Dissolved organic matter (DOM) originates from a large pool of organisms of diverse trophic levels, with flows shown as the open lines, and is consumed by heterotrophic bacteria (e). The "viral shunt" is the flow of DOM from bacteria themselves, lysed from viral infection, back into the DOM pool. The solid arrows show predator–prey relations. Larval fish (a) eat copepods (b). Copepods eat large phytoplankton (f) such as dinoflagellates and diatoms as well as small microzooplankton such as tintinnids (c). The autotrophic nanoflagellates (g), and the heterotrophic nanoflagellates (d), are eaten by the microzooplankton. The heterotrophic nanoflagellates feed on autotrophic bacteria (h), such as *Synechococcus*, as well heterotrophic bacteria. The "microbial loop" proper is the production of DOM, its flow to heterotrophic bacteria transforming it into bacterial biomass, and the subsequent transfer of this bacterial biomass up the food web through whose activities, in part, supply the DOM pool thus closing the loop.

morphological data (e.g. Kim et al. 2010; Agatha & Strüder-Kypke 2007, 2012; Santoferrara et al. 2012a, 2012b; Xu et al. 2012; Bachy et al. 2012), an approach that appears especially valuable for tintinids (see Chapters 2 and 3).

1.5 TINTINNIDS AS MODEL ORGANISMS FOR MARINE PLANKTON

Hutchinson posed the question, "How can so many species co-exist in a relatively homogenous environment?"; his "Paradox of the Plankton" is now over 50 years old (Hutchinson 1961). So far, no explanation, including his own, has been generally accepted and new mechanisms continue to be proposed (e.g., Shoresh et al. 2008; Fox et al. 2010). Furthermore, the diversity Hutchinson described was based on observations of morphologically defined species and this diversity is now known to be dwarfed by the genetic diversity revealed by sequencing the DNA of natural plankton communities. Genetic data suggests that thousands of species of protists can be found in a few liters of seawater (see, for example, Edgcomb et al. 2011). Such an astounding diversity is difficult to explain. Complicating any attempt is the fact that for most species-rich groups of planktonic organisms, phylogeny, morphology, and ecology are not easily related to one another. For example, harpacticoid copepods include both planktonic and benthic species, forms that are herbivorous, others that are carnivorous, or still others that are parasitic. Similarly, gymnodinid dinoflagellates can be benthic, planktonic, phototrophic, mixotrophic, heterotrophic, or parasitic. In contrast to such taxa, tintinnid ciliates represent a singular exception of a coherent group of organisms.

Tintinnid ciliates constitute a single suborder of the ciliate order Choreotrichidae and are thus phylogenetically united. In terms of morphology, all are characterized by the possession of a shell (lorica) whose architecture forms the basis of classic taxonomic schemes. In this species-rich group (over 1000 morphotypes have been described, see Chapter 3), virtually all are restricted to the marine plankton. A few species are known from freshwater plankton, but no tintinnids are benthic. In marine plankton assemblages they are ubiquitous and the great majority are consumers of nanoplankton, including the smaller size-fractions of phytoplankton. Distinct species have long been distinguished using light microscopy. Consequently, compared with most other groups of planktonic organisms, especially other planktonic protists, there exists a very substantial and relatively detailed literature, both modern and historical. Tintinnids are a phylogenetically coherent group (e.g., Lynn 2008) of morphologically and ecologically similar species (e.g., Dolan 2010). Thus, they are the ideal group of planktonic organisms to examine questions of variability in time, space, and composition.

Tintinnid ciliates most likely originated in the marine plankton from a shell-less, oligotrich-like ciliate (see Chapter 3). Tintinnids are often attributed one of the most extensive and continuous fossil records for zooplanktonic organisms, dating back to the Ordovician period (Fig. 1.11). Fossils resembling tintinnid loricae have been described as fossil tintinnid species and continue today to be described as tintinnids (see, for example, Bignot & Poignant 2010). Thus, in principle, tintinnids are good model organisms for reconstructing paleo-environments. Unfortunately, the fossil record of tintinnids is, in reality, ambiguous. This is because many fossil remains reported to be tintinnids, for example calpionellids, most likely are not (see Chapter 8). The fossil loricae of calpionellids are formed of calcite and as no known ciliates form external structures of calcite, calpionellids were probably

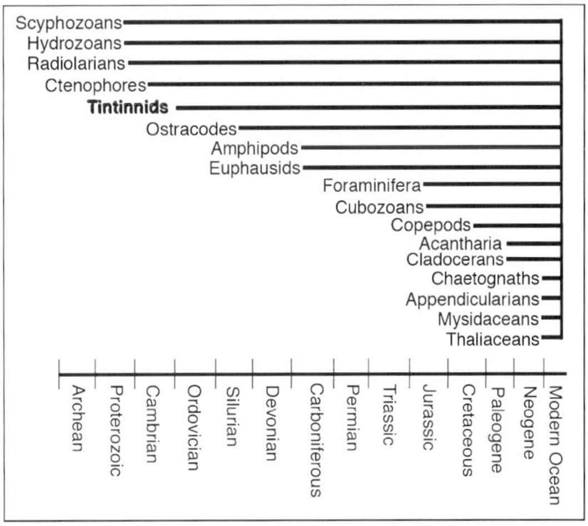

Fig. 1.11 The fossil record of major zooplankton taxa according to Rigby and Milsom (2000). The record attributed to tintinnids is typically one of the oldest among contemporary taxa of the plankton. However, many fossils described as tintinnids probably are not (see Chapter 8). Based on data from Rigby & Milsom (2000).

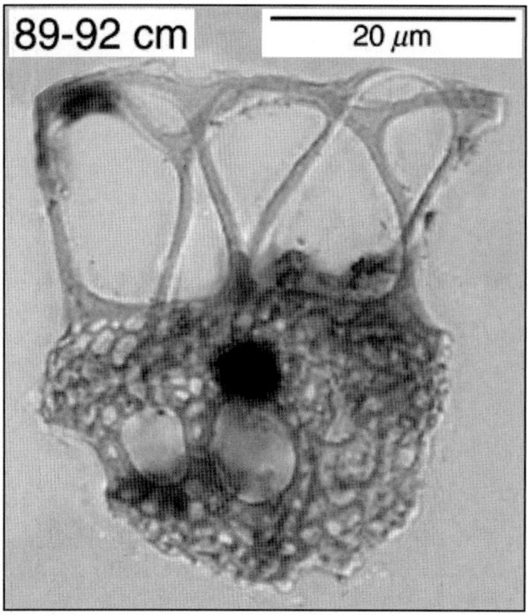

Fig. 1.12 A lorica of *Dicytocysta lepida* from a section of a sediment core dated to 3000–6000 years before present. Sediment core samples from the Orca Basin (2249 m depth) of the Gulf of Mexico contained tintinnid loricae and crustacean remains in the section corresponding to 89–92 cm depth in the core. Reprinted from Earth and Plantary Science Letters, 272, Meckler, A.N., Schubert, C.J., Hochuli, P.A., Plessen, B., Birgel, D., Flower, B.P., Hinrichs, K.-U. & Haug, G.H. Glacial to Holocene terrigenous organic matter input to sediments from Orca Basin, Gulf of Mexico. A combined optical and biomarker approach, 251–263. Copyright (2008) with permission from Elsevier.

not even ciliates, much less tintinnids (Remane 1963, 1998). Some authorities then attribute a fossil record to tintinnids which begins only in early Jurassic period (e.g., Armstrong & Brasier 2005).

Although the fossil record is unsettled, it is clear that the remains of tintinnids can provide evidence of relatively recent climatic change. Very well preserved lorica have been found in late Holocene deposits (Fig. 1.12). Loricae are among the microfossils that can be used as indicators of ecological and hydrographic shifts. For example, changes in an Antarctic lake were inferred using the sediment record of tintinnids and other planktonic organisms by Cromer et al. (2005). Likewise paleo-hydrographic changes in waters surrounding the Faroe Islands were investigated by analyzing the occurrences of dinoflagellate cysts, acritarchs, and tintinnid loricae in sediment cores (Roncaglia 2004).

Tintinnids are probably the group of planktonic protists with best potential for analysis using automated or semi-automated imaging systems. They are relatively large and the geometry of the lorica is simple. Furthermore, some characteristics of the simple geometry, specifically the diameter of the lorica oral opening, is both of taxonomic significance in distinguishing species (see Laval-Peuto & Brownlee 1986) and correlates with some ecological characteristics, such as prey size (Fig. 1.13) or maximum growth rate (see Chapter 4). Among tintinnids, morphological diversity is correlated with species diversity and morphological differences reflect ecological differences. Therefore, data in the simple form of morphological descriptors can provide information on species diversity and ecological characteristics in tintinnids. Unfortunately, existing imaging systems are not suitable for analyzing most natural assemblages in their entirety. The size range of tintinnid loricae (50–400 µm length, 15–200 µm diameter) is incompatible with either the size range or resolving capacity of existing instruments such as the Flowcam (e.g., Zarauz et al. 2009) or the ZooScan (e.g., Gorsky et al. 2010). However, the near future, like the recent past, will likely see great progress in instruments for analyzing plankton populations (see Erickson et al. 2011).

Although studying tintinnids may not provide a general explanation for "the paradox of the plankton", any attempt to explain the paradox will have to account for co-existence of dozens of tintinnid species, all of roughly similar morphology and ecology, in a few liters of seawater. Does the answer lie in the partitioning of food resources among tintinnid species, variability in competition from other microzooplankters, differential mortality, or life-cycle differences among species? The question of what regulates or rules the composition of assemblages of tintinnids is part of the question of what regulates planktonic assemblages. This question is far from trivial or only of academic interest. Predicting the consequences of climate change on marine systems requires an understanding the functioning of planktonic food webs, which in turn requires an understanding of what regulates their composition.

As a concluding note, it is worth noting that outside the realm of science, tintinnids can be considered as models. In the art world, tintinnids have appeared as both models of artistic form in nature as well as subjects. The best known are perhaps those in Haeckel's "Art Forms of Nature" seen in Plate 3, Ciliata (Haeckel 1904). Several of the species shown are from the plates of his 1873 paper describing forms he found in the Canary Islands and Naples (Fig. and Plate 1.14).

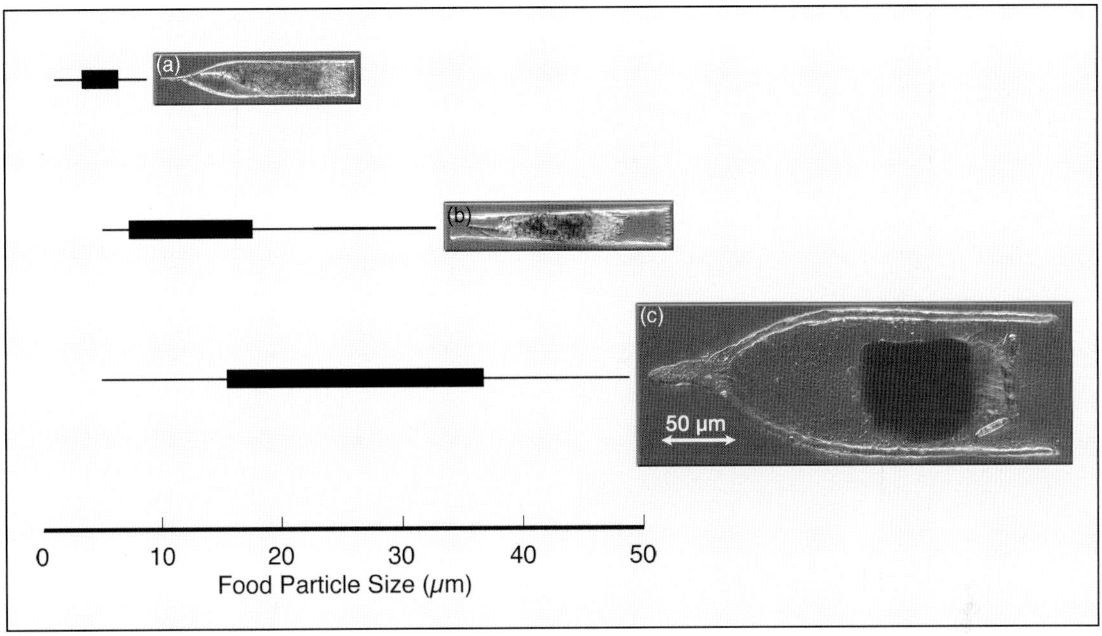

Fig. 1.13 The size range of the food items ingested by three tintinnid species ((a) *Helicostomella subulata*, (b) *Eutintinnus pectinus*, (c) *Favella ehrenbergii*) in Danish coastal waters varies with the size of the oral diameters of the loricae. Based on data from Fenchel (1987), fig. 7.1.

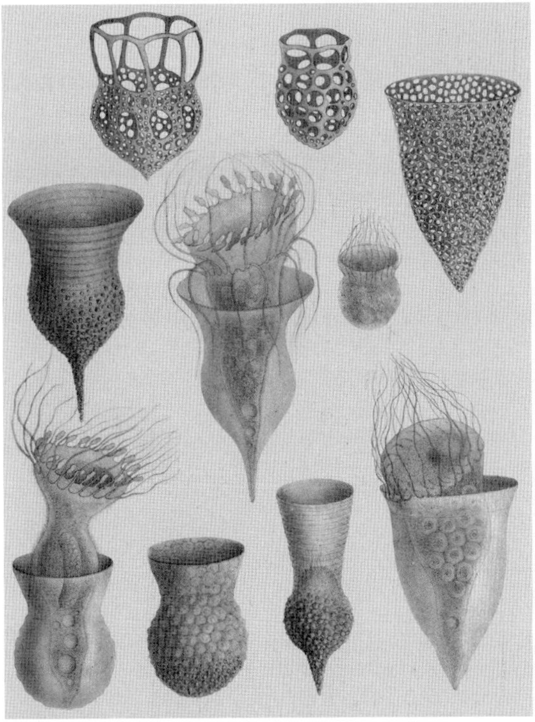

Fig. 1.14 Tintinnids of Ernst Haeckel. Restored drawings from Ernst Haeckel's 1873 paper, the first scientific article devoted to tintinnid ciliates. Image by J.R. Dolan. For color version, see Plate 1.14.

Fig. 1.15 "La Boheme: A Portrait of Today's Oceans in Peril", the 2012 sculpture by Mara G. Haseltine depicts two tintinnids ensnarled in plastic. Dimensions: 91 cm × 122 cm × 185 cm (length × width × height). Materials: uranium-infused blown glass and plastic. By permission of Mara Heseltine. For color version, see Plate 1.15.

A recent example of tintinnids in the art world is in Mara Haseltine's sculpture "La Boheme", which depicts tintinnids entangled in microscopic pieces of ultraviolet-degraded plastic (Fig. and Plate 1.15).

1.6 KEY POINTS

1. Tintinnids are ciliate protists, constituting a suborder of the order Choreotrichidae, with the common characteristic of having a lorica. They are all planktonic and the overwhelming majority are marine. Based on differences in lorica architecture, literally hundreds of species have been described since the first tintinnid was noted over 230 years ago.
2. The role of tintinnids in the pelagic ecosystems is that of a component of the microzooplankton, grazers at the base of the food web. They are a species-rich group of organisms which are ecologically united as planktonic ciliates, mainly consumers of nanoplankton, and are morphologically as well as phylogenetically united as "shelled", choreotrich ciliates.
3. Tintinnid ciliates represent an excellent group of organisms to examine questions of what regulates or rules the composition of planktonic assemblages.

ACKNOWLEDGMENTS

The comments and ideas of all the chapter authors as well as Richard Christan, Tom Fenchel, Dave Kirchman, Charles Bachy, and Puri Lopez-Garcia are gratefully acknowledged. Funding for this work was provided by the French Agence National de la Recherche and Pole Mer PACA through the ANR-Biodiversité AQUA-PARADOX project.

Chapter 2

THE TINTINNID LORICA

Sabine Agatha[1], *Michèle Laval-Peuto*[2], *and Paul Simon*[3]

[1]University of Salzburg, Salzburg, Austria
[2]Chemin des Campons 144, La Colle-sur-Loup, France
[3]Max Planck Institute for Chemical Physics of Solids, Dresden, Germany

The Biology and Ecology of Tintinnid Ciliates: Models for Marine Plankton, First Edition. Edited by John R. Dolan, David J.S. Montagnes, Sabine Agatha, D. Wayne Coats, and Diane K. Stoecker.

2.1 INTRODUCTION

Tintinnids are unique among planktonic ciliates in building solid shells. These shells, called loricae (singular lorica; Latin, "corselet"), are minute works of art. Sometimes they are simply tube- or vase-shaped, sometimes incredibly elaborate, but the builders are only unicellular organisms (Figs 1.4, 1.5, 2.1, 2.2, 3.16, and 3.17).

Because our knowledge about the inhabitants is rather scarce, the taxonomic classification of the tintinnids is still commonly based on features of the loricae (Chapter 3). Therefore, we should look closely at the factors that influence the appearance of the

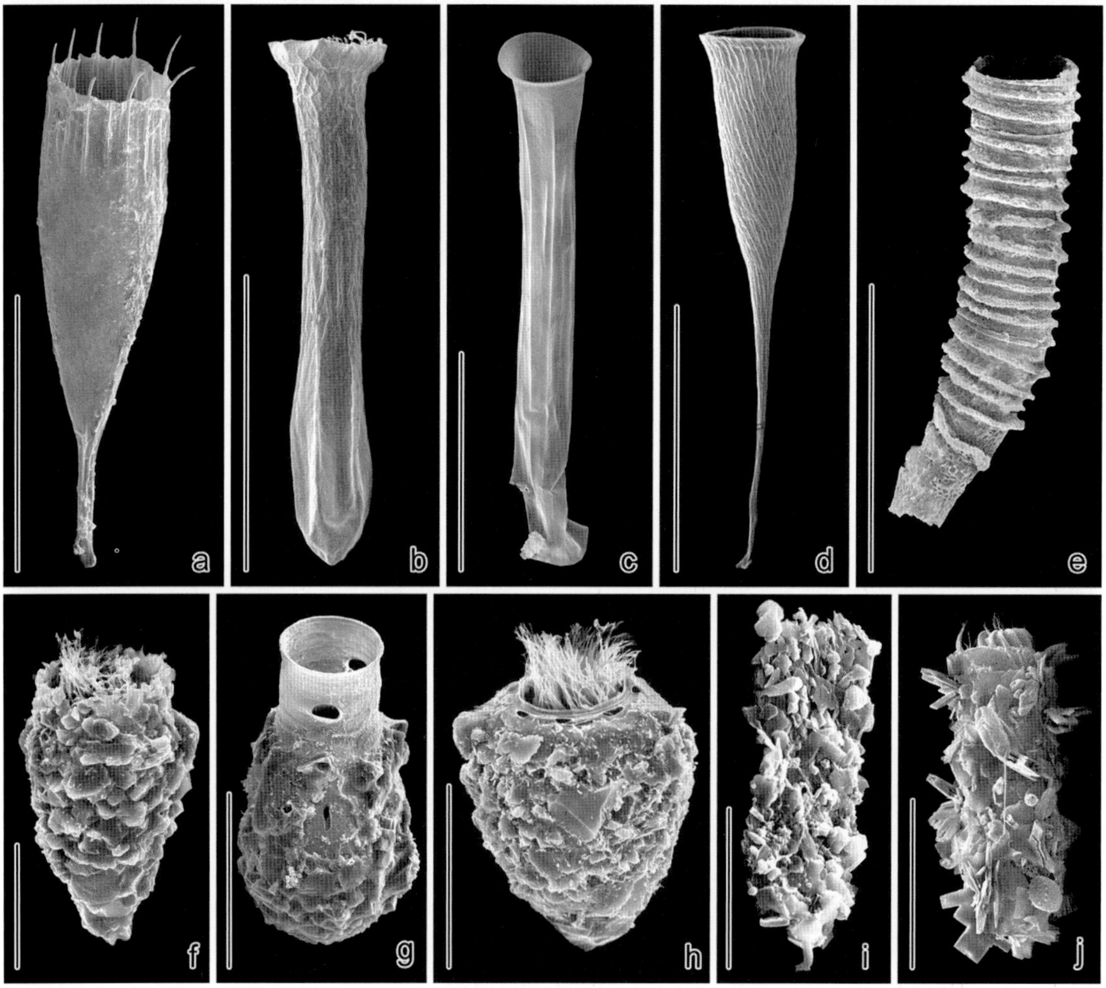

Fig. 2.1 Hyaline loricae (a–e), a hard, agglomerated lorica (f), loricae composed of a hyaline collar and an agglomerated bowl (g, h), and soft, agglomerated loricae (i, j) in the scanning electron microscope (a–e, g, originals of S. Agatha; f, adapted from Agatha 2010b; h, adapted from Agatha, S. & Tsai, S. (2008) Redescription of the tintinnid *Stenosemella pacifica* Kofoid and Campbell, 1929 (Ciliophora, Spirotricha) based on live observation, protargol impregnation, and scanning electron microscopy. Journal of Eukaryotic Microbiology, 55, 75–85; i, j, adapted from Foissner et al. 1999). (a) *Dadayiella ganymedes*. (b) *Steenstrupiella* species. (c) *Eutintinnus brandti*. (d) *Rhabdonella spiralis*. (e) *Climacocylis elongata*. (f) *Tintinnopsis parvula*. (g) *Codonellopsis schabi*. (h) *Stenosemella pacifica*. (i, j) *Tintinnopsis cylindrata*. The empty lorica in Fig. 2.1i collapsed in the scanning electron microscope, as it is soft. Scale bars 50 μm (a, b), 100 μm (c), 150 μm (d, e), 20 μm (f), 50 μm (g, i, j), and 30 μm (h).

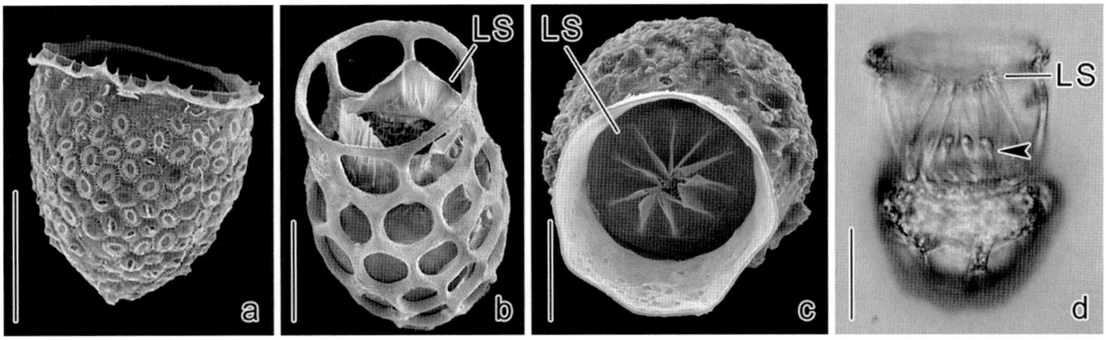

Fig. 2.2 Lorica of *Acanthostomella* species covered with coccoliths (a; calcium carbonate plates formed by coccolithophorids, unicellular algae) and specimens of *Dictyocysta mitra* (b), *D. lepida* (d), and *Codonella aspera* (c) showing a lorica sac. Arrowhead in Figure 2.2d denotes tentaculoids (pin-shaped cytoplasmic extensions containing tintinnid extrusomes). Micrographs from the scanning electron microscope (a–c; a, adapted from Young & Geisen 2002; b, c, adapted from Agatha, S. (2010a) A Light and scanning electron microscopic study of the closing apparatus in tintinnid ciliates (Ciliophora, Spirotricha, Tintinnina): a forgotten synapomorphy. Journal of Eukaryotic Microbiology, 57, 297–307) and a live cell (d; adapted from Agatha 2010a). LS, lorica sac. Scale bars 10 µm (a) and 20 µm (b–d).

loricae, which are (1) the amount and quality of the lorica material, (2) the environmental conditions during formation, and (3) the cell cycle. Astonishingly, many of these unicellular organisms are able to construct not only a single type of lorica, but several, which are occasionally so dissimilar that they were affiliated with different families. After the death of the ciliate or when the lorica is abandoned, it sediments, transporting chemical compounds to deeper water layers and possibly to the bottom of the sea or lake. As tintinnids occasionally dominate the microzooplankton (heterotrophic organisms of the pelagic zone 20–200 µm in size), the material flux can be considerable.

2.2 DIVERSITY, FORMATION, AND VARIABILITY OF LORICAE

Diversity of loricae

The loricae are often much longer than the cell proper (Figs 1.4, 1.5, and 2.10) and they are thus able to harbor the entire disturbed ciliate retracted by its contractile peduncle. In extended state, the anterior cell portion with the adoral membranelles (fan-like ciliary structures used for swimming and feeding; Chapter 4) protrudes from the lorica. The lorica might be a protection against the numerous predators (mainly copepods and invertebrate larvae), but other functions are also possible (Chapter 1).

Just after division and before it starts the construction of its own lorica, the proter (anterior division product) is naked, whereas the opisthe (posterior division product) keeps the old lorica. Tintinnids that lost or abandoned their loricae are also naked, until they form new loricae. Reports of naked tintinnids from the plankton are not available, probably because the lorica formation is generally a rapid process. An occasional misidentification with aloricate choreotrichids of the genus *Strombidinopsis* can also not be ruled out, considering that usually fixed material was investigated. Furthermore, naked tintinnids are slow swimming, relatively small cells that could easily be caught by numerous predators. In cultures, however, naked tintinnids can be observed.

According to the famous monographs on extant (non-fossil) tintinnids by Kofoid & Campbell (1929, 1939), the vast majority of genera have hyaline loricae (42 genera in 10 families), which are remarkably diverse in morphology (Figs 1.5 and 2.1a–e). The loricae range from small, smooth, and rounded forms resembling bowls (e.g., in *Undella* and *Proplectella*) and tube-shaped loricae (e.g., in *Eutintinnus*) to large bowls with everted collars (e.g., in *Petalotricha*; Fig. 2.4) and chalice-shaped loricae (e.g., in *Epiplocylis*). The anterior lorica portions can be quite elaborate with denticulations, flaring collars, and portions of different structure or sculptured appearance (Figs 2.1 and 2.2). The second major type of lorica is hard and has entirely, or only in the posterior portion, particles of mineral or biogenic origin agglomerated. These loricae are found

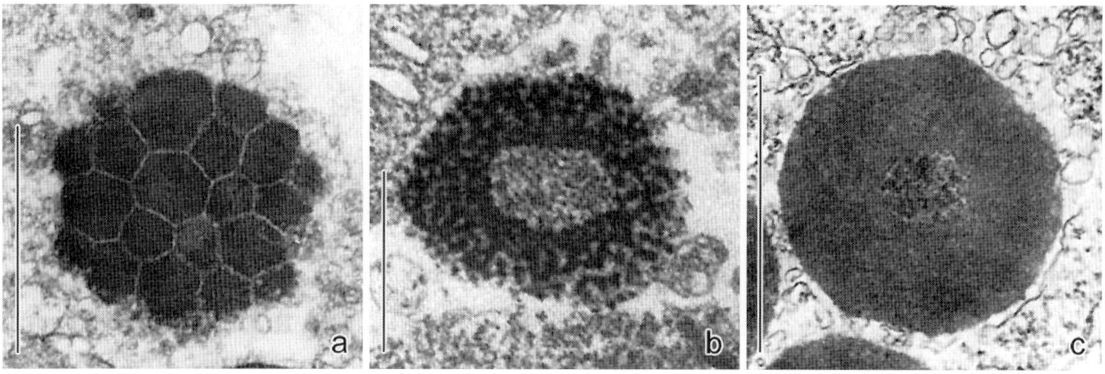

Fig. 2.3 Transmission electron micrographs of lorica forming granules. (a) Morula-like aggregate of granules in *Petalotricha ampulla* (adapted from Laval-Peuto 1994); similar structures occur in *Cymatocylis convallaria*, *Cyttarocylis brandti*, and *Parafavella denticulata*. (b, c) Granular and compact types of granules in *Parafavella gigantea* (adapted from Hedin 1975b). Scale bars 2 μm (a) and 1 μm (b, c).

in 11 genera belonging to four families (Figs 1.4, 2.1f–j, 2.2a, c, d, and 2.12). A few agglomerated species have soft loricae; they belong to the family Tintinnidiidae, which comprises three genera.

A unique character, the closing apparatus, was recently re-investigated, using light and scanning electron microscopy (Agatha 2010a). The first observations on this structure were performed about 130 years ago by Fol (1883; for details, see Agatha 2010a). Two kinds of closing apparatus exist. A membranous closing apparatus, which shuts the lorica opening by 8–24 folds, usually 12, in disturbed and thus retracted ciliates, is found in four genera, which were until recently affiliated with three families (Chapter 3; Fig. 2.2b–d): *Codonella*, *Codonaria* (hard, agglomerated loricae), *Codonellopsis* (loricae composed of a hyaline collar and an agglomerated bowl), and *Dictyocysta* (loricae entirely hyaline or composed of a hyaline collar and an agglomerated bowl). Owing to a similar morphology and functionality, the foldable closing apparatus was regarded as synapomorphy (shared derived character) of these genera (Chapter 3). This kind of closing apparatus merges posteriorly into a lorica sac, which is difficult to recognize as it completely lines the lorica bowl (broadened lorica portion). Even if not hermetically sealed, the lorica sac and its closing apparatus might be a valuable protection against contact predators, who use feeding tubes (dinoflagellates) and pseudopodia (radiolaria), especially, in tintinnids with fenestrated loricae, for example *Dictyocysta*. The second type of closing apparatus is diaphragm-like and occurs in the genera *Salpingacantha* and *Salpingella* (Tintinnidae; with hyaline loricae). Because the diaphragm-like closing apparatus shows neither folds nor closing movements and is not associated with a lorica sac, a homology with the foldable apparatus is less likely. Therefore, the genera *Salpingacantha* and *Salpingella* are probably not closely related to the four genera mentioned above. This is supported by phylogenetic analysis of the small subunit ribosomal RNA (SSU rRNA) gene (Chapter 3).

Because hyaline loricae lack agglomerated particles obscuring their content and texture (see below), those tintinnids were preferentially studied. Even with a very simple microscope at low magnification, Müller (1779) was able to observe the ciliate, the contractility of its peduncle, and some stages of cell division (see Box 1.1 in Chapter 1). To avoid impediments by the agglomerated particles, the first sections of tintinnid cells to study the cell structures were mainly conducted in species with hyaline loricae (Entz Jr 1909b; Merkle 1909). Differences in the structure of the lorica wall and sometimes its texture were described, using light microscopy (Brandt 1906, 1907). Later, ultrathin sections for transmission electron microscopy provided new data on the cytology of tintinnids and the texture of the lorica wall (Laval 1972; Laval-Peuto 1975, 1980). Several types of texture are distinguishable (Figs 2.13 and 2.14): monolaminar and compact (e.g., in *Eutintinnus*), monolaminar with alveoli (small chambers; e.g., in *Climacocylis*), trilaminar with different tubules in the three layers (e.g., in

Petalotricha), or monolaminar and tubular (e.g., in *Dictyocysta*). The lorica wall can have windows (e.g., in *Codonellopsis*, *Dictyocysta*, and *Stenosemella*; Figs 2.1g, h and 2.2b, d), perforations of different sizes (e.g., in *Epiplocylis* and *Cyttarocylis*), or pores (e.g., in *Rhabdonella* and *Schmidingerella*; Figs 2.14c and 3.17e). The outer lorica surface may show reticulate, longitudinal, or spiraled ribs or folds (e.g., in *Climacocylis*, *Daturella*, *Rhabdonella*, and *Schmidingerella*; Figs 2.1d, e, 2.14a, c, and 3.17e). Furthermore, the wall can have a spiraled structure in most of the lorica (e.g., in *Climacocylis*; Fig. 2.1e) or only in the collar (e.g., in *Codonellopsis*; Fig. 2.1g).

The texture of the lorica wall is distinguished from the lorica structure (for details, see Laval-Peuto & Brownlee 1986; Laval-Peuto 1994). Continuous and spiraled lorica walls are occasionally found in the same species. They are the result of different construction modes. The wall texture, however, depends on the chemical composition of the lorica material. Despite some variability in the height of the surface ridges and the abundance of pores, the wall texture appears invariably monolaminar with alveoli in *Schmidingerella arcuata* (compare Fig. 2.7a with 3.17). Likewise, the monolaminar texture with alveoli is apparently constant during the cell cycle of *Favella ehrenbergii* and *F. panamensis* (Laval-Peuto 1981; S. Agatha, unpublished observations). Therefore, the wall texture seems to be a rather reliable taxonomic character. Accordingly, the trilaminar texture of the lorica wall with a tubular middle layer in *Cyttarocylis* and *Petalotricha* indicates a close relationship, contradicting their current classification in different families.

General aspects of lorica formation

Tintinnids can create new loricae on different occasions. After each cell division, the proter forms a new lorica, whereas the opisthe keeps the old one, which might be slightly modified during its trophic life. If the lorica is lost or abandoned by the ciliate, a new one is built that might be different from the preceding one. So, the same species possibly displays different forms (phenotypes) of loricae. As damaged, deformed, or excessively modified loricae are very rarely found in plankton samples, it seems likely that loricae are usually not older than two or three cell divisions.

Two conditions are required for lorica construction: the presence of lorica material and the activity of the somatic ciliature (assemblage of body cilia; Fig. 1.2; Chapter 3). The latter apparently moulds the secreted lorica material, as soon as it is extruded; only large quantities of lorica material allow a rapid and complete construction of a lorica. The behavior and division rate of the species (Chapter 4) and the environmental conditions (e.g., temperature, salinity, and particle availability) during the construction probably influence the final size, shape, and structure of the lorica. Therefore, the final morphology of a lorica reflects the history of its construction.

Lorica material

The lorica material is an organic substance produced by the ciliate mainly before cell division. The first detailed observations of this material were probably from Campbell (1927), who named the cytoplasmic state "lorica forming granules". Transmission electron microscopic studies revealed the existence of different types of osmiophilic granule and arrangement: morula-like bodies composed of densely packed granules in *Cymatocylis*, *Cyttarocylis*, and *Petalotricha* (Fig. 2.3a; Laval 1972; Laval-Peuto 1975, 1994; Wasik & Mikołajczyk 1992); single granular and compact types in *Parafavella* (Fig. 2.3b, c; Hedin 1975b) and *Tintinnopsis* (Laval-Peuto et al. 1979); single granules 1–1.5 μm across in *Favella ehrenbergii*; and an aggregate of granules, forming a single mass in *Dictyocysta* (Laval-Peuto 1994). Even within a single genus, different granule types were recorded, namely granular and compact types in *Parafavella gigantea* (Fig. 2.3b, c; Hedin 1975b), whereas there are morula-like aggregates in *P. denticulata* (Sokolova & Gerassimova 1984; Sokolova et al. 1986); possibly, the first two types represent only developmental stages. The different morphologies of the granules likely reflect distinct chemical compositions and might be correlated with the final texture of the lorica wall. Possibly, the lorica material is differentiated and stored without any limiting membrane inside the vesicular reticulum, cortical cavities that communicate with each other and the pericellular space underneath the perilemma (an additional membrane covering the entire cell; Laval-Peuto 1975; Sokolova & Gerassimova 1984; Sokolova et al. 1986).

The process of granule accumulation was studied in live specimens with hyaline loricae (Fauré-Fremiet 1924), for example in *Petalotricha* (Fig. 2.4), and described after application of various histological

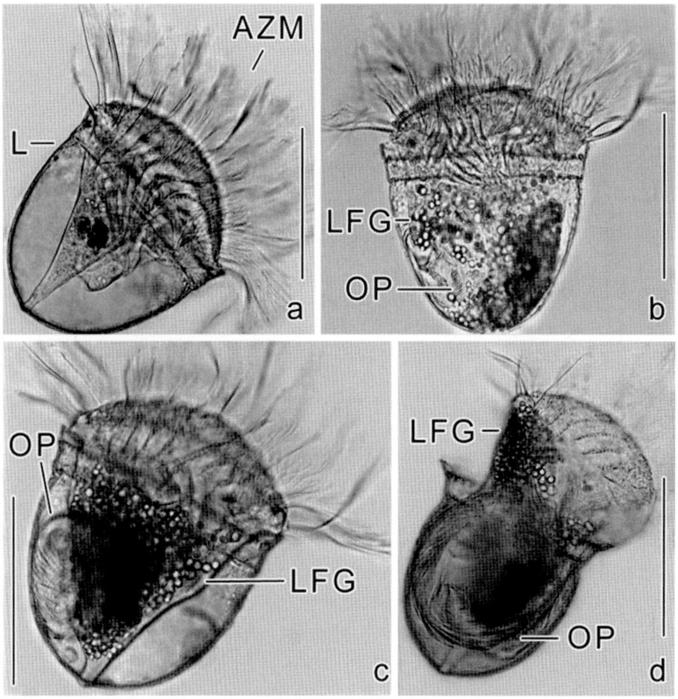

Fig. 2.4 Trophont (a) and dividers (b–d) of *Petalotricha ampulla* (live cells; originals of M. Laval-Peuto). (a) Very young trophont in its newly formed lorica. It possesses only few lorica forming granules. (b) Older trophont with some recently formed refractile granules in the posterior cell portion (stage II in Fig. 2.5). The opisthe's future oral apparatus starts to originate. (c) In the middle division stage, the already abundant granules are found in almost the whole cell (stage III in Fig. 2.5). (d) Just before separation, the granules accumulate in the anterior cell portion above the opisthe's future oral apparatus, namely in the proter (stage IV in Fig. 2.5). AZM, adoral zone of membranelles; L, lorica; LFG, lorica forming granules; OP, oral primordium. Scale bars 100 μm.

stains (Campbell 1927; Biernacka 1965). In particular, the detailed observations of a *Favella* species stained with protargol (which reveals the basal bodies, macronucleus nodules, and micronuclei) demonstrated this process during morphogenesis (reproduction of the pre-existing form during cell division; Fig. 2.5; Laval-Peuto 1994). The lorica granules gradually increase in number in the posterior portion of the ciliate during the early division stages (stages II and III). During completion of the opisthe's oral ciliature, the granules migrate into the anterior cell half, namely into the future proter (stage IV). With the separation of the two division products, the opisthe is provided with the new oral apparatus and keeps the old lorica, whereas the naked proter with the old oral apparatus is equipped with a sufficient number of granules for the rapid and complete formation of a new lorica.

The secretion of the lorica material by the proter was studied in live *Favella* specimens by Laval-Peuto (1981). Her findings are supported by scanning electron micrographs (Figs 2.6 and 2.7), which show just extruded globules of lorica material in the anterior ventral cell portion at the level of the somatic ciliature (Fig. 2.6). First, the posterior lorica portion including posterior processes is formed by the mergence of the material globules with those secreted before. Then, the lorica grows anteriorly by attaching a band of lorica material to the opening rim (Fig. 2.7).

In tintinnids, locomotion is provided by the adoral zone of membranelles, specifically by the large collar membranelles inserting on the elevated peristomial rim around the circular peristomial field (Fig. 1.2; Chapter 4). Swimming is a permanent activity of these planktonic ciliates. Even, when they divide or conjugate, the membranellar movements only stop for a short period. Just before separation of the division products, the opisthe moves to the lorica opening and pushes the proter outside. Still connected by a thin cytoplasmic thread, the opisthe directs the locomotion by its vigorously beating adoral membranelles, while the proter with the original adoral zone of membranelles swims actively, but not in phase with the opisthe (Fig. 2.10a–d; Laval-Peuto 1981).

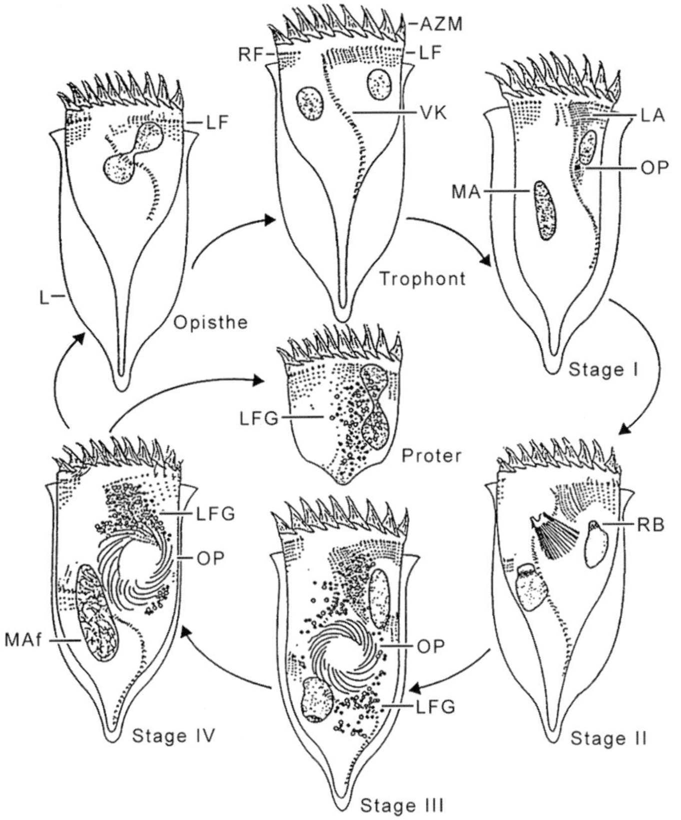

Fig. 2.5 Morphogenesis of a *Favella* species that probably belongs to the recently established genus *Schmidingerella* [Chapter 3; adapted from Laval-Peuto (1994); after protargol staining]. The production of the lorica forming granules occurs during all stages; however, lorica formation depletes the material reserve. Stage I: first signs of the oral primordium (for details on the somatic ciliary pattern and its development, see Chapter 3). Stage II: the adoral membranelles start to form, and replication bands traverse the macronucleus nodules. Stage III: the lorica forming granules become numerous and migrate anteriorly above the oral primordium. Stage IV: the macronucleus nodules fuse before division. Now, most of the granules are in the anterior cell portion. After fission, the opisthe receives the new adoral zone of membranelles and keeps the old lorica, whereas the naked proter with the old adoral zone and the reserve of granules swims away and starts the construction of a new lorica (Fig. 2.10a–d). AZM, adoral zone of membranelles; L, lorica; LA, lateral ciliary field; LF, left ciliary field; LFG, lorica forming granules; MA, macronucleus nodules; MAf, fused macronucleus nodules; OP, oral primordium; RB, replication bands (place of DNA duplication); RF, right ciliary field; VK, ventral kinety.

Role of the somatic cilia

The somatic ciliature, except for the anteriormost, often elongated cilia, is hidden by the lorica. As in the related oligotrichids and aloricate choreotrichids, the somatic cilia of tintinnids are not involved in locomotion, but their arrangement in specialized fields and rows indicates other functions. Laval-Peuto & Brownlee (1986) suggested a role in (1) the construction of the lorica (Hofker 1931a), (2) the general maintenance and cleaning of the interior space (e.g., the removal of faeces from the lorica; Entz Jr 1909b), (3) the positioning of the ciliate within the lorica, and (4) sensory activities. Live observations on the movement of the somatic cilia are rare. In *Eutintinnus*, with its hyaline lorica, the elongated anteriormost cilia are rather immobile, grasping the upper rim of the lorica while swimming (Fauré-Fremiet 1924). In *Tintinnopsis* (Fig. 1.2), *Stenosemella*, *Codonella*, and *Codonellopsis*, especially, the elongated anteriormost cilia contribute to the lorica formation by carrying foreign particles to the newly forming lorica portions (Fauré-Fremiet 1924; Hofker 1931a). However, a correlation between the somatic ciliary pattern and the lorica structure is not evident (Chapter 3).

Lorica phenotypes formed during the cell cycle

The phenotypic plasticity of the loricae can only be demonstrated in cultured species by following several generations or gene-based diagnosis. The first culture studies provided interesting, but brief and incomplete,

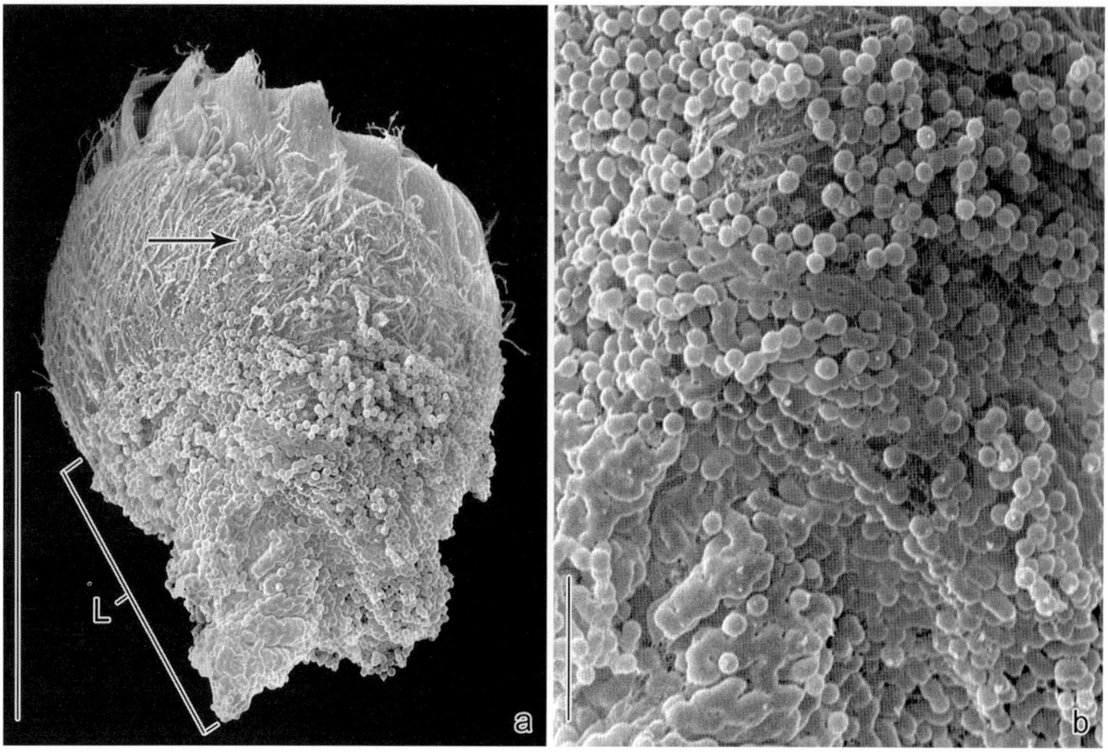

Fig. 2.6 Lorica formation in *Favella panamensis* from field material (scanning electron micrographs; originals of S. Agatha). (a) Oblique posterior polar view. The globules of lorica material are apparently arranged in a single oblique stripe on the ciliated anterior ventral cell portion (arrow), while the material already forms an obconical cap covering the whole unciliated posterior portion. This indicates that the lorica material is extruded in the anterior cell portion and transported to the growing lorica margin by ciliary movements. (b) Detail showing the mergence of the material globules and an underpinning network, which might represent the first stage in the merging process. L, just forming lorica. Scale bars 30 µm (a) and 5 µm (b).

observations of species with hyaline or translucent loricae (Entz Sr 1885; Schweyer 1905, 1909; Entz Jr 1909b; Busch 1925). They suggested the existence of distinct stages in lorica construction and a certain variability in lorica morphology. Unfortunately, Kofoid & Campbell (1929) neglected these studies and elevated many lorica variants to species rank. Findings on preserved samples from natural populations also indicated an intraspecific variability in lorica morphology (e.g., Davis 1978, 1981). For several decades, only the investigations by Hofker (1931a, b) and Biernacka (1952, 1965) gave some insight into lorica formation, describing briefly the early stages passed by the proter in the genus *Tintinnopsis*.

The first comprehensive study was conducted by Laval-Peuto (1981) on cultures of *Favella ehrenbergii*. She demonstrated that a single species is able to produce a variety of lorica morphologies during its cell cycle and that tintinnid behavior during lorica formation is highly complex (Fig. 2.8). In addition, biometric analyses of loricae from natural populations verified that the diameter of the lorica opening, specifically, the internal diameter (correlates with the diameter of the adoral zone of membranelles) is the only reliable lorica feature for identification. This was confirmed by conjugants (specimens performing a reciprocal fertilization) with different lorica morphologies, but similar lorica opening diameters (Laval-Peuto 1983).

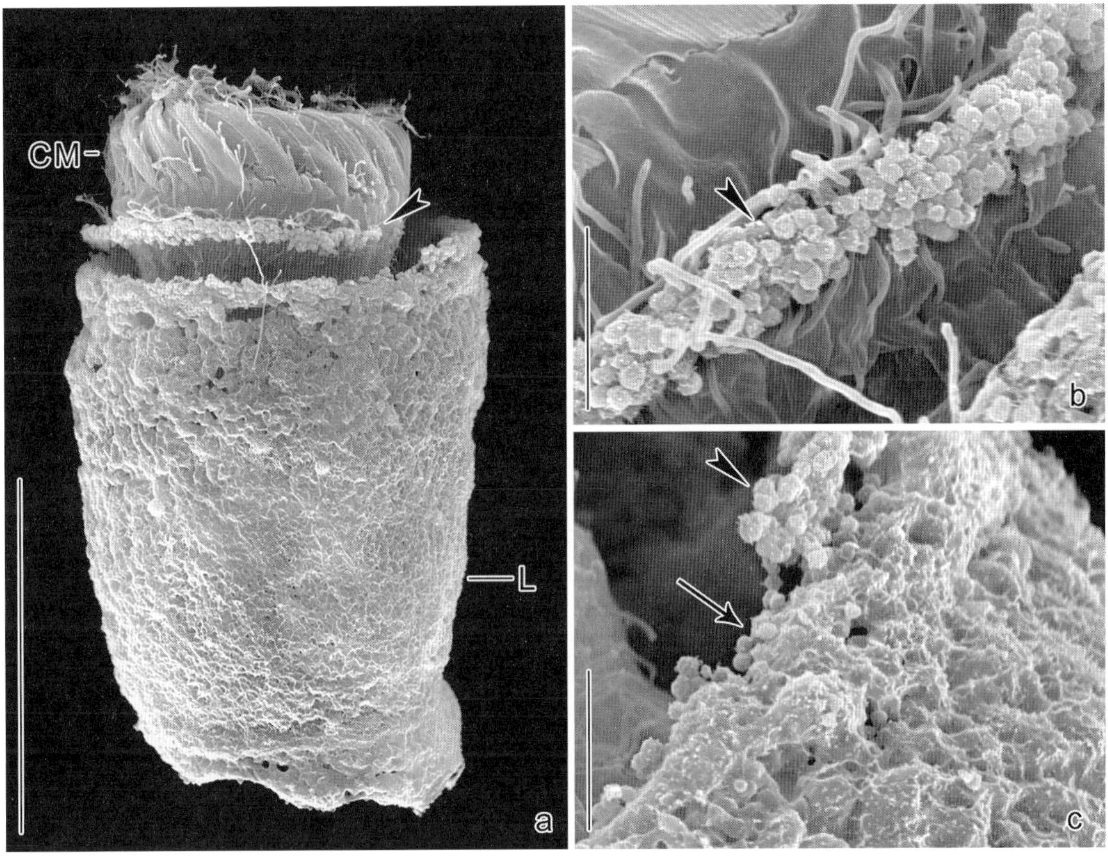

Fig. 2.7 Formation of an abnormal (posteriorly truncate) lorica in *Schmidingerella arcuata* from field material (scanning electron micrographs; originals of S. Agatha). (a) Lateral view showing the ciliate with a band of just secreted lorica material (arrowhead). It is uncertain whether the band was detached from the lorica rim by movements of the cell during the preparation procedure or is attached to the lorica rim only after reaching a certain length. (b) Detail showing the band of just secreted globules of lorica material (arrowhead). (c) Detail showing the recently secreted globules of lorica material (arrowhead) that started to merge with the already existing lorica portion (arrow). CM, collar membranelles; L, lorica. Scale bars 50 µm (a) and 5 µm (b, c).

Laval-Peuto & Brownlee (1986) introduced a terminology for the different types of lorica formed during a cell cycle (Fig. 2.9), as follows. (1) Protolorica (Greek prefix *protos*, first). The first lorica built by the naked proter just after division, using an intracellular reserve of lorica material. (2) Paralorica (Greek prefix *para*, beside). The replacement lorica built by a naked trophont, which has lost or abandoned its lorica; it might be of *coxliella* form (Fig. 2.8). (3) Epilorica (Greek prefix *epi*, upon). The part secondarily added by a trophont to the anterior end of a protolorica or paralorica; it is a kind of collar and often spiraled. Several successive epiloricae are possible.

Variability of loricae during the cell cycle in hyaline species

The occurrence of different lorica phenotypes in cultures or natural populations depends on the conditions

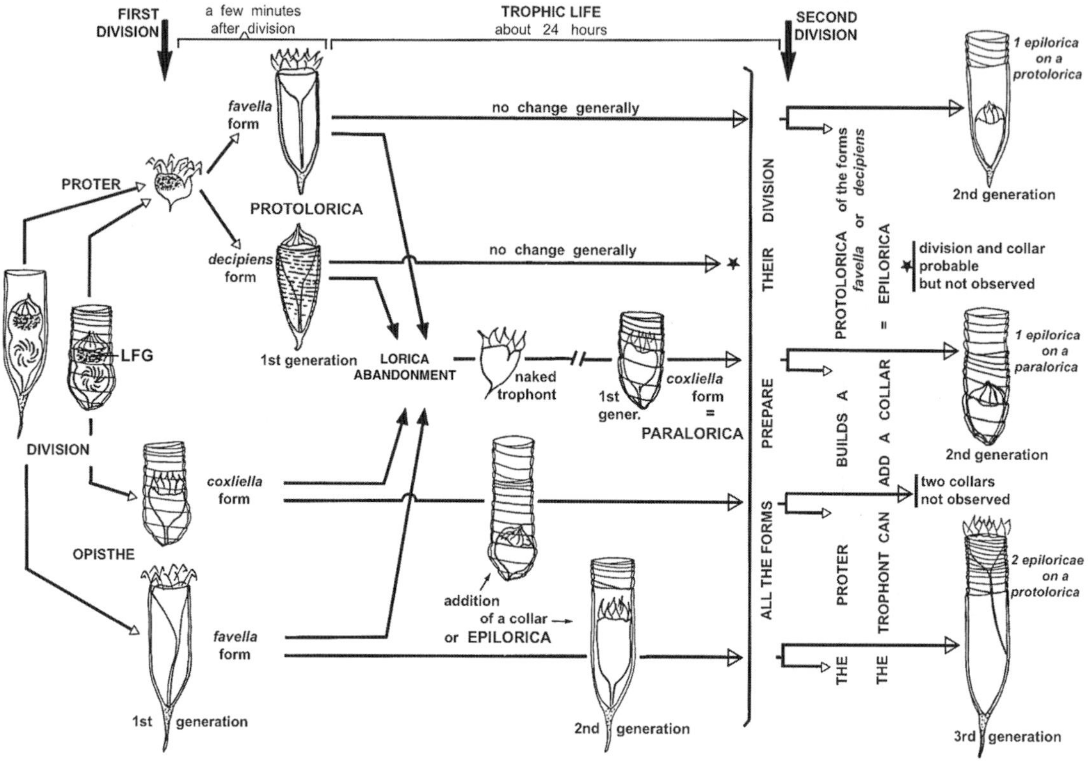

Fig. 2.8 Main lorica phenotypes formed by *Favella ehrenbergii* during two cell cycles (adapted from Laval-Peuto 1981). The protolorica is rapidly built by the proter just after cell division, using the reserve of lorica material; the *favella* form is the common phenotype, whereas the *decipiens* form and other types (abnormalities) are rarer. The opisthe keeps the old lorica. A paralorica is produced by a trophont which lost or abandoned its lorica; the *coxliella* form, slowly built, is an entirely spiraled phenotype. Epiloricae are one or more spiraled collars that are added to any phenotypes during the trophic life. In all phenotypes, the length depends on the amount of available lorica material, whereas their spiraled structure is caused by a slow secretion of the material. LFG, lorica forming granules.

encountered by the ciliate during its cell cycle, specifically during the construction of the protolorica and paralorica and the addition of one or more epiloricae (Figs 2.10 and 2.11). The most detailed studies were preformed on *Favella* by Laval-Peuto (1981, 1983, 1994).

The length of the constructed lorica depends on the amount of lorica material available. Accordingly, the bowl and posterior process might be more or less long, and the posterior process and lorica closure might even be lacking. Although the wall texture is invariable, the structure of the protolorica wall might vary. The wall is continuous and hyaline in the *favella* form of *Favella ehrenbergii*, *F. markusovszkyi*, *F. adriatica*, and *F. panamensis* representing the most common phenotype in natural populations (Figs 2.8, 2.10e, f, and 3.16a, c).

Protoloricae of the *favella* form are constructed within 2–10 minutes. The process apparently stops when the stored lorica material is used up. Abnormally short and deformed loricae may come from a shortage of material and its reduced plasticity. This is shown by ciliates that lose part of their lorica material owing to the adverse conditions under the light microscope; they are unable to complete lorica construction, pro-

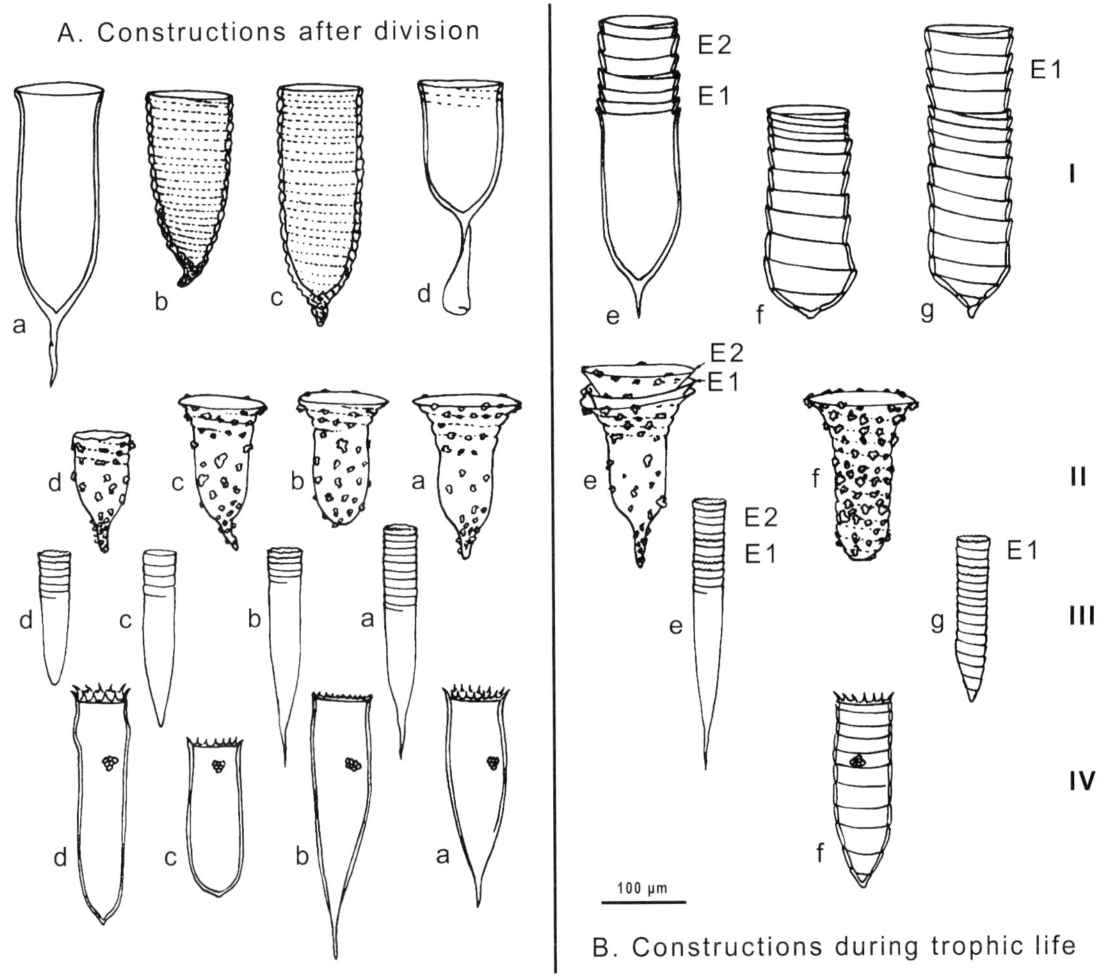

Fig. 2.9 Various phenotypes of loricae built by tintinnids after division (A) and during their trophic life (B; adapted from Laval-Peuto & Brownlee 1986). I *Favella ehrenbergii*; II *Tintinnopsis campanula*; III *Helicostomella subulata*; IV *Parafavella denticulata*. (a–c) The most common protoloricae; (d) abnormalities; (e) protoloricae with successively added epiloricae (E1, E2) or "collars"; (f) paraloricae (*coxliella* form); (g) paraloricae with added epiloricae. Some of the phenotypes were considered distinct species until their cell cycles were known, e.g., *Coxliella decipiens* (I b, c), *Coxliella annulata* (I f), and *Tintinnopsis bütschlii* (II b).

ducing ring-shaped loricae, which are promptly abandoned (Fig. 2.10j, k).

The *decipiens* form is named after *Coxliella decipiens*, which represents a further, less common lorica phenotype in the cell cycle of *Favella ehrenbergii* (Figs 2.8 and 2.10l–o). It is generally shorter than the *favella* form and has a stout twisted horn. The wall of the *decipiens* form is translucent and irregular because of both the compaction of its texture and its spiraled structure. In nature, the *decipiens* form is probably caused by less abundant and more viscous lorica material and certain environmental conditions; in the laboratory, it might be the result of elevated temperatures under the light microscope.

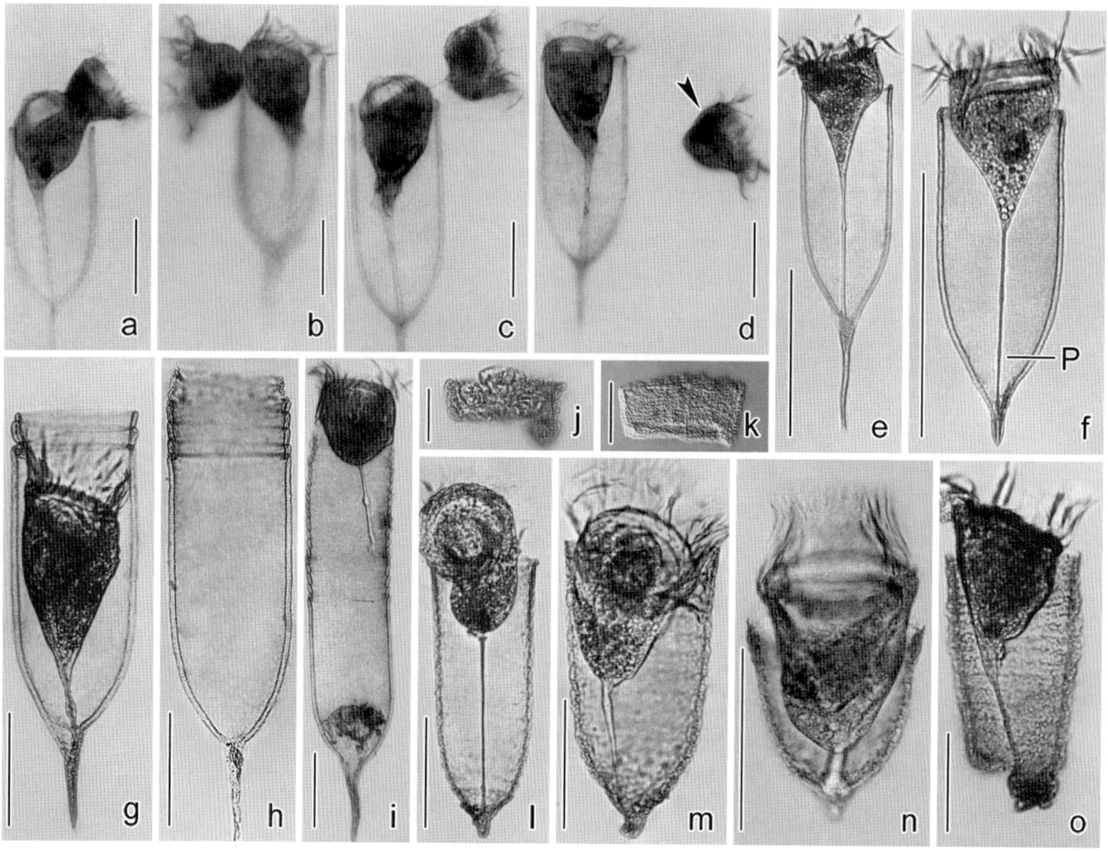

Fig. 2.10 Phenotypes produced by *Favella ehrenbergii* during the cell cycle (live cultured specimens; adapted from Laval-Peuto 1981). (a–d) Very late dividers (a–c) and postdividers (d). The arrowhead denotes the reserve of lorica material used by the proter to build a new lorica. (e, f) Protoloricae of *favella* form showing the variability in length. (g–i) Protoloricae of *favella* form with epiloricae of different length. Cultured specimen (i) shows exceptionally long epilorica not known from plankton samples. (j, k) Abnormal ring-shaped protoloricae. (l–o) Protoloricae of *decipiens* form showing the variability in length (l–n) and form of the posterior end (o). P, peduncle. Scale bars 100 μm (a–d, g–i, l–o), 200 μm (e, f), and 50 μm (j, k).

Besides the two types of protoloricae, a third distinct lorica phenotype is produced by *F. ehrenbergii*, namely a paralorica of *coxliella* form; the name refers to *Coxliella annulata* (Fig. 2.11). The construction of this replacement lorica can only occur when the granules of lorica material are being formed (perhaps only during the second part of the vegetative period). In contrast to the rapid production of the protoloricae, the secretion of the material for a paralorica lasts several hours and its spiraled accretion yields the helical wall structure. The paralorica is again hyaline and usually closed posteriorly, but it lacks a posterior process. It is quite similar in length and opening diameter to the bowls of the protoloricae. The same is true for the *coxliella* form of *Favella panamensis* (Fig. 3.16g).

Brandt (1907) and Entz Jr (1909b) already assumed that loricae of the *coxliella* form are part of the cell cycle in *Favella ehrenbergii*. Thus, the genus *Coxliella* is artificial; however, it cannot be disbanded as the species affiliations of its members are unknown (Chapter 3).

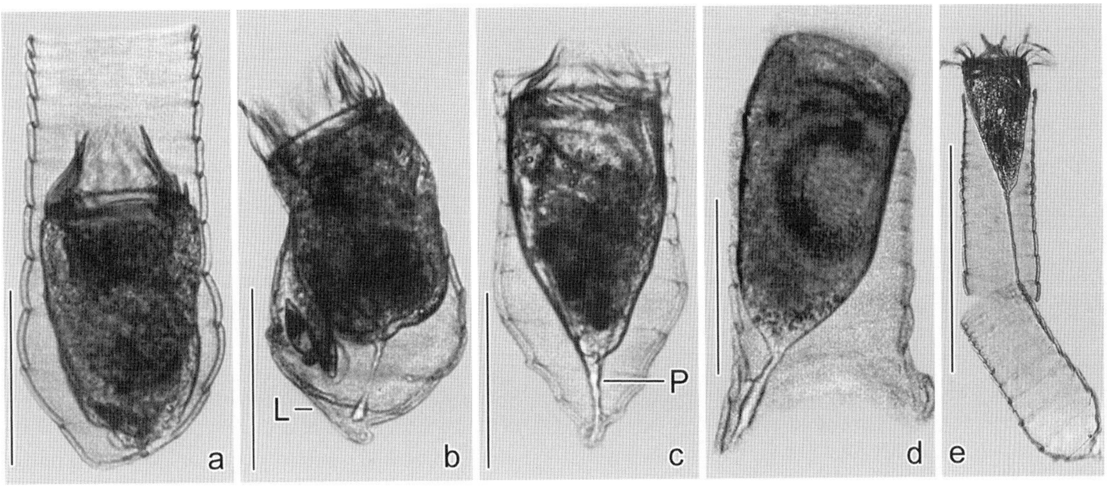

Fig. 2.11 Phenotypes produced by *Favella ehrenbergii* during the cell cycle (live cultured specimens; adapted from Laval-Peuto 1981). Paraloricae of *coxliella* form showing the variability in length (a–c) and an open posterior end (d). Cultured specimen (e) with an exceptionally long epilorica not known from plankton samples. L, lorica; P, peduncle. Scale bars 100 µm (a–d) and 300 µm (e).

Epiloricae may be added anteriorly to the protoloricae and paraloricae of *F. ehrenbergii* (Figs 2.8, 2.10g–i, and 2.11e). They form one or more additional collars with a helical structure, indicating a slow formation owing to the concomitant production of the lorica material by the trophont, like for paraloricae. When a lorica becomes very long because of the addition of epiloricae, the extended cell is no longer able to reach the lorica opening, and the peduncle is thus attached more anteriorly. However, extraordinarily long loricae as observed in cultures (Figs 2.10i and 2.11e) have not been found in natural populations, probably because they are cumbersome for locomotion. Accordingly, such long loricae are perhaps abandoned and a new lorica is built.

Both protoloricae and paraloricae are built from the posterior to the anterior end by the proter and the trophont, respectively. This contradicts previous hypotheses suggesting a backward construction and the assistance of the opisthe (Kofoid 1930), but it matches the observations by Biernacka (1952) on *Tintinnopsis*, which forms agglomerated loricae.

A phenotypic plasticity of the loricae, as found in *Favella ehrenbergii*, is also supposed for several further tintinnids, in which helical structures occur at least in parts of their hyaline loricae (Fig. 2.9; Laval-Peuto & Brownlee 1986). On the other hand, several other species apparently lack helical structures in their hyaline loricae, for example *Petalotricha* with its bowl-shaped loricae or *Salpingella* with its extremely long tubular loricae. Gold & Morales (1975b) suggested that the lorica of *Parafavella* is layered down by the cell in a yet-to-be described manner, involving the cell membrane as template, because no growth region is recognizable. These suggestions indicate that there might be further modes of lorica formation and different cell cycles in tintinnids with hyaline loricae; however, further studies, especially, on cultured species, are required for verification.

Formation of hard, agglomerated loricae

For species with agglomerated loricae (Figs 1.4 and 2.1f–j), it is the capability to agglutinate particles and the implicated complex behavior, which received most attention. However, the steps of cell division and lorica formation are not as easily observed as in hyaline forms. The principal question addressed has been the origin of the mineral or biogenic particles, which cover the entire lorica (e.g., in *Tintinnopsis* and *Leprotintinnus*) or only the bowl (e.g., in *Codonellopsis* and *Stenosemella*).

Loricae exclusively covered with diatoms or coccoliths (Fig. 2.2a), those with large particles on the bowl and small ones around the lorica opening (Fig. 2.1f), as well as loricae composed of an agglomerated bowl and a hyaline collar (Fig. 2.1g, h) indicate some kind of particle selection and a certain behavior of the ciliate during the lorica construction. Different sources of the particles have been suggested: biogenic particles from their food remnants (Hofker 1931a; Gold & Morales 1977; Takahashi & Ling 1984), suspended particles from the water column (Gold & Morales 1976a, c), and particles from the sediment (Gold & Morales 1976a, c; Rassoulzadegan 1980). Because tintinnids occasionally divide more than once a day (Chapter 9) and the distances between the sediment and the upper water layers with the tintinnid prey are large relative to tintinnid swimming speeds (Chapter 4), even in relatively shallow waters, the use of particles from the sediment is unlikely. Concerning particle selectivity, the findings are contradictory: Winter et al. (1986) suggested that tintinnids are not strongly selective, but restricted by particle size, whereas Takahashi & Ling (1984) and Wasik et al. (1996) assumed a capability to select biogenic particles, as at least some tintinnids do not single out the dominant ones. The chemical composition of the particles seems to be irrelevant (Gold 1979). In species with hyaline collars, the agglutination stops after construction of the agglomerated bowl probably because of a change in the ciliate's behavior, the habitat, or the stickiness of the lorica material; in *Stenosemella*, a second agglomerated collar might occur (Agatha & Tsai 2008).

The inconsistent results indicate that there is no single way in tintinnids to build an agglomerated lorica, but several. The experiments by Gold (1979) demonstrated that the lorica material loses its ability to agglutinate particles after hardening.

The construction of agglomerated loricae is probably not fundamentally different from that in species with hyaline loricae. This assertion is based on observations of species, such as *Tintinnopsis campanula*, from plankton samples (Daday 1887; Margalef & Durán 1953) and laboratory maintenance (Fig. 2.12). The rather translucent lorica of *T. campanula* allowed observations of the ciliate building the everted spiraled collar of its protolorica, to which a second everted collar (epilorica) may be added (Fig. 2.9); additionally, the species is able to produce an entirely spiraled paralorica (Fig. 2.9). A similar phenotypic plasticity is also expected in other species with agglomerated loricae. Because the attached particles often obscure the matrix, the spiraled structures, especially in the paraloricae, might, however, not be recognizable. It cannot be excluded that besides the mode of lorica formation described above, others exist.

Soft, agglomerated loricae

Soft, agglomerated loricae characterize the family Tintinnidiidae, which occurs in marine, brackish, and freshwater habitats. In lakes, the Tintinnidiidae are the dominating tintinnid group. Only two freshwater species with hard, agglomerated loricae are known: *Codonella cratera* and *Stenosemella lacustris* (Foissner & Wilbert 1979; Foissner & O'Donoghue 1990; Foissner et al. 1999).

Abiogenic and biogenic particles are incorporated in the jelly-like, structureless matrix of the soft loricae (Foissner & Wilbert 1979; Bernatzky et al. 1981). Observations of cell division and lorica construction are even rarer than in species with hyaline or hard, agglomerated loricae. Although these few studies reported the common process of lorica formation by the proter, they also indicated a second, very different mode by which the proter might obtain a lorica. Entz Sr (1885) noticed a bipartition of the lorica in *Tintinnidium fluviatile*. The rotation of the opisthe resulted in a twisted middle of the lorica. Finally, this narrowed portion broke, producing a posteriorly open lorica for the proter and an anteriorly shortened lorica for the opisthe. Similarly, Reck (1988) described a process named "lorica splitting" in *Tintinnidium pusillum*; here, the "co-operation" of both opisthe and proter caused the break of the twisted lorica. Reck considered "lorica splitting" an adaptive strategy against selective pressure in temperate lakes and assumed it also for coastal and estuarine species. In samples from marine systems, pairs of apparently recently divided *Tintinnidium* ciliates can be found in a single lorica (M. Laval-Peuto, unpublished observations). The prerequisite for lorica splitting is, in any case, the possession of a soft lorica.

Problems of lorica typological taxa

The tintinnid classification established by Kofoid & Campbell (1929, 1939) was based exclusively on char-

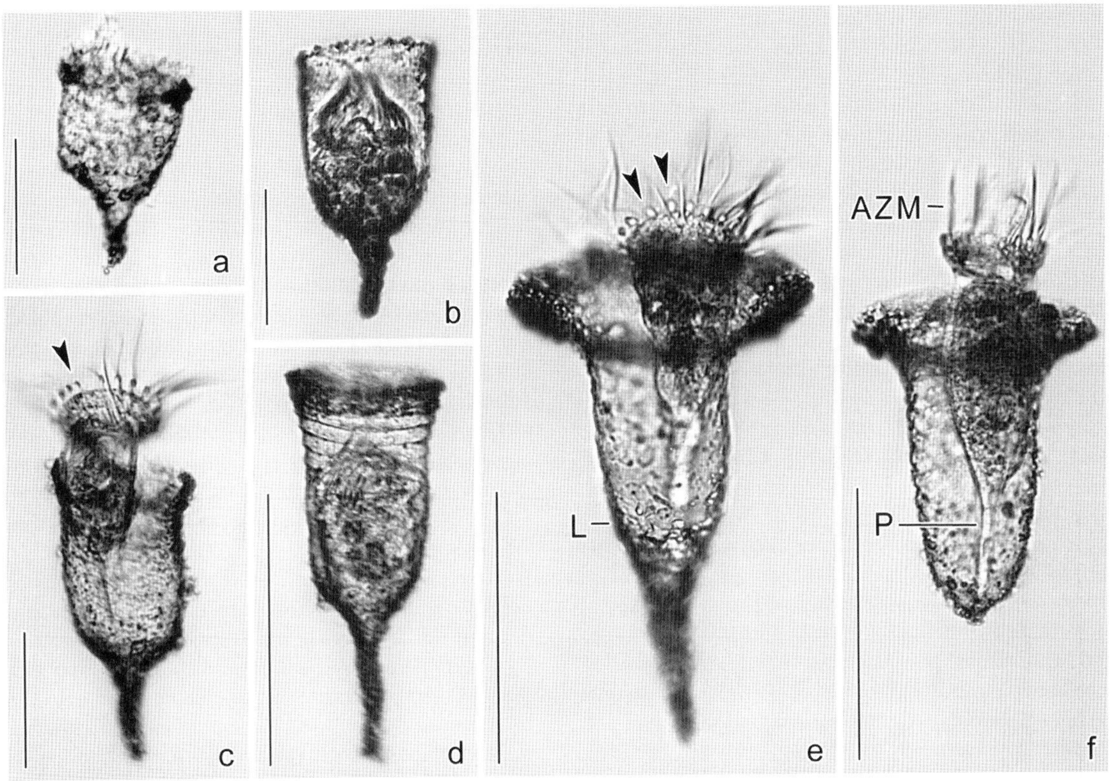

Fig. 2.12 Lorica formation in *Tintinnopsis campanula* from field material (from live specimens; originals of M. Laval-Peuto). The loricae are translucent owing to the only sparsely agglomerated particles. Arrowheads mark pin-shaped tentaculoids. First, the posterior lorica portion with a process is formed, and the ciliate attaches to the bottom of the lorica by its contractile peduncle (a, b). Next, the lorica elongates at the anterior end (c, d). The spiraled anterior lorica portion (d) indicates a slow formation. Finally, a flaring collar is formed, which has particles agglomerated also on its inner wall (e). Owing to the lack of a posterior process (f), this phenotype was regarded as a distinct species (*T. bütschlii*) by Kofoid & Campbell (1929). AZM, adoral zone of membranelles; L, lorica; P, peduncle. Scale bars 40 µm (a–c) and 80 µm (d–f).

acteristics of the loricae. By focusing on relatively small variations in the morphology of this extracellular product as a means to distinguish species, many forms and variants were raised to species rank (Chapter 3). Kofoid and Campbell cited, but neglected, previous excellent publications, for example that of Brandt (1907), who stated that species circumscriptions solely based on lorica features are arbitrary, depending on the recognition of variants as distinct species. Brandt even emphasized that only by the thorough study of the cell features as well as the lorica structure and texture, the real species limits are recognizable and a natural classification can be established. Despite early criticisms and ongoing critiques, Kofoid and Campbell's classifications have remained nearly intact as the system used by protistologists and ecologists, because the robust loricae are more easily collected, preserved, and thus studied than the fragile ciliate cells. The cytological features are recognizable only during patient live observation and after application of special staining techniques (Chapter 3). However, several cytological features are promising for the establishment of a natural tintinnid classification and the circumscription of species (for details, see Chapter 3). Differences in the phenotypic plasticity of the loricae, the chemical composition of the lorica material, the wall texture, and the morphology of the resting cysts are possibly also of taxonomic significance (Chapter 3; Laval-Peuto & Brownlee 1986).

Even nowadays, our knowledge about tintinnids is still insufficient to modify significantly the concept of Kofoid & Campbell (1929, 1939). Molecular methods combined with taxonomic investigations of the cell and the lorica, applying modern methods, will finally lead to a far-reaching revision of tintinnid classification (Chapter 3). Nevertheless, the monographs of Kofoid and Campbell are precious inventories of the early literature and nearly complete catalogues, showing the remarkable diversity of tintinnid loricae.

2.3 ULTRASTRUCTURE OF LORICAE

Although tintinnid loricae are often separated into two major groups, namely the "agglomerated" and "hyaline" ones, there are actually four general lorica types. The first, the supposedly primitive one (Kofoid & Campbell 1939), is soft and agglomerated. It occurs in *Tintinnidium*, *Membranicola*, and *Leprotintinnus* of the family Tintinnidiidae (Fig. 2.1i, j). The second type is also agglomerated, but hard, as in species of the genus *Tintinnopsis* (Fig. 2.1f). The loricae of the third type are composed of a bowl (broadened lorica portion) and collar of distinct compositions: the bowl is hard and agglomerated, the collar hyaline. This type is seen, for example, in species of the genera *Codonellopsis* and *Stenosemella* of the families Dictyocystidae and Stenosemellidae, respectively (Fig. 2.1g, h). The fourth lorica type is entirely hyaline and found, for instance, in *Climacocylis*, *Dadayiella*, *Eutintinnus*, *Favella*, *Rhabdonella*, *Schmidingerella*, and *Steenstrupiella* of the families Metacylididae, Ptychocylididae, Rhabdonellidae, and Tintinnidae (Figs 2.1a–e, 3.16a, c, g, and 3.17a, b). The cladistic and genetic analyses indicate that the hard, agglomerated loricae and the hyaline ones do not represent separate taxonomic entities (Chapter 3).

Within these types, investigations of the ultrastructure revealed distinct subtypes. This is apparent in the variety of textures reported for different species of the genus *Tintinnopsis*, all with hard, agglomerated loricae. So far, four distinct types of lorica wall matrixes have been described: (1) a monolaminar texture with alveoli is found in *T. lobiancoi* (Fig. 2.13b; Wasik et al. 1997a; Wasik 1998); (2) a matrix with a variable number of discontinuous alveolar layers in *T. parva* grown in particle-free cultures (Laval-Peuto et al. 1979); (3) a monolaminar and compact matrix found in a *Tintinnopsis* species (Laval-Peuto 1980); and (4) a monolaminar and fibrous matrix in *T. parvula* grown in almost particle-free cultures (Fig. 2.14f; Agatha 2010b). Besides the somatic ciliary pattern, the matrix texture thus represents a promising feature for a reliable subdivision of the paraphyletic genus *Tintinnopsis* (Chapter 3). In the case of *T. parva* and *T. parvula*, however, cultivation artifacts cannot be excluded, namely the textures of specimens from particle-free cultures might deviate from that of specimens from field material (Agatha 2010b).

The soft, agglomerated lorica matrix in *Tintinnidium fluviatile* appears structureless in the scanning electron microscope (Foissner & Wilbert 1979; Bernatzky et al. 1981). It resembles that of the freshwater species *Tintinnopsis cylindrata* (Fig. 2.1i, j; Bernatzky et al. 1981), whose close relationship to the genus *Tintinnidium* has also been inferred from its somatic ciliary pattern (Chapter 3). In loricae of a marine *Tintinnidium* species formed in an almost particle-free culture, a thin film covers several layers of fibrous globules held together by anastomosing fibers. Because these globules are similar to the just extruded lorica forming unites in other tintinnids (Figs 2.6 and 2.7), this texture possibly represents a cultivation artifact (Gold & Morales 1976c).

Loricae composed of an agglomerated bowl and a hyaline collar also show diverse wall textures, and differences might even occur within a single genus. For instance, a monolaminar texture with alveoli was found in *Codonellopsis gaussi* (Wasik et al. 1997a), whereas a possibly continuous organic matrix overlaid with fibrous material was recognizable in *C. americana* after removing the agglomerated particles (Gold & Morales 1977). These fibers resemble those composing the in vitro-formed loricae of *Tintinnopsis parvula* (Fig. 2.14f; Agatha 2010b), in which, however, an inner organic layer was not found. The lorica matrix of *Stenosemella* is again monolaminar with alveoli (Fig. 2.14e; own data of S. Agatha).

Three main types of wall texture are found in hyaline loricae. A monolaminar texture with alveoli occurs in the loricae of *Favella* (Figs 2.14b and 3.16d; Hedin 1975b; Laval-Peuto 1994), *Rhabdonella* (Fig. 2.14c), *Schmidingerella* (Figs 2.14a and 3.17d), *Xystonella* (S. Agatha, unpubl. observations), *Helicostomella*

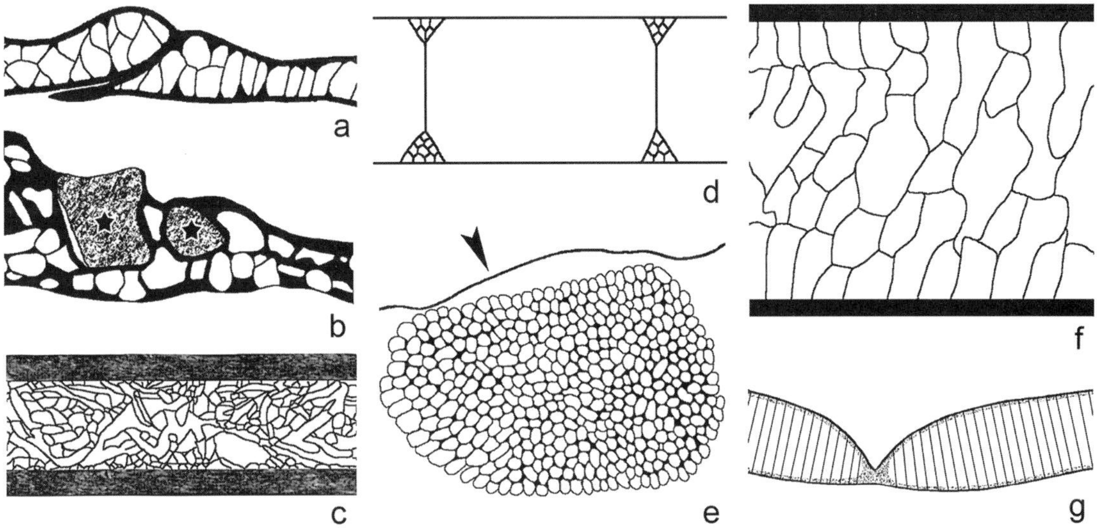

Fig. 2.13 Schemes of lorica wall textures based on transmission electron micrographs of field material (a, b, based on data from Wasik et al. 1997a; c, e, f, based on data from Laval-Peuto 1994; d, based on data from Hedin 1975b; g, based on a micrograph kindly provided by W. Coats). (a) Monolaminar texture of the overlapping spirals in the about 0.9 µm thick, hyaline collar wall of *Helicostomella subulata*. (b) Monolaminar texture with alveoli and embedded particles (asterisks) in the about 1.5 µm thick, agglomerated wall of *Tintinnopsis lobiancoi*. (c) Trilaminar texture with three layers of different tubules in the about 1.2 µm thick, hyaline wall of *Petalotricha ampulla*. (d) Monolaminar and alveolar texture and triangles with minute alveoli in the about 1.8 µm thick, hyaline wall of *Parafavella gigantea*. (e) Monolaminar and tubular texture probably of a mullion (about 1.8 µm × 2.7 µm) in *Dictyocysta elegans*. Arrowhead marks the lorica sac lining the lorica. (f) Trilaminar texture in the about 2.8 µm thick, hyaline wall of an *Undella* species. (g) Fusion of two cross-striated circular platelets in the monolaminar and compact, about 0.5 µm thick, hyaline wall of *Eutintinnus pectinis*.

(Fig. 2.13a), *Laackmanniella* (Wasik et al. 1997a; Wasik 1998), and *Parafavella* (Fig. 2.13d; Hedin 1975b; Sokolova & Gerassimova 1984; Wasik et al. 1997a). In *Eutintinnus*, the lorica wall is monolaminar and compact, as probably in *Amphorides*, *Salpingella*, and several further genera. Transmission electron micrographs showed that the *Eutintinnus* lorica is apparently composed of cross-striated platelets stuck together with their narrowed margins (Figs 2.13g, 6.7a; Laval-Peuto 1994).

A trilaminar texture with thick inner and outer layers enclosing a tubular middle layer was found in the hyaline loricae of *Cyttarocylis* and *Petalotricha* (Fig. 2.13c; Laval 1972; Laval-Peuto 1994), whereas the middle layer is alveolar in *Proplectella* and *Undella* (Fig. 2.13f; Laval-Peuto 1980, 1994). Different wall textures occur in the genus *Cymatocylis*. In *C. drygalskii* and *C. vanhöffeni*, the wall is monolaminar with alveoli (Wasik et al. 1997a; Wasik 1998). In contrast, *C. convallaria* has a trilaminar wall texture with alveolar inner and outer layers and a tubular middle layer, which is similar to that found in *Petalotricha*. Furthermore, the wall texture changes towards the anterior lorica end in *C. convallaria* owing to a reduction of the middle layer and its final disappearance in the collar (Wasik & Mikołajczyk 1992). In some hyaline species, the alveolar pattern becomes more complex by a branching of the alveoli septa (*Cymatocylis*, *Helicostomella*; Wasik et al. 1997a; Wasik 1998) or the formation of triangles with minute alveoli at their corners (*Parafavella*; Fig. 2.13d; Hedin 1975b; Sokolova & Gerassimova 1984; Wasik et al. 1997a).

A tubular texture occurs in the frames of the fenestrated lorica in *Dictyocysta* (Figs 2.13e, 2.14d;

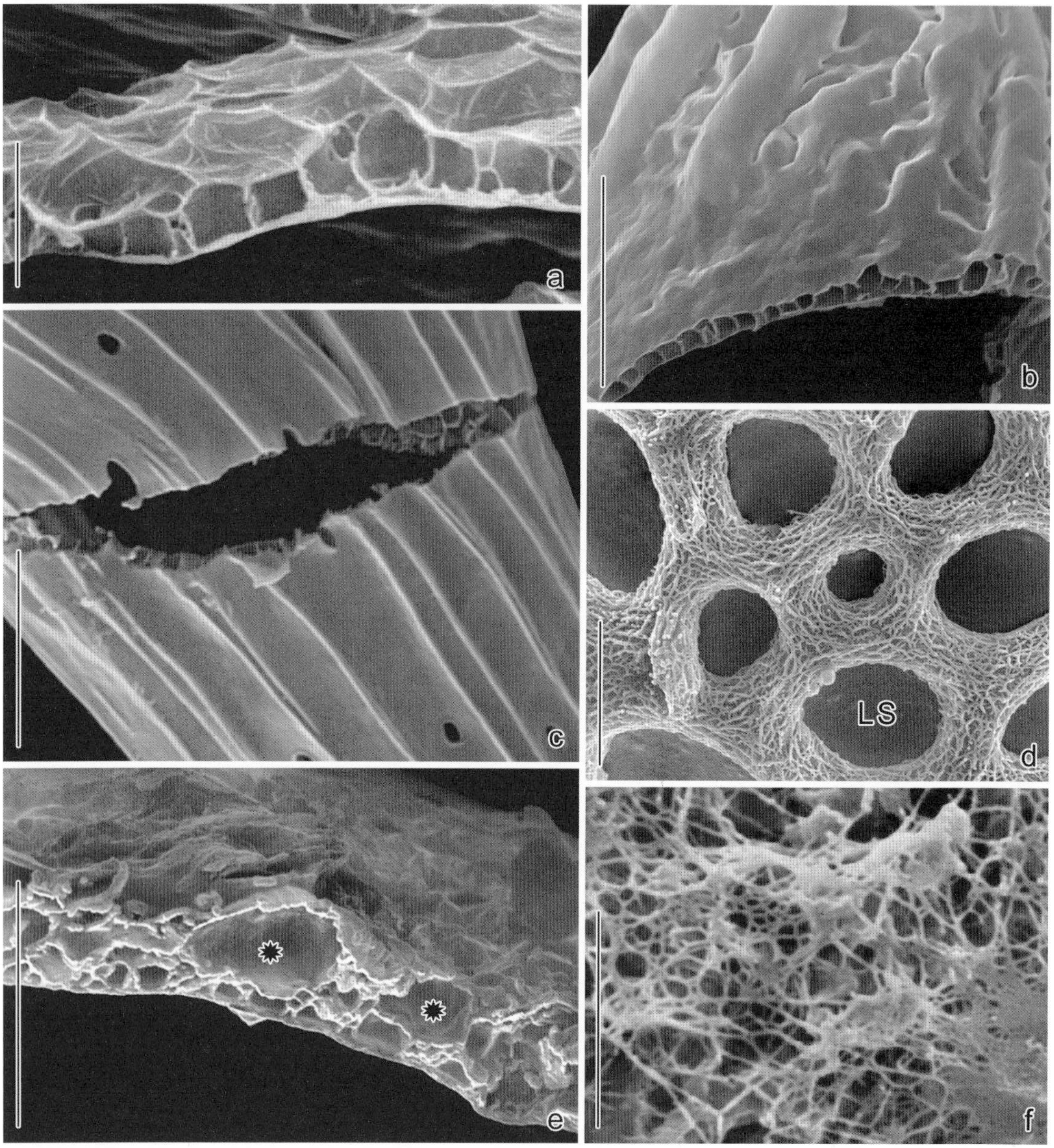

Fig. 2.14 Rupture surfaces (a–c, e) and surface (d, f) views showing the wall textures in loricae from field material (a–e) and a particle-free culture (f) in the scanning electron microscope (originals of S. Agatha). (a) Monolaminar texture with alveoli and reticulate ridges on the outer surface in *Schmidingerella arcuata* (compare with Fig. 3.17). (b) Monolaminar texture with alveoli and smooth outer surface in *Favella panamensis* (compare with Fig. 3.16). (c) Monolaminar texture with alveoli, spiraled surface ribs, and pores in *Rhabdonella spiralis*. (d) Tubular texture in *Dictyocysta mitra*. The lorica is lined by a lorica sac (LS). (e) Monolaminar texture with alveoli in *Stenosemella ventricosa*. Owing to the rupture, some formerly embedded particles were lost (asterisks). (f) Monolaminar texture with fibers in *Tintinnopsis parvula* from a particle-free culture (possibly an artifact). Scale bars 2 μm (a, f), 10 μm (b), and 5 μm (c–e).

Laval-Peuto 1994; Agatha 2010a). A further texture type was found in the hyaline lorica of *Dadayiella ganymedes*. In this species, the lorica wall shows a complex pattern of longitudinal and dextrally spiraled lines and its outer lorica surface exhibits irregularly arranged globules of different sizes and ridges formed by whorls overlapping the posterior portion of the next whorl (Lecal 1967).

Overall, it is obvious that the lorica wall texture in tintinnids varies considerably among the basic lorica types and occasionally even among species of the same genus. Thus, the similarities in lorica morphology that unite species in a genus or genera in higher-level taxa are not necessarily reflected in the texture of the loricae. However, the current classification of tintinnids does not account for differences in wall texture. So, the wall texture may finally turn out to be of taxonomic significance in some groups. For instance, in the paraphyletic genus *Tintinnopsis*, the different wall textures might be characteristic of the separate branches in the morphologic and gene trees (Figs 3.2 and 3.13).

2.4 CHEMICAL COMPOSITION OF LORICAE

History of chemical studies

Several protists are able to construct tests, shells, or loricae, which are often quite elaborate. The study of these fascinating loricae concerns not only their morphology and formation process, but also their chemical composition. In foraminifera, the test walls and matrixes sticking together the agglomerated particles are of proteinaceous nature (Hedley 1963; Pierce et al. 1968; Hedley & Rudall 1974; Bowser & Bernhard 1993); the material was often called "tectin" or "pseudochitin" (Hedley 1963). Likewise, the organic tests of Amoebozoa (Moraczewski 1970, 1971a, b) and filose amoebae (Hedley 1960) consist of proteins. Chitin was detected in the loricae of the peritrich ciliate *Cothurnia* sp. and the heterotrich ciliate *Parafolliculina violacea* as well as in the resting cysts of the genera *Blepharisma, Bursaria, Climacostomum, Cyclogramma, Euplotes, Fabrea, Nassula, Nassulopsis, Phacodinium, Pseudomicrothorax, Telotrochidium* (Bussers & Jeuniaux 1974), and *Meseres* (a rather close relative of the tintinnids: see Chapter 3 and Foissner et al. 2007). In other ciliate species, the resistant resting cysts contain other polysaccharides, proteins, and/or lipids (Bussers & Jeuniaux 1974); however, proteins are usually among the main components.

In tintinnids, there is a long history of investigations into the chemical composition of their loricae, dating back to Fol (1881). The most comprehensive studies were conducted by Daday (1887), Entz Jr (1909a, b), and Hofker (1931a). Usually, a chitinous nature of the lorica wall or matrix was inferred from their resistance against strong bases. However, Entz Jr (1909b) and Bussers & Jeuniaux (1974) excluded chitin, at least for some species, and Entz Jr (1909b) suspected a proteinaceous substance. Later studies, even using energy-dispersive X-ray spectroscopy (EDX analysis; Wasik et al. 1997b) or further histochemical methods (Gold 1968, 1980; Gold & Morales 1975b) failed to identify clearly the composition of tintinnid loricae. Consequently, Agatha & Simon (2012) recently readdressed the subject, applying previous techniques and new methods, for example enzymatic digestion and high-resolution transmission electron microscopy. Their results, along with those from previous studies, are summarized below (for details on the species studied and on the histochemical and proteolytic methods, see Agatha & Simon 2012).

Cytochemical stains and enzymatic experiments

Most studies were performed on loricae of marine tintinnids. Thus, most data are available on hyaline and hard, agglomerated loricae, whereas the soft, agglomerated loricae of the species mainly occurring in freshwater have rarely been studied.

Former investigations inferred from the resistance of the loricae against strong bases a chitinous material. Actually, the loricae demonstrated an astonishing resistance, withstanding potassium hydroxide at 160 °C for 40 minutes, in the experiments of Agatha & Simon (2012). However, the Van-Wisselingh test did not reveal chitin in the hyaline lorica of *Rhabdonella spiralis* and the hard, agglomerated ones of *Codonella aspera* and *Stenosemella ventricosa*. Furthermore, a dissolution of the loricae took place after an elongated exposure to the strong, hot base (Bussers & Jeuniaux 1974; Agatha & Simon 2012). So, the alkaline treatment was probably insufficient (too short exposure periods and/or too low temperatures) in former studies to observe the disintegration of the loricae, and the chitin-specific staining was not applied. The failure of the chitinase digestion in the hard, agglomerated loricae of *Codonella aspera*, *Tintinnopsis* cf. *cylindrica*, and *T. compressa*, and the hyaline ones of *Eutintinnus brandti* and *Rhabdonella spiralis*, also excluded chitin as main component of the lorica walls and matrixes (Agatha & Simon 2012). Accordingly, the inference of a chitinous lorica material in the former studies was probably based on inadequate experiments.

The presence of various other polysaccharides was excluded by the periodic acidic Schiff reaction in the hard, agglomerated lorica of *Tintinnopsis* cf. *cylindrica* and the hyaline one of *Rhabdonella spiralis* (Agatha & Simon 2012). After application of iodine alcohol or Lugol's solution, the hard, agglomerated loricae of *Codonella* and *Tintinnopsis beroidea* and the hyaline ones of *Petalotricha*, *Proplectella*, and *R. spiralis* became yellow–brown (Entz Jr 1909b), indicating glycogen. This demonstrates a variability in the occurrence of glycogen in *R. spiralis* and the genus *Tintinnopsis*. Based on the data available, it can be concluded that polysaccharides are probably never the main component in tintinnid loricae.

A proteinaceous, horny, or keratin-like substance in the lorica walls and matrixes was assumed by Kent (1880–1882), Entz Jr (1909b), and Laval-Peuto & Barria de Cao (1987). Because keratin is easily soluble in potassium hydroxide, its presence can be excluded in the hyaline loricae of *Amphorides*, *Cyttarocylis*, *Petalotricha*, *Rhabdonella*, and *Undella*, the soft, agglomerated lorica of *Tintinnidium*, and the hard, agglomerated loricae of *Codonella*, *Codonellopsis*, *Dictyocysta*, and *Tintinnopsis*. Furthermore, keratin stains with Millon's solution (it detects some phenols, especially the amino acid tyrosine), but most loricae did not color (Fol 1881; Daday 1887; Fauré-Fremiet 1908; Entz Jr 1909a, b; Bussers & Jeuniaux 1974), and hair keratin produced a much higher sulfur peak in the EDX analysis than the loricae (Agatha & Simon 2012).

Some proteins are, however, known to persist in hot and concentrated potassium hydroxide for a while and they did not stain with Millon's reagent (Krishnan 1954; Moraczewski 1971a). The findings of Agatha & Simon (2012) actually indicated a proteinaceous composition of the lorica walls and matrixes: (1) the proteinase K digestion of the hyaline loricae of *Climacocylis* and *Eutintinnus* and the hard, agglomerated one of *Tintinnopsis levigata*; (2) the detection of nitrogen in the EDX analyses of the hyaline loricae in *Eutintinnus* and *Rhabdonella* and the hard, agglomerated ones of *Codonella*, *Codonellopsis*, *Stenosemella*, and *Tintinnopsis* (Fig. 2.15 and Plate 2.1); (3) the crystal lattice revealed by high-resolution transmission electron microscopy

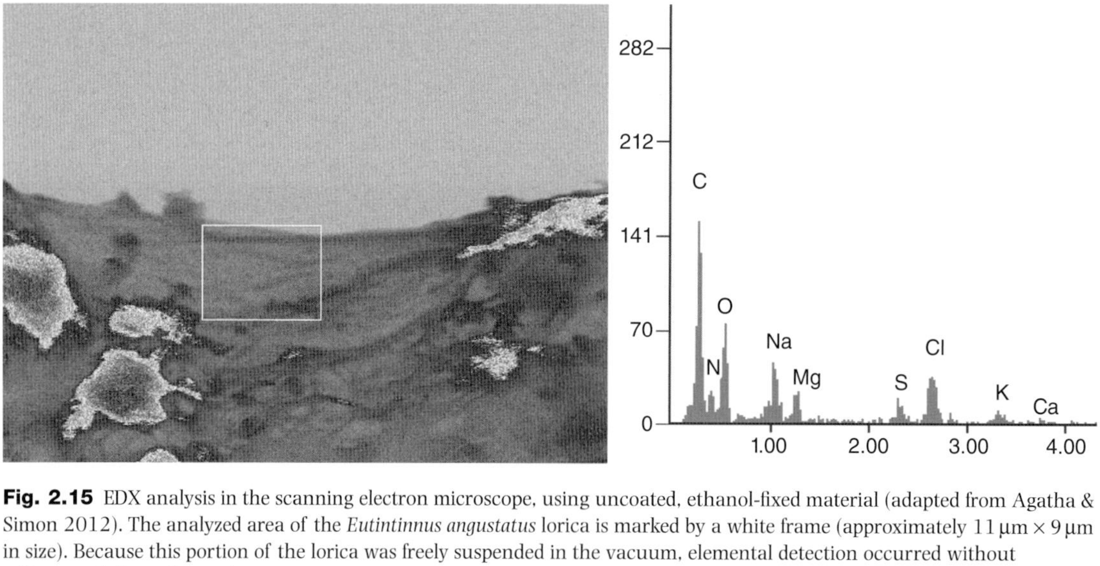

Fig. 2.15 EDX analysis in the scanning electron microscope, using uncoated, ethanol-fixed material (adapted from Agatha & Simon 2012). The analyzed area of the *Eutintinnus angustatus* lorica is marked by a white frame (approximately 11 μm × 9 μm in size). Because this portion of the lorica was freely suspended in the vacuum, elemental detection occurred without influence of the carbon substrate.

in the hyaline *Eutintinnus* lorica, which resembles the proteinaceous surface layer (S-layer) of bacteria and archaea (Fig. 2.16b–d); and (4) the striation recognizable in sections of the hyaline *Eutintinnus* lorica wall (Figs 2.13g and 6.7a), which is similar to that produced by crystalline proteins in extruded trichocysts and kinetodesmal fibers of ciliates.

By means of EDX analysis, Wasik et al. (1997b) analyzed the hard, agglomerated loricae of *Codonellopsis gaussi*, *Laackmanniella naviculaefera*, and *Tintinnopsis lobiancoi* and the hyaline ones of *Cymatocylis affinis/convallaria*, *Helicostomella subulata*, and *Parafavella denticulata*. The EDX spectrogram was only shown for *L. naviculaefera* without a detailed description of the findings. Interestingly, the highest EDX peaks observed corresponded to copper and zinc, indicating that the lorica of *L. naviculaefera* consisted of brass, which is less likely; thus, the composition of the electron microscopic stub was probably analyzed. Recently, Kaulich et al. (2009) performed low-energy X-ray fluorescence microscopy on the hard, agglomerated lorica of *Tintinnopsis radix*. All loricae analyzed by Agatha & Simon (2012) revealed carbon, nitrogen (not mentioned by Kaulich et al. 2009), and oxygen, indicating organic lorica walls and matrixes (Fig. 2.15 and Plate 2.1). Inorganic components, such as sodium, potassium, magnesium, chlorine, and sulfur, which stem from sea water traces in the samples, were also recorded. In agglomerated loricae or lorica portions, peaks of further elements occurred, namely of aluminum and silicon (Plate 2.1), owing to the presence of mineral particles (Gold & Morales 1975b; Agatha & Simon 2012).

In contrast to proteins, the polysaccharides cellulose, starch, and glycogen do not exhibit nitrogen in EDX analyses, and chitin with its *N*-acetyl groups can be excluded to be a lorica component by the Van-Wisselingh test and/or the failure of the chitinase digestion. Therefore, the data indicate a proteinaceous material in the lorica walls and matrixes.

Staining with mercuric bromophenol blue was conducted by Gold (1968), Gold & Morales (1975b), Wasik et al. (1997b), and Agatha & Simon (2012). A positive reaction was found in the hyaline loricae of *Climacocylis*, *Cymatocylis*, *Eutintinnus*, *Parafavella*, and *Rhabdonella*, and the hard, agglomerated ones of *Codonella*, *Codonellopsis*, and *Laackmanniella* (Plate 2.2a–f). An intrageneric variability was observed in *Tintinnopsis* with a coloration of *T. lobiancoi*, but not of *T. compressa* and *T.* cf. *cylindrica*. For *Helicostomella subulata*, inconsistent results were obtained, possibly because of the application of different fixatives. Specimens of *Rhabdonella spiralis* varied in their staining intensity even in the same experiment. Although quite often applied, the significance of this stain concerning the detection of proteins is doubtful (Baker 1958; Kanwar 1960). Acid fuchsin, clearly indicating proteins, and phosphor-molybdenum acid, indicating proteins and alkaloids, gave positive results in *Tintinnopsis* (Merkle 1909).

Further stains, which were more or less specific for proteins or certain amino acids, produced inconsistent results. Intrageneric variations were observed, especially, in the hard, agglomerated loricae of *Codonella*, *Codonellopsis*, and *Tintinnopsis* and the hyaline lorica of *Favella* (for details, see Agatha & Simon 2012). Intraspecific differences were recorded in the freshwater species *Codonella cratera* with its hard, agglomerated lorica (Entz Jr 1909a, b; Foissner & Wilbert 1979). Even within a single lorica, deviations in the staining properties occurred (*Stenosemella ventricosa*; Plate 2.2g–i; Agatha & Simon 2012).

The first high-resolution transmission electron microscopic study was performed on the hyaline, monolaminar, and compact lorica of *Eutintinnus angustatus* (Agatha & Simon 2012). This mode of transmission electron microscopy allowed distinctly deeper insights into the composition of lorica by displaying crystallographic structures. Possibly because of the desiccation of the ethanol-fixed loricae, the walls were slightly wrinkled and ruptured (Fig. 2.16b). Minute details of the lorica surface became recognizable by high-resolution transmission electron microscopy, namely a crystal lattice (Fig. 2.16b–d) and electron-dense aggregates of bitter salt and spots of sodium and potassium chloride (Fig. 2.16b). The ultrastructure was identical in different regions of the lorica wall. By means of a computer algorithm, the fast Fourier transform, periodic structures in the image gave rise to sharp spots in the resulting diffraction pattern. After application of Fourier filtering and at maximum magnification, the lorica surface revealed a hexagonal symmetry of the crystal lattice with a periodicity of about 23.7 nm; the finest structural details of the crystal hold a size of approximately 3 nm. Three primitive rhombohedral unit cells each form the hexagonal structures recognizable in the Fourier-filtered micrograph (Fig. 2.16c). In the deduced simple model of the crystal structure, the main motifs consist of triangles with a diameter of approximately 18 nm. They are interconnected on

Fig. 2.16 Transmission electron micrographs of the hyaline lorica in *Eutintinnus angustatus* at different magnifications (adapted from Agatha & Simon 2012). (a) Overview of right lorica half. The 130 μm × 80 μm-sized lorica lies almost horizontally on the electron transparent holey carbon substrate. The black rectangular structures are the copper bars of the transmission electron microscopic grid. (b) At higher magnification, a crystal lattice is recognizable with some dark spots (arrow) caused by sodium and potassium chloride. (c) At the highest magnification and after application of Fourier filtering, hexagonal structures are recognizable in the crystal lattice. (d) Scheme of the crystal structure (for details, see text).

each side by tiny tubes approximately 7 nm in diameter. The empty spaces between the triangles and tubes, respectively, appear dark (Fig. 2.16d).

The crystal lattice recognizable in the surface view of the *Eutintinnus angustatus* lorica is very similar to the proteinaceous surface layers (S-layers) of bacteria and archaea (Schuster & Sleytr 2000; Sleytr et al. 2001). Because only a single high-resolution transmission electron microscopic study of a tintinnid lorica is currently available, it is unknown whether the crystalline composition is typical of tintinnids, namely shared by the monolaminar textures with alveoli and the trilaminar textures and all other compact, monolaminar walls.

Conventional transmission electron microscopy of longitudinal and transversal wall sections of the hyaline lorica in *Eutintinnus* shows a cross striation due to alternating thin, electron-dense and broad, electron-lucent stripes (Fig. 2.13g; Laval-Peuto 1994). A similar striation and periodicity of the stripes (approximately 35 nm compared with 50–55 nm) is found in the "body" of extruded trichocysts in *Paramecium* and *Frontonia*, which consist of crystalline proteins (Hausmann 1978; Rosati & Modeo 2003); both the extruded lorica material and the extrusomes remain insoluble upon secretion and undergo a rapid change of state, resulting in an extended extracellular form.

Enzyme digestion experiments provided conflicting results, revealing a proteolytic resistance in the hyaline loricae of *Schmidingerella arcuata* and *Rhabdonella spiralis* and the hard, agglomerated loricae of *Stenosemella pacifica*, *Tintinnopsis* cf. *cylindrica*, and *T. compressa* (Agatha & Simon 2012). However, these results do not necessarily reject the idea of proteinaceous lorica material, as proteolytic-resistant proteins are known from prions (McKinley et al. 1983; Basu et al. 2007), bacteria (Butler et al. 1991), and the surface layer (S-layer) of archaea (Kaminski et al. 2010).

Despite intrageneric and even intraspecific inconsistencies in the staining and proteolytic properties, data in the literature indicate proteins in 13 of the 24 tintinnid genera studied, namely in those representing the four main lorica types: *Climacocylis*, *Codonella*, *Codonellopsis*, *Dictyocysta*, *Eutintinnus*, *Favella*, *Laackmanniella*, *Leprotintinnus*, *Parafavella*, *Rhabdonella*, *Stenosemella*, *Tintinnidium*, and *Tintinnopsis*. In the remaining genera, proteins were not or not unequivocally detected, probably because the tests were not specific for proteins or were restricted to certain groups of proteins or amino acids. Furthermore, the loricae displayed a rather uncommon proteolytic resistance, resembling that of prions, bacteria, and archaea. The most parsimonious assumption is a single origin of the loricae (Chapter 3) and a common kind of organic macromolecule. Therefore, it seems likely that all loricae consist of proteins, which are responsible for the resistance of the loricae against strong, hot bases. Only the protein composition and the composition of the additional substances (e.g., carbohydrates, lipids) vary, causing the observed differences between and within the genera and occasionally even within a single species.

Acid treatments again displayed deviations between and within genera; thus, differences in the solubility were apparently not correlated with the lorica structure (for details, see Agatha & Simon 2012). Hydrochloric acid not only dissolved the calcium carbonate particles (coccoliths) in hard, agglomerated loricae (Gold & Morales 1977), but at least occasionally also the loricae themselves. On the other hand, hydrofluoric acid merely removed the siliceous particles from the hard, agglomerated loricae of *Codonellopsis*, *Stenosemella*, and *Tintinnopsis rapa*, whereas the lorica matrixes or hyaline loricae were not influenced (Gold & Morales 1977; Gold 1980; Agatha & Simon 2012); only in *Tintinnopsis wangi*, the lorica completely broke down, probably because of a very low content of resistant matrix substance (Gold 1980).

The lorica forming granules have rarely been studied ultrastructurally and histochemically. The magnification of the available transmission electron micrographs is insufficient for the recognition of crystalline structures (Fig. 2.3). However, the osmiophilic nature of the granules, indicating lipids, unsaturated –C=C– bonds, and/or proteins (especially, those with lysine, arginine, tryptophane, histidine, and cysteine), is in accordance with the assumption of a proteinaceous lorica material.

Taxonomically significant differences in the chemical composition of the loricae were predicted by Kofoid (1930). Although histochemical and enzymatic experiments actually demonstrated a variability in the reactions, general differences between the hyaline, the soft, agglomerated, and the hard, agglomerated loricae as well as between freshwater and marine species are not obvious. Correlations between the wall texture, the chemical composition of the lorica material (Laval-Peuto 1994), and the granule types are also not evident. Distinctly more data are required to elucidate the taxonomic significance of these features.

Despite the number of investigations, identification of the chemical elements provided by the mother cell and incorporated by the proter into the different kinds of lorica (hard, agglomerated; soft, agglomerated; hyaline; composed of a hyaline collar and an agglomerated bowl) is still lacking completely. However, these data are needed to assess properly the role of the tintinnids and their sedimenting loricae in energy flux as well as in distribution and fate of the organic compounds in marine and freshwater ecosystems (Boltovskoy et al. 1993; Peulvé et al. 1996).

2.5 LORICA SEDIMENTATION

Because tintinnids are less abundant and diverse in freshwater, most studies have dealt with loricae of marine species. In laboratory experiments with empty loricae, sinking rates were estimated. For the hyaline loricae of *Acanthostomella norvegica*, *Schmidingerella taraikaensis*, *Helicostomella subulata*, *Parafavella* species, and *Ptychocylis obtusa*, the sinking velocities were calculated to be 0.25–2.08 m d^{-1} (Suzuki & Taniguchi 1995). Not surprisingly, agglomerated loricae sink faster; their sinking rates were estimated to be about 5.7 m d^{-1} (66 ± 4.1 µm s^{-1} in *Tintinnopsis lobiancoi*; Jonsson 1989) and 1.9–15.9 m d^{-1} (*Tintinnopsis beroidea*, *T. ampla*; Suzuki & Taniguchi 1995).

Interestingly, bulk community sedimentation rates appear to vary within a relatively narrow range (a factor of 5) among quite distinct systems, as indicated by sediment trap material. In the northeast Pacific, rates of up to 500×10^3 loricae $m^{-2} d^{-1}$ were recorded from traps at about 200 m depth (Price & Pospelova 2011). In the Norwegian Sea, a maximum value of about 300×10^3 loricae $m^{-2} d^{-1}$ was found at 500 m depth (Bathmann et al. 1990), which is similar to the rate of $221 \times 10^3 \pm 142 \times 10^3$ loricae $m^{-2} d^{-1}$ at 100 m depths estimated for a Norwegian fjord (González et al. 1994). In the southeast Pacific, a rate of 250×10^3 loricae $m^{-2} d^{-1}$ at 2300 m depth was estimated (González et al. 2004), which is quite similar to the figure of about 201×10^3 loricae $m^{-2} d^{-1}$ at 2195 m depth for the northeast tropical Atlantic (Boltovskoy et al. 1996). These rates are comparable to the 150×10^3 loricae $m^{-2} d^{-1}$ found at 100 m depth in the subtropical northeast Atlantic (Passow & Peinert 1993) and the 125×10^3 loricae $m^{-2} d^{-1}$ at 853 m depth in the eastern equatorial Atlantic (Boltovskoy et al. 1993).

Gilron & Lynn (1989a) estimated the lorica carbon content of hard, agglomerated loricae in four species and of hyaline loricae in six species, using the regression equation of Verity & Langdon (1984). The carbon content of their loricae ranged from 1569 pg in *Dadayiella ganymedes* to 73 671 pg in *Ascampbelliella urceolata* with an average of 15 828 pg carbon per lorica, and the loricae contributed 50–79%, on average 63%, to the total carbon content of the tintinnids. Using the maximum sedimentation rate of 500×10^3 loricae $m^{-2} d^{-1}$ recorded by Price & Pospelova (2011) and the average carbon content of the loricae estimated by Gilron & Lynn (1989a), about 7.9 mg carbon per day and per square metre might have sedimented in the northeast Pacific.

The chemical composition of the tintinnid loricae affects their decomposition during sedimentation and at the bottom of the sea or lake. The loricae are rather resistant against various chemicals and enzymes, as also indicated by fossilized and intact loricae in copepod fecal pellets (Fig. 5.3; Turner & Anderson 1983; Turner 1984b). However, loricae are apparently rare in copepod fecal pellets compared with the actual abundances and production rates of tintinnids, constituting 1.6–13.7% in subarctic waters and 3.0–15.6% in subtropical waters. Thus, most loricae probably sediment without passing a copepod digestive tract (Suzuki 2000).

Sediment traps from about 3800 m depth in the Panama Basin collected both intact hyaline and hard, agglomerated loricae with a dominance of the latter type (Ling 1992), and about 1000 loricae m^{-3} were recorded in 7000 m depth (Suzuki & Taniguchi 1995). These loricae were not decomposed during sedimentation, which might correspond to the proteolytic resistance observed in some species (Agatha & Simon 2012). Peulvé et al. (1996) concluded that the N-containing compounds in the material from sediment traps in 1000 m and 2000 m depth provide an intermediate stability of the organic nitrogen, namely that their macromolecular form and/or structure prevent them from degradation and thus may lead to compounds acting as a substantial nitrogen source for the benthos.

However, less than 1% of the standing crop is collected in the deep traps, as concluded by Boltovskoy et al. (1996) based on tintinnid abundances in surface waters and their reproduction rates. Decomposition during sedimentation was also indicated by a comparison of data from traps at 853 m depth in the eastern equatorial Atlantic and at 2195 m depth in the northeast tropical Atlantic. The percentage of intact loricae was lower in the deeper trap (78% compared with 63%; Boltovskoy et al. 1993, 1996). Actually, Boltovskoy et al. (1993) observed signs of bacterial degradation in loricae, and Suzuki & Taniguchi (1995) found a half-value period of lorica decomposition of 0.3–1.0 year at 5 °C in non-sterile GF/F filtered sea water. Using these data, Suzuki & Taniguchi (1995) calculated that the sedimentation velocity is too low and the decomposition velocity is too high to explain the occurrence of loricae in 7000 m depth. Possibly, these loricae were incorporated into marine snow or fecal pellets, which sink faster. But, more likely, they represent the proteolytic-resistant tintinnid loricae, whereas those with a different chemical composition were successively digested by bacterial enzymes during sedimentation.

2.6 KEY POINTS

1. Lorica formation. After cell division, the proter produces its own (proto-)lorica, whereas the opisthe keeps the old lorica. A tintinnid that lost or abandoned its lorica is able to form a paralorica. Both the protolorica and the paralorica can be elongated anteriorly by the addition of one or more usually spiraled epiloricae. The somatic ciliature is probably involved into the formation of the lorica. However, a correlation between the somatic ciliary pattern and the lorica structure is currently not evident.

2. Lorica structure. The lorica structure largely depends on the formation process, namely slowly formed loricae or lorica parts are often spiraled, whereas the quickly formed ones are usually continuous. The capability to agglutinate biogenic and/or abiogenic particles is a valuable taxonomic feature at genus and usually at family level, separating the hyaline and the agglomerated loricae. Under the last type soft, agglomerated loricae, hard, agglomerated loricae, and loricae that are composed of a hyaline collar and a hard, agglomerated bowl are often subsumed.

3. Lorica texture. The lorica texture depends on the chemical composition of the lorica material and its hardening properties; thus, it is not very variable. Five main types of textures are distinguished: the monolaminar and compact texture; the monolaminar and alveolar texture; the trilaminar texture with an alveolar middle layer; the trilaminar texture with a tubular middle layer; and the tubular texture. Some further textures are known, but might represent culture-dependent artifacts.

4. Chemical composition. The assumption of a chitinous lorica material was inferred from the resistance of the loricae against strong bases. However, the exposure periods were too short and the temperatures too low to observe the dissolution of the loricae. This disintegration in hot, strong bases, the absence of chitin-specific stainings, and the failure of digestion by chitinase exclude a chitinous nature in the loricae tested. In fact, the lorica material seems to consist of proteins in 13 of the 24 genera investigated because of the proteolytic properties and staining reactions. In the remaining genera, proteins were not or not unequivocally detected, probably because they are rather uncommon, resembling those of prions, bacteria, and archaea in their reactions.

5. Lorica sedimentation. Tintinnid loricae were collected in sediment traps at up to 7000 m depth. Differences in the abundances and species compositions between the plankton communities and the trap material, however, indicated the loss of the less proteolytic-resistant loricae by bacterial digestion during sedimentation.

6. Future efforts. By means of culture experiments and fluorescence in situ hybridization (which detects signature sequences in the DNA), the intraspecific phenotypic plasticity of the tintinnid loricae should be studied. Knowledge about the different loricae in a species' cell cycle is necessary for a far-reaching revision of tintinnid classification at the genus and occasionally the family levels. A quantification of the elemental composition of the lorica material is required to assess the role of tintinnids and their sedimenting loricae in the pelagic and benthic energy flux. The taxonomic significance of the lorica texture, the types of lorica forming granules, and the chemical composition of the lorica material should be elucidated by further histochemical, enzymatic, and ultrastructural studies.

ACKNOWLEDGMENTS

We thank the other chapter authors for comments and ideas, Wayne Coats for providing a transmission electron micrograph of a *Eutintinnus* lorica, and the Austrian Science Fund (FWF; Project P20461-B17) for the financial support of S. Agatha.

Chapter 3

SYSTEMATICS AND EVOLUTION OF TINTINNID CILIATES

Sabine Agatha[1] *and Michaela C. Strüder-Kypke*[2]

[1]University of Salzburg, Salzburg, Austria
[2]University of Guelph, Guelph, ON, Canada

The Biology and Ecology of Tintinnid Ciliates: Models for Marine Plankton, First Edition. Edited by John R. Dolan, David J.S. Montagnes, Sabine Agatha, D. Wayne Coats, and Diane K. Stoecker.

3.1 INTRODUCTION

Taxonomy (from Ancient Greek: *taxis* "arrangement" and *nomos* "law") is the science of describing and classifying organisms, and systematics (from Ancient Greek: *systēmatikós* "ordered") investigates the phylogenetic relationships among organisms. The systematics and taxonomy of the extant tintinnids are the topic of the present chapter.

In his "scala naturae" (ladder of life), the Greek philosopher Aristotle (384–322 BC) already classified organisms according to their perfectness, namely, from simply to highly organized ones, to order the otherwise bewildering diversity of species. The modern hierarchical taxonomic system and binominal nomenclature was devised by Carolus Linnaeus (1758) in his "Systema Naturae". Although initially classification was simply a useful means of cataloguing the diversity of life and providing unambiguous names for the organisms, the view of taxonomy was changed distinctly by Charles Darwin's (1859) famous book "On the Origin of Species". He provided convincing evidence that all organisms are descendents of a common ancestor, i.e. an ancestral species gave rise to daughter species through lineage splitting.

Trees convey our hypotheses of the histories of individual organisms, populations, taxa, genes, proteins, or morphological characters and serve to organize and summarize where and how information has changed. Supposing all similarities and differences among organisms are the result of cladogenesis (lineage splitting) and anagenesis (character change), the inference of phylogenetic trees through cladistic (using derived characters; Mayr & Bock 2002) and genetic analyses is a very useful tool. Theodosius Dobzhansky (1964) was correct in stating "nothing in biology makes sense except in the light of evolution". Taxonomic classification should reflect this evolutionary history and the natural relationships of organisms. To reconstruct the unobserved evolution, all available data on the organisms (mainly about morphology, ontogenesis, ultrastructure, and fossils) are compiled and analyzed.

By investigating the extant species, we develop ideas about morphology and further features in their ancestors, inferring a close relationship of organisms based on their similarity due to inherited traits from their common ancestor. However, Hennig (1966) demonstrated that overall similarity might be misleading and only the shared derived homologous characters (synapomorphies) indicate the true relationships. The first step in phylogenetic analyses is the detection of homologous features and the separation of analogous features (homoplasies), which evolved in unrelated species independently/convergently, for example as an adaptation to the environment. By comparison with an outgroup (close relatives of the monophyletic group whose relationships shall be analyzed), the plesiomorphic (ancestral) and apomorphic (derived) states are defined.

Although gene sequence analyses use a variety of algorithms to infer phylogenetic relationships by means of computer programs (the interested reader is referred to Felsenstein 2004), cladistic analyses only use the method of maximum parsimony. The parsimony principle, also called the principle of Occam's razor, dates back to the medieval and was apparently established by William von Ockham (about 1285–1347). It suggests that the best estimate is the one with the fewest assumptions, or in other words, the true phylogeny is the one that requires us to postulate the fewest evolutionary changes. The trees based on cladistic and gene sequence analyses represent hypotheses that can be tested by the addition of features and taxa.

A taxonomic classification, which uses the groupings of the inferred tree, allows predictions about properties of the organisms and provides insights into the processes of speciation, adaptation, and maintenance of diversity. In this chapter, extant tintinnids are used as model to demonstrate how ciliate evolution can be reconstructed with morphologic, morphogenetic (reproduction pattern of the pre-existing form; Lynn 2008), ultrastructural, and genetic data.

In tintinnids, there is also a fossil record of the loricae, which one might think would be useful in reconstructing phylogenies. However, developmental studies (Chapter 2), cladograms (trees built on cladistic analyses; Mayr & Bock 2002) as well as molecular trees refute the idea of inferring phylogenetic relationships from lorica types. In the following, we will show that some features or character states are well suitable for inferring relationships, whereas others are not, as they represent homoplasies.

First, a short sketch of the changeful history of tintinnid taxonomy and classification is given, showing the strong influence of method developments on ciliate taxonomy (light microscopy, staining techniques, and electron microscopy). Then, we follow the evolution of tintinnids as inferred from cladistic analyses, discussing some of the features considered. Using one tintinnid genus as example, we demonstrate how information on a taxon can be inferred from the cladogram and its databases, even if some features were not studied. Next, the methods and results of gene sequence analyses are briefly described and a molecular tree is compared with the cladogram. The new insights into tintinnid evolution by recent cladistic and genetic analyses are incorporated into a revised tintinnid classification. Admittedly, it is not the far-reaching revision that is necessary. However, the present state of knowledge allows only slight modifications. Subsequently, the most important similarities and differences in the morphology of tintinnids and their relatives as well as the main features for the reconstruction of the evolution in the halteriids, oligotrichids, and aloricate choreotrichids are extracted. This will elucidate whether the constraints of the pelagic zone caused similar adaptations in these mainly marine ciliate groups. The gaps in our knowledge about tintinnids and future fields of research are the focus of the last section.

3.2 HISTORY OF TINTINNID TAXONOMY AND SYSTEMATICS

The beginning

The first tintinnid was described by Müller (1776) as *Trichoda inquilinus* (now *Tintinnus inquilinus*) from marine waters. Hence, tintinnids belong to the first ciliates ever described. From the beginning, they fascinated amateur microscopists and scientists with their ability to form wonderful vase- or tube-shaped loricae (shells), which are carried through the pelagic zone. During the early period of light microscopy, only a few features of these planktonic unicellular organisms could be recognized, as the magnification and resolution of the instruments were comparatively low. Nevertheless, the optics allowed the investigation of the lorica, which is easily collected by nets and preserved and remains intact after the inhabitant has died. Even at low magnification, different lorica types could be distinguished: (1) hyaline loricae with transparent walls; (2) hard, agglomerated loricae with biogenic and abiogenic particles attached to the entire wall; (3) soft, agglomerated loricae with biogenic and abiogenic particles attached to the whole wall; and (4) hard loricae composed of a hyaline collar and an agglomerated bowl (broadened lorica portion; Chapter 2).

The "inhabitants" of the loricae were, however, rarely studied (Claparède & Lachmann 1858; Haeckel 1873; Fol 1881; Entz Sr 1884; Daday 1887), possibly because they move vigorously with and within their loricae, are fragile and thus die soon under the light microscope, and do not show many features after preservation. The only cytological feature sufficiently studied at that time was the adoral zone of membranelles (fan-like ciliary structures) at the apical cell end, the only portion of the ciliate that extends out of the lorica. The huge collar membranelles are arranged on an elevated rim around the peristomial field with the eccentric buccal cavity and move strongly, generating water currents for locomotion and suspension feeding (Chapter 4). Accordingly, the descriptions of tintinnid species focused on lorica characteristics and the presence of a conspicuous oral ciliature, especially as the oral structures were the main taxonomic feature for the classification of ciliates in general during the ages of discovery and exploitation. The somatic (body) ciliature was comparatively poorly known.

The first ciliate classifications appear strange seen from the present state of knowledge. For instance, Kent (1880–1882) placed the tintinnid genera *Codonella*, *Tintinnidium*, *Tintinnopsis*, and *Tintinnus* and the aloricate choreotrichid genus *Strombidinopsis* together with the peritrich genus *Vasicola* in the order Heterotricha (now belonging to the Postciliodesmatophora) owing to the membranellar structures at the apical cell end and a rather dense somatic ciliature. The tintinnid genera *Dictyocysta* and *Petalotricha*, the halteriid genus *Halteria*, the oligotrichid genera *Strombidium*, *Acarella*, and *Arachnidium* (the last two are nomina oblita, forgotten names; Corliss 1979), as well as the haptorid genera *Mesodinium* and *Didinium* were, however, assigned to the order Peritricha (now mainly sessile ciliates) based on a ciliature almost restricted to an apical "ciliary wreath". With an improved quality of the optics, the descriptions of ciliate species became more detailed, but only few tintinnid taxonomists studied the cells. Although Daday (1887), Brandt (1907), Entz Jr (1909b), and Hofker (1931a) denounced the lorica-based classification of tintinnids as likely artificial and emphasized the usefulness of cytological features, most tintinnid studies did not follow their recommendations.

When Kofoid & Campbell (1929) revised this huge group of ciliates, most species were only known by their loricae. Moreover, the authors established many new species, using minute deviations in lorica shape and size. Finally, they regarded 706 species, 51 genera, and 12 families of extant (non-fossil) tintinnids as valid. In a subsequent revision, which focused on material from the eastern tropical Pacific (Chapter 1), they added a further family, 11 genera, and 28 species (Kofoid & Campbell 1939). A few years later, material from the Carnegie cruise yielded two additional genera and 14 new species (Campbell 1942). Consequently, there were about 750 species in 64 genera and 13 families known in the 1940s. Kofoid & Campbell (1939) regarded the genera *Tintinnidium* and *Leprotintinnus*, with their soft, agglomerated loricae, as the most primitive tintinnids, followed by the genera *Tintinnopsis* and *Codonella*, with their hard, agglomerated loricae. They also suggested that hyaline, monolaminar lorica walls, as in the genus *Eutintinnus*, characterize the most highly evolved tintinnids. Almost concurrently, Kahl (1932) assigned the following three families to the suborder Oligotricha (order Spirotricha), which he distinguished from the heterotrich ciliates by a reduced somatic ciliature: the Tintinnidae (including the aloricate choreotrichid genus *Strombidinopsis*), the Strobilidiidae (*Strobilidium*, *Lohmanniella*, and *Ciliospina*), and the Halteriidae (*Meseres*, *Halteria*, and the oligotrichid genus *Strombidium*).

In the following years, there were a few notable contributions of further species descriptions and redescriptions, based on studies of lorica features, for example Enrique Balech (1968 and references therein), Miguel Durán (1965 and references therein), and Yoshine Hada (1970 and references therein). Currently, we thus know about of approximately 1000 non-fossil tintinnid species descriptions (Fig. 3.1), virtually all solely based on lorica characteristics.

Fauré-Fremiet (1970) distinguished the tintinnids from the aloricate taxa, although he recognized that tintinnids shared a circular adoral zone of membranelles with the aloricate strobilidiids. Corliss (1979) maintained this separation in his valuable book on ciliated protozoa. His two suborders, Tintinnina and Oligotrichina, were united in the order Oligotrichida based on the common characteristic of a reduced somatic ciliature and an apical adoral zone of membranelles.

The introduction of staining techniques

In the 1950s, the use of histological staining procedures caused major changes in ciliate classification. By revealing the ciliary patterns, the stains allowed the characterization of the somatic ciliature and its use for taxonomy and systematics. Notably, the protargol (silver proteinate) stain was used, revealing the basal bodies (basal termination of the cilia) and the nuclear apparatus. In 1979, Foissner and Wilbert stained the

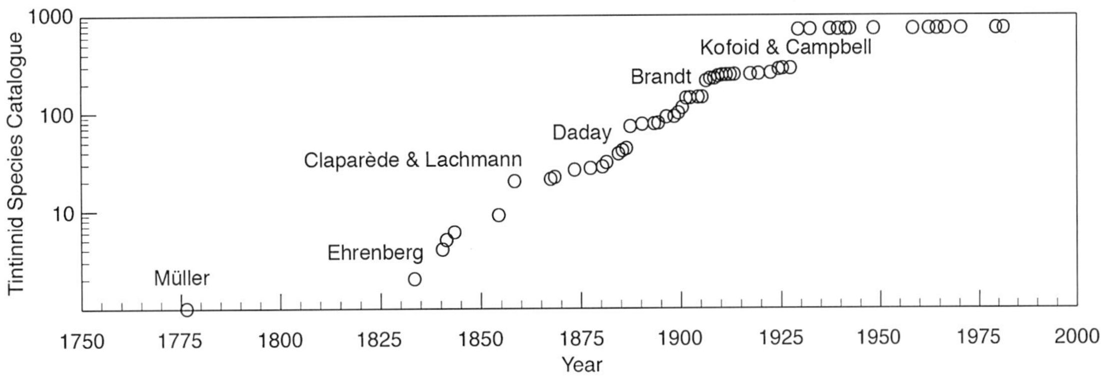

Fig. 3.1 Cumulative number of species descriptions in tintinnid ciliates since their discovery in 1776 by O.F. Müller plotted on a log-scale. Distinct increases can be attributed to comprehensive taxonomic works by certain authors. Note that during the past 20 years, relatively few new tintinnid species have been described.

first four tintinnid species, all collected from freshwater (Foissner & Wilbert 1979).

The era of electron microscopy

The introduction of electron microscopy provided further new insights in cell morphology; now, cell organelles and minute surface structures could be studied. Transmission electron microscopy revealed, for instance, fibers associated with the basal bodies. They were found to be taxonomically important because they are conserved compared with oral structures, which are exposed to high evolutionary pressure due to their role in food uptake and thus in fitness. In their famous revision, Small & Lynn (1985) used the circular and C-shaped arrangements of the adoral membranelles as main distinguishing feature to separate the choreotrichids (tintinnids plus aloricate taxa, for example *Strobilidium* and *Strombidinopsis*) and the oligotrichids (including the halteriids); these two orders constituted the subclass Choreotrichia within the class Spirotrichea.

Cladistic analysis

Protargol staining allowed detailed studies of dividing tintinnids, in particular the development of the somatic and oral ciliature for the opisthe (posterior division product) – a developmental sequence that previously could only be studied by observing living cells. The developmental characteristics of halteriids (*Halteria*; Fauré-Fremiet 1953; Fernandez-Leborans 1983), oligotrichids (*Strombidium*; Fauré-Fremiet 1912, 1953; Kormos & Kormos 1958; Deroux 1974), and freshwater tintinnids were combined with morphologic features in a cladistic analysis by Petz & Foissner (1992, 1993).

In terms of development, they found that halteriids share with the oligotrichids and choreotrichids a unique inverse orientation of the opisthe during cell division, termed an enantiotropic division mode. This is in contrast to most ciliates, in which the proter (anterior division product) and opisthe have the same orientation, termed a homeotropic division mode, such as in the related hypotrich (e.g., *Euplotes*) and stichotrich (e.g., *Oxytricha*, *Stylonychia*) spirotrichs. However, there are differences distinguishing the developmental pattern of halteriids from that of oligotrichids and choreotrichids. The opisthe's oral ciliature develops on the cell surface in halteriids, as usual for most ciliates, while this occurs in an invagination in oligotrichids and choreotrichids. Therefore, Petz & Foissner (1992) separated the halteriids (*Halteria*, *Meseres*) from the oligotrichids based on their distinct morphogenetic patterns, although the two groups have a C-shaped adoral zone of membranelles.

A separation of oligotrichids and choreotrichids was also indicated by a developmental characteristic. The two groups differ with regard to the shape of the invagination in which the new mouth develops during cell division. In oligotrichids, the invagination is tube-shaped, whereas in choreotrichids, the invagination is pouch-shaped. Based on the internal development of the oral apparatus, the oligotrichids and choreotrichids formed a cluster, which was the sister group to the halteriids, the closest relatives to the stichotrich ciliates. This topology was maintained even after including further morphologic, morphogenetic, and ultrastructural features in a cladistic analysis by Agatha (2004b). In the classification by Lynn & Small (2002), however, the choreotrichids and oligotrichids (including the halteriids) are not united by shared derived characters, but simply represent two of the five subclasses within the spirotrich ciliates.

Gene sequence analysis

A new era of phylogenetic analyses started with the introduction of gene sequencing. In particular, the small subunit ribosomal RNA (SSU rRNA) gene is frequently used to infer relationships. In the most recent general classification of ciliates by Lynn (2008), genetic data were considered. Based on SSU rRNA genealogies, Lynn (2008) affiliated the halteriids with the stichotrich spirotrichs. This point is controversally discussed as the close relationship with the stichotrichs is neither substantiated by the α-tubulin tree (Snoeyenbos-West et al. 2002) nor by morphologic and morphogenetic (e.g., the enantiotropic division mode) characteristics. Recently, Paiva et al. (2009) concluded from statistical tests on competing phylogenetic scenarios that the resolution of the SSU rRNA gene alone is insufficient to define the position of the halteriids unambiguously.

The molecular data currently available indicate that the oligotrichids and choreotrichids are closely related, forming a sister group to the stichotrich ciliates, which include the halteriids (see Snoeyenbos-West et al. 2002; Strüder-Kypke & Lynn 2003, 2008; Agatha & Strüder-Kypke 2007, 2012; Gao et al. 2009; Li Z. et al. 2009; Kim S.Y. et al. 2010). These relationships are only partially reflected in the classification of Lynn (2008). At the same time, Berger (2008) suggested a new taxon, the Perilemmaphora, that unites the stichotrichs, halteriids, oligotrichids, and choreotrichids based on the presence of a perilemma (additional membrane covering the entire cell). Here, for the sake of simplicity, vernacular names are used, i.e. stichotrichs, halteriids, oligotrichids, choreotrichids (= tintinnids plus aloricate choreotrichids).

For tintinnids, characteristics of the lorica have remained the primary means of distinguishing species. It has often been stated that lorica features might show considerable intraspecific variability depending on environmental conditions and the cell cycle (Biernacka 1965, 1968; Gold 1974; Gold & Morales 1974, 1975c; Bakker & Phaff 1976; Bernatzky et al. 1981; Boltovskoy et al. 1990) and this has been demonstrated in cultures (see Chapter 2; Laval-Peuto 1977, 1981). However, the cytological characteristics of relatively few tintinnid species are known. At this time, we know the cell features of merely 24 out of the more than 1000 species in the literature (Table 3.1). Consequently, Lynn (2008) again provided a lorica-based classification. He and Loeblich Jr & Tappan (1968) compiled 75 extant valid genera, whereas Laval-Peuto (1994) considered 67 genera as valid.

Species descriptions and redescriptions

Although tintinnids have been known for more than 230 years, their classification, based solely on lorica characteristics, has remained largely unchanged since the revision by Kofoid & Campbell (1929). This clearly demonstrates the need for thorough species descriptions and redescriptions, using live observation, protargol staining, and electron microscopy. Although the use of protargol staining is now widely acknowledged, the value of careful live observations is underappreciated. Many features can be observed only in live specimens, or they are better recognized in live than in stained ones, for example the length of the somatic cilia and adoral membranelles, the position of the cytopyge (cell anus) as well as the structures containing capsules (probably extrusive organelles), namely the striae (beaded strands on the collar membranelles), tentaculoids (possibly contractile, finger- or pin-shaped cytoplasmic extensions between the collar membranelles), and accessory combs (distinct ridges between the collar membranelles). Because these characters are probably of taxonomic significance, there is a need for detailed studies, drawings, and micrographs of live specimens. Even the lorica requires more attention. Apart from the structure of the lorica wall (hyaline or agglomerated), light microscopes equipped with interference contrast optics often clearly show its texture, for example small chambers (alveoli; Chapter 2).

Whenever possible, gene sequences (mainly SSU rRNA but also other molecular markers) should be provided in a species description (Lynn & Simpson 2009). These methods are nowadays routinely used and have been refined so that only few specimens are needed for extraction and amplification. Predominantly in cases where morphologic data alone are not sufficient to separate species, or are obscured, molecular analyses can help to define species boundaries and determine taxonomic placement.

Monographs and identification keys

The monographs by Kofoid & Campbell (1929, 1939) are still foremost concerning the number of included species. Therefore, it is understandable that these monographs are generally used for tintinnid identification. Although these books are outstanding, they should, however, only represent the first step in species determination, followed by the study of the original description. Why? Kofoid & Campbell (1929, 1939) did a wonderful job; however, (1) they changed the species circumscriptions by synonymizations that might be not justified; (2) they provided only one line drawing per species, which was often not from the original description and might show an uncommon specimen; (3) the lorica sizes can often only be estimated from the minute single illustrations and thus the size ranges are unknown, especially those of the type population; and (4) similar species and distinguishing features are rarely mentioned. Additionally, some mistakes occurred (for instance, illustration 18 shows *Stenosemella ventricosa* instead of *Tintinnopsis major*; Kofoid & Campbell 1929), which is not surprising, considering the enormous number of species and illustrations

Table 3.1 Tintinnid species studied after silver impregnation. Five lorica types are distinguished (A–E). (A) Soft, agglomerated. (B) Soft, agglomerated and with subterminal membrane. (C) Hard, agglomerated. (D) With hyaline collar and agglomerated bowl. (E) Entirely hyaline.

Species	Reference
Codonella cratera (Leidy, 1877) Imhof, 1885; (C)	Foissner & Wilbert (1979)
Codonellopsis glacialis (Laackmann, 1907) Kofoid & Campbell, 1929; (D)	Petz et al. (1995)
Cymatocylis calyciformis (Laackmann, 1907) Laackmann, 1910; (E)	Petz et al. (1995)
Cymatocylis convallaria Laackmann, 1910; (E)	Petz et al. (1995), Wasik & Mikołajczyk (1994b)
Eutintinnus angustatus (Daday, 1887) Kofoid & Campbell, 1939; (E)	Choi et al. (1992)
Eutintinnus pectinis (Kofoid & Campbell, 1929) Kofoid & Campbell, 1939; (E)	Choi et al. (1992)
Eutintinnus tenuis (Kofoid & Campbell, 1929) Kofoid & Campbell, 1939; (E)	Choi et al. (1992)
Favella ehrenbergii (Claparède & Lachmann, 1858) Jörgensen, 1924; (E)	Kim S.Y. et al. (2010)
Membranicola tamari Foissner, Berger & Schaumburg, 1999; (B)	Foissner et al. (1999)
Nolaclusilis bicornis Snyder & Brownlee, 1991; (E)	Snyder & Brownlee (1991)
Nolaclusilis hudsonicus Sniezek, Capriulo, Small & Russo, 1991; (E)	Sniezek et al. (1991)
Schmidingerella arcuata (Brandt, 1906) Agatha & Strüder-Kypke, 2012; (E)	Agatha & Strüder-Kypke (2012)
Stenosemella lacustris Foissner & O'Donoghue, 1990; (C)	Foissner & O'Donoghue (1990)
Stenosemella pacifica Kofoid & Campbell, 1929; (D)	Agatha & Tsai (2008)
Tintinnidium fluviatile (Stein, 1863) Kent, 1881; (A)	Foissner & Wilbert (1979)
Tintinnidium pusillum Entz, 1909; (A)	Foissner & Wilbert (1979)
Tintinnidium semiciliatum (Sterki, 1879) Kent, 1881; (A)	Blatterer & Foissner (1990), Song & Wilbert (1989)
Tintinnopsis brasiliensis Kofoid & Campbell, 1929; (C)	Cai et al. (2006)
Tintinnopsis campanula (Ehrenberg, 1840) Daday, 1887; (C)	S. Agatha, unpublished observations
Tintinnopsis cylindrata Kofoid & Campbell, 1929; (A)	Foissner & Wilbert (1979)
Tintinnopsis cylindrica Daday, 1887; (C)	Agatha & Riedel-Lorjé (2006)
Tintinnopsis fimbriata Meunier, 1919; (C)	Agatha (2008)
Tintinnopsis parvula Jörgensen, 1912; (C)	Agatha (2010b)
Tintinnopsis tubulosoides Meunier, 1910; (C)	S. Agatha, unpublished observations

included. Although these books are still indispensable in tintinnid identification, the original descriptions should thus always be consulted.

Keys distinctly facilitate species determination. Those that guide to the tintinnid species are either incomplete or restricted to particular geographic areas (Daday 1887; Hada 1938; Cosper 1972; Rampi & Zattera 1982; Gilron et al. 1990; Strüder-Kypke & Montagnes 2002). The most recent key for tintinnids by Lynn & Small (2002) guides only to representative genera. So, there is need for a complete key considering all extant species (about 1000), but such a project is unrealistic as merely 2–3% of the lorica-based species have been characterized cytologically. Consequently, synonymies are impossible to identify with certainty.

3.3 EVOLUTION OF TINTINNIDS

In this section, we will track the evolution of the tintinnids as shown by the cladistic analysis of extant taxa. Currently, we have information on 11 groups of characters: (1) the cell shape, (2) the life style, (3) the oral ciliature, (4) the somatic ciliature, (5) the nuclear apparatus, (6) special organelles and cell structures, (7) features of the cell cortex, (8) the lorica, (9) particulars of cell division, (10) the structure of the resting cysts (Chapter 7), and (11) the particulars of conjugation (reciprocal fertilization, a unique feature of the ciliates; Agatha & Strüder-Kypke 2007, 2012). Certainly, the number of features will further increase with our knowledge about tintinnids.

Two types of tree result from cladistic analyses: (1) a cladogram built by the Hennigian argumentation based on the discussion of each character and the use of synapomorphies and the parsimony principle (Fig. 3.2) and (2) trees generated by maximum parsimony computer programs calculating the tree statistics and confidence estimates of the branches or divisions (Fig. 3.12; for methods and interpretation, see below). In the following, only the most important characters and their changes are discussed, but all characters and their states are summarized in Table 3.2 and their distribution among the taxa is shown in Table 3.3 (for details of the computer methods and the characters omitted here, see Agatha & Strüder-Kypke 2007, 2012).

Kinetal maps

Protargol stains the single (monokinetids) and paired (dikinetids) basal bodies as well as the linked basal bodies in the polykinetids (e.g., the bases of the membranelles). The patterns displayed are complicated and difficult to understand in tintinnids owing to the globular or conical cell shape and the curvature of the kineties (ciliary rows; Figs 3.3, 3.4, 3.7, and 3.10). To facilitate portrayal and comparison of the somatic ciliary patterns, kinetal maps were introduced (Foissner & Wilbert 1979; Choi et al. 1992). These maps depict the ciliary patterns of a protargol-stained trophont (the non-dividing, feeding stage) in two dimensions, namely the cell cortex is drawn as cut longitudinally on dorsal side. The adoral zone of membranelles consists of two parts: large collar membranelles used for locomotion and suspension feeding and small buccal membranelles leading to the eccentric cytostome (cell mouth). Horizontal bars in the kinetal maps symbolize the large collar membranelles, whereas the diagonal bars symbolize those membranelles that are partly or entirely in the buccal cavity, namely, the elongated collar membranelles and the small buccal membranelles. Occasionally, the somatic cilia are symbolized by horizontal lines.

Before we consider evolution within the tintinnids, some earlier relationships resulting in the loricate choreotrichids should be described. In particular those of the stichotrichs (e.g., *Oxytricha*, *Stylonychia*), halteriids (e.g., *Halteria*, *Meseres*), oligotrichids (e.g., *Strombidium*, *Tontonia*), and aloricate choreotrichids (e.g., *Strobilidium*, *Strombidinopsis*), which evolved before the tintinnids as indicated by SSU rRNA gene trees (Fig. 3.13) and cladistic analyses (Fig. 3.2).

The ancestor of the stichotrichs, halteriids, oligotrichids, and choreotrichids

The common ancestor of the stichotrichs, halteriids, oligotrichids, and choreotrichids presumably was a dorsoventrally flattened benthic ciliate with a conspicuous C-shaped adoral zone of membranelles used for suspension feeding and some ventral cirri (ciliary bristles) used for walking (Foissner et al. 2004). So, it resembled what we today would call a stichotrich ciliate. Opposite the adoral zone of membranelles that extended along the left margin of the peristome, the ancestor had two monostichomonad (single-rowed) membranes, often called undulating membranes. The endoral membrane extended in the buccal cavity and the paroral membrane extended on the buccal lip bordering the buccal cavity on the right (Petz & Foissner 1992 and references therein). The dorsal somatic ciliature usually comprised three to nine longitudinal kineties (Curds & Wu 1983; Berger 1999, 2006). As already recognized by Small & Lynn (1985), the structure of the somatic kinetids (basal body/ies plus associated fibrillar structures) is important for reconstructing evolution, but only Agatha (2004a, b) and Agatha & Strüder-Kypke (2007) used it in cladistic analyses of the oligotrichids and choreotrichids, respectively. They hypothesized that the

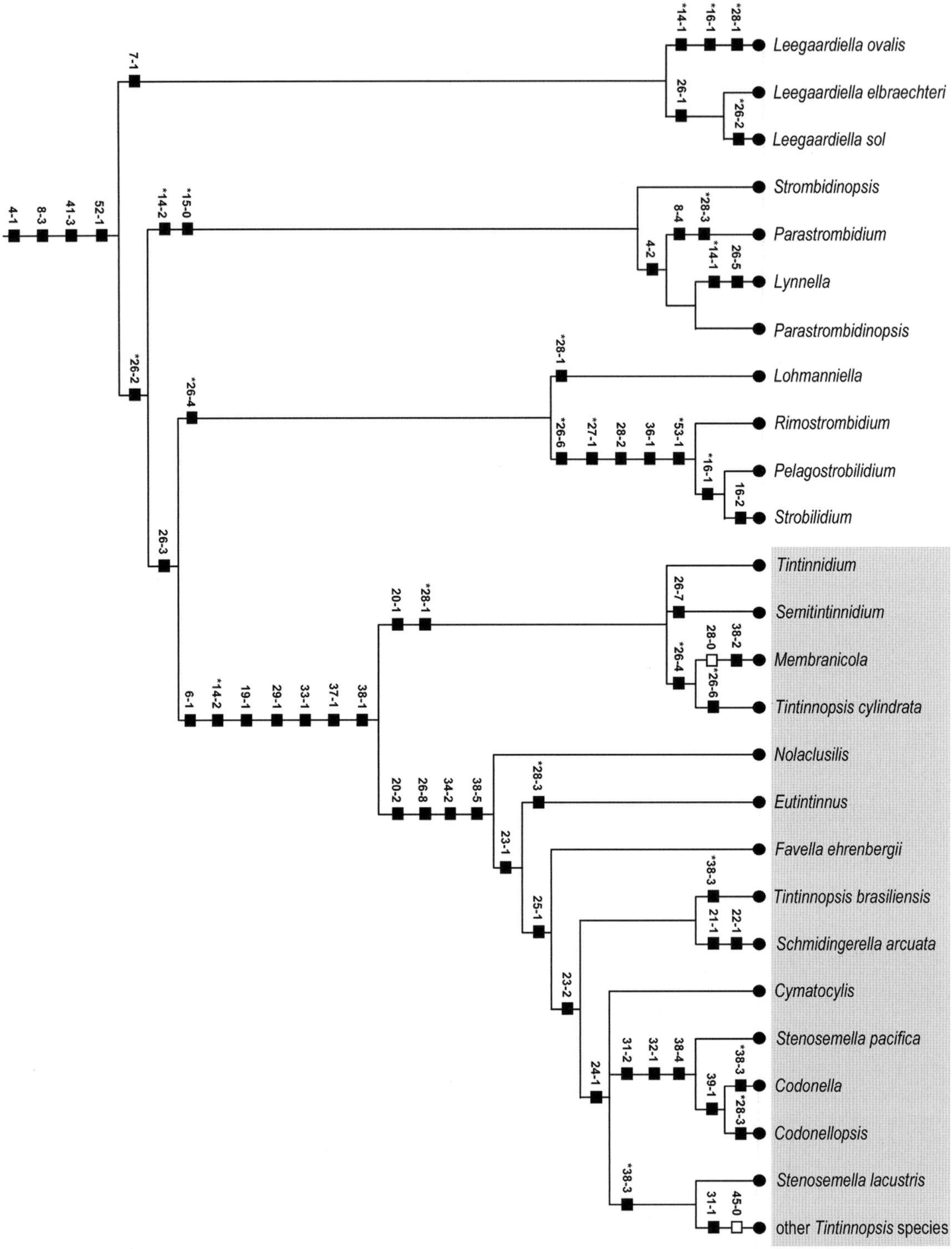

Fig. 3.2 Cladogram generated by the Hennigian argumentation method (adapted from Agatha, S. & Strüder-Kypke, M.C. (2012) Reconciling cladistic and genetic analyses in choreotrichid ciliates (Protists, Spirotrichea, Oligotrichea). Journal of Eukaryotic Microbiology, in press). For character coding, see Table 3.2. Black squares mark apomorphies. Asterisks denote homoplasies. *Semitintinnidium*, subgenus *Semitintinnidium* of genus *Tintinnidium* with *T. semiciliatum*; *Tintinnidium*, subgenus *Tintinnidium* of genus *Tintinnidium* with *T. fluviatile* and *T. pusillum*.

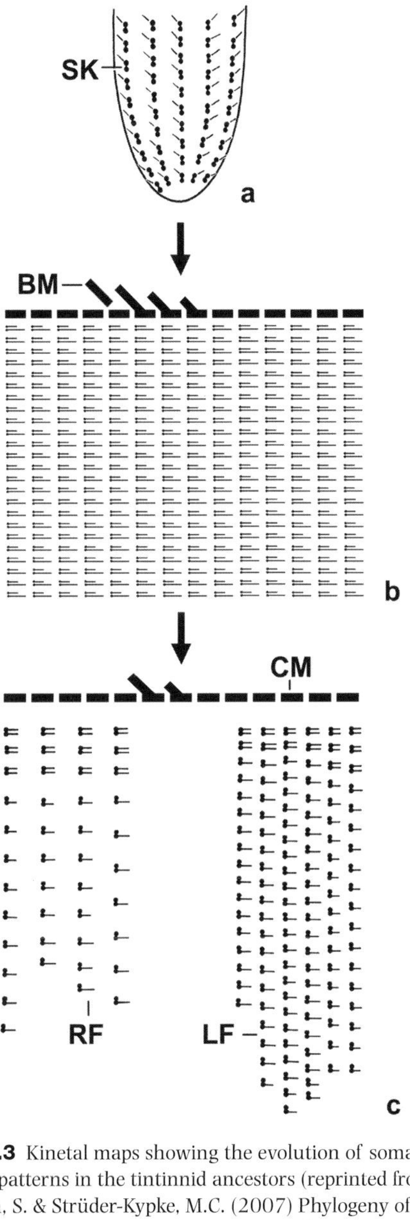

Fig. 3.3 Kinetal maps showing the evolution of somatic ciliary patterns in the tintinnid ancestors (reprinted from Agatha, S. & Strüder-Kypke, M.C. (2007) Phylogeny of the order Choreotrichida (Ciliophora, Spirotricha, Oligotrichea) as inferred from morphology, ultrastructure, ontogenesis, and SSrRNA gene sequences. European Journal of Protistology, 43, 37–63: Fig. 1a, with permission from Elsevier; protargol stain). (a) Ancestor of stichotrichs, halteriids, oligotrichids, and choreotrichids. (b) *Strombidinopsis*. The somatic dikinetids have two cilia. (c) Hypothetical tintinnid ancestor. BM, buccal membranelle; CM, collar membranelles; LF, left ciliary field; RF, right ciliary field; SK, somatic kineties.

ancestor had longitudinal somatic ciliary rows composed of dikinetids bearing only one cilium, which is associated with the anterior basal body (Fig. 3.3a). During cell division, the opisthe's oral ciliature developed on the cell surface, as it is common in ciliates, and originated *de novo*, i.e. without involving previously existing ciliature (epiapokinetal stomatogenesis; Foissner 1996). Simultaneously, a replication band – the site of DNA duplication – traversed each macronucleus nodule before division (Deroux 1974; Salvano 1974, 1975; Raikov 1982).

The ancestor of the halteriids, oligotrichids, and choreotrichids

Entering the pelagic zone and obtaining a planktonic life style apparently coincided with some serious morphologic changes, a "dorsalization". Foissner et al. (2004) suggested that the flattened ventral side of the ancestor with the cirri was reduced, whereas the dorsal side with the longitudinal somatic kineties became enlarged (character 2). As the result of this "dorsalization" process, the ciliates were no longer dorsoventrally flattened but rather had a globular, ellipsoidal, or obconical cell shape (character 1). The cells developed an apically located adoral zone of membranelles (character 3), which became also responsible for swimming, as the somatic cilia (the dorsal cilia of the ancestor) were minute and possibly had only a sensory function (Fauré-Fremiet & Ganier 1970) or became involved in lorica construction (Chapter 2; Laval-Peuto & Brownlee 1986). Possibly, the "dorsalization" process also resulted in the enantiotropic division mode (character 40) because the ventral side comprising the area for the opisthe's oral apparatus was reduced (Foissner 1996; Agatha 2004b; Agatha & Strüder-Kypke 2007). Therefore, the voluminous oral ciliature probably obtained an inverse orientation in the unciliated posterior cell portion, as in halteriids, and finally invaginated in oligotrichids and choreotrichids (see below).

Although the membranelles gradually decrease in size towards the cytostome in stichotrichs, a bipartition of the adoral zone of membranelles into a portion with large collar membranelles and small buccal membranelles characterized the ancestor of the halteriids, oligotrichids, and choreotrichids (character 8). Simultaneously, the somatic kineties that extended longitudinally between the adoral zone of membranelles and the posterior cell end in stichotrichs were shortened posteriorly and/or anteriorly (character 15). This was the time when the ancestor of the halteriids branched off according to cladistic analyses. The large number of special features indicates a high evolution rate in halteriids (Agatha 2004b; Agatha & Strüder-Kypke 2007; Agatha & Foissner 2009).

The ancestor of the oligotrichids and choreotrichids

In the ancestor of the oligotrichids and choreotrichids, the generation of the opisthe's C-shaped adoral zone of membranelles was transferred into a tube-shaped invagination (hypoapokinetal stomatogenesis; character 41). Additionally, the paroral membrane on the outer surface of the buccal lip was lost (character 10; Agatha 2004a).

The ancestor of the choreotrichids

The ancestor of the aloricate choreotrichids and tintinnids was probably characterized by a circular arrangement of the large collar membranelles on an elevated rim around the peristomial field containing the eccentric buccal cavity (character 4). Some collar membranelles, which are equally sized in the halteriids and oligotrichids, became proximally elongated, plunging into the buccal cavity (character 8). The small buccal membranelles became inconspicuous as they were transferred into the funnel-shaped buccal cavity. A single-rowed endoral membrane with rather long cilia extended across the peristomial field, entering the buccal cavity. Although the endoral membrane is usually not visible owing to the perilemma cover, its movements are occasionally recognizable (Fig. 4.2). The adoral zone of membranelles of the opisthe developed in an invagination, which became pouch-shaped (character 41). This change was probably correlated with the creation of a circular adoral zone.

Two feature complexes were found to be most valuable in the reconstruction of evolution within the choreotrichids, and especially the tintinnids (Section 3.5): the structure of the somatic kinetids (monokinetids or dikinetids) and the somatic ciliary patterns. In the following, we will see that the differentiation between both kinetid states is important, but the states are often difficult to distinguish owing to the minuteness of the organelles: monokinetids are about 0.26 μm across, dikinetids about 0.5 μm × 0.26 μm in size. A further problem is caused by the different staining characteristics of ciliated and unciliated basal bodies, namely those basal bodies that bear a cilium stain more intensely. Thus, occasionally only the ciliated basal body of dikinetids is distinct, suggesting the presence of monokinetids. Atypical observations should hence be verified by transmission electron microscopy. Furthermore, it is important which dikinetidal basal bodies are ciliated: both, only the anterior, or only the posterior. Unfortunately, these characters were not always mentioned in species descriptions.

During the first phase of choreotrichid evolution, primarily the structure of the somatic kinetids changed, whereas the second phase was mainly characterized by an increase in complexity of the somatic ciliary patterns. The ancestor of the choreotrichids probably had ten or more somatic kineties composed of dikinetids (character 14). In contrast to the stichotrichs, halteriids, and oligotrichids, which have cilia only at each anterior dikinetidal basal body (for details, see Agatha 2004a, b, 2011), the choreotrichid ancestor probably had cilia associated with both basal bodies (character 26; Agatha & Strüder-Kypke 2007, 2012). Such a diciliated kinetid structure is still found in the extant aloricate genera *Strombidinopsis*, *Parastrombidinopsis*, and *Parastrombidium* (Fig. 3.3b) and represents a keystone for understanding the evolution not only within the aloricate choreotrichids (Section 3.7), but also in the tintinnids.

The early evolution within the tintinnids

The main and most conspicuous synapomorphy of the tintinnids is, of course, the lorica, the vase-shaped or tube-shaped house (character 38). As we have already seen, it has been used to define tintinnids for a long time. The ciliate is attached to the inner lorica wall by a contractile peduncle (character 33). Contractility is a rare feature in the taxa under consideration. It occurs only in the tail of the oligotrichid tontoniids, such as *Tontonia appendiculariformis* (Fauré-Fremiet 1924). However, the ultrastructure of the contractile tail in tontoniids is very different from the myonemes in tintinnids; thus, the tontoniid tail probably represents a homoplasy (Laval 1972; Hedin 1976b; Greuet et al. 1986; Wasik & Mikołajczyk 1992). The somatic ciliary pattern of the tintinnid ancestor probably comprised more than ten somatic kineties arranged in two ciliary fields on the right and left cell sides, separated by an unciliated ventral stripe (character 19). Both fields differed in the spacing of their ciliary rows, i.e. the kineties of the left ciliary field were more closely spaced than those of the right ciliary field (Figs 3.3c, 3.4a, and 3.7a; Snyder & Brownlee 1991; Agatha & Riedel-Lorjé 2006). The anterior cilia were reduced in the posterior dikinetids (character 26).

Capsules represent a further synapomorphy (character 29). They are membrane-bound, exocytotic cell organelles (extrusomes). However, to our knowledge, ejection has never been observed; so, their extrusive character was merely inferred from their location on or between the oral ciliature (see below) and the invariable orientation of the ampoule-shaped organelles with the tip directed to the cell membrane. Extrusomes are widely distributed in protists (Hausmann 1978). They are attached to the inner side of the cell membrane and are extruded in fractions of a second. Some of the extrusome types are used in defence or to attack prey organisms. They act like mini-harpoons or the nematocysts of Cnidaria (which are not cell organelles, but whole cells). The ultrastructure of tintinnid capsules differs from possibly extrusive granules in some stichotrichs and aloricate choreotrichids and the needle-shaped extrusomes (trichites) in the closely related oligotrichids (Agatha & Strüder-Kypke 2007, Agatha 2004b, and references therein). Both trichites and capsules are regarded as apomorphies.

Five types of tintinnid capsule have been distinguished based on characteristics revealed by transmission electron microscopy (Laval-Peuto & Barria de Cao 1987), but only two types occur in the species cladistically analyzed here: type I capsule in the genus *Tintinnopsis* and type II in the genera *Codonella*, *Codonellopsis*, *Dictyocysta*, and *Stenosemella* (character 31). The capsules are assumed to be always associated with organelles recognizable in live cells and scanning electron micrographs: striae, accessory combs, and tentaculoids (Agatha & Strüder-Kypke 2007, 2012, Agatha & Tsai 2008, and references therein). The tentaculoids were observed in several tintinnids (character 34; Figs 2.2d, 2.12c, e, and 4.2b, c; for details, see Agatha & Strüder-Kypke 2007, 2012). Their movement roughly follows that of the collar membranelles, which extend almost perpendicularly to the main cell axis in swimming specimens and bend towards the peristomial field in disturbed cells; according to Laval-Peuto (1994), the tentaculoids are contractile. Observations concerning the presence of capsules, striae, accessory combs, and tentaculoids are occasionally contradictory, possibly because their occurrence is correlated with food availability (Chapter 4; Entz Jr 1909b; Capriulo et al. 1986).

The vesicular reticulum probably represents a cortical structure characteristic of tintinnids (character 37; Sokolova & Gerassimova 1984). It was found in *Cyttarocylis* and *Parafavella*, in which it comprises cortical cavities that communicate with each other and the pericellular space underneath the perilemma (Laval-Peuto 1975; Sokolova & Gerassimova 1984; Sokolova et al. 1986).

Descending from the tintinnid ancestor, two lineages diverged: one comprises the mainly freshwater genera *Tintinnidium* and *Membranicola* and the freshwater species *Tintinnopsis cylindrata*, the other comprises the remaining, mainly marine tintinnids (Agatha & Strüder-Kypke 2007). We will follow the track of evolution in the two lineages separately (see below).

Freshwater species

Cladistic analyses show an early branching of the genera *Tintinnidium* and *Membranicola* and the species *Tintinnopsis cylindrata* (Fig. 3.2). In their ancestor, two ventral organelles were probably introduced on the ventral side between the right and left ciliary fields (Fig. 3.4b; character 20). These ventral organelles are composed of dikinetids and have a special origin, namely they apparently develop *de novo* (without contact to parental ciliary structures) during morphogenesis (Foissner & Wilbert 1979; Petz & Foissner 1993; Foissner et al. 1999), whereas somatic kineties develop by intrakinetal proliferation of basal bodies in stichotrichs and oligotrichids (for the *de novo* origin and reorganization of the somatic ciliature in halteriids, see below). The loricae are soft and have agglomerated some biogenic and abiogenic particles (mineral particles, diatom frustules or their fragments, etc.; character 38; Chapter 2); thus, they resemble the mineral envelope of the aloricate choreotrichid *Strombidinopsis minima*, which is, however, directly attached to the cell surface (Agatha 2003 and references therein). The first somatic ciliary pattern produced in this branch was probably similar to that of *Tintinnidium fluviatile* and *T. pusillum* (Figs 3.4b and 3.5a–c).

Subsequently, two lineages split within the tintinnids with soft loricae by (1) the partial loss of the unciliated anterior basal bodies in the posterior dikinetids, creating a pattern found in *Tintinnidium semiciliatum* (Figs 3.4c and 3.5d, e), and (2) the loss of all anterior cilia

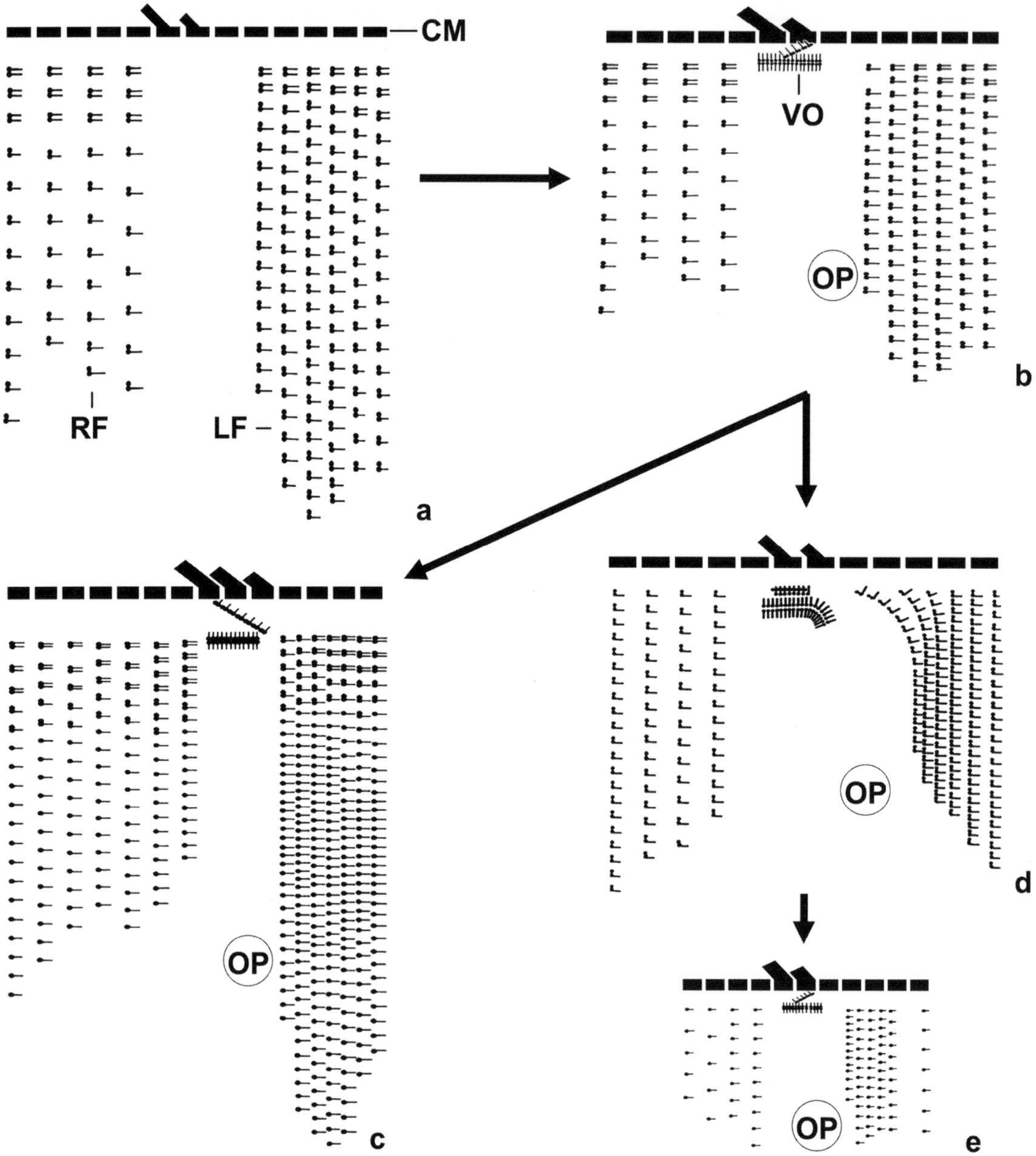

Fig. 3.4 Kinetal maps showing the evolution of somatic ciliary patterns in tintinnids with two dikinetidal ventral organelles (a, reprinted from Agatha, S. & Strüder-Kypke, M.C. (2007) Phylogeny of the order Choreotrichida (Ciliophora, Spirotricha, Oligotrichea) as inferred from morphology, ultrastructure, ontogenesis, and SSrRNA gene sequences. European Journal of Protistology, 43, 37–63: Fig. 1e, with permission from Elsevier; b, e, adapted from Foissner, W. & Wilbert, N. (1979) Morphologie, Infraciliatur und Ökologie der limnischen Tintinnina: *Tintinnidium fluviatile* Stein, *Tintinnidium pusillum* Entz, *Tintinnopsis cylindrata* Daday und *Codonella cratera* (Leidy) (Ciliophora, Polyhymenophora). Journal of Protozoology, 26, 90–103: Figs 2 and 6; c, adapted from Blatterer, H. & Foissner, W. (1990) Beiträge zur Ciliatenfauna (Protozoa: Ciliophora) der Amper (Bayern, Bundesrepublik Deutschland). Archiv für Protistenkunde, 138, 93–115: Fig, 4b, with permission from Elsevier; d, adapted from Foissner et al. 1999; protargol stain). (a) Hypothetical tintinnid ancestor. (b) Two ventral organelles were introduced, resulting in the pattern of *Tintinnidium* (*Tintinnidium*). (c) The unciliated anterior dikinetidal basal bodies were partly lost in the posterior parts of the kineties, giving rise to the pattern of *Tintinnidium* (*Semitintinnidium*). (d) The anterior cilia of all dikinetids were lost, producing the *Membranicola* pattern. (e) The unciliated anterior dikinetidal basal bodies were lost, creating the pattern of *Tintinnopsis cylindrata*. CM, collar membranelles; LF, left ciliary field; OP, oral primordium; RF, right ciliary field; VO, ventral organelles.

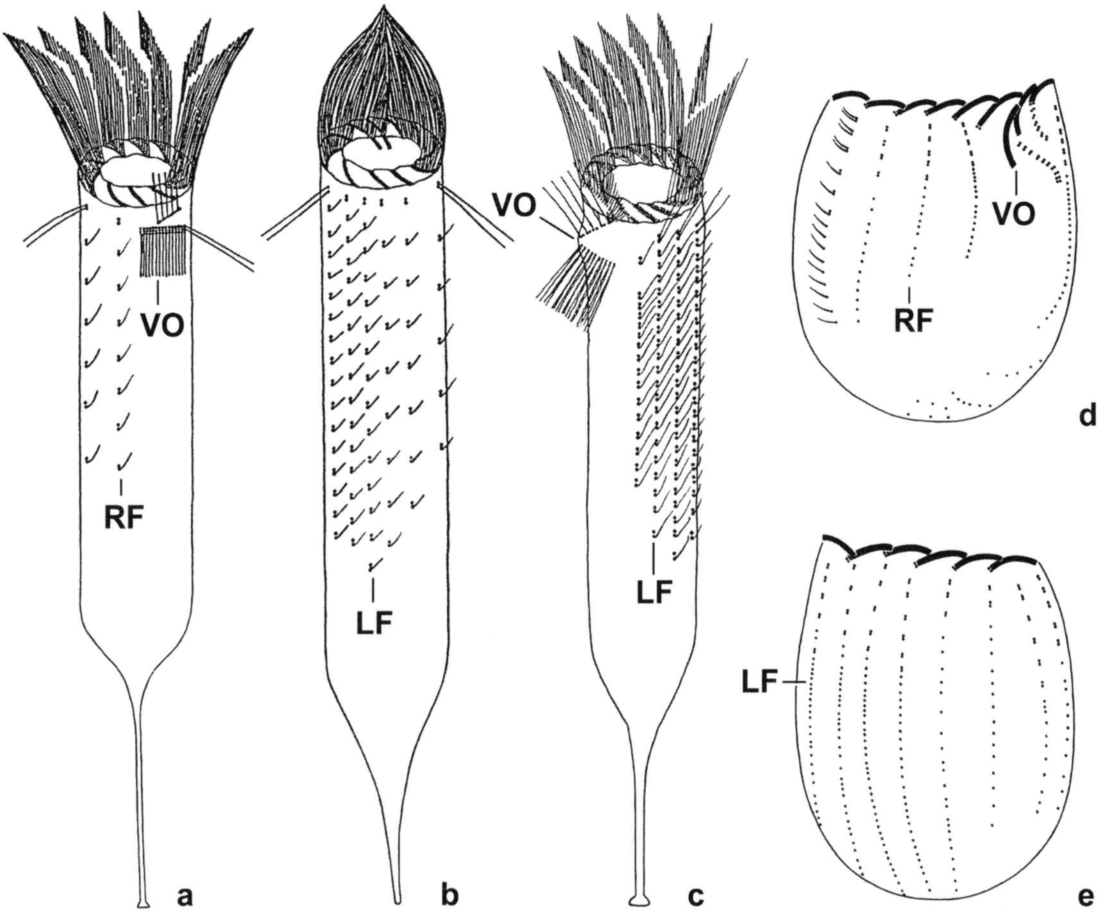

Fig. 3.5 Tintinnids with two dikinetidal ventral organelles (a, c, adapted from Foissner, W. & Wilbert, N. (1979) Morphologie, Infraciliatur und Ökologie der limnischen Tintinnina: *Tintinnidium fluviatile* Stein, *Tintinnidium pusillum* Entz, *Tintinnopsis cylindrata* Daday und *Codonella cratera* (Leidy) (Ciliophora, Polyhymenophora). Journal of Protozoology, 26, 90–103: Figs 1 and 3; b, from Foissner et al. (1991) © Bayerisches Landesamt fur Umwelt; d, e, reprinted from Blatterer, H. & Foissner, W. (1990) Beiträge zur Ciliatenfauna (Protozoa: Ciliophora) der Amper (Bayern, Bundesrepublik Deutschland). Archiv für Protistenkunde, 138, 93–115: Figs 4c, d, with permission from Elsevier; protargol stain). (a, b) *Tintinnidium* (*Tintinnidium*) *pusillum*, 155 µm, 160 µm. (c) *Tintinnidium* (*Tintinnidium*) *fluviatile*, 210 µm. (d, e) *Tintinnidium* (*Semitintinnidium*) *semiciliatum*, 50 µm. The unciliated anterior dikinetidal basal bodies were partly lost in the posterior parts of the kineties. LF, left ciliary field; RF, right ciliary field; VO, ventral organelles.

in the dikinetids, producing the *Membranicola* pattern (Figs 3.4d and 3.6a, b; character 26). The differences in the somatic ciliary patterns of *Tintinnidium fluviatile* and *T. pusillum* on the one hand and *Tintinnidium semiciliatum* on the other were used by Agatha & Strüder-Kypke (2007) to establish two subgenera within the genus *Tintinnidium*. The rather recently been described freshwater genus *Membranicola* has a unique subterminal membrane in its lorica, to which the ciliate's peduncle is attached (character 38; Foissner et al. 1999). The *Membranicola* pattern apparently gave rise to the *Tintinnopsis cylindrata* pattern with its monokinetidal somatic ciliary rows (Figs 3.4e and 3.6c, d). Mainly because of the presence of two ventral organelles, the somatic ciliary pattern of *T. cylindrata* differs from that of its congeners (Fig. 3.10c) and

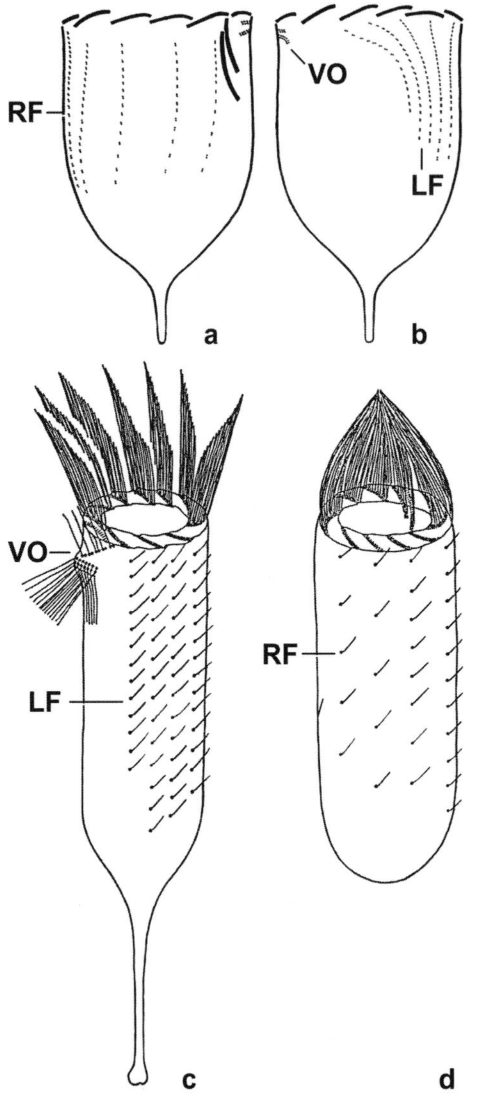

Fig. 3.6 Tintinnids with two dikinetidal ventral organelles (a, b, adapted from Foissner et al. 1999; c, adapted from Foissner, W. & Wilbert, N. (1979) Morphologie, Infraciliatur und Ökologie der limnischen Tintinnina: *Tintinnidium fluviatile* Stein, *Tintinnidium pusillum* Entz, *Tintinnopsis cylindrata* Daday und *Codonella cratera* (Leidy) (Ciliophora, Polyhymenophora). Journal of Protozoology, 26, 90–103: Fig. 5; d, from Foissner et al. (1991) © Bayerisches Landesamt fur Umwelt; protargol stain). (a, b) *Membranicola tamari*, 42 µm. Merely the posterior dikinetidal basal bodies are ciliated. (c, d) *Tintinnopsis cylindrata*, 97 µm, 97 µm. The somatic kineties are composed of ciliated monokinetids. LF, left ciliary field; RF, right ciliary field; VO, ventral organelles.

resembles that of the *Tintinnidium* species; furthermore, its lorica is soft. Despite these discrepancies, the generic affiliation of *Tintinnopsis cylindrata* cannot be changed at the moment, as the ciliary pattern of the type species of the genus *Tintinnopsis*, *T. beroidea*, is unknown (Agatha & Riedel-Lorjé 2006).

Marine species

In the second lineage originating from the tintinnid ancestor, we find all remaining taxa, namely, all marine tintinnids and the freshwater species *Stenosemella lacustris* and *Codonella cratera*. The generic affiliations of the last two species are doubtful (Agatha & Tsai 2008; Agatha 2010a). The main synapomorphy of this lineage is the introduction of a specialized monokinetidal ventral kinety between the right and left ciliary fields (Fig. 3.7b; character 20). In contrast to the ventral organelles of the previous lineage, this ciliary row originates as common, namely, by intrakinetal proliferation of basal bodies. Simultaneously, the dikinetids of the right and left ciliary fields, except for the anteriormost ones, became ciliated monokinetids (character 26), producing a pattern found in the extant genus *Nolaclusilis* (Figs 3.7b and 3.8a, b). The members of this genus have unique protrusions near the adoral zone of membranelles that are attached to the bell-shaped loricae and used to close the lorica opening. The lorica walls are hyaline and apparently monolaminar and compact (character 38; Sniezek et al. 1991; Snyder & Brownlee 1991).

Next, two dorsal kineties evolved between the right and left ciliary fields (character 23). A monokinetidal or dikinetidal structure of the dorsal kineties and cilia at the anterior, posterior, or even at both dikinetidal basal bodies were reported (Foissner & O'Donoghue 1990; Choi et al. 1992; Agatha & Tsai 2008); electron microscopic studies are required to confirm these conflicting observations. The pattern created matches that of *Eutintinnus* species (Figs 3.7c and 3.8c–h; Choi et al. 1992). In this genus, some variation is observed in the position and shape of the ventral kineties. The loricae of *Eutintinnus* species are hyaline and tube-shaped with an anterior and posterior opening (Figs 2.1c and 2.13g); the wall is monolaminar and compact.

By the introduction of a lateral ciliary field (character 25), the complexity of the somatic ciliary pattern further increased. In contrast to the right and left ciliary fields with their anterior dikinetids, this field

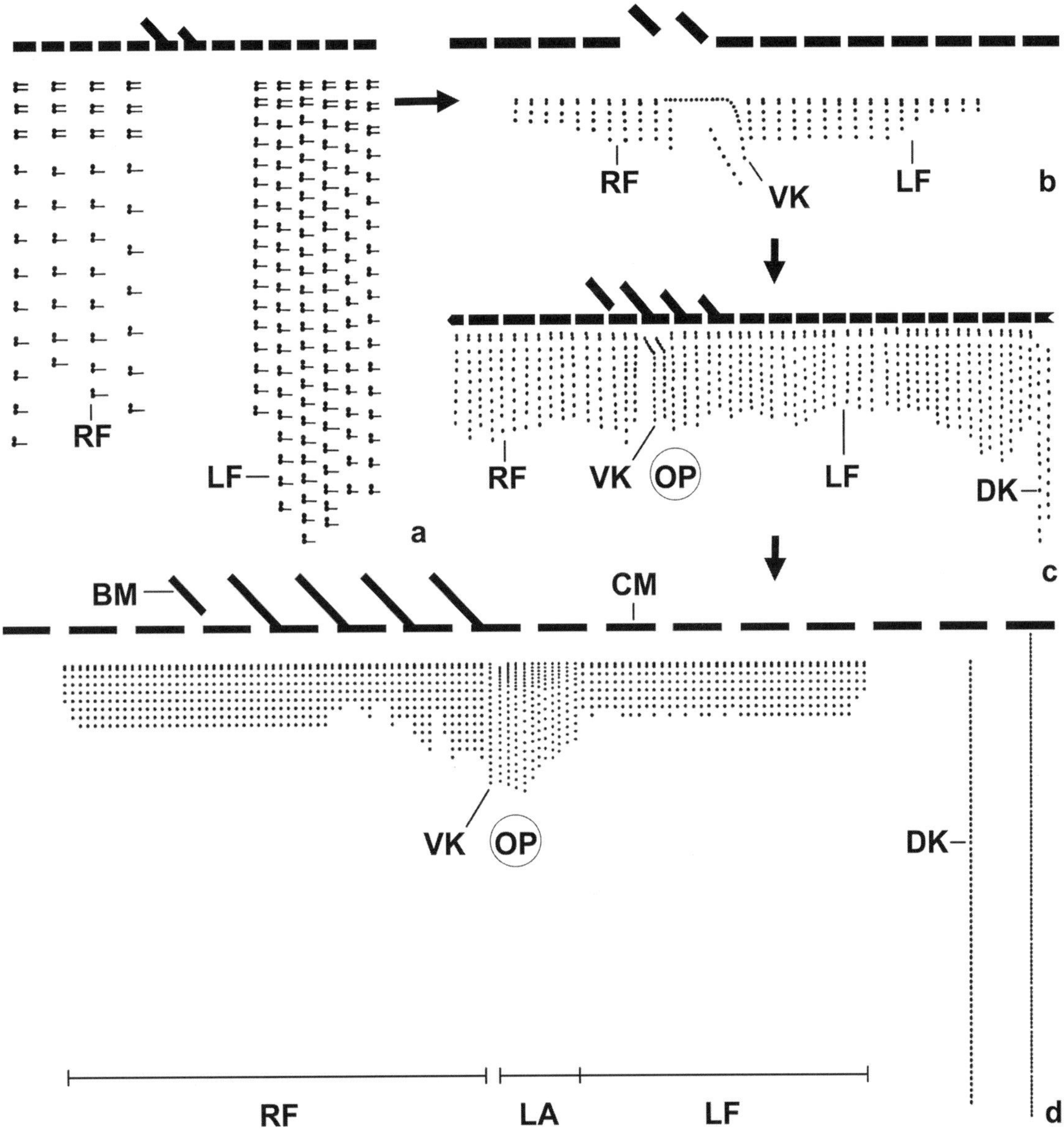

Fig. 3.7 Kinetal maps showing the evolution of somatic ciliary patterns in tintinnids with a monokinetidal ventral kinety (a, adapted from Agatha, S. & Strüder-Kypke, M.C. (2007) Phylogeny of the order Choreotrichida (Ciliophora, Spirotricha, Oligotrichea) as inferred from morphology, ultrastructure, ontogenesis, and SSrRNA gene sequences. European Journal of Protistology, 43, 37–63: Fig. 1e; b, adapted from Snyder, R.A. & Brownlee, D.C. (1991) *Nolaclusilis bicornis* n. g., n. sp. (Tintinnina: Tintinnidiidae): a tintinnine ciliate with novel lorica and cell morphology from the Chesapeake Bay estuary. Journal of Protozoology, 38, 583–589: Fig. 4; c, adapted from Choi, J.K., Coats, D.W., Brownlee, D.C. & Small, E.B. (1992) Morphology and infraciliature of three species of *Eutintinnus* (Ciliophora; Tintinnina) with guidelines for interpreting protargol-stained tintinnine ciliates. Journal of Protozoology, 39, 80–92: Fig. 3c; d, adapted from Kim, S.Y., Yang, E.J., Gong, J. & Choi, J.K. (2010) Redescription of *Favella ehrenbergii* (Claparède and Lachmann, 1858) Jörgensen, 1924 (Ciliophora: Choreotrichia), with phylogenetic analyses based on small subunit rRNA gene sequences. Journal of Eukaryotic Microbiology, 57, 460–467: Fig. 13; protargol stain). (a) Hypothetical tintinnid ancestor. (b) *Nolaclusilis*. A ventral kinety was created and the dikinetids, except for the anteriormost ones, transformed into monokinetids. (c) *Eutintinnus*. Two dorsal kineties developed. (d) *Favella ehrenbergii*. A monokinetidal lateral ciliary field was introduced. BM, buccal membranelle; CM, collar membranelles; DK, dorsal kineties; LA, lateral ciliary field; LF, left ciliary field; OP, oral primordium; RF, right ciliary field; VK, ventral kinety.

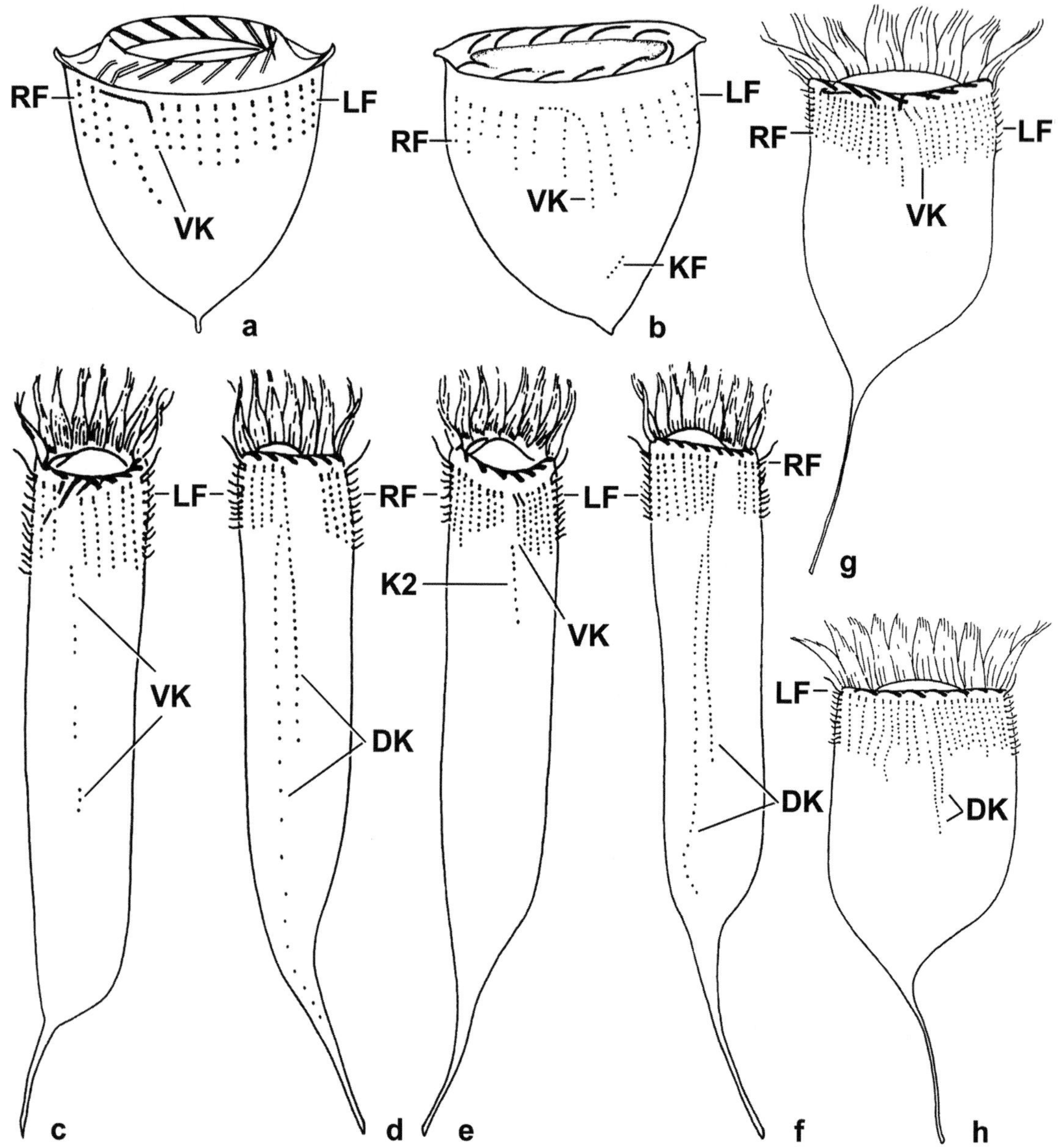

Fig. 3.8 Tintinnids with a monokinetidal ventral kinety (a, adapted from Snyder, R.A. & Brownlee, D.C. (1991) *Nolaclusilis bicornis* n. g., n. sp. (Tintinnina: Tintinnidiidae): a tintinnine ciliate with novel lorica and cell morphology from the Chesapeake Bay estuary. Journal of Protozoology, 38, 583–589: Fig. 2; b, adapted from Sniezek, J.H., Capriulo, G.M., Small, E.B. & Russo, A. (1991) *Nolaclusilis hudsonicus* n. sp. (Nolaclusiliidae n. fam.) a bilaterally symmetrical tintinnine ciliate from the lower Hudson River estuary. Journal of Protozoology, 38, 589–594: Fig. 1; c–h, adapted from Choi, J.K., Coats, D.W., Brownlee, D.C. & Small, E.B. (1992) Morphology and infraciliature of three species of *Eutintinnus* (Ciliophora; Tintinnina) with guidelines for interpreting protargol-stained tintinnine ciliates. Journal of Protozoology, 39, 80–92: Fig. 4a; protargol stain). (a, b) *Nolaclusilis bicornis* and *N. hudsonicus*, 26 μm, 38 μm. (c–h) *Eutintinnus pectinis* (c, d), *E. tenuis* (e, f), and *E. angustatus* (g, h) have two dorsal kineties, 88 μm, 88 μm, 125 μm, 130 μm, 93 μm, 90 μm. DK, dorsal kineties; K2, second kinety (first row of RF; which commences often more posteriorly and/or with more than one dikinetid); KF, kinety fragment; LF, left ciliary field; RF, right ciliary field; VK, ventral kinety.

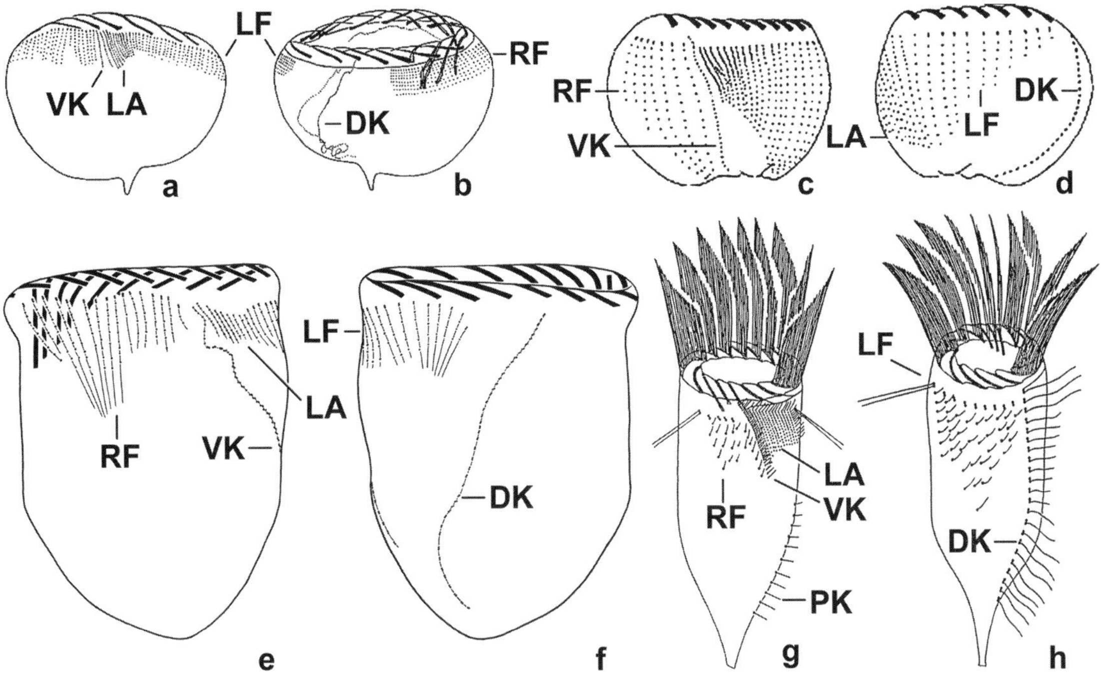

Fig. 3.9 Tintinnids with ventral kineties, lateral ciliary fields, and dorsal kineties (a, b, adapted from Kim, S.Y., Yang, E.J., Gong, J. & Choi, J.K. (2010) Redescription of *Favella ehrenbergii* (Claparède and Lachmann, 1858) Jörgensen, 1924 (Ciliophora: Choreotrichia), with phylogenetic analyses based on small subunit rRNA gene sequences. Journal of Eukaryotic Microbiology, 57, 460–467: Fig. 15; c, d, adapted from Cai et al. 2006; e, f, adapted from Agatha, S. & Strüder-Kypke, M.C. (2012) Reconciling cladistic and genetic analyses in choreotrichid ciliates (Protists, Spirotrichea, Oligotrichea). Journal of Eukaryotic Microbiology (in press). g, h, adapted from Foissner, W. & Wilbert, N. (1979) Morphologie, Infraciliatur und Ökologie der limnischen Tintinnina: *Tintinnidium fluviatile* Stein, *Tintinnidium pusillum* Entz, *Tintinnopsis cylindrata* Daday und *Codonella cratera* (Leidy) (Ciliophora, Polyhymenophora). Journal of Protozoology, 26, 90–103: Fig. 9; protargol stain). (a, b) *Favella ehrenbergii*, 142 μm. (c, d) *Tintinnopsis brasiliensis*, 58 μm. (e, f) *Schmidingerella arcuata* (Brandt, 1906) Agatha & Strüder-Kypke, 2012, 75 μm. (g, h) *Codonella cratera*, 70 μm. DK, dorsal kinety/ies; LA, lateral ciliary field; LF, left ciliary field; PK, posterior kinety; RF, right ciliary field; VK, ventral kinety.

is exclusively composed of monokinetids. Even if the structure of the anterior kinetids is not clearly recognizable in protargol preparations, the closer distances of the kinetids and ciliary rows in the lateral ciliary field facilitate its separation from the left ciliary field. The pattern generated corresponds to that of *Favella ehrenbergii* and *F. panamensis* (Figs 3.7d and 3.9a, b; Kim S.Y. et al. 2010; Agatha & Strüder-Kypke 2012). Like the previous two genera, *F. ehrenbergii* has a hyaline lorica, but its wall is monolaminar with alveoli (Figs 2.14b and 3.16d).

Next, one of the formerly two dorsal kineties was lost (character 23). The ancestor of the remaining tintinnids, thus, possessed a right, left, and lateral ciliary field as well as a ventral and dorsal kinety. This pattern is found in *Tintinnopsis brasiliensis* (Fig. 3.9c, d; Cai et al. 2006). As indicated by the generic affiliation, the species has a hard, agglomerated lorica, which represents the first time this lorica type is formed by tintinnids. Besides *T. brasiliensis*, two further lineages probably developed after the loss of one dorsal kinety.

In the second lineage, the right ciliary field separated from the ventral kinety, which got a unique dikinetidal posterior portion, producing the pattern of *Schmidingerella arcuata* (characters 21, 22; Agatha & Strüder-Kypke 2012). Although the loricae of the genus *Favella* and *S. arcuata* are very similar at first glance owing to their hyaline, monolaminar walls with alveoli, careful live observation at 1000× magnification and scanning electron micrographs revealed

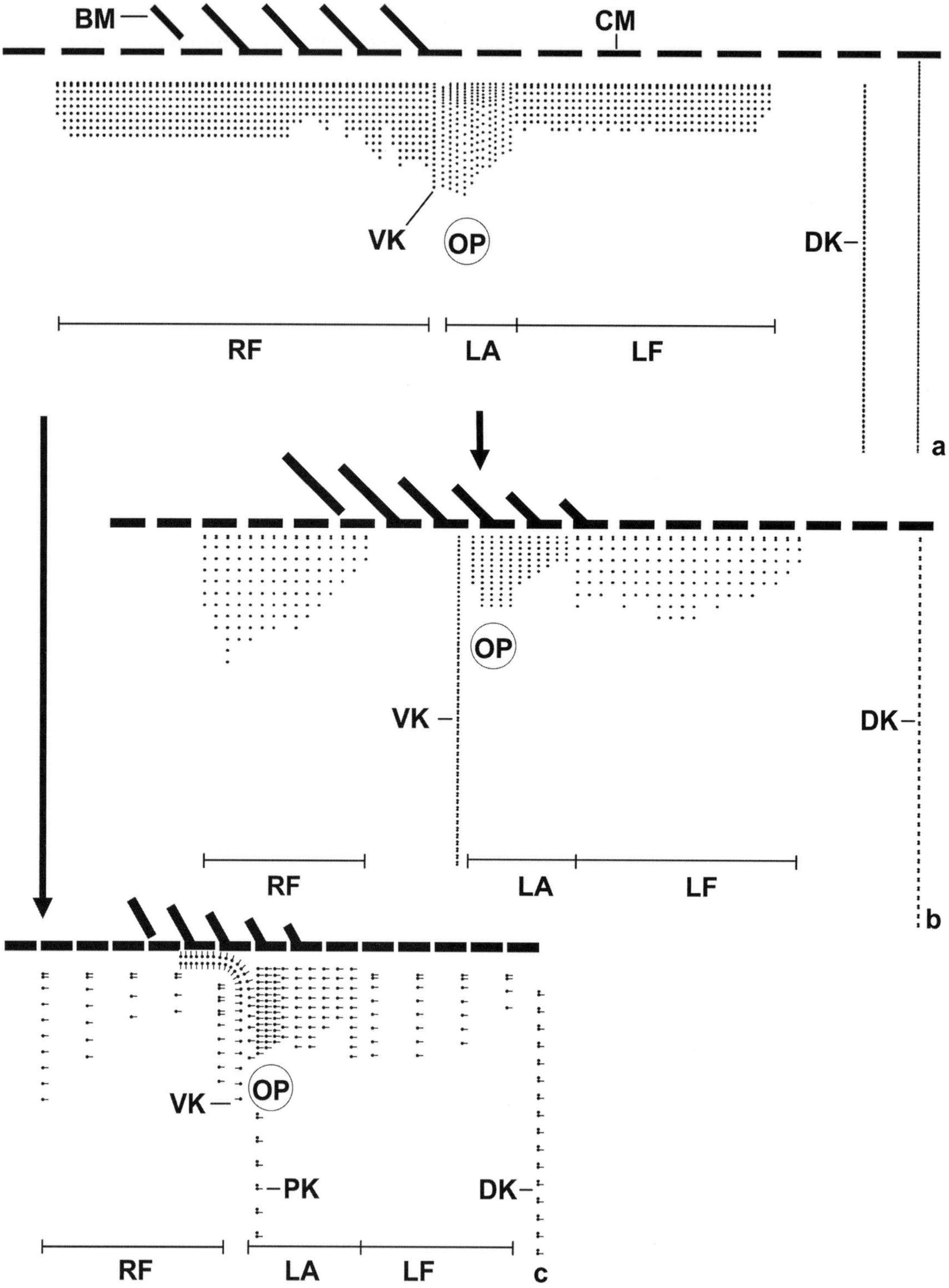

Fig. 3.10 Kinetal maps of tintinnids with ventral and dorsal kineties and lateral ciliary fields (a, adapted from Kim, S.Y., Yang, E.J., Gong, J. & Choi, J.K. (2010) Redescription of *Favella ehrenbergii* (Claparède and Lachmann, 1858) Jörgensen, 1924 (Ciliophora: Choreotrichia), with phylogenetic analyses based on small subunit rRNA gene sequences. Journal of Eukaryotic Microbiology, 57, 460–467: Fig. 13; b, adapted from Agatha, S. & Strüder-Kypke, M.C. (2012) Reconciling cladistic and genetic analyses in choreotrichid ciliates (Protists, Spirotrichea, Oligotrichea). Journal of Eukaryotic Microbiology (in press); c, adapted from Agatha 2010b; protargol stain). (a) *Favella ehrenbergii*. (b) *Schmidingerella arcuata*. (c) *Tintinnopsis parvula*. Most complex pattern with a posterior kinety. BM, buccal membranelle; CM, collar membranelles; DK, dorsal kinety/ kineties; LA, lateral ciliary field; LF, left ciliary field; OP, oral primordium; PK, posterior kinety; RF, right ciliary field; VK, ventral kinety.

deviations: the wall has pores and surface ridges in *S. arcuata* (Figs 2.14a and 3.17d, e), absent in *Favella* (Figs 2.14b and 3.16c). The differences in the lorica walls and the somatic ciliary patterns and the distant positions in the gene trees (Figs 3.9e, f, 3.10b, and 3.13) justify not only a separation of the species on genus level, but also an affiliation of the genera with different tintinnid families, namely, the Rhabdonellidae and Ptychocylididae.

In the third lineage, a posterior kinety was introduced, creating the most complex tintinnid ciliary pattern so far known (Fig. 3.10c; character 24), comprising a right, left, and lateral ciliary field as well as a ventral, dorsal, and posterior kinety. A monokinetidal or dikinetidal structure of the posterior kinety and cilia at the anterior, posterior, or even at both dikinetidal basal bodies were reported (Foissner & Wilbert 1979; Foissner & O'Donoghue 1990; Agatha & Tsai 2008). Electron microscopic studies are needed to verify these conflicting observations. The complex somatic ciliary pattern occurs with minute deviations in a wide variety of taxa: in the genera *Codonella* (Fig. 3.9g, h) and *Tintinnopsis* (Fig. 3.11i–l; for *T. cylindrata*, see page 55) with hard, agglomerated loricae; in the genera *Codonellopsis* (Fig. 3.11g, h) and *Stenosemella* (Fig. 3.11m, n; for *S. lacustris*, see below) with loricae composed of a hyaline collar and an agglomerated bowl, and the genus *Cymatocylis* (Fig. 3.11a–d) with hyaline loricae. The subtle deviations in the somatic ciliary patterns concern the following: (1) the course of the ventral kinety; (2) the number of dikinetids at the anterior end and the position of the first and second kinety in the right ciliary field (numbering in clockwise direction in top view; Chatton et al. 1931); (3) the course of the last kinety of the lateral ciliary field; (4) the structure of the dorsal kinety; (5) the structure, position, and course of the posterior kinety; and (6) the occurrence of fragmented kineties. Unfortunately, most studies so far have ignored these differences although they probably represent taxonomically relevant features.

The capsule types

Among the tintinnids with the most complex ciliary pattern, the cladogram shows a grouping of *Stenosemella pacifica* with the genera *Codonella* and *Codonellopsis*. They are united by the same types of capsules and mucocysts (membrane-bound extrusome with a paracrystalline structure that forms a mucus-like mass after ejection; Laval-Peuto & Barria de Cao 1987).

The lorica sac

The lorica sac represents a recently discovered synapomorphy of the genera *Codonella*, *Codonellopsis*, *Codonaria*, and *Dictyocysta* (Fig. 2.2b–d; Agatha 2010a). The last two genera were, however, not cladistically analyzed owing to the scarcity of cytological data. In these four genera, the ciliate cell is not only surrounded by the lorica, but also by a membranous sac that lines the lorica. Apically, the sac merges into a foldable closing apparatus that shuts the lorica opening in disturbed/contracted specimens. On the other hand, the diaphragm-like apparatuses in the genera *Salpingacantha* and *Salpingella* are probably not homologous to the foldable ones (Agatha 2010a). As mentioned earlier, the generic affiliation of *Codonella cratera* is doubtful, as a lorica sac was not reported in this comparatively well-known freshwater species (Section 3.6; Foissner et al. 1999; Agatha 2010a). Likewise, the generic assignment of *Stenosemella lacustris* is incorrect because the lorica collar is not hyaline but agglomerated and lacks windows (Foissner & O'Donoghue 1990; Agatha & Tsai 2008). The hard, entirely agglomerated lorica thus unites *S. lacustris* (Fig. 3.11e, f) with those *Tintinnopsis* species that have the most complex somatic ciliary pattern (the somatic ciliary pattern of *Tintinnopsis cylindrata* resembles that of the *Tintinnidium* species and *Tintinnopsis brasiliensis* lacks a posterior kinety).

Cladistic analysis

The Hennigian argumentation involved 56 characters, mainly those of morphology, ultrastructure, cell division, conjugation, and resting cysts (Table 3.2). The apomorphic character states were used to infer relationships, applying the parsimony principle, by which the simplest solution with the fewest homoplasies and evolutionary steps is preferred [primers for this approach are found in Lipscomb (1998) and Futuyma (2009); details are provided by Felsenstein (2004)]. Although the ancestral and derived character states are simply coded in the Hennigian argumentation method (first or only code in Table 3.2), the computer programs require a special procedure for character state trees. The bifurcations during character development are translated into several binary codes [second codes in Table 3.2; for coding of multistate characters, see Lipscomb (1992)]. Accordingly, these character state trees form several columns in Table 3.3 and not

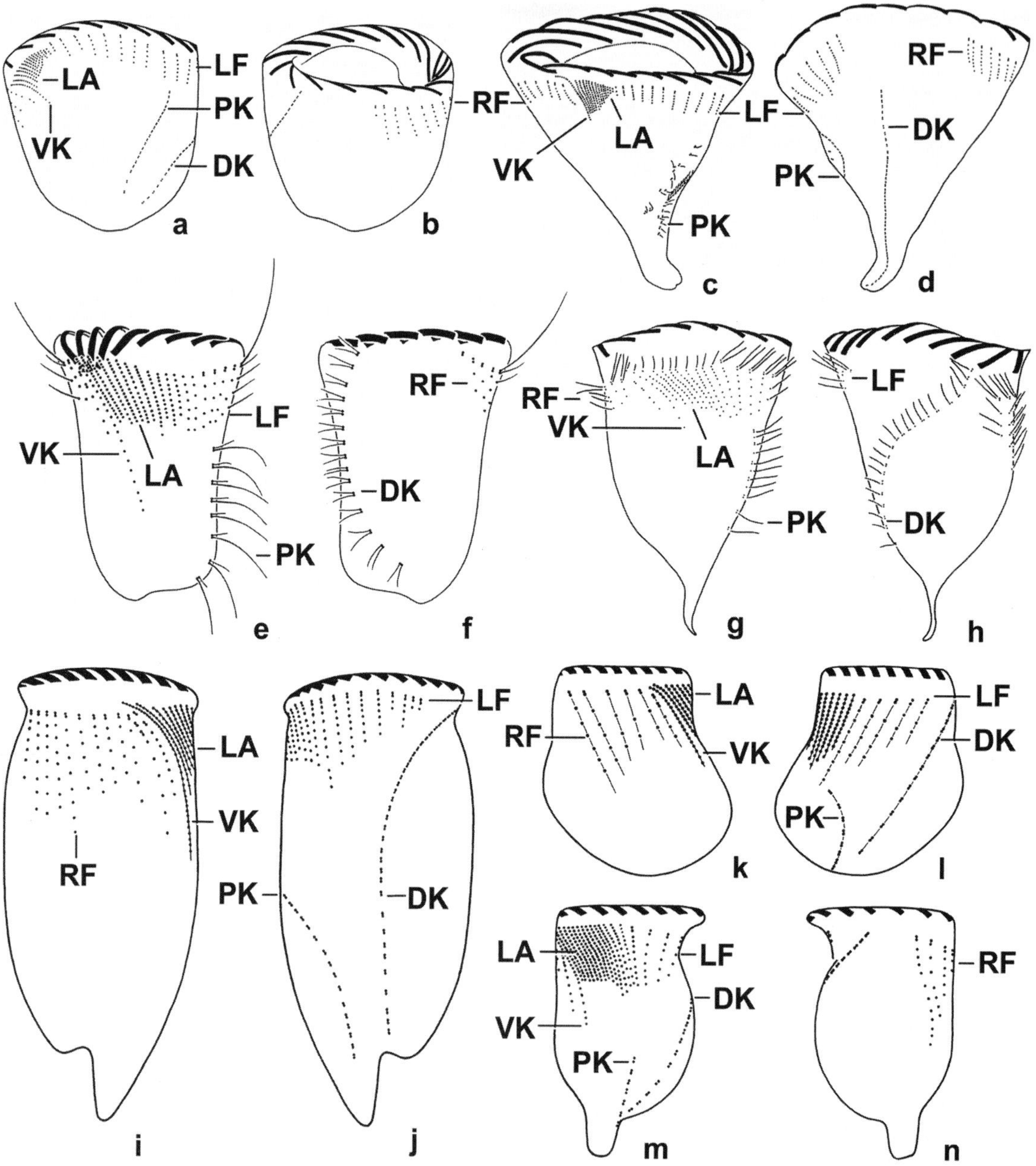

Fig. 3.11 Tintinnids with the most complex somatic ciliary pattern, comprising a ventral, dorsal and posterior kinety as well as a right, left, and lateral ciliary field (a–d, g, h, adapted from Petz et al. 1995; e, f, adapted from Foissner & O'Donoghue 1990, CSIRO Publishing, http://www.publish.csiro.au/nid/120/paper/IT9890661.htm; i, j, adapted from Agatha & Riedel-Lorjé 2006; k, l, adapted from Agatha 2008; m, n, adapted from Agatha, S. & Tsai, S.-F. (2008) Redescription of the tintinnid *Stenosemella pacifica* Kofoid and Campbell, 1929 (Ciliophora, Spirotricha) based on live observation, protargol impregnation, and scanning electron microscopy. Journal of Eukaryotic Microbiology, 55, 75–85: Fig. 67; protargol stain). (a, b) *Cymatocylis convallaria*, 74 µm. (c, d) *Cymatocylis calyciformis*, 125 µm. (e, f) *Stenosemella lacustris*, 32 µm. (g, h) *Codonellopsis glacialis*, 57 µm. (i, j) *Tintinnopsis cylindrica*, 85 µm. (k, l) *Tintinnopsis fimbriata*, 24 µm. (m, n) *Stenosemella pacifica*, 36 µm. DK, dorsal kinety; LA, lateral ciliary field; LF, left ciliary field; PK, posterior kinety; RF, right ciliary field; VK, ventral kinety.

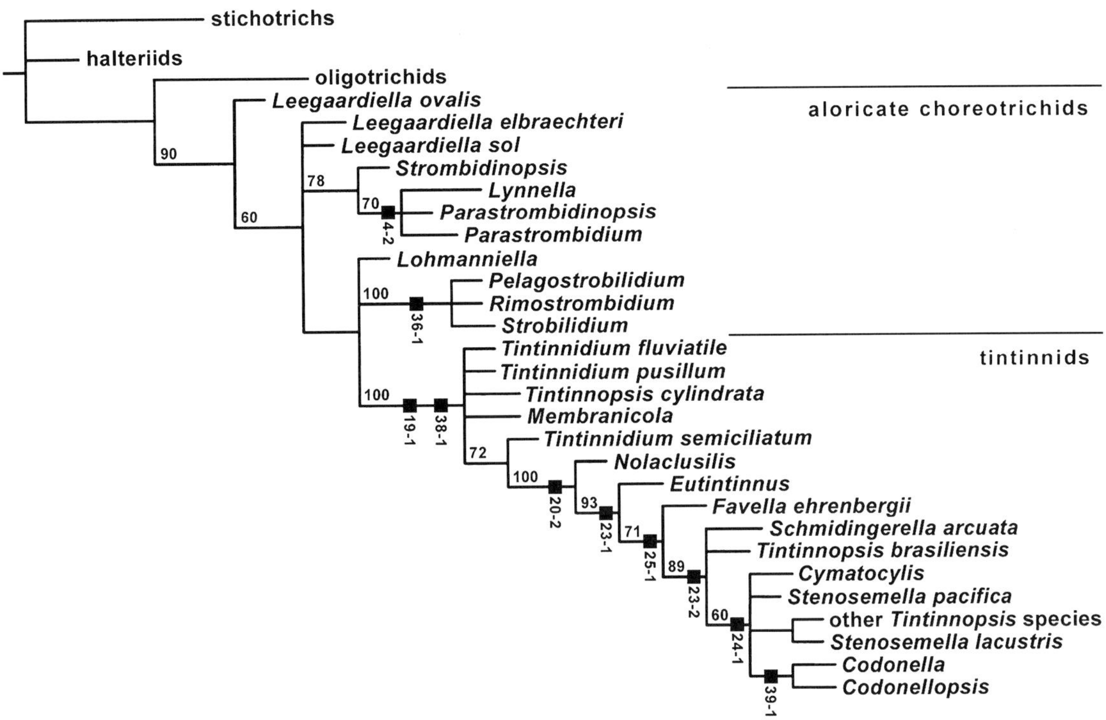

Fig. 3.12 The monophyletic branch of the choreotrichid ciliates in the maximum parsimony tree of the Oligotrichea computed with Hennig86, using the stichotrichs as outgroup (adapted from Agatha, S. & Strüder-Kypke, M.C. (2012) Reconciling cladistic and genetic analyses in choreotrichid ciliates (Protists, Spirotrichea, Oligotrichea). Journal of Eukaryotic Microbiology, in press). For character coding, see Tables 3.2 and 3.3. The tree is the 50% majority-rule consent of 100 trees (length = 164, consistency index = 75, retention index = 93). Numbers on the branches are the bootstrap values (percentage out of 10 replicates) for the congruent internal nodes calculated with PAUP* version 4.0b10 (Swofford 2002). Black squares mark the main apomorphies.

only a single one as typical. Hence, the data matrix for computer analyses comprises 76 characters (= columns). The computer program first searches for the shortest trees, which are according to the parsimony principle the most likely ones. Next, a consensus tree is calculated displaying only those branches occurring in all or most trees (Fig. 3.12). The final tree thus shows besides the unquestionable bifurcations unresolved relationships in form of polytomies (multiple branching of a common ancestor).

The cladogram (Fig. 3.2) clearly demonstrates the separation of tintinnids with two dikinetidal ventral organelles (e.g., *Tintinnidium*) from those with a monokinetidal ventral kinety (e.g., *Eutintinnus*, *Favella*), which is not recognizable in the computed tree (Fig. 3.12). Likewise, a clustering of the genera *Codonella*, *Codonellopsis*, and *Stenosemella* (excluding *S. lacustris*) owing to the capsule type is not recognizable.

Conclusions

The cladistic analysis shows the following.

1. Most freshwater tintinnids are characterized by a somatic ciliary pattern comprising a right and left ciliary field and two *de novo* originating dikinetidal ventral organelles.

2. The structure of the somatic kinetids is highly variable in tintinnids with two ventral organelles, but the complexity of the patterns is low.

3. The structure of the somatic kinetids apparently changed rapidly and then remained constant in tintinnids with a ventral kinety, whereas the complexity increased distinctly by the introduction of specialized kineties and ciliary fields.

4. Hyaline and hard, agglomerated loricae do not form distinct evolutionary lineages, whereas soft, agglomerated loricae are characteristic of most freshwater tintinnids, and loricae composed of a hyaline collar and an agglomerated bowl group rather well together.

5. The capsule types, the lorica sac with its closing apparatus, and the hypothesis about somatic ciliary pattern development, represent the most important features/feature complexes for the reconstruction of evolution among tintinnids.

6. The genus *Tintinnopsis* is paraphyletic (not all descendents of the last common ancestor fall in this taxon).

3.4 HOW TO READ THE TINTINNID CLADOGRAM

Cladograms are the graphic result of a cladistic analysis, which is again based on the interpretation of many morphologic, ontogenetic, and ultrastructural data and their distribution among the taxa analyzed. Although the pure tree only reflects the relationships of the taxa, the combination of the cladogram with the databases provides the reader with information about the occurrence and changes of character states during the evolution of the organisms. To show how to read a cladogram and how to infer this additional information, we use the tintinnid tree shown in Fig. 3.2 as well as the Tables 3.2 and 3.3. Each of the black squares on the branches marks an apomorphy or synapomorphy; the associated number indicates the character, followed by its state as given in Table 3.2.

On our way through the cladogram starting with the genus *Codonellopsis*, for instance, we pass several black squares. The most important apomorphies will be discussed. The first square has associated the number 28-3 and is additionally marked by an asterisk indicating a homoplasy. Using Table 3.2, we see that *Codonellopsis* has more than two macronucleus nodules, a character state that independently developed in *Eutintinnus*. Furthermore, we recognize that two macronucleus nodules are the plesiomorphic (ancestral) character state and that there are two further apomorphic character states, namely the presence of an ellipsoidal and a C-shaped macronucleus

Table 3.2 Character states and coding used for the construction of a cladogram built by the Hennigian argumentation (first or only code; Fig. 3.2) and for computer analyses (only or second code; Fig. 3.12). The coding is mainly based on the outgroup comparison with the stichotrichs. If not stated otherwise, the characters are ordered/additive (the states have a certain sequence; Wagner/Farris optimization).

	Character states	
	Apomorphy	**Plesiomorphy**
1	Cell shape usually globular to obconical (coded 1)	Cell shape usually distinctly dorsoventrally flattened (coded 0)
2	Usually planktonic (coded 1)	Usually benthic (coded 0)
3	Adoral zone of membranelles mainly apical (coded 1)	Adoral zone of membranelles mainly ventral (coded 0)
4	Adoral zone of membranelles circular (coded 1) or secondarily with minute ventral gap (coded 2)	Adoral zone of membranelles C-shaped (coded 0)
5	30–50% (coded 1) or 0% (coded 2) of adoral polykinetids composed of four rows of basal bodies	>90% of adoral polykinetids composed of four rows of basal bodies (coded 0)
6	Postciliary and transverse microtubules absent in adoral membranelles (coded 1)	Postciliary and transverse microtubules present in adoral membranelles (coded 0)
7	Collar polykinetids bipartite (coded 1)	Collar polykinetids continuous (coded 0)

Continued

Table 3.2 *Continued*

	Character states	
	Apomorphy	**Plesiomorphy**
8[b]	Adoral zone bipartite in buccal membranelles with small polykinetids and short cilia and collar membranelles with broad polykinetids and long cilia (coded 1; coded 1000), buccal membranelles absent (coded 2; coded 1100), last collar membranelles proximally elongated (coded 3; coded 1010), last collar membranelles proximally and distally elongated (coded 4; coded 1011)	Polykinetids and cilia of adoral membranelles gradually decrease in size towards the cytostome (coded 0; coded 0000)
9	Undulating membrane(s) often diplostichomonad (two parallel rows of basal bodies) or polystichomonad (more than two parallel rows of basal bodies; coded 1)	Undulating membrane(s) monostichomonad (single row of basal bodies; coded 0)
10	Paroral membrane absent (coded 1), paroral and endoral membranes absent (coded 2)	Usually endoral and paroral membranes present (coded 0)
11	Cyrtos-like (conspicuously strong) pharyngeal fibres (coded 1)	Common pharyngeal fibres (coded 0)
12	Cirri absent (coded 1)	Cirri present (coded 0)
13[a]	Somatic kinetids unciliated (coded 1) or with clavate cilia (coded 2)	Somatic kinetids with rod-shaped or fusiform cilia (coded 0)
14[a]	Usually one or two somatic kineties (coded 1), usually ten or more somatic kineties (coded 2)	Usually 3–9 somatic kineties (coded 0)
15	≥40% of unspecialized somatic kineties shortened or entirely reduced (coded 1)	Unspecialized somatic kineties extend from adoral zone of membranelles to posterior cell end (coded 0)
16[a]	Some unspecialized somatic kineties distinctly curved (coded 1) or forming a posterior spiral (coded 2)	Unspecialized somatic kineties longitudinal (coded 0)
17[a]	Oligotrichid ventral kinety erected (coded 1), usually indistinct or absent (coded 2)	Oligotrichid ventral kinety dextrally spiraled (coded 0)
18[b]	Oligotrichid girdle kinety: dextrally spiraled with posterior end inversely orientated (coded 1; coded 100000000); horizontally orientated with oral primordium posteriorly (coded 2; coded 010000000); sinistrally spiraled (coded 3; coded 011000000); horizontally orientated with dorsal gap (coded 4; coded 010100000); horizontally orientated with oral primordium anteriorly (coded 5; coded 010010000); Ω-shaped with oral primordium posteriorly (coded 6; coded 010001000); Ω-shaped with oral primordium anteriorly (coded 7; coded 000000100); in two inverted U-shaped fragments (coded 8; coded 000000110); in several mostly clockwise inclined fragments (coded 9; coded 000000101)	Oligotrichid girdle kinety dextrally spiraled (coded 0; coded 000000000)
19	Somatic kineties arranged in a right and left ciliary field (coded 1)	Somatic kineties more or less equidistantly arranged (coded 0)
20[a]	Two ventral organelles (coded 1) or one specialized tintinnid ventral kinety (coded 2)	Specialized ventral organelles or tintinnid ventral kinety absent (coded 0)
21	Tintinnid ventral kinety composed of a monokinetidal anterior and a dikinetidal posterior portion (coded 1)	Tintinnid ventral kinety monokinetidal (coded 0)
22	Right ciliary field and tintinnid ventral kinety separated by a broad unciliated stripe (coded 1)	Right ciliary field abuts on tintinnid ventral kinety (coded 0)

Continued

Table 3.2 *Continued*

	Character states	
	Apomorphy	**Plesiomorphy**
23	Two specialized dorsal kineties (coded 1) or one dorsal kinety (coded 2)	Specialized dorsal kinety/kineties absent (coded 0)
24	Specialized posterior kinety present (coded 1)	Specialized posterior kinety absent (coded 0)
25	Lateral ciliary field present (coded 1)	Lateral ciliary field absent (coded 0)
26[b]	Unspecialized somatic kineties: some dikinetids with cilia only at the anterior basal bodies, other dikinetids with two cilia (coded 1; coded 10000000); all dikinetids with two cilia (coded 2; coded 11000000); most dikinetids with cilia only at the posterior basal bodies, few dikinetids with two cilia (coded 3; coded 11100000); all dikinetids with cilia only at the posterior basal bodies (coded 4; coded 11110000); some dikinetids with cilia only at the posterior basal bodies, some ciliated monokinetids (coded 5; coded 11111000); ciliated monokinetids (coded 6; coded 11111100); mostly ciliated monokinetids, some dikinetids with two cilia, some dikinetids with cilia only at the posterior basal bodies (coded 7; coded 11100010); mostly ciliated monokinetids, some dikinetids with two cilia (coded 8; coded 11100011)	Unspecialized somatic kineties composed of dikinetids, each has a distinct cilium associated only with the anterior basal body (coded 0; coded 00000000)
27	Somatic kinetids condensed (coded 1)	Somatic kinetids distinctly separate (coded 0)
28[a]	Majority of members with one ellipsoidal macronucleus nodule (coded 1), one C-shaped macronucleus (coded 2), or more than two macronucleus nodules (coded 3)	Majority of members with two macronucleus nodules (coded 0)
29[a]	Tintinnid extrusomes (capsules) and/or structures usually associated with tintinnid extrusomes (coded 1) or oligotrichid extrusomes (trichites; coded 2) present	Tintinnid and oligotrichid extrusomes absent (coded 0)
30	Stripe of extrusome attachment sites distinctly apart from the oligotrichid girdle kinety (coded 1)	Stripe of extrusome attachment sites directly anterior to the oligotrichid girdle kinety (coded 0)
31[a]	Capsule type I (coded 1) or type II (coded 2) present	Capsules absent (coded 0)
32	Tintinnid mucocyst type A present (coded 1)	Mucocysts absent or different (coded 0)
33[a]	Contractility of peduncle (coded 1) or tail (coded 2)	Posterior cell portion acontractile (coded 0)
34[a]	Anterior cell portion with contractile tentacles (coded 1) or tentaculoids (coded 2)	Anterior cell portion without cytoplasmic appendages (coded 0)
35	Polysaccharidic cortical platelets present (coded 1)	Cortical platelets absent (coded 0)
36	Kinetal lips covering the bases of the somatic cilia present (coded 1)	Kinetal lips covering the bases of the somatic cilia absent (coded 0)
37	Vesicular reticulum present (coded 1)	Vesicular reticulum absent (coded 0)
38[a]	Lorica: soft, agglomerated (coded 1); soft, agglomerated, and with subterminal membrane (coded 2); hard, agglomerated (coded 3); posterior portion agglomerated, anterior hyaline (coded 4); entirely hyaline (coded 5)	Majority of taxa without lorica (coded 0)
39	Lorica sac and foldable closing apparatus present (coded 1)	Lorica sac and foldable closing apparatus absent (coded 0)

Continued

Table 3.2 *Continued*

	Character states	
	Apomorphy	**Plesiomorphy**
40	Enantiotropy (coded 1)	Homeotropy (coded 0)
41[b]	Stomatogenesis hypoapokinetal in transient tube (coded 1; coded 100), permanent tube (coded 2; coded 110), or transient pouch (coded 3; coded 101)	Stomatogenesis epiapokinetal (coded 0; coded 000)
42	Undulating membranes originate *de novo* (coded 1)	Undulating membranes originate from oral primordium or cirral anlagen (coded 0)
43	Unspecialized somatic kineties originate *de novo* (coded 1)	Unspecialized somatic kineties originate usually by intrakinetal proliferation of basal bodies (coded 0)
44	Reorganization of somatic kineties present (coded 1)	Reorganization of somatic kineties absent (coded 0)
45	Preformed emergence pore of resting cyst closed with a plug (coded 1)	Preformed emergence pore and plug absent in resting cyst (coded 0)
46	Ectocyst (outer cyst layer) bipartite and granular (coded 1)	Ectocyst comprises a single microfibrillar or membranous layer (coded 0)
47	Wall of resting cyst with inorganic layers (coded 1)	Wall of resting cyst without inorganic layers (coded 0)
48[a]	Lepidosomes: with tubular fine structure (coded 1); with fibrous fine structure and conical/spine-like shape (coded 2); with fibrous fine structure and globular shape (coded 3)	Lepidosomes absent (coded 0)
49	"Curious structures" in cytoplasm of resting cyst (coded 1)	Cytoplasm of resting cyst without "curious structures" (coded 0)
50	Cyst wall precursors of halteriid type (coded 1)	Cyst wall precursors of stichotrich type (coded 0)
51	Inner cyst membrane that encloses the ciliate emerging from the cyst absent (coded 1)	Inner cyst membrane that encloses the ciliate emerging from the cyst present (coded 0)
52	Pycnosis of vegetative macronucleus without fragmentation (coded 1)	Fragmentation of vegetative macronucleus before pycnosis (coded 0)
53[a]	Interlocking arrangement (coded 1) or oblique arrangement (coded 2) of conjugants	Parallel arrangement of conjugants (coded 0)
54	Transient dimorphism of conjugants (coded 1)	Isomorphic conjugants (coded 0)
55	Conjugants share membranelles (coded 1)	Conjugants do not share membranelles (coded 0)
56	Single derivative of first maturation division performs second division (coded 1)	All derivatives of first maturation division participate in second division (coded 0)

[a] Non-additive (unordered) character states, i.e. each state can change into any other state by one step.
[b] Binary coding of character state trees (first code for Hennigian argumentation scheme; second code for computer analyses).

Table 3.3 Distribution of character states over the taxa cladistically analyzed with the computer program Hennig86. Note that the character state trees of character 8 (bipartition of the adoral zone of membranelles), character 18 (oligotrichid girdle kinety), character 26 (structure of the somatic kinetids), and character 41 (stomatogenesis) were converted into additive binary coding (Table 3.2).

	10	20	30	40	50	60	70
Stic	0000000000	010000000-	----------	------0000	0000000---	000000-000	0001000100	002110
Mese	1110100100	000010000-	----------	------0000	0000010---	000000-100	0111010311	0?1111
Halt	1110100100	000010010-	----------	------0000	0000110---	000000-100	011101021?	?01111
Phal	1110100100	000010010-	----------	------0000	0000110---	000000-100	01110?021?	??1111
Cyrt	1110200110	0-21101111	010100000-	------0000	00000120--	001000-110	01001?12??	??000?
Plst	1110200100	0010111112	010000000-	------0000	00000120--	001000-111	0100111 2??	??000?
Limn	1110200100	0010121111	010010000-	------0000	00000120--	001000-111	01001?12??	??000?
Labo	1110200100	0010101112	011000000-	------0000	00000320--	001000-110	01001?12??	??000?
Noap	1110200100	0010101111	000000000-	------0000	00000021--	001000-110	01001?12??	??000?
Nosi	1110200100	0010101111	000000000-	------0000	00000020--	001000-110	01001?12??	??000?
Omeg	1110200100	0010101111	000000100-	------0000	00000120--	001000-110	01001?12??	??000?
Opis	1110200100	0010101111	010010000-	------0000	00000120--	001000-110	01001?12??	??000?
Fois	1110200100	0010101111	010010000-	------0000	00000021--	001000-110	01001?12??	??000?
Plle	1110200100	0010101110	000000000-	------0000	00000120--	001000-110	01001?12??	??000?
Spir	1110200100	0010101111	100000000-	------0000	00000120--	001000-110	01001?12??	??000?
Stro	1110200100	0010101111	010000000-	------0000	00000120--	001000-110	01001?12??	1?000?
Pato	1110200100	0010101111	010001000-	------0000	00000320--	201000-110	01001?12??	??000?
Psto	1110200100	0010101111	010000000-	------0000	00000320--	211000-110	01001?12??	??000?
Spto	1110200100	0010101112	011000000-	------0000	00000320--	201000-110	01001?12??	??000?
Tont	1110200100	0010101112	000000000-	------0000	00000320--	201000-110	01001?12??	??000?
Apos	1110200100	0010101112	000000110-	------0000	00000120--	001000-110	01001?12??	??000?
Vari	1110200100	0010101112	000000101-	------0000	00000120--	001000-110	01001?12??	??000?
Lelb	1111201101	001010010-	----------	------1000	0000000---	000000-110	11001?10??	?1?000
Lova	1111201101	001010111-	----------	------0000	0000010---	000000-110	11001?10??	?1?000
Lsol	1111201101	001010010-	----------	------1100	0000000---	000000-110	11001?10??	?1?000
Lohm	1111200101	001010010-	----------	------1111	0000010---	000000-110	11001?10??	?1?000
Lynn	1112200101	001010100-	----------	------1111	1000000---	000000-110	11001?10??	?1?000
Pstb	1111200101	001010011-	----------	------1111	1100120---	000100-110	11001?10??	?11000
Rimo	1111200101	001010010-	----------	------1111	1100120---	000100-110	11001?10??	?11000
Strb	1111200101	001010012-	----------	------1111	1100120---	000100-110	11001?10??	?11000
Parp	1112200101	001010200-	---------0	0--0001100	0000000---	000000-110	11001?10??	?1?000
Parb	1112200101	101010200-	---------0	0--0001100	0000030---	000000-110	11001?10??	?1?000
Spsi	1111200101	001010200-	---------0	0--0001100	0000000---	000000-110	11001?10??	?1?000
semi	1111210101	001010210-	---------1	1--0001110	0010011-??	1?00110110	1100??10??	?10000
fluv	1111210101	001010210-	---------1	1--0001110	0000011-??	1?00110110	1100??10??	?10000
pusi	1111210101	001010210-	---------1	1--0001110	0000011-??	1?00110110	1100??10??	?10000
cyli	1111210101	001010210-	---------1	1--0001111	1100011-??	1?00110110	1100??10??	?10000
Memb	1111210101	001010210-	---------1	1--0001111	000000?-??	1?00120110	1100??10??	?10000
Nola	1111210101	001010210-	---------1	2000001110	0011001-??	1200150110	1100??10??	?10000
Euti	1111210101	001010210-	---------1	2001001110	0011031-??	1200150110	1100??10??	?10000
Fehr	1111210101	001010210-	---------1	2001011110	0011000--?	1200150110	1100??10??	?10000
Sarc	1111210101	001010210-	---------1	2112011110	0011001-??	1200150110	11001?10??	?10000
bras	1111210101	001010210-	---------1	2002011110	001100?-??	1200130110	1100??10??	?10000
Codo	1111210101	001010210-	---------1	2002111110	0011001-21	1200131110	1100??10??	?10000

Continued

Table 3.3 *Continued*

	10	20	30	40	50	60	70	
Cpsi	1111210101	001010210-	---------1	2002111110	0011031-21	1200141110	1100??10??	?10000
Cyma	1111210101	001010210-	---------1	2002111110	0011001-??	1200150110	1100??10??	?10000
Otsp	1111210101	001010210-	---------1	2002111110	0011001-1?	1200130110	11000?10??	?10000
lacu	1111210101	001010210-	---------1	2002111110	001100?-??	1?00130110	1100??10??	?10000
paci	1111210101	001010210-	---------1	2002111110	0011001-21	1200140110	1100??10??	?10000

Apos, *Apostrombidium*; *bras*, *Tintinnopsis brasiliensis*; *Codo*, *Codonella*; *Cpsi*, *Codonellopsis*; *cyli*, *Tintinnopsis cylindrata*; *Cyma*, *Cymatocylis*; *Cyrt*, *Cyrtostrombidium*; *Euti*, *Eutintinnus*; *Fehr*, *Favella ehrenbergii*; *fluv*, subgenus *Tintinnidium* with *T. fluviatile*; *Fois*, *Foissneridium*; *Halt*, *Halteria*; *Labo*, *Laboea*; *lacu*, *Stenosemella lacustris*; *Lelb*, *Leegaardiella elbraechteri*; *Limn*, *Limnostrombidium*; *Lohm*, *Lohmanniella*; *Lova*, *Leegaardiella ovalis*; *Lsol*, *Leegaardiella sol*; *Lynn*, *Lynnella*; *Memb*, *Membranicola*; *Mese*, *Meseres*; *Noap*, *Novistrombidium apsheronicum*; *Nola*, *Nolaclusilis*; *Nosi*, *Novistrombidium sinicum*; *Omeg*, *Omegastrombidium*; *Opis*, *Opisthostrombidium*; Otsp, other *Tintinnopsis* species; *paci*, *Stenosemella pacifica*; *Parb*, *Parastrombidium*; *Parp*, *Parastrombidinopsis*; *Pato*, *Paratontonia*; *Phal*, *Pelagohalteria*; *Plle*, *Parallelostrombidium*; *Plst*, *Pelagostrombidium*; *Pstb*, *Pelagostrobilidium*; *Psto*, *Pseudotontonia*; *pusi*, subgenus *Tintinnidium* with *T. pusillum*; *Rimo*, *Rimostrombidium*; *Sarc*, *Schmidingerella arcuata*; *semi*, subgenus *Semitintinnidium* with *Tintinnidium semiciliatum*; *Spir*, *Spirostrombidium*; *Spsi*, *Strombidinopsis*; *Spto*, *Spirotontonia*; Stic, stichotrichs; *Strb*, *Strobilidium*; *Stro*, *Strombidium*; *Tont*, *Tontonia*; *Vari*, *Varistrombidium*.

nodule. The cladogram (Fig. 3.2) is only shown for the choreotrichids, but by means of Table 3.3, we can see that the tontoniid oligotrichids also possess more than two macronucleus nodules. Admittedly, Table 3.3 is more difficult to read than Table 3.2, owing to the binary coding of character state trees, which is provided by the second code in Table 3.2. Character 28 is marked as non-additive in Table 3.2, i.e. the character states are unordered and each state can turn into each other by one step.

For parsimony reasons, the introduction of an entirely hyaline lorica was assumed in the ancestor of tintinnids with a monokinetidal ventral kinety (character 38-5). The anterior cell portion of *Codonellopsis* bears tentaculoids (character 34-2); similar, but probably analogous structures are found in an tontoniid oligotrichid (character 34-1; Skovgaard & Legrand 2005). The unspecialized somatic kineties (dorsal, posterior, and ventral kineties are excluded) of *Codonellopsis* are composed of monokinetids and few dikinetids with two cilia (character 26-8). The complex changes in the structure of the somatic kineties are translated into binary codes, forming eight columns in Table 3.3. And we see in the cladogram that the evolution of the kinety structure passed two previous character states (26-3 and 26-2). The parsimony principle again necessitated the introduction of a soft, agglomerated lorica at the beginning of the tintinnid evolution (character 38-1). Although not studied in the genus *Codonellopsis*, some of its character states can simply be predicted owing to its affiliation with the tintinnids: the presence of a vesicular reticulum (character 37-1), the lack of postciliary and transverse microtubules at the adoral membranelles (associates of basal bodies extending posteriorly and leftwards, respectively; character 6-1), and a disintegration of the macronucleus nodules without a preceding fragmentation during conjugation (character 52-1; with fragmentation in stichotrichs and halteriids; Agatha & Foissner 2009 and references therein). The last feature probably requires some explanation.

During conjugation, partners of probably different mating types fuse in their anterior cell portions. The somatic macronucleus disintegrates with or without a preceding fragmentation, while the generative diploid micronucleus performs a meiosis comprising a first and second maturation division; usually, three of the haploid derivatives disintegrate. The surviving fourth derivative divides mitotically, generating a stationary and a migratory pronucleus (Fig. 4.6). As indicated by the names of the pronuclei, the stationary remains in each partner, while the migratory pronuclei are exchanged between the partners. Finally, each conjugant has its own haploid stationary pronucleus and the haploid migratory pronucleus of its partner. Subsequently, the pronuclei fuse, forming the diploid synkaryon. Taxon-specific divisions of the synkaryon produce the micronuclei, the macronucleus nodules, and disintegrating derivatives.

The affiliation of the genus *Codonellopsis* with the tintinnid family Dictyocystidae in the revised classification (Table 3.4) reflects its evolutionary history and natural relationships at least partly (Section 3.6). This assignment is based on the presence of the lorica sac with its foldable closing apparatus, a feature that is shared by the genera *Codonaria*, *Codonella*, and *Dictyocysta*.

3.5 MOLECULAR ANALYSIS AND COMPARISON WITH MORPHOLOGIC DATA

Here, we present a short introduction into gene sequencing, followed by the results of the SSU rRNA analyses and a comparison with the hypothesis of tintinnid evolution inferred from morphologic data.

Gene sequence analysis in general

The SSU rRNA is the gene usually analyzed to infer phylogenetic relationships in ciliates, as its mutation rate is not too low (conserved) or high (multiple nucleotide substitutions at the same site obscure the true evolution). In some studies, the internal transcribed spacer (ITS 1 and 2) sequences are analyzed to infer the evolution at and below species level and the phylogeography, as they have elevated mutation rates. In recent years, mitochondrial DNA, in particular the cytochrome-*c* oxidase subunit 1 (*cox1*) gene, has been suggested as a possible molecular marker for species distinction (Hebert et al. 2003a, b) owing to its higher resolution at species level. Initial studies in ciliates showed the great potential of this gene to distinctly separate closely related species (see Barth et al. 2006; Lynn & Strüder-Kypke 2006; Strüder-Kypke & Lynn 2010). The *cox1* gene is, however, still comparatively rarely studied in ciliates, and data for tintinnids are not available.

Although gene sequence analysis was expensive and time-consuming a few years ago, now extraction, polymerase chain reaction (PCR) amplification, and sequencing are routine. The nucleotide sequences are subsequently aligned with homologous sequences obtained from a databank (e.g., GenBank), and the alignment is further refined based on the secondary structure of the SSU rRNA molecule. The result is a data matrix similar to that shown in Table 3.3; instead of numbers, presumably homologous nucleotides are listed. The amount of characters provided by SSU rRNA gene sequencing is much higher than that of cladistic analyses, as about 1600 or even more nucleotides remain after trimming the ends of the sequences and deleting highly variable regions. However, we should keep in mind that the morphologic characters are often the product of more than one gene and that gene trees and species trees are not the same (Nichols 2001).

Next, the phylogenetic relationships can be inferred by various analytical methods, including algorithms based on the parsimony principle (Maximum Parsimony, MP). Because each method has its advantages and disadvantages, usually more than one computer program is run to analyze the data [for an introduction to the methods, see Holder & Lewis (2003); details are provided by Felsenstein (2004)]. Some of them, first calculate the relative probability of transitions (A < > G, C < > T) versus transversions (A < > C, G < > T) or the degree to which the rate of evolution differs across the nucleotide sites based on the given data set, and then implement the obtained substitution model (e.g., the Kimura-2-parameter or General-Time-Reversible models) into the computer program (e.g., Maximum Likelihood). Others merely calculate the evolutionary distances between sequences, namely, the number of different nucleotides (Neighbor-Joining). Finally, the confidence of the obtained trees is calculated by, for example, bootstrapping procedures.

Data acquisition

In August 2011, there were 66 SSU rRNA gene sequences of tintinnid ciliates available from public databases (GenBank, EMBL). These 66 sequences can be assigned to 37 genetically distinct entities (operational taxonomic units). However, 11 of these taxa remain unidentified, because they result from environmental sequencing or because the genus and/or species were not determined during sampling. A major shortcoming in many molecular studies is uncertain species identification, which is generally based on lorica morphology only. Therefore, in many cases, the identity of sequenced tintinnid taxa is unknown or remains doubtful. Compared with the number of lorica-based species, the percentage of species with molecular data is extremely small, and only 14 of the assumed 75 genera (Table 3.4) are represented by SSU rRNA gene sequences (about 19%). Thirteen of the SSU rRNA sequences in the phylogenetic tree (Fig. 3.13) are assigned to the genus *Tintinnopsis*, which we regard as a taxon incertae sedis (with uncertain placement; Section 3.6) within the order Tintinnida until more morphologic

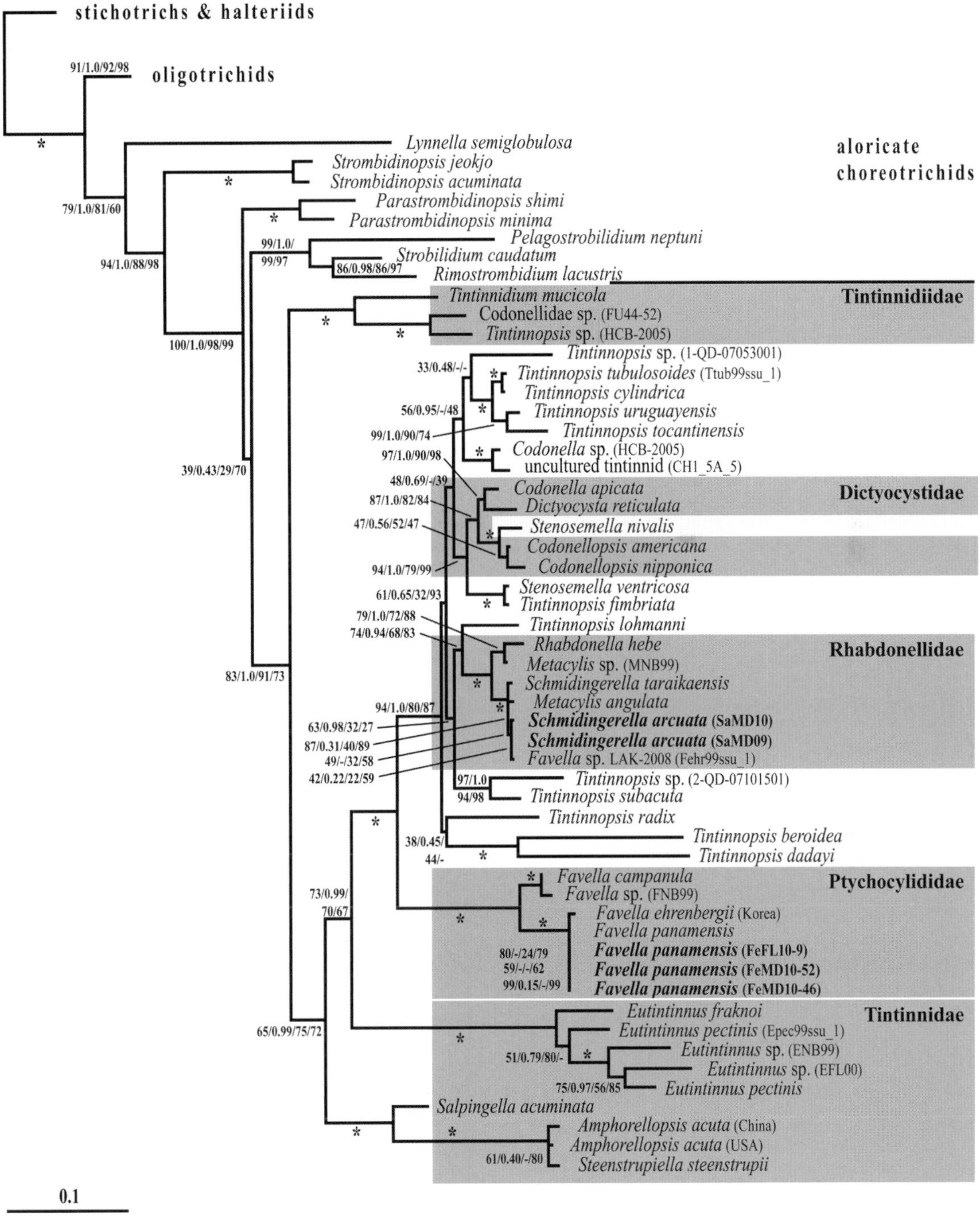

Fig. 3.13 Maximum likelihood tree inferred from SSU rRNA gene data of representative tintinnid species, computed with PhyML (Guindon et al. 2010) and based on the general time-reversible (GTR) model with gamma distribution and an estimate of invariable sites (adapted from Agatha, S. & Strüder-Kypke, M.C. (2012) Reconciling cladistic and genetic analyses in choreotrichid ciliates (Protists, Spirotrichea, Oligotrichea). Journal of Eukaryotic Microbiology, in press). The numbers at the nodes represent the support values for the maximum likelihood, Bayesian inference, maximum parsimony, and neighbor joining analyses, respectively. Asterisks mark full support in all analyses, whereas dashes indicate support values of less than 20% (or 0.2 posterior probability). The scale bar represents 10 substitutions per 100 nucleotides.

and molecular data become available. Therefore, any results of molecular phylogenetic analyses can only be regarded as preliminary, although some patterns begin to emerge, as more sequences are included.

SSU rRNA phylogenies

All published phylogenetic trees show the following topology (Fig. 3.13).

1. The basal branch within the tintinnids represents the family Tintinnidiidae (*Tintinnidium*; soft, agglomerated loricae; right and left ciliary fields, ventral organelles).

2. The family Tintinnidae (*Eutintinnus*, *Salpingella*, *Amphorellopsis*, *Steenstrupiella*, all with hyaline, monolaminar and compact lorica walls) is paraphyletic with a branch representing the genus *Eutintinnus* (right and left ciliary fields and ventral and dorsal kineties) and another for the remaining genera. Sequences from additional genera of this family are required to confirm whether the deep divergence is real or owing to undersampling (species filling the gap have not been sequenced).

3. The family Ptychocylididae branches next, highly supported, but only represented by the genus *Favella* (hyaline, monolaminar lorica walls with alveoli; right, left, and lateral ciliary fields, ventral and dorsal kineties). All remaining taxa display a less statistically supported topology, which varies depending on the sequences analyzed and the algorithm used. However, two clusters are consistently supported: the family Dictyocystidae (lorica sac and capsule type II) and the family Rhabdonellidae (*Rhabdonella* and *Schmidingerella* with some former *Favella* species; hyaline, monolaminar lorica walls with alveoli, surface ridges, and pores).

4. The genus *Stenosemella* is placed within the family Dictyocystidae. The lack of a lorica sac contradicts this affiliation, but *Stenosemella* shares the capsule type II and its lorica is composed of a hyaline collar and an agglomerated bowl like those in *Codonellopsis* and most *Dictyocysta* species.

5. The genus *Metacylis* is placed within the family Rhabdonellidae. The apparently monolaminar lorica wall with alveoli, surface ridges, and pores supports this assignment at least for *M. angulata*. Nevertheless, additional molecular and morphological evidence is required to confirm the close phylogenetic relationships of *Stenosemella* and *Metacylis*, respectively.

In taxonomy, a species name is a hypothesis, and a species diagnosis represents a compilation of the distinguishing features, whereas the description includes all data. Likewise, molecular biologists search for nucleotide differences that characterize certain taxa. Actually, a detailed analysis of the gene sequences revealed some signature nucleotides at species, genus, or family level (Figs 3.14 and 3.15). The deep divergence in the family Tintinnidae is based on several nucleotide differences, shown in Fig. 3.14a. On the other hand, there are also several unique nucleotide positions that unite the genera *Eutintinnus*, *Amphorellopsis*, *Steenstrupiella*, and *Salpingella* and separate them from the genera *Favella*, *Codonella*, *Tintinnopsis*, and *Rhabdonella*. Some positions are shared among the family Tintinnidae and *Favella panamensis* (family Ptychocylididae), which is reflected in the topology of the phylogenetic tree that strongly separates these basal taxa from the remaining species. The genera *Rhabdonella*, *Metacylis*, and *Schmidingerella* (with some former *Favella* species) cluster as one branch in the phylogenetic tree. The two sequences of *Metacylis* group with *Rhabdonella* and *Schmidingerella*, respectively. Several signature nucleotides support this division of the genus *Metacylis* (Fig. 3.14b). However, three signature nucleotides are unique to the two *Rhabdonella* species and are absent in *Metacylis*. The analysis of additional sequences is required to determine the level of sequence variation among these genera and to resolve their taxonomic assignment.

The genus *Favella* (family Ptychocylididae) is separated by a long branch from the remaining tintinnid clusters, indicating many nucleotide changes and a high mutation rate. Accordingly, many signature positions exist, which separate this genus from the other tintinnid taxa (Fig. 3.14b).

The genus *Tintinnopsis* is paraphyletic; probably, it consists of several genera characterized by different somatic ciliary patterns (Fig. 3.2) and lorica wall textures (Chapter 2). Thus, it is not surprising that the genus as a whole displays no signature nucleotides. Nevertheless, some clusters in the phylogenetic tree reveal unique signatures, for example the branch comprising *Tintinnopsis cylindrica*, *T. tubulosoides*, *T. tocantinensis*, and *T. uruguayensis* (Fig. 3.15).

Several signatures distinguish the family Dictyocystidae plus *Stenosemella nivalis* and *S. ventricosa* from the remaining tintinnid taxa. Within this grouping the two branches consisting of (1) *Codonellopsis* and *Stenosemella nivalis* and (2) *Dictyocysta* and *Codonella* reveal some distinct signatures. The relationship of the *Stenosemella* species to the Dictyocystidae still needs to be resolved (see above).

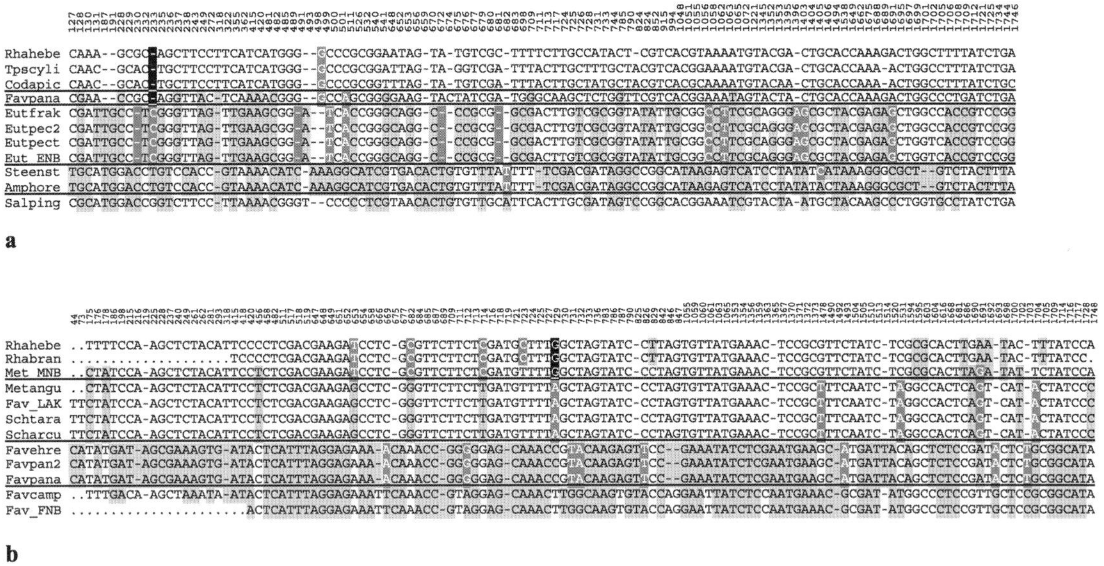

Fig. 3.14 Signature nucleotide positions of supported clusters in the order Tintinnida. (a) Family Tintinnidae. (b) Families Ptychocylididae and Rhabdonellidae. The numbers above the nucleotides refer to the positions in the original alignment. A background in different gray tints defines the distinguishing nucleotides; in cases where several signature nucleotides are present at the same position, darker and lighter tints are used to differentiate between them. The shading does not relate to any evolutionary state. The horizontal lines separate the supported clusters of the phylogenetic tree. Dots present missing data. Species abbreviations: Amphore, *Amphorellopsis acuta*; Codapic, *Codonella apicata*; Eut_ENB, *Eutintinnus* sp. (ENB99); Eutfrak, *Eutintinnus fraknoi*; Eutpec2, *Eutintinnus pectinis* (Epec99ssu_1); Eutpect, *Eutintinnus pectinis*; Fav_FNB, *Favella* sp. (FNB99); Fav_LAK, *Favella* sp. (Fehr99ssu_1); Favcamp, *Favella campanula*; Favehre, *Favella ehrenbergii* (Korea); Favpan2, *Favella panamensis* (FeMD10-46); Favpana, *Favella panamensis*; Met_MNB, *Metacylis* sp. (MNB99); Metangu, *Metacylis angulata*; Rhabran, *Rhabdonella brandti*; Rhahebe, *Rhabdonella hebe*; Salping, *Salpingella acuminata*; Scharcu, *Schmidingerella arcuata*; Schtara, *Schmidingerella taraikaensis*; Steenst, *Steenstrupiella steenstrupii*; Tpscyli, *Tintinnopsis cylindrica*.

Comparison of SSU rRNA phylogenies and cladograms with former hypothesis

A comparison of the gene trees with cladograms is currently impeded by differences in the species analyzed. Actually, only eight genera were analyzed cladistically and genetically (Figs 3.2 and 3.13), and Kim S.Y. et al. (2010) published the first study, in which both molecular and detailed morphological data were described for the same tintinnid species, namely *Favella ehrenbergii*, the type species of the genus *Favella*. Therefore, the following comparison cannot be very detailed.

It should be noted that for tintinnids the gene tree and cladogram match rather well. This is in marked contrast to some other ciliate groups, in which the phylogenies obtained by both methods deviate distinctly, necessitating a reconsideration of homology and the investigation of further morphologic and/or molecular data (for colpodid ciliates, see Foissner et al. 2011; for halteriids, see Agatha & Foissner 2009). In tintinnids, the two approaches both show the following features.

1. The early branching of the genus *Tintinnidium* followed by *Eutintinnus*, *Favella*, and a diverse group of highly developed tintinnids (compare Figs 3.2 and 3.12 with Fig. 3.13).

2. A grouping of the genera with a lorica sac and foldable closing apparatus.

3. A paraphyly of the genus *Tintinnopsis*.

The two *Favella* clusters found in previous gene trees, could be confirmed by morphologic features (Sections 3.3 and 3.6), justifying the establishment of the genus *Schmidingerella* for *Favella* species clustering with *Rhabdonella*. Although the phylogenetic relationship of the genus *Stenosemella* with taxa possessing a lorica sac requires some further investigations, the presence of this structure warranted a far-reaching revision of the tintinnid families

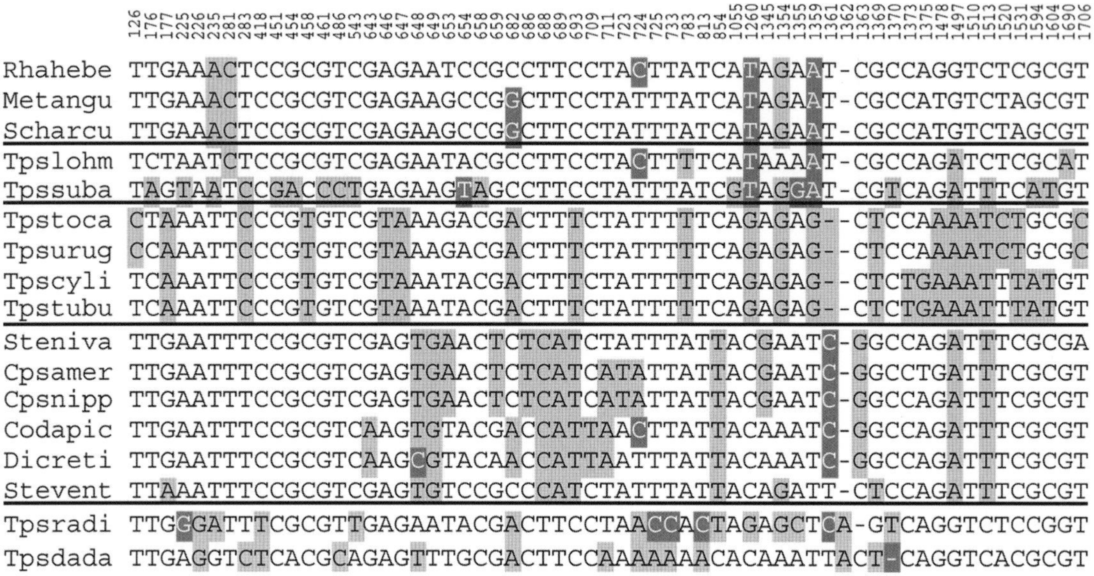

Fig. 3.15 Signature nucleotide positions of supported clusters in the order Tintinnida. Families Dictyocystidae and Rhabdonellidae, and various *Tintinnopsis* species. The numbers above the nucleotides refer to the positions in the original alignment. A background in different gray tints defines the distinguishing nucleotides; in cases where several signature nucleotides are present at the same position, darker and lighter tints are used to differentiate between them. The shading does not relate to any evolutionary state. The horizontal lines separate the supported clusters of the phylogenetic tree. Dots present missing data. Species abbreviations: Cpsamer, *Codonellopsis americana*; Cpsnipp, *Codonellopsis nipponica*; Codapic, *Codonella apicata*; Dicreti, *Dictyocysta reticulata*; Metangu, *Metacylis angulata*; Rhahebe, *Rhabdonella hebe*; Scharcu, *Schmidingerella arcuata*; Steniva, *Stenosemella nivalis*; Stevent, *Stenosemella ventricosa*; Tpscyli, *Tintinnopsis cylindrica*; Tpsdada, *Tintinnopsis dadayi*; Tpslohm, *Tintinnopsis lohmanni*; Tpsradi, *Tintinnopsis radix*; Tpssuba, *Tintinnopsis subacuta*; Tpstoca, *Tintinnopsis tocantinensis*; Tpstubu, *Tintinnopsis tubulosoides*; Tpsurug, *Tintinnopsis uruguayensis*.

Codonellidae, Codonellopsidae, and Dictyocystidae. The differences in the somatic ciliary pattern within the genus *Tintinnopsis* as well as its paraphyly in the gene trees necessitate a reasonable split, which cannot be performed until the cell morphology and SSU rRNA gene sequence of its type species, *T. beroidea*, are known and confirmed.

The phylogenetic relationships inferred from the SSU rRNA gene and morphology (Figs 3.2 and 3.13; Snoeyenbos-West et al. 2002; Strüder-Kypke & Lynn 2003; Agatha & Strüder-Kypke 2007; Strüder-Kypke & Lynn 2008; Gao et al. 2009; Li Z. et al. 2009; Agatha & Strüder-Kypke 2012) markedly contradict the hypothesis of tintinnid evolution and the lorica-based classification by Kofoid & Campbell (1939). The most obvious disparity is the fact that in the phylogenetic trees, species with hyaline and agglomerated loricae are not separate taxonomic entities. On the contrary, species with hyaline and agglomerated loricae are interrelated, and an evolutionary trend from agglomerated to hyaline loricae – as postulated by Kofoid & Campbell (1939) – is not evident (Strüder-Kypke & Lynn 2003). Only the family Tintinnidiidae was predicted by Kofoid & Campbell (1939) to be primitive, and the genus *Tintinnidium* actually branches at the base of the tintinnids in the cladograms and gene trees. Both analyses reflect the increase in complexity of the somatic ciliary patterns in tintinnids with a ventral kinety. The most highly developed pattern occurs in a diverse group of tintinnids and is associated with different lorica types: hard, agglomerated loricae, hyaline loricae, and loricae composed of a hyaline collar and an agglomerated bowl. The same type of lorica can also be associated with different patterns of the somatic ciliature, as shown for the hyaline loricae by *Nolaclusilis*, *Eutintinnus*, *Favella*, *Schmidingerella*, and *Cymatocylis*. Likewise, hard, agglomerated loricae occur in genera with a right, left, and lateral ciliary field as well as a ventral and dorsal kinety, which differ in the presence of a posterior kinety (absent in *Tintinnopsis brasiliensis*, while present in most *Tintinnopsis* species, *Codonella*, and *Stenosemella lacustris*). Accordingly, the different lorica types are not necessarily correlated with different somatic ciliary patterns, as suggested by Kofoid & Campbell (1939).

Laval-Peuto & Brownlee (1986) expected that the most primitive tintinnids would have many somatic kineties per cell circumference (high kinetal density index), long ciliary rows, and no specialized kineties and ciliary fields. Although the last features are also regarded as plesiomorphic in the present analysis (Sections 3.3 and 3.4), the consideration of the number of kineties resulted in a rather different idea of tintinnid evolution. The position of *Tintinnidium* near the base of the tintinnids was also proposed by the model of Laval-Peuto & Brownlee (1986), but the sequence of the remaining taxa was in general agreement with the hypothesis of Kofoid & Campbell (1939) and thus contradicts the recent findings.

The quintessence of the genetic and cladistic analyses is that the lorica types (hyaline; soft, agglomerated; hard, agglomerated; composed of a hyaline collar and an agglomerated bowl) can hardly be used to infer phylogenetic relationships because they indicate natural groupings only at the genus level, while rarely at higher ranks (e.g., Tintinnidiidae). The comparison of both analyses indicates that the development of the somatic ciliary patterns appropriately reflects evolution in tintinnids. For a higher resolution among the genera with the most complex ciliary pattern, data on the subtle differences and further features are required. Admittedly, the investigation of taxonomic features and their cladistic analysis are more time-consuming in ciliates than genetic studies. Nevertheless, the cladistic results are more informative, as they represent a testable hypothesis about character evolution and adaptive processes. On the other hand, molecular analyses might indicate homoplasious characters. So, combined analyses of morphologic and genetic data are strongly recommended for future species descriptions (Lynn & Simpson 2009) and investigations of phylogenetic relationships.

3.6 SYSTEMATICS

Ideal taxonomic classifications (systematics) are the translation of the evolutionary history and natural relationships of the taxa. In most tintinnids, only a small part of the organism, the lorica, has been used in species descriptions, whereas cytological data are not available. The lorica, however, shows a high intraspecific variability due to environmental conditions and the cell cycle (Chapter 2). On the other hand, genetic analyses of the internal transcribed spacer regions indicated high levels of genetic heterogeneity in similar tintinnid morphotypes (morphologically defined units; Snoeyenbos-West et al. 2002). Therefore, it is not surprising that cladistic and molecular analyses also demonstrated the inadequacy of the lorica types for inferring phylogenetic relationships (Sections 3.3 and 3.5).

Although tintinnids have been studied for more than 230 years, we are still unable to circumscribe most species properly and thus to provide even a rough estimate of how many tintinnid species exist. Therefore, a reliable list of valid species cannot be presented here. The scarcity of cytological data about these exciting plankton ciliates also prevents a far-reaching revision beyond those by Loeblich Jr & Tappan (1968), Corliss (1979), Small & Lynn (1985), Laval-Peuto (1994), Aescht (2001), and Lynn & Small (2002), as argued by Lynn (2008). Nevertheless, recent cladistic (Section 3.3) and molecular (Section 3.5) analyses allow partial improvements of the lorica-based classification (Agatha & Strüder-Kypke 2012). Because the relationships among the tintinnids are far from been fully understood, the families and the genera within the families are merely alphabetically listed (Table 3.4). The improvements of the tintinnid classification described below concern (1) the revision of the families Codonellidae, Codonellopsidae, and Dictyocystidae, (2) the affiliation of the recently established genus *Schmidingerella* with the family Rhabdonellidae, and (3) a preliminary revision of the genera. These kinds of taxonomic action are ruled by the International Code of Zoological Nomenclature (ICZN 1999).

Recent changes in the classification

The interpretation of the lorica sac (Fig. 2.2b–d) as synapomorphy of the genera *Codonaria*, *Codonella*, *Codonellopsis*, and *Dictyocysta* was supported by the close relationship of the genera in SSU rRNA phylogenies (Fig. 3.13; Agatha 2010a). Far-reaching changes were thus necessary, especially as three of them are type genera (name-bearing types) of the families Codonellidae, Codonellopsidae, and Dictyocystidae (Agatha & Strüder-Kypke 2012). The four genera were united in the oldest family, the Dictyocystidae, and the genera *Wangiella* and *Laackmanniella* were affiliated incertae sedis. In *Wangiella*, the prismatic flaps of the inner collar described by Nie (1934) might correspond to a closing apparatus, and the genus thus possibly represents a junior synonym of the genus *Dictyocysta*. In the genus *Laackmanniella*, a lorica sac and closing apparatus are unknown; hence, its assignment to the Dictyocystidae is uncertain, although indicated by a *Codonellopsis*-like, posteriorly open lorica. The generic affiliation of the comparatively well-known freshwater species *Codonella cratera* is probably incorrect, as a closing apparatus and lorica sac were never mentioned (Foissner et al. 1999; Agatha 2010a). Because the cytological features of the type species *Tintinnopsis beroidea* are insufficiently known (Agatha & Riedel-Lorjé 2006), *C. cratera* is not affiliated with the genus *Tintinnopsis*, but its somatic ciliary pattern is not any longer representative of the genus *Codonella*. However, Petz et al. (1995) found in *Codonellopsis glacialis*, another member of the Dictyocystidae, the same complex somatic ciliary pattern (a ventral, dorsal, and posterior kinety plus a lateral, right, and left ciliary field). Accordingly, it is assumed that this pattern is characteristic of the entire family.

After the transfer of the genus *Codonellopsis* to the family Dictyocystidae, there was only one extant genus left, namely *Stenosemella*, whose diagnosis was improved by Agatha & Tsai (2008). This genus is type of the subfamily Stenosemellinae Campbell, 1954, which was raised to family rank by Agatha & Strüder-Kypke (2012). The affiliation of the many fossil genera of the family Codonellopsidae mentioned by Lynn (2008) is questionable. Agatha & Tsai (2008) synonymized the genus *Luminella* with *Stenosemella*, as they recognized that the type species *S. ventricosa* possesses small windows in its hyaline collar, the distinguishing feature of the genus *Luminella*.

Three genera lost their home, the family Codonellidae, after the transfer of the type genus *Codonella* to the family Dictyocystidae: *Codonopsis*, *Poroecus*, and *Tintinnopsis*. No alternative family name exists for these taxa and only the lorica features are known of the type species. Because these features are insufficient for placement in a natural tintinnid classification, no new families are established, and the three genera remain for the present as incertae sedis in the tintinnids.

Previously published SSU rRNA analyses displayed a paraphyly of the genus *Favella*, forming two distantly related clusters: one cluster branched just after the *Eutintinnus* species, the other was closely related to *Metacylis* and *Rhabdonella* (Strüder-Kypke & Lynn 2003, 2008; Gao et al. 2009; Li Z. et al. 2009; Kim S.Y. et al. 2010). The paraphyly was also recognizable in the cladogram. Thus, it is now possible to relate genetic grouping with morphologic features. This justified the establishment of two distinct genera, namely *Favella* with *F. ehrenbergii* as type and the new genus *Schmidingerella* with *Favella arcuata* as type (Agatha & Strüder-Kypke 2012). The distinguishing morphologic features concern the monolaminar lorica wall with alveoli and the somatic ciliary pattern. *Favella* has a smooth wall and a monokinetidal ventral kinety abutting on the right ciliary field (Figs 2.14b, 3.9a, b, 3.10a, and 3.16a–g). *Schmidingerella arcuata* has a wall with surface ridges and pores, and its ventral kinety is separated from the right ciliary field by an unciliated stripe and composed of a monokinetidal anterior and dikinetidal posterior portion (Figs 2.14a, 3.9e, f, 3.10b, and 3.17a–f).

Although *Favella ehrenbergii* remains affiliated with the family Ptychocylididae, *Schmidingerella arcuata* was assigned to the family Rhabdonellidae owing to a similar lorica wall and the close relationship with the genus *Rhabdonella* in the SSU rRNA gene trees (Fig. 3.13; Gao et al. 2009; Li Z. et al. 2009; Kim S.Y. et al. 2010; Agatha & Strüder-Kypke 2012). The relationship of *Metacylis angulata* with *Rhabdonella* and *Schmidingerella* in the SSU rRNA genealogies is supported by the monolaminar lorica wall with alveoli, pores, and folds (Lackey & Balech 1966). Further genetic and, especially, morphologic data about the cell and lorica are required to decide on a transfer of *Metacylis*, which is the type of the family Metacylididae, to the family Rhabdonellidae.

Additional changes

Further minor changes concern the families Ascampbelliellidae, Metacylididae, Petalotrichidae, Undellidae, and Xystonellidae.

Family Ascampbelliellidae. Laval-Peuto (1994) speculated about a synonymy of the genus *Luxiella* with *Acanthostomella*, although both differ mainly in a subapical row of small windows (present compared with

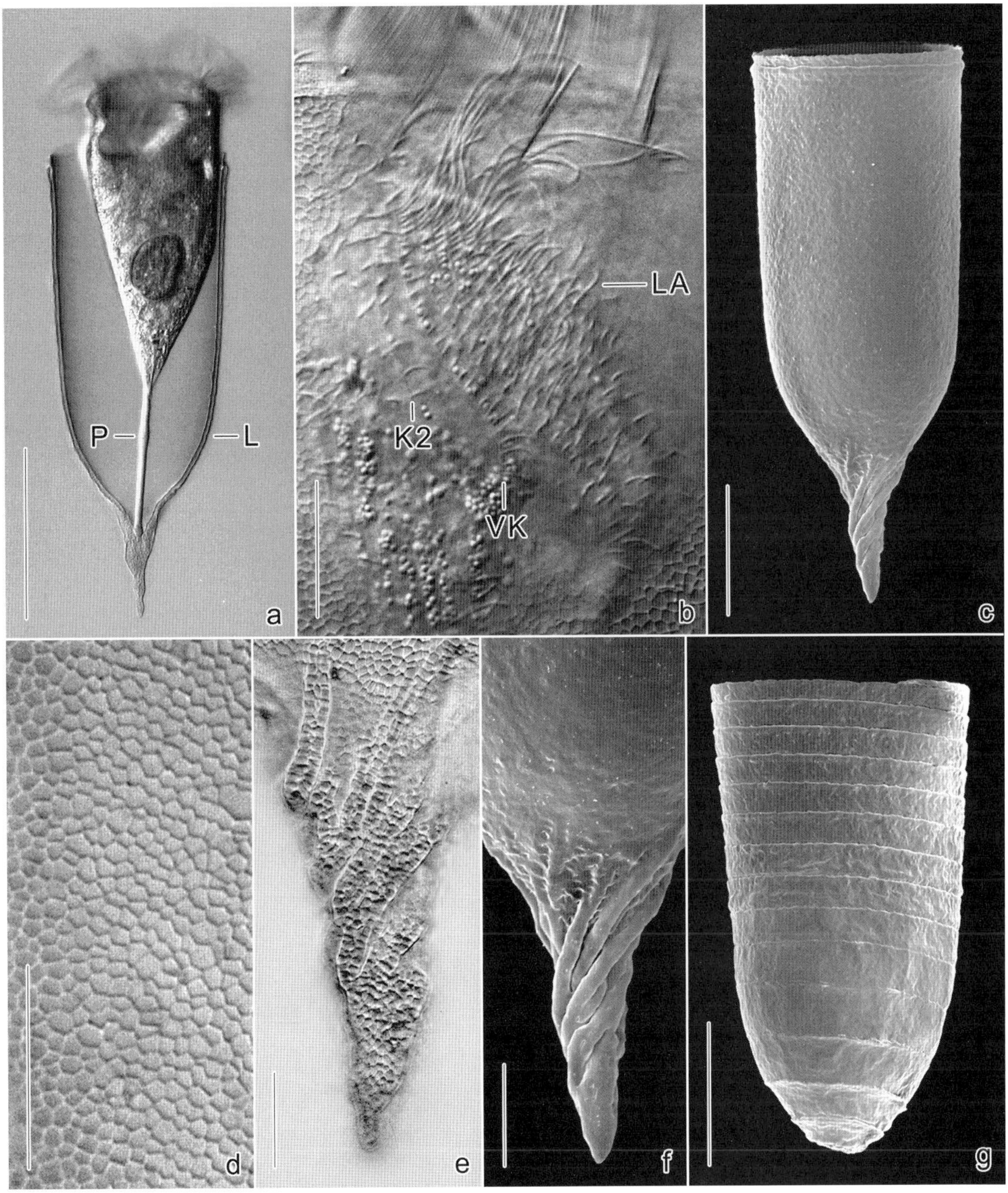

Fig. 3.16 *Favella panamensis* (adapted from Agatha, S. & Strüder-Kypke, M.C. (2012) Reconciling cladistic and genetic analyses in choreotrichid ciliates (Protists, Spirotrichea, Oligotrichea). Journal of Eukaryotic Microbiology, in press; a, b, d, e, from life; c, f, g, scanning electron micrographs). (a) Longitudinal optical section. (b) Anterior ventral cell portion. Kinety 2, first row of the right ciliary field, abuts on the ventral kinety. (c) Lateral view of lorica showing the smooth surface. (d) Alveolar texture of the lorica wall. (e, f) Posterior lorica processes showing twisted ribs. (g) Lorica of *coxliella* form. Note the smooth surface of the whorls. K2, first kinety of right ciliary field; L, lorica; LA, lateral ciliary field; P, peduncle; VK, ventral kinety. Scale bars 100 μm (a), 50 μm (c, g), 20 μm (b, e, f), and 10 μm (d).

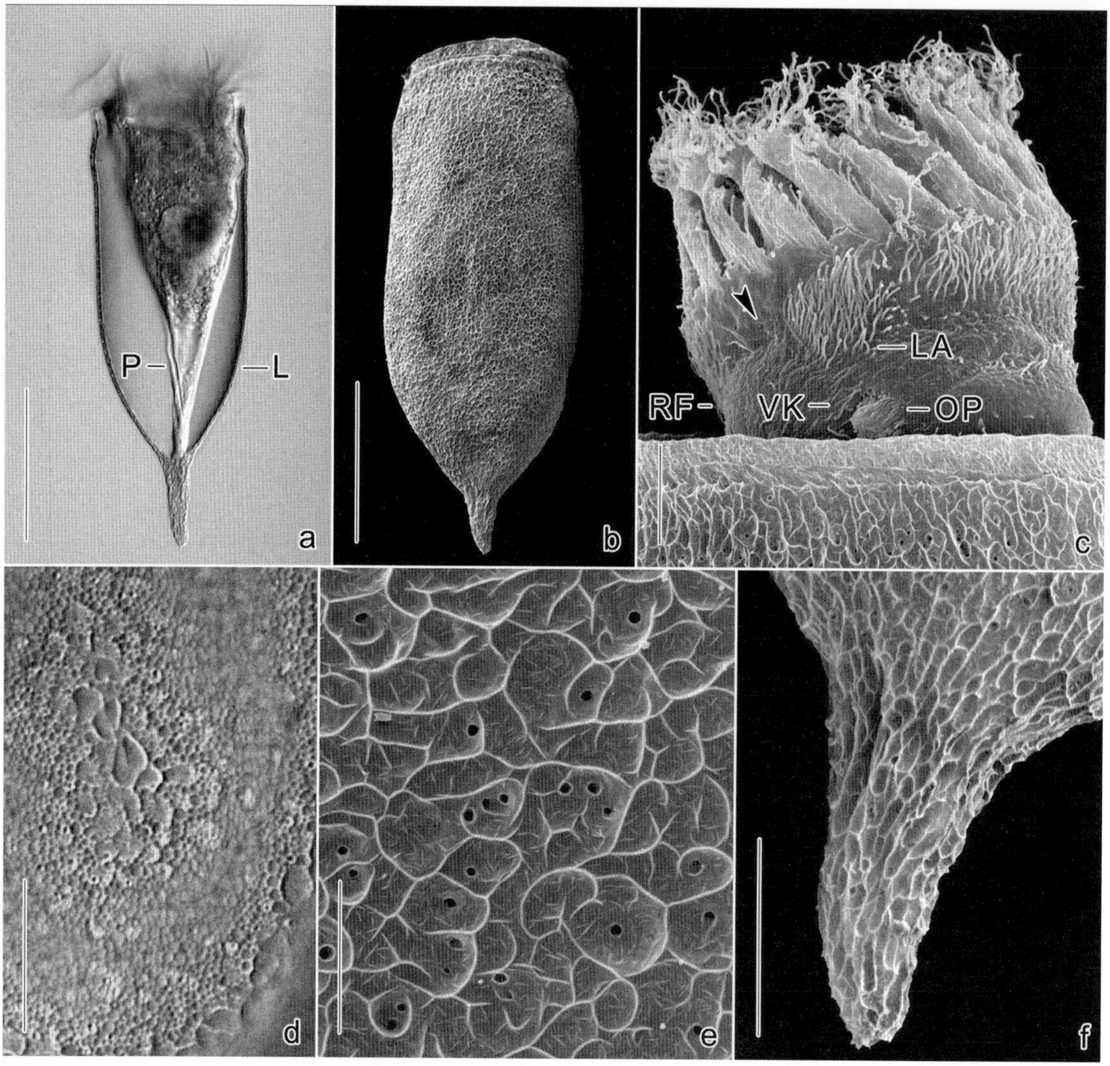

Fig. 3.17 *Schmidingerella arcuata* (adapted from Agatha, S. & Strüder-Kypke, M.C. (2012) Reconciling cladistic and genetic analyses in choreotrichid ciliates (Protists, Spirotrichea, Oligotrichea). Journal of Eukaryotic Microbiology, in press; a, d, from life; b, c, e, f, scanning electron micrographs). (a, b) Longitudinal optical section and lateral view. Note the rough lorica surface (b). (c) Anterior ventral cell portion of a middle divider showing an unciliated stripe (arrowhead) separating the right ciliary field and the ventral kinety. (d) Two kinds of reticulation are recognizable in the lorica wall: one caused by the small alveoli, the other caused by the reticulate rims on the outer surface. (e) Outer lorica surface showing a reticulation and pores. (f) Posterior lorica process showing a reticulate surface. L, lorica; LA, lateral ciliary field; OP, oral primordium; P, peduncle; RF, right ciliary field; VK, ventral kinety. Scale bars 100 μm (a), 50 μm (b), 20 μm (d), 10 μm (c, f), and 5 μm (e).

absent; Lecal 1953). Because the cell cycles are unknown, we keep the genera separate. Likewise, Laval-Peuto (1994) speculated about a synonymy of the genus *Niemarshallia* with its single species and the genus *Ascampbelliella*. Because *Niemarshallia aperta* is distinguished by a posterior lorica opening from the *Ascampbelliella* species (Marshall 1934), we keep the genera separate. *Luxiella* and *Niemarshallia* are incertae sedis in the family Ascampbelliellidae, following Lynn (2008).

Family Metacylididae. The genus *Rhizodomus* was established with *R. tagatzi* as type species by monotypy (it included only a single species when established; Strelkow & Wirketis 1950; Aescht 2001). As it might have a sparsely agglomerated lorica and thus resembles some *Tintinnopsis* species (especially *T. corniger* Hada, 1964; Chapter 10), its assignment to the family Metacylididae is questionable. Likewise, the genus *Stylicauda* with its single species *S. platensis* resembles *Tintinnopsis* species, except for the posterior process lacking agglomerated particles (Cunha & Fonseca 1917). Because a future split of the paraphyletic genus *Tintinnopsis* is very likely and distinct branches might finally be assigned to the two genera, we refrain from a synonymization and affiliate *Rhizodomus* and *Stylicauda* incertae sedis with the tintinnids.

It is well known that the genus *Coxliella* might actually have no reality, as loricae of *coxliella* form belong to the cell cycles of *Favella* (Chapter 2) and probably several further distantly related genera. Accordingly, Corliss (1979) regarded the genus as questionable and replaced the former familial name Coxliellidae by Metacylididae. Because it is impossible at the current state of knowledge to synonymize all members of the genus *Coxliella* with species of other genera, we still keep it here as nomen inquirendum (name under enquiry).

Family Petalotrichidae. Laval-Peuto (1994) regarded the genus *Parapetalotricha* Hada, 1970 as a synonym of the genus *Petalotricha* Kent, 1881. According to Aescht (2001), the type species, *Parapetalotricha meridiana*, was fixed owing to monotypy of the genus. However, Hada (1970) affiliated simultaneously a second species, namely *Parapetalotricha entzi*. Hence, the type species (representative of the genus) is uncertain and the genus name unavailable.

Family Undellidae. The genus *Rotundocylis* was established together with its single species, *R. arcellopsis*, by Kufferath (1950), but it was not considered by Loeblich Jr & Tappan (1968) in their index to the tintinnids. The double contour of the lorica opening indicates a small collar and the bowl might have attached some granules. Thus, the genus is rather similar to *Stenosemella*, whose bowl is, however, densely agglomerated. Laval-Peuto (1994) considered the genus a synonym of *Undella*, as Kufferath (1950) also described unnamed specimens with loricae lacking "ornaments" (apparently interpreted as agglomerated particles). However, Kufferath mentioned that these specimens might represent a distinct species. Without additional information, *Rotundocylis* thus cannot properly be assigned to any family; therefore, it is placed incertae sedis in the tintinnids.

The genera *Amplectellopsis* and *Cricundella* were distinguished by Kofoid & Campbell (1929) from *Amplectella* owing to the absence of a buckling of the lorica walls and a broadened bowl, respectively. Laval-Peuto (1994) apparently regarded these distinguishing features as insufficient and synonymized the genera with *Amplectella*. Because the cells might provide features that support the separation, the synonymizations seem premature, and the genera are kept separate as in the classification by Lynn (2008).

The genus *Proplectella* was established by Kofoid & Campbell (1929) for species that are distinguished from *Undella* species by a subapical inner projection of the lorica wall. Laval-Peuto (1994) synonymized both genera, apparently attributing this difference to the intrageneric variability of *Undella*. As it is unknown whether deviating cytological features accompany this admittedly small difference, we follow Lynn (2008) in keeping the genera separate.

The genus *Micrundella* was established by Busch (1948) with three new species, but he omitted to fix a type. This was remedied by Loeblich Jr & Tappan (1968), who made the genus name available (ICZN 1999; Aescht 2001). The descriptions of the genus and the affiliated species did not provide many details and were based on very few specimens. Apparently, Busch used the distinctly smaller lorica length (less than 20 μm long) and the presence of perhaps a single macronucleus nodule to distinguish the genus from *Undella*. Because macronucleus nodules might fuse to a single mass before cell division in tintinnids, the occurrence of a single nodule and the smaller lorica size seem insufficient for a generic separation. Hence, we follow Laval-Peuto (1994) in regarding the genus *Micrundella* as a synonym of the genus *Undella*.

Family Xystonellidae. *Spiroxystonella* was established by Kofoid & Campbell (1939) as a subgenus of *Xystonella*. Although the lorica outline is typical of *Xystonella*, the spiraled ridges resemble those of *Climacocylis* (compare Fig. 2.1e). Following Loeblich Jr & Tappan (1968), Lynn (2008) regarded *Spiroxystonella* as a distinct, monotypic genus within the family Xystonellidae. Corliss (1979), however, used the similarity with *Climacocylis* for its affiliation with the Metacylididae. According to Laval-Peuto (1994), *Spiroxystonella scandens* might represent a paralorica (replacement

Table 3.4 Revised classification of extant tintinnid ciliates.

- Ascampbelliellidae Corliss, 1960
  - *Acanthostomella* Jörgensen, 1927
  - *Ascampbelliella* Corliss, 1960
    - Incertae sedis
    - *Luxiella* Lecal, 1953
    - *Niemarshallia* Corliss, 1960
- Cyttarocylididae Kofoid & Campbell, 1929
  - *Cyttarocylis* Fol, 1881
- Dictyocystidae Haeckel, 1873
  - *Codonaria* Kofoid & Campbell, 1939
  - *Codonella* Haeckel, 1873
  - *Codonellopsis* Jörgensen, 1924
  - *Dictyocysta* Ehrenberg, 1854
    - Incertae sedis
    - *Laackmanniella* Kofoid & Campbell, 1929
    - *Wangiella* Nie, 1934
- Epiplocylididae Kofoid & Campbell, 1939
  - *Epicancella* Kofoid & Campbell, 1929
  - *Epiplocylis* Jörgensen, 1924
  - *Epiplocyloides* Hada, 1938
- Metacylididae Kofoid & Campbell, 1929
  - *Climacocylis* Jörgensen, 1924
  - *Helicostomella* Jörgensen, 1924
  - *Metacylis* Jörgensen, 1924
  - *Pseudometacylis* Balech, 1968
- Nolaclusiliidae Sniezek, Capriulo, Small & Russo, 1991
  - *Nolaclusilis* Snyder & Brownlee, 1991
- Petalotrichidae Kofoid & Campbell, 1929
  - *Petalotricha* Kent, 1881
- Ptychocylididae Kofoid & Campbell, 1929
  - *Cymatocylis* Laackmann, 1910
  - *Favella* Jörgensen, 1924
  - *Protocymatocylis* Kofoid & Campbell, 1929
  - *Ptychocylis* Brandt, 1896
  - *Wailesia* Kofoid & Campbell, 1939
- Rhabdonellidae Kofoid & Campbell, 1929
  - *Epirhabdonella* Kofoid & Campbell, 1939
  - *Protorhabdonella* Jörgensen, 1924
  - *Rhabdonella* Brandt, 1906
  - *Rhabdonellopsis* Kofoid & Campbell, 1929
  - *Schmidingerella* Agatha & Strüder-Kypke, 2012
- Stenosemellidae Campbell, 1954
  - *Stenosemella* Jörgensen, 1924
- Tintinnidae Claparède & Lachmann, 1858
  - *Albatrossiella* Kofoid & Campbell, 1929
  - *Amphorellopsis* Kofoid & Campbell, 1929
  - *Amphorides* Strand, 1928
  - *Brandtiella* Kofoid & Campbell, 1929
  - *Bursaopsis* Kofoid & Campbell, 1929
  - *Buschiella* Corliss, 1960
  - *Canthariella* Kofoid & Campbell, 1929
  - *Clevea* Balech, 1948
  - *Dadayiella* Kofoid & Campbell, 1929
  - *Daturella* Kofoid & Campbell, 1929
  - *Epicranella* Kofoid & Campbell, 1929
  - *Eutintinnus* Kofoid & Campbell, 1939
  - *Odontophorella* Kofoid & Campbell, 1929
  - *Ormosella* Kofoid & Campbell, 1929
  - *Proamphorella* Kofoid & Campbell, 1939
  - *Prostelidiella* Kofoid & Campbell, 1939
  - *Rhabdosella* Kofoid & Campbell, 1929
  - *Salpingacantha* Kofoid & Campbell, 1929
  - *Salpingella* Jörgensen, 1924
  - *Salpingelloides* Campbell, 1942
  - *Steenstrupiella* Kofoid & Campbell, 1929
  - *Stelidiella* Kofoid & Campbell, 1929
  - *Tintinnus* Schrank, 1803
- Tintinnidiidae Kofoid & Campbell, 1929
  - *Leprotintinnus* Jörgensen, 1900
  - *Membranicola* Foissner, Berger & Schaumburg, 1999
  - *Tintinnidium* Kent, 1881
- Undellidae Kofoid & Campbell, 1929
  - *Amplectella* Kofoid & Campbell, 1929
  - *Amplectellopsis* Kofoid & Campbell, 1929
  - *Cricundella* Kofoid & Campbell, 1929
  - *Proplectella* Kofoid & Campbell, 1929
  - *Undella* Daday, 1887
  - *Undellopsis* Kofoid & Campbell, 1929
- Xystonellidae Kofoid & Campbell, 1929
  - *Parafavella* Kofoid & Campbell, 1929
  - *Parundella* Jörgensen, 1924
  - *Spiroxystonella* Kofoid & Campbell, 1939
  - *Xystonella* Brandt, 1906
  - *Xystonellopsis* Jörgensen, 1924
- Incertae sedis in tintinnids
  - *Codonopsis* Kofoid & Campbell, 1939
  - *Poroecus* Cleve, 1902
  - *Rhizodomus* Strelkow & Wirketis, 1950
  - *Rotundocylis* Kufferath, 1950
  - *Stylicauda* Balech, 1951
  - *Tintinnopsis* Stein, 1867
- Nomen inquirendum
  - *Coxliella* Brandt, 1906

lorica) with *coxliella* form of a *Xystonella* species. Accordingly, she suggested a synonymization of *Spiroxystonella* with *Xystonella*. Admittedly, spiraled structures often occur in paraloricae (Chapter 2), but they are usually not combined with elevated ridges as in *Spiroxystonella*. Until we know the cell features and the cell cycles of the genera *Spiroxystonella*, *Climacocylis*, and *Xystonella*, the former genus is kept separate within the family Xystonellidae.

3.7 COMPARISON WITH THE EVOLUTION OF RELATED PLANKTONIC CILIATES: THE ALORICATE CHOREOTRICHIDS, OLIGOTRICHIDS, AND HALTERIIDS

The freshwater plankton comprises ciliates from many different classes (Foissner et al. 1999), while the marine and brackish water plankton is dominated by close relatives of the tintinnids: the oligotrichids (e.g., *Strombidium*, *Tontonia*), the aloricate choreotrichids (e.g., *Pelagostrobilidium*, *Strombidinopsis*), and rarely the halteriids (e.g., *Pelagohalteria*). Most of these about 140 species were described or redescribed, using protargol staining, and the SSU rRNA genes have been sequenced in about thirty species (Kim Y.-O. et al. 2010). Accordingly, our knowledge about this important group of marine plankton ciliates is much better than that about tintinnids. Nevertheless, we again face the problem of poor overlap in the taxa included in the cladistic and genetic analyses. The aim of this section is to show interesting similarities and differences in the evolution of tintinnids and their relatives.

The planktonic life style

The globular, obconical, or ellipsoidal cell shapes found in these taxa are common in eupelagic (always living in the pelagic zone) ciliates, whereas dorsoventrally flattened taxa, for example the stichotrichs and hypotrichs, might only occasionally occur in the pelagic zone of shallow waters. The high percentage of halteriids, oligotrichids, and choreotrichids, which is not found in the soil or marine interstitial, is used to name the planktonic ciliate community in freshwater "Oligotrichetea". The large collar membranelles of the adoral zone generate water currents for locomotion and suspension feeding. The somatic ciliature is not involved in locomotion, except for the halteriids. Merely two shortened ciliary rows occur in the oligotrichids, whereas more than ten kineties are usually found in the aloricate choreotrichid genera *Strombidinopsis*, *Parastrombidinopsis*, and *Parastrombidium* (Lynn et al. 1991; Kim et al. 2005; Xu et al. 2007) and the tintinnids. The enantiotropic division mode is the most important synapomorphy of the halteriids, oligotrichids, and choreotrichids. The plesiomorphic state, i.e. the epiapokinetal stomatogenesis, is only found in the halteriids, whereas the remaining taxa form the new oral apparatus in an invagination (hypoapokinetal stomatogenesis). The assumption by Kahl (1932) that the subsurface development of the new oral apparatus became necessary when the membranelles became responsible for the cell's movement cannot be supported. The epiapokinetal stomatogenesis in the planktonic halteriids and the hypoapokinetal stomatogenesis in the benthic hypotrichs even contradict Kahl's hypothesis.

The somatic ciliature

In contrast to the halteriids, oligotrichids, and aloricate choreotrichids, the somatic ciliature of the tintinnids is covered with a lorica. The selective forces that affect the somatic ciliature are thus probably different in loricate and aloricate taxa. However, even among the aloricate taxa the length and function of the somatic cilia is different: a mainly sensory function of the stubby cilia is assumed in the oligotrichids and aloricate choreotrichids (Fauré-Fremiet & Ganier 1970), whereas the conspicuously long and closely arranged cilia of the halteriids are probably involved in the jumping movements. In tintinnids, the somatic cilia participate in lorica construction and remove the fecal material (Chapter 2).

In the halteriids and oligotrichids, the plesiomorphic state of the kinetid structure persists, i.e. only the anterior dikinetidal basal bodies are associated with cilia; specialized ciliary rows and fields are unknown. The evolution of the kinetid structure is not restricted to the lorica-forming tintinnids, but already started in the aloricate choreotrichids (character 26). The general trend was to form a second cilium at the posterior dikinetidal basal body, to lose the cilium at the anterior basal body, and finally to lose the unciliated anterior basal body. The reasons for and the advantages of these structural changes are unknown.

Accordingly, we find dikinetids with only an anterior cilium (*Leegaardiella ovalis*; Lynn & Montagnes 1988), dikinetids with two cilia (e.g., *Strombidinopsis*; Agatha 2003), dikinetids with only a posterior cilium (e.g., *Lohmanniella*; Agatha & Strüder-Kypke 2007), and monokinetids (e.g., *Pelagostrobilidium*; Agatha et al. 2005) in the aloricate choreotrichids and tintinnids (character 26).

Although the oligotrichids are more conservative than the choreotrichids in the structure of the somatic kinetids and special ciliary components, they are able to create various, occasionally very complex, patterns with their two ciliary rows, especially with their girdle kinety (character 18; Agatha 2004a, 2011). Again, the reasons for and the advantages of these different patterns are unknown. Like in tintinnids, the cladogram and gene tree of oligotrichids roughly follow the evolution of the somatic ciliary patterns, demonstrating its importance for inferring the phylogenetic relationships (Agatha 2004a, b, 2011; Agatha & Strüder-Kypke 2007, 2012).

Resting cysts

The analysis of resting cyst data is difficult owing to their scarcity (Chapter 7). A preformed emergence pore closed by a plug occurs in the resting cysts of oligotrichids (*Strombidium*; Agatha et al. 2005), aloricate choreotrichids (*Rimostrombidium*, *Strombidinopsis*; S. Agatha, unpublished observations on cysts provided by A. Saage, University of Kiel, Germany; Ichinomiya et al. 2004), and the tintinnid genus *Schmidingerella* (McManus & Katz 2009; Agatha & Strüder-Kypke 2012). The cysts of stichotrichs and halteriids, however, lack a preformed emergence pore and the ciliates thus leave the cyst through rupture of the wall (character 45; Foissner et al. 2005, 2007). Likewise, emergence pores and plugs are absent in the ellipsoidal or globular structures, which probably represent the cysts of the tintinnids *Acanthostomella* (Davis 1985), *Coxliella* (Laackmann 1908), *Helicostomella* (Laackmann 1908; Meunier 1919; Paranjape 1980), *Leprotintinnus* (Meunier 1910; Reid & John 1978), *Parafavella* (Busch 1920; Reid & John 1978), *Parundella* (Paranjape 1980), and marine *Tintinnopsis* species (Plate 7.1; S. Agatha, unpublished observations; Meunier 1919; Biernacka 1952). The most parsimonious assumption is that a cyst plug represents a synapomorphy of the oligotrichids and choreotrichids, whereas some tintinnids have secondarily lost this feature.

The cyst surface is covered by often globular or spine-shaped structures (lepidosomes) in some stichotrichs, halteriids, oligotrichids, and aloricate choreotrichids, in which they probably developed independently (character 48; Foissner et al. 2007 and references therein), whereas lepidosomes are apparently lacking in tintinnids (Chapter 7; Van Breemen 1905; Laackmann 1908; Meunier 1910, 1919; Biernacka 1952; Margalef & Durán 1953; Reid & John 1978; Paranjape 1980; Davis 1985; McManus & Katz 2009).

The halteriids, an enigmatic ciliate group

The halteriids are an interesting group of planktonic ciliates, as they show some unique morphologic features and because their phylogenetic relationships are uncertain. The occurrence of these special characters might cause a long-branch attraction artefact (phenomenon in phylogenetic analyses, mainly maximum parsimony, incorrectly showing rapidly evolving lineages to be closely related), resulting in an affiliation of the halteriids with the stichotrichs. Distinguishing features of the halteriids are the *de novo* origin of the somatic kineties during morphogenesis (character 43) accompanied by a resorption of the old somatic ciliary rows (character 44; Petz & Foissner 1992; Song 1993; Agatha & Strüder-Kypke 2007).

Further special features concern the conjugation of the halteriids. Although they fuse with their anterior ventral cell portions like the oligotrichids and choreotrichids (character 53; Agatha & Foissner 2009 and references therein), the formerly globular conjugation partners form an extraordinary ventral concavity and finally obtain an interlocking arrangement by inserting the left cell side into the partner's furrow. Such a curious arrangement of the partners is otherwise found only in the aloricate choreotrichid *Pelagostrobilidium* (Ota & Taniguchi 2003; Agatha & Foissner 2009) and might represent a convergent adaptation to the jumping escape movements performed by the two taxa. Further peculiarities of the halteriid conjugation emphasize the uniqueness of this taxon; for details and references, see Agatha & Foissner (2009).

Gene sequence analyses

In the recently published SSU rRNA gene trees of the oligotrichid ciliates, most basal nodes were statistically not well supported, many polytomies could not be resolved, and paraphylies were recognizable (Gao et al. 2009; Kim Y.-O. et al. 2010; Zhang et al. 2010; Liu et al. 2011). The paraphyletic oligotrichid genus *Strombidium* might, as the genus *Tintinnopsis*, actually comprise several genera, but their separation is currently impossible owing to the lack of appropriate morphologic features. Other branches of the gene trees are well supported by morphologic characters, especially synapomorphies of the somatic ciliary pattern. The aloricate choreotrichids are paraphyletic in cladograms and gene trees. Both genealogies differ, however, in their basal topologies, probably because genetic data on the genera *Leegaardiella*, *Lohmanniella*, and *Parastrombidium* are not available (Agatha & Strüder-Kypke 2012).

Despite differences in the evolution of halteriids, oligotrichids, and choreotrichids, the SSU rRNA gene trees and the cladograms roughly reflect in all taxa the development of the somatic ciliature, which is thus the most important feature complex in phylogenetic reconstructions. In contrast, the oral ciliature is rather conserved, except for small changes in the aloricate choreotrichids (character 8).

3.8 KEY POINTS

1. Taxonomic value of the lorica. The phylogenetic relationships inferred from cladistic and genetic analyses contradict the lorica-based hypothesis of tintinnid evolution and classification, namely species with hyaline and agglomerated loricae are not separate taxonomic entities. Furthermore, an evolutionary trend from agglomerated to hyaline loricae is not evident. However, soft, agglomerated loricae are apparently restricted to the basal branch of tintinnid evolution comprising most freshwater species. Taxa with loricae composed of a hyaline collar and an agglomerated bowl also group rather well together.

2. Reliable morphologic characters for phylogenetic inference. The cladistic analysis of morphologic characters and the sequence analysis of the small subunit ribosomal RNA gene produce rather similar trees of the monophyletic tintinnids. In particular, the hypothesis about somatic ciliary pattern development, the capsule types, and the lorica sac with its foldable closing apparatus seem to be valuable features/feature complexes for the reconstruction of tintinnid evolution. In the closely related oligotrichids and aloricate choreotrichids, the morphologic and molecular trees also widely reflect the development of the somatic ciliary patterns.

3. Origin of the tintinnids. The tintinnids most probably originated from marine planktonic ciliates, as their supposed aloricate ancestor, resembling the genus *Strombidinopsis*, was restricted to the marine pelagic zone and most tintinnid species occur in this habitat. Tintinnids apparently entered freshwater only twice, namely at the beginning of their evolution with species possessing soft, agglomerated loricae (*Tintinnidium* and *Membranicola*) and later with species characterized by the most complex ciliary pattern (*Codonella cratera* and *Stenosemella lacustris*).

4. Shortcomings. A taxonomic classification should reflect the evolutionary history and natural relationships of organisms. However, the number of tintinnid species sequenced and studied with modern taxonomic methods is currently insufficient for a revision of the lorica-based sorting, especially as the intraspecific variability is unknown. Furthermore, a reconciliation of the information provided by gene and morphologic trees is problematic, as the sequenced species usually were not identified, applying appropriate methods, and most species descriptions based on modern morphological methods lack gene sequence data.

5. Future efforts. A simultaneous acquisition of both genetic and morphologic data and their phylogenetic analysis are recommended. Because the taxonomic classification requires distinguishing features for taxa diagnoses, detailed morphologic studies on the loricae and cells should be performed, applying live observation, protargol staining, and preferably electron microscopy; permanent slides should be deposited in acknowledged collections. Additionally, the intraspecific lorica variability should be studied by cultivation experiments and fluorescence in situ hybridization (which detects species-specific gene sequences). Finally, the real number of tintinnid species might be estimated. One of the most important tintinnids to be redescribed is the type species *Tintinnopsis beroidea*, whose gene sequence has also to be confirmed.

Once the distinguishing features of most/all species are known, keys for identification can be established.

Reliable species determination is in turn the base for the investigation of biogeographic patterns as well as ecological and evolutionary processes, which might help to understand speciation and the maintenance of diversity in the pelagic realm. To attain these objectives, synergistic approaches combining the expertise of morphological and molecular taxonomists, ecologists, and physiologists will be indispensable.

ACKNOWLEDGMENTS

The comments and ideas of the other chapter authors are gratefully acknowledged. Sabine Agatha was supported by the Austrian Science Fund (FWF; Project P20461-B17).

Chapter 4

ECOPHYSIOLOGY AND BEHAVIOR OF TINTINNIDS

David J.S. Montagnes
University of Liverpool, Liverpool, UK

The Biology and Ecology of Tintinnid Ciliates: Models for Marine Plankton, First Edition. Edited by John R. Dolan, David J.S. Montagnes, Sabine Agatha, D. Wayne Coats, and Diane K. Stoecker.

4.1 INTRODUCTION

Planktonic ciliates are a key component of pelagic food webs and are of substantial ecological importance (see, Chapter 9). Although tintinnids are rarely the main planktonic ciliates (Chapter 9), owing to their ease of capture and identification (Chapter 1), historically, they have been extensively studied. In this Chapter, I provide an indication of our understanding of their behavior and ecophysiology, primarily from laboratory studies, as they provide controlled, reliable measurements. Much of these data are then synthesized to reveal general and specific traits and indicate trends that may then be applicable to planktonic ciliates and protists in general. Finally, I contextualize the above observations by indicating how our knowledge of tintinnids may be incorporated into autecological and synecological models. In this way, it is revealed that tintinnids may provide useful tools to address larger issues. Note that in this chapter the species names are those as reported in the literature.

4.2 FEEDING

An overview of tintinnid feeding

Feeding by tintinnids, as for most protozoa, can be divided into discrete, linked steps (Fig. 4.1) that may be governed by chemical, physical, or biological cues (Montagnes et al. 2008). These steps could, of course, be infinite in number, but developing a series of measurable points provides a framework for descriptive purposes. Thus, to provide structure for ecophysiological issues that will follow, I also present a generalized tintinnid, as a consumer of prey (Fig. 4.2).

As is described in more detail elsewhere in this book (Chapter 1), tintinnids possess a ring (or zone) of adoral polykinetids or membranelles surrounding an oral cavity or infundibulum, the "floor" of which is at least in part covered by the endoral membranelle. Within the oral cavity is an excentrically placed cytostomal region, at the base of which vacuoles form (Fig. 4.2). In general, but described in more detail below, the cilia of the adoral membranelles beat, creating currents that bring prey to the cytostomal region (Fig. 4.2d), as well as propelling the ciliate; thus, locomotion and grazing are linked in tintinnids. Note, the somatic cilia are covered by the lorica and do not participate in locomotion but seem to help the cell move within the lorica (see Chapter 3 for details).

Searching, a term extensively used in the literature on tintinnids and, therefore, adopted here, per se, is a misnomer, as planktonic grazers rarely "search"; rather through random swimming they encounter prey. Regardless, "searching" for prey by tintinnids will be influenced by their swimming speed and the frequency of turning, as these will alter the volume of water encountered. Such behaviors will be affected by the distribution, abundance, and type of prey and will be discussed in more detail later in this section and in Section 4.4, but first I will turn to contact, the next step in feeding.

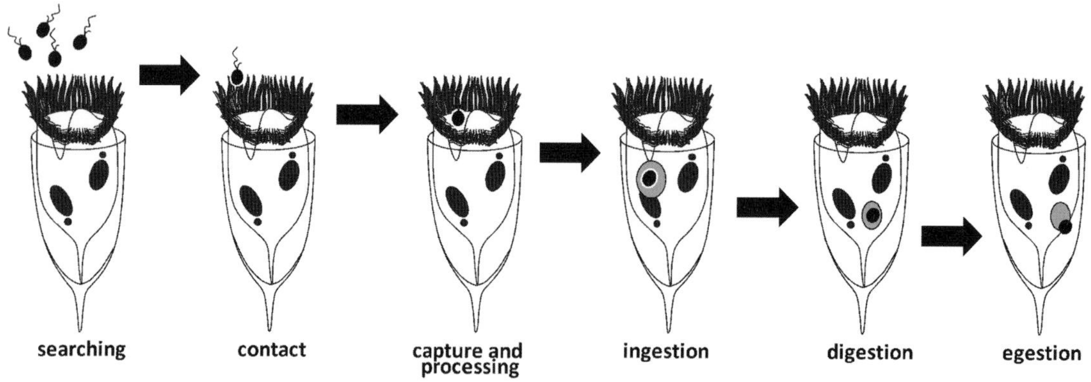

Fig. 4.1 Feeding steps of tintinnids. As detailed in the text, the ciliate swims, processing water and searching for prey. Contact is then made, often with the cilia. Capture will then occur, but prey may be rejected at this point. Once captured, prey are moved into the cytostomal region where they are processed into a food vacuole. Next, prey are digested, and any indigestible part is egested through the cell anus (or cytopyge, see Fig. 4.2).

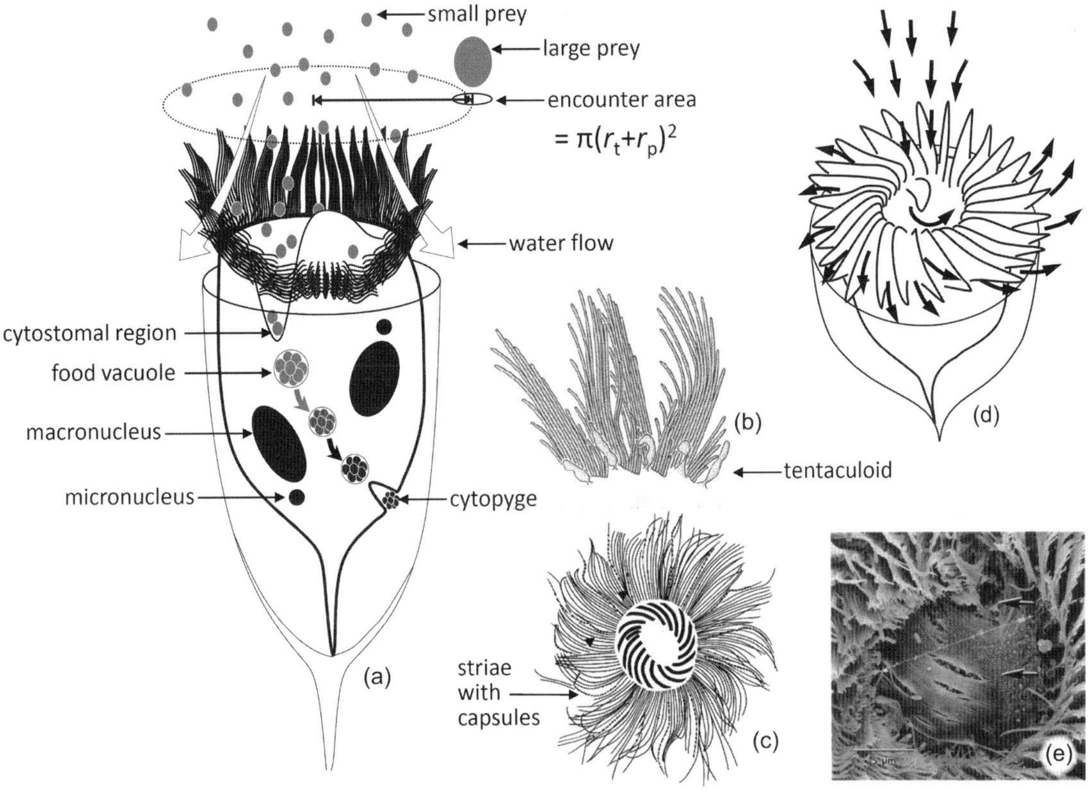

Fig. 4.2 The generalized tintinnid consumer. (a) An indication of how a tintinnid will capture and process prey; white arrows represent the general flow of water; dark arrows indicate the processing of food in a vacuole within the cell; the dotted lines at the top of the ciliate represent the area of encounter that can be approximated by the equation (r_t, tintinnid radius; r_p, prey radius). (b, c) Indications of tentaculoids and striae with permission, redrawn from Agatha (2009). (d) A prediction of the flow of water through the membranelles of a tintinnid, based on work on the naked oligotrich *Lohmanniella spiralis* (Jonsson 1986). (e) An indication of the structure of the endoral membranelle, lying on the base of the cytosomal region (note this scanning electron micrograph is from a naked ciliate, not a tintinnid).

Contact probability between the tintinnid and its prey will increase with both increasing predator and prey size; note "prey size" may be an ambiguous term as it typically refers to a diameter, but prey are rarely spherical; still this is the simplified practice of the discipline, which is adopted here. Following very simple mechanistic principles, initially ignoring prey size if they are small, and considering only physical properties, the maximum encounter area by a feeding tintinnid will be equal to the area of its oral region, although this may include the oral cilia if they are sensory (Fenchel 1986; Fig. 4.2), and therefore they too may need to be measured. Furthermore, when prey are large, their size must also be considered, as the area of encounter includes the transactional region around the oral area, defined by the radius of the prey (Fig. 4.2). Thus, larger prey might be assumed to be encountered and, consequently, ingested with a higher probability, but see Sections on selectivity on pages 96 and 98. Note also that chemical signals will increase this encounter area beyond the cell and cilia, as the prey may exude material that the ciliate may sense (see Section "Selectivity other than size" on page 98). Furthermore, video observations of *Favella* sp. indicate that different regions of the oral zone can preferentially retain particles (Stoecker et al. 1995), so in fact treating the entire oral surface as uniform is undoubtedly simplistic, although it may be appropriate for most applications. Furthermore, prey that collide with the oral cilia are often not captured (Blackbourn 1974), so efficiency is not 100% (see Box 4.1). Finally, ciliates may sense movement of the prey through mechnoreception (Fenchel 1987; Hausmann et al. 2003), but the extent to which such behavior occurs in tintinnids is undetermined. Clearly, the behavior of these ciliates is complex!

A second physical property that will influence contact is the speed at which predator and prey move. In work that examined *Tintinnopsis dadayi* feeding on a range of prey, Verity (1991a) modeled the effect of swimming speed and predator encounter area on encounter rate, using Equation 4.1 (from Gerritsen & Strickler 1977), where encounter rate (E) is a function of prey concentration (P), prey speed (u), predator speed (v), and predator encounter radius (r),

$$E = \frac{\pi r^2 P}{3} \times \frac{u^2 + 3v^2}{v} \qquad (4.1)$$

Based on estimated rates of a range of prey, Verity (1991a) concluded that prey speed had little influence on encounter (predator speed was not considered), whereas encounter radius of the tintinnid did. At least, this work suggests we should focus on tintinnid and prey size when considering contact, although as we will see later in this chapter, the swimming behavior of the tintinnid will alter prey capture.

The flow of water past a ciliate's feeding structure is greater when it is attached to a surface, and thus the probability of contact with prey will be greatest for attached ciliates when prey are at saturating levels; when prey are scarce searching will play a greater role in contact, and thus attachment would be a disadvantage (Fenchel 1987). Although tintinnids are not generally permanently attached, residing in a lorica could reduce their speed, relative to naked oligotrichs. We might then conclude that, when prey are scarce, the lorica reduces encounter rate by reducing swimming speed but increases the flow of water past the oral region (i.e., a greater amount of water is processed). Thus, by acting as a "sea anchor" the lorica might provide a selective advantage over naked (i.e., aloricate) ciliates when prey are abundant, and searching is not a priority but processing water is. This is an entertaining argument for the function of the lorica. However, in Section 4.4 I indicate that tintinnids seem to swim as fast, or faster, than their aloricate relatives. Furthermore, tintinnids are common in oligotrophic waters, where prey are sparsely distributed (see, for example, Chapter 9). Therefore, this argument for the lorica acting as an adaptive trait to improve feeding when prey are abundant lacks veracity (see Chapter 1 for a further discussion of the purpose of the lorica). Note though that there is some support for the argument that the lorica can help to act as a sea anchor. Jonsson et al. (2004) indicate that when *Eutintinnus inquilinus* uses its lorica to attach to particles in the water column it can increase its feeding rate by up to 80%, so the concept bears further exploration.

Swimming behavior will be examined in more detail in Section 4.4, but it is noteworthy that tintinnids feed and grow on a range of prey, including non-motile diatoms and highly motile dinoflagellates (Table 4.1); thus, although prey motility may, in theory, increase the rate of contact, it is far from essential. Another mechanism that should increase encounter rate is turbulence in the water column (Shimeta et al. 1995), as it effectively increases the speed of both ciliates and their prey and may affect organisms the size of larger tintinnids (Kiørboe 2008). However, the work on tintinnids that has assessed the effect of turbulence on feeding (Shimeta et al. 1995) suggests either no influence on *Favella* sp. or a reduced clearance rate (for a definition of clearance rate see Box 4.1) of *Helicostomella* sp. The latter observation was attributed to feeding inhibition through some mechanical disturbance (see Kiørboe 2008 for an overview of such inhibitory processes). Similar, negative effects on feeding, attributed to turbulence-induced behavioral changes, occur for the naked oligotrich *Strombidium* (Dolan et al. 2003). Thus, the sparse data available on turbulence suggest more negative than positive impacts on feeding in planktonic ciliates.

Once contact has occurred, the prey must be *captured* and *processed* or, possibly, *rejected*. Although there are no good estimates of water flow through the oral structures of tintinnids, there has been work conducted on *Lohmanniella spiralis* (= *Strobilidium spiralis*, *Pelagostrobilidium spirale*) a closely related aloricate oligotrich that is structurally and functionally similar in feeding behavior to tintinnids (Jonsson 1986). In *L. spiralis* the adoral membranelles are orientated so that their beating propels water between them and out of from the peristomal infundibulum, transporting water and food particles into the oral region from the anterior of the cell (Fig. 4.2). Jonsson (1986) estimated the filtering area of *L. spiralis* to be approximately $3.7 \times 10^3\,\mu m^2$ with a current velocity through the membranelles of $1.6 \times 10^3\,\mu m\,s^{-1}$, suggesting a flow rate of $21\,\mu l\,h^{-1}$, a theoretical value near his observed

Table 4.1 An indication of prey types that do (+) and do not (–) support tintinnid growth in monocultures. Note these data do not consider the concentration of potential prey or the age of the prey cultures, both of which may alter the ability for tintinnids to grow. Furthermore, prey that do not support growth may be consumed. Prey sizes (in micrometers) are approximate linear values, based on averages reported in the literature and should not be considered as accurate; they are provided to indicate a rough range. The interested reader is advised to examine the cited works for details. *, β-chitin threads on diatoms were removed by shaking; ☠, a prey that was considered either toxic or potentially toxic to aquatic organisms; "clone" refers to a case where one clone of a species supported growth, whereas another did not; "poor" refers to growth but at a very low level. References: (1) Apple et al. 2011; (2) Clough & Strom 2005; (3) Fulco 2007; (4) Graham & Strom 2010; (5) Hansen 1989; (6) Hansen 1995; (7) Hansen et al. 1992; (8) Jakobsen et al. 2001; (9) Kamiyama 1997a; (10) Kamiyama 1997b; (11) Kamiysma & Arima 1997; (12) Kamiyama & Suzuki 2006; (13) Kamiyama et al. 2005; (14) Rosetta & McManus 2003; (15) Stoecker & Evans 1985; (16) Stoecker et al. 1981; (17) Taniguchi & Takeda 1988; (18) Verity 1991a; (19) Verity & Stoecker 1982; (20) Verity & Villareal 1986.

Class	Prey	Size	Tintinnid	Toxicity	Growth	Reference
Bacillariophyceae	*Chaetoceros* sp.	10	*Tintinnopsis vasculum*		–	(20)
	Cyclotella caspia	5	*Tintinnopsis vasculum*		–/+*	(20)
	Cyclotella meneghiniana	7	*Tintinnopsis vasculum*		–/+*	(20)
	Minutocellus polymorphus	3	*Tintinnopsis vasculum*		+	(20)
	Thalassiosira constricta	7	*Tintinnopsis vasculum*		–/+*	(20)
	Thalassiosira oceanica	4	*Tintinnopsis vasculum*		–/+*	(20)
	Thalassiosira pseudonana	4	*Tintinnopsis vasculum*		–/+ *	(20)
	Thalassiosira sp.	4	*Tintinnopsis vasculum*		–/+*	(20)
Bicoecida	*Bicoea* cf. *maris*	5	*Tintinnopsis dadayi*		+	(18)
	Cafeteria sp.	3	*Tintinnopsis dadayi*		+	(18)
	Pseudobodo cf. *tremulans*	4	*Tintinnopsis dadayi*		+	(18)
Chlorophyceae	*Dunaliella tertiolecta*	8	*Tintinnopsis dadayi*		+	(18)
Chroococcales	*Synechococcus* sp.	1	*Eutintinnus* sp.	☠	–	(1)
	Synechococcus sp.	1	*Metacylis* sp.	☠	–	(1)
Chrysophyceae	*Paraphysomonas* sp.	7	*Tintinnopsis dadayi*		+	(18)
Dinophyceae	*Alexandrium ostenfeldii*	30	*Favella ehrenbergii*	☠	–	(7)
	Alexandrium tamarense	30	*Eutintinnus* sp.	☠	–	(3)
	Alexandrium tamarense	30	*Favella taraikaensis*	☠	+	(13)

Continued

Table 4.1 *Continued*

Class	Prey	Size	Tintinnid	Toxicity	Growth	Reference
	Alexandrium tamarense	30	*Favella taraikaensis*	☠	+	(12)
	Alexandrium tamarense	30	*Favella taraikaensis*	☠	–	(3)
	Alexandrium tamarense	30	*Favella ehrenbergii*	☠	+/– clone dependent	(5)
	Amphidinium carterae	15	*Favella ehrenbergii*		–	(16)
	Gonyaulax polyedre	35	*Favella ehrenbergii*		+	(16)
	Gonyaulax tamarensis	30	*Favella ehrenbergii*		+	(16)
	Gyrodinium aureolum	25	*Favella ehrenbergii*	☠	–	(6)
	Heterocapsa circularisquama	15	*Favella taraikaensis*	☠	+	(10)
	Heterocapsa circularisquama	15	*Favella taraikaensis*	☠	+	(11)
	Heterocapsa circularisquama	15	*Favella azorica*	☠	+	(9)
	Heterocapsa sp.	25	*Favella ehrenbergii*		+	(16)
	Heterocapsa triquetra	20	*Favella taraikaensis*		+	(10)
	Heterocapsa triquetra	20	*Favella azorica*		+	(10)
	Heterocapsa triquetra	20	*Favella ehrenbergii*		+	(6)
	Heterocapsa triquetra	20	*Favella* sp.		+	(4)
	Katodinium rotundatum	7	*Tintinnopsis dadayi*		+	(18)
	Prorocentrum mariaelebouriae	15	*Favella ehrenbergii*		poor+	(16)
	Prorocentrum minimum	15	*Eutintinnus pectinis*		–	(14)
	Prorocentrum minimum	15	*Eutintinnus pectinis*	☠	–	(14)
	Prorocentrum minimum	15	*Favella ehrenbergii*	☠	+	(14)
	Prorocentrum minimum	15	*Favella taraikaensis*		+	(17)
	Prorocentrum minimum	15	*Metacylis angulata*	☠	+	(14)
	Scrippsiella trochoidea	25	*Favella ehrenbergii*		+	(16)
	Scrippsiella trochoidea	20	*Eutintinnus* sp.		+	(3)
	Scrippsiella trochoidea	20	*Favella taraikaensis*		+	(3)
	Heterocapsa pygmaea	10	*Favella* sp.		+	(19)

Table 4.1 *Continued*

Class	Prey	Size	Tintinnid	Toxicity	Growth	Reference
Pelagophyceae	*Aureoumbra lagunensis*	5	*Amphorides* sp. *quadrilineata*	☠	–	(8)
Prymnesiophyceae	*Isochrysis galbana*	5	*Amphorides* sp. sp.*quadrilineata*		+	(8)
	Isochrysis galbana	5	*Metacylis* sp.		+	(4)
	Isochrysis galbana	3	*Tintinnopsis dadayi*		+	(18)
	Isochrysis galbana	3	*Tintinnopsis* sp. sp.*tubulosoides*		+	(19)
	Isochrysis galbana	3	*Eutintinnus pectinis*		+	(14)
	Isochrysis galbana	3	*Tintinnopsis vasculum*		+	(20)
	Prymnesium parvum	10	*Eutintinnus pectinis*	☠	–	(14)
	Prymnesium parvum	10	*Favella ehrenbergi*	☠	+	(14)
	Prymnesium parvum	10	*Metacylis angulata*	☠	+	(14)
Pyrenomonadaceae	*Rhodomonas lens*	10	*Eutintinnus pectinus*		+	(14)
	Rhodomonas sp.	10	*Coxliella* sp.		–	(2)
	Rhodomonas sp.	10	*Eutintinnus* sp.		+	(2)
	Rhodomonas sp.	10	*Favella taraikaensis*		–	(17)
	Rhodomonas sp.	10	*Metacylis* sp.		–	(2)
Raphidophyceae	*Heterosigma akashiwo*	15	*Coxliella* sp.	☠	–	(2)
	Heterosigma akashiwo	15	*Eutintinnus* sp.	☠	+/–	(2)
	Heterosigma akashiwo	15	*Favella* sp.	☠	–	(4)
	Heterosigma akashiwo	15	*Favella taraikaensis*	☠	–	(17)
	Heterosigma akashiwo	15	*Metacylis* sp.	☠	–	(4)
	Heterosigma akashiwo	15	*Metacylis* sp.	☠	–	(2)
	Olisthodiscus luteus	10	*Favella* sp.	☠	–	(19)
	Olisthodiscus luteus	10	*Tintinnopsis dadayi*		–	(18
	Olisthodiscus luteus	10	*Tintinnopsis* sp. sp.*tubulosoides*	☠	–	(19)
Ciliate	*Balanion* sp.	33	*Favella* sp.			(15)

maximum clearance rate for this species of 26 μl h^{-1}; note, on suboptimal particles, where capture efficiency was lower, the apparent maximum clearance was an order of magnitude lower. Although slightly higher, these values generally agree with estimates for tintinnids (see Table 4.2). It is an appealing notion to consider the oral structure as a simple filtration system, but as will be seen below, there may be substantial "handling" of prey before it is ingested.

Of all the planktonic ciliates, tintinnids best lend themselves to the observation of prey capture and rejection, as they may be immobilized without damaging the cell; thus they act as useful models for other ciliates, especially the numerically dominant naked oligotrichs and choreotrichs (Chapter 9), which they are closely related to (Chapter 3). For instance, *Favella taraikaensis* was held stationary by trapping its lorica in petroleum jelly on a microscope slide (Taniguchi & Takeda 1988), and *Favella* sp. was held by a complex method using among other compounds egg albumen, to tether the lorica to a slide (Stoecker et al. 1995). Also, as the lorica appears to prevent tintinnids from rapidly jumping, a trait exhibited by many of their close, naked relatives (see Gilbert 1994; Jakobsen 2001), this allows researchers to follow them under a microscope.

Blackbourn (1974) made careful observations of free-swimming tintinnids and indicated that when small (4–10 μm) particles were ingested, the ciliate underwent a slight "shudder", whereas ingestion of larger items resulted in an interruption of motion, often with the outer adoral membranelles bent inward, possibly to prevent prey escaping and the inner membranelles forcing the prey into the oral region. This parallels observations by Stoecker et al. (1995) who noted cilia reversal aided in prey capture. Blackbourn (1974) also suggested that the "oral plug", a structure at the base of the oral region is used as a "pump", somehow aiding in prey processing; this oral plug is likely to be the endoral membranelle (Fig. 4.2), although it could also be a contractile portion of the cell within the oral cavity. In general, for *Favella taraikaensis*, processing of prey takes approximately 10–100 s (Taniguchi & Takeda 1988) and between 2–8 s for some other tintinnids (Blackbourn 1974); the extent to which this applies to tintinnids is unknown.

Prey may be rapidly *rejected* from the oral region (Taniguchi & Takeda 1988). For instance, particles may be expelled from the cytopharyngeal region by reversal of cilia (Blackbourn 1974; Stoecker et al. 1995). My own observations also suggest that rejection of prey may occur by rapid motion of the endoral membranelle, analogous to a tongue that flicks prey away. In contrast, others indicate that reversal of adoral membranelles may result in prey removal (Taniguchi & Takeda 1988; Stoecker et al. 1995). Regardless of the specific mechanism, it is clear that tintinnids are capable of selecting against prey that have been captured, and are thus far from passive filter- or suspension-feeders (Stoecker 1988).

Ciliary action may not be the only method of prey *capture*. Extrusomes are a structure common to many ciliates, some of which contain toxins to paralyze prey and others contain adhesive substances to retain prey (Lynn 2008; Hausmann et al. 2003). Some tintinnids possess specialized extrusomes, called "capsules", contained within thread-like "striae", which then have a beaded appearance (Fig. 4.2c; Laval-Peuto et al. 1979). Tintinnids may also possess tentaculoids, flap-shaped cytoplasmic extensions inserting between the adoral membranelles that contain extrusomes (Fig. 4.2b). By observing fed and unfed tintinnids, Capriulo et al. (1986) revealed that tintinnids possessed significantly more striae when fed, suggesting this is a mechanism for increasing prey capture; i.e., the ciliate expends energy on striae production when prey are abundant. Again, the implication is that tintinnids are not simple filter feeders. We might extend this physically observable method of prey capture and predict that like other protozoa, tintinnids also use unobservable (molecular) methods, such as ligands, to bind to and capture prey (see Roberts et al. (2011) for a review on this subject); although no data exist to support this for tintinnids, there is clear evidence for chemosesory abilities (see section "Selectivity other than size" on page 98), and it might be assumed such mechanisms exist.

Besides direct observations of prey rejection, several studies have experimentally indicated that tintinnids exhibit prey preference (Table 4.1). For instance, *Tintinnopsis tubulosa* prefers yeast to carbon particles (Spittler 1973); *Eutintinnus pectinis* prefers one prey flagellate over another (Heinbokel 1978a); *Favella ehrenbergii* prefers dinoflagellates over other similarly sized prey (Stoecker et al. 1981); *Favella* sp. prefers live prey over latex polystyrene beads of similar size, beads over heat killed dinoflagllates, and shows other prey preferences, including a preference for non-toxic species over toxic ones (Stoecker 1988; Strom and

Loukos 1998; Graham & Strom 2010, respectively). Finally, Verity (1991a) indicated preference for prey, between similar-sized flagellates and through rather elegant analysis suggested that this preference was based on rejection, rather than encounter. Thus, there is rather conclusive evidence that tintinnids can be particular in their diet.

Once captured, tintinnids *ingest* prey through a defined cytostomal region and form a vacuole around either single (Taniguchi & Takeda 1988) or multiple (Stoecker 1984) prey; as for many ciliates, the formation of the food vacuole, from existing membrane, may ultimately be the limiting step in feeding when food is abundant (see Box 4.1 and Fig. 4.2; Hausmann et al. 2003); although it must be recognized that this is a complex, and poorly examined, process, and other limiting steps may exist in tintinnids. *Digestion* is then within food vacuoles, by hydrolytic enzymes that reduce pH and allow nutrient absorption (see Hausmann et al. (2003) and Hausmann & Bradbury (1996) for a more detailed general description). *Egestion* (the removal of undigested material) follows, with fecal material, at times composed of the products of more than one prey, being within the same size range as the prey; at least this occurred when *Favella* sp. consumed the thecate dinoflagellate *Heterocapsa triquetra* (Stoecker 1984). Older reports on tintinnids indicate that they possess a defined region where egestion occurs, the so-called "cell anus", or cytopyge (see, for example, Campbell 1927; Fig. 4.2); the presence of a cytopyge has been more recently observed on several tintinnids (S. Agatha, personal communication). Egested material is generally so small (~10 µm) that it will not sink quickly and is undoubtedly remineralized by planktonic bacteria; thus egested nutrients are maintained suspended in the upper water column, rather than being exported to below the thermocline. Similarly, excreted material (the end-products of metabolized food), such as nitrogenous wastes, will be released by diffusion and used by bacteria or autotrophs. Thus, tintinnids act as re-mineralizers of nutrients within their local environment (see Chapter 9).

Functional response

Possibly the best studied and potentially most important ecophysiological response associated with tintinnids is their change in ingestion rate with prey abundance, the *functional response*, as it both reveals behavioral traits and is central to incorporating ciliates into food-web models (see section "Incorporation of tintinnids into models" on page 118). The above description of tintinnid prey capture and processing (see section "An overview of tintinnid feeding" on page 86) reveals two discrete steps: prey-encounter (searching–contact) and prey-processing (capture–egestion); these lend support for the argument that tintinnids, like many other organisms, exhibit a typical type II functional response (Kiørboe 2008), homologous in its fundamental mechanism to Michaelis–Menton enzyme kinetics, with the contact rate increasing at a decreasing rate with prey abundance, until it becomes constant at satiation (Fig. 4.3); this process can be mechanistically modeled by Equation 4.2 (Box 4.1), where predator ingestion rate (I, prey per predator per day) varies with prey concentration (P, prey per milliliter), I_{max} is the maximum ingestion rate (the reciprocal of handling or processing time), and $k_{½}$ is the prey concentration that elicits half I_{max} ($k_{½} = I_{max}$/maximum clearance rate); for a description of this process associated with protozoa in general, see Fenchel (1986).

$$I = \frac{I_{max} P}{k_{1/2} + P} \tag{4.2}$$

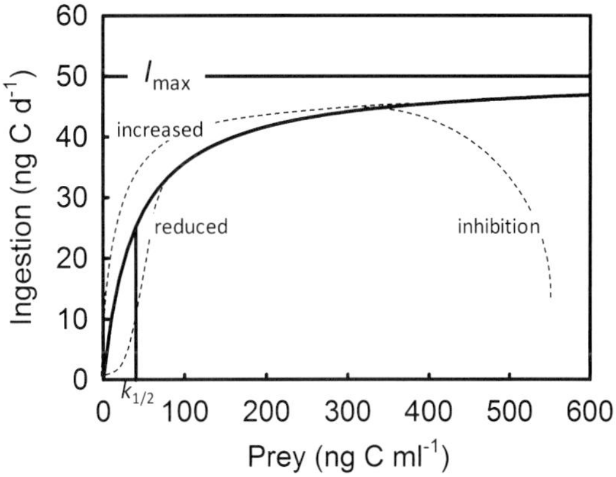

Fig. 4.3 The functional response. Ingestion rate increases to the asymptote I_{max} in the typical response (solid line), with the response reaching half the asymptotic level at $k_{½}$. However, behavioral changes (see text) with prey abundance may "increase" or "reduce" the response closer to the origin. Also, high prey abundance may "inhibit" feeding.

Box 4.1 Derivation of the functional response

Let us assume that there are two activities involved when a tintinnid consumes a prey: *searching* for the prey and *handling* of the prey. In tintinnids, handling is the time to capture a prey item and to form a food vacuole around it, and ultimately the limiting step is likely the re-processing of vacuole membrane to allow new vacuole formation (Fig. 4.2). We will also assume (although this may not be so, and possibly this should be tested), that the tintinnid cannot actively search for prey when it is processing the prey; i.e., the two processes are mutually exclusive, so that the total time (T_{total}) for the tintinnid to capture prey is $T_{total} = T_{search} + T_{handle}$. However, we are generally interested in the prey consumed (P_c) over the total time (T_{total}) so that we can determine an ingestion rate (prey per unit time).

The total time associated with handling (T_{handle}) will be the product of the time to handle a single prey (T_h) and the number of prey consumed (P_c) over a defined time: P_cT_h. Also, a tintinnid will consume several prey (P_c) over a defined searching period (T_{search}). Assuming the tintinnid has a constant searching rate (a, the volume of water processed per time), contact will depend on the prey abundance (P, prey per volume). This can be expressed as $P_c = aPT_{search}$, or $T_{search} = P_c/aP$. Note, in this simple case, we assumed that if contact is made, then the prey is captured; i.e., there is 100% efficiency in capture. However, the relationship holds if capture efficiency is constant and less than 100%; in contrast, if it varies with prey abundance, then the shape of the functional response (Fig. 4.3) will be altered.

The above information can now be substituted into $T_{total} = T_{search} + T_{handle}$, and $T_{total} = P_c/aP + P_cT_h$; this can be rearranged to provide Equation B4.1, representing the type II functional response (Kiørboe 2008).

$$P_c = \frac{aPT}{1 + T_h aP} \tag{B4.1}$$

The inverse of handling time is the maximum ingestion rate (I_{max}, i.e., assuming the tintinnid has ample food available, the only process limiting food capture is the time taken to process a prey item); thus $1/T_h = I_{max}$. Also, T_h and a can be combined, where $k_{½} = 1/T_h a$, which is the half-saturation constant, the prey abundance that elicits $0.5I_{max}$. Equation B4.1 can then be rewritten as,

$$P_c = \frac{I_{max}P}{k_{1/2} + P} \tag{B4.2}$$

Thus, by determining the functional response, by measuring tintinnid ingestion at a range of prey abundances, Equation B4.2 can be fit to the data; e.g., by nonlinear curve-fitting methods (see Montagnes & Berges (2004)), and I_{max} and $k_{½}$ can be determined. Both of these parameters have biological meaning. The inverse of I_{max} is the time required by a tintinnid to process a prey item (T_h). Also, $I_{max}/k_{½} = a$, which is the amount of water processed by a tintinnid per unit time (assuming the 100% capture efficiency); this is also referred to as the "maximum clearance rate".

Here I would like to add a word of caution. Much of the older (and some of the newer) literature on tintinnids and other micro- and mesozooplankton provides data on the relationship between "clearance rate" and prey abundance (Fig. B4.1); in this case, "clearance rate" is an artificial construct, and in my view this relationship should not be reported. In all cases for tintinnids, ingestion rate is the variable that is directly measured, even when the loss of prey from a system is measured rather than direct uptake. The change in clearance rate with prey abundance is then inappropriately determined by dividing ingestion (prey per time) by prey abundance (prey per volume); this is homologous to the heinous error of autocorrelation by plotting $1/x$ versus x (Berges 1997). The only useful term that could be obtained from the plot of clearance rate versus prey is the y-intercept, which is maximum clearance rate (a). However, this can be more readily and accurately determined from $a = I_{max}/k_{½}$, which can then be used in a relative manner to determine the capture efficiency of various prey. In conclusion, given that a can be obtained as described above from Equation B4.2, there is no need to present clearance rate versus prey concentration (e.g., Fig. B4.1), and the literature that does so only adds to confusion, rather than recording trends.

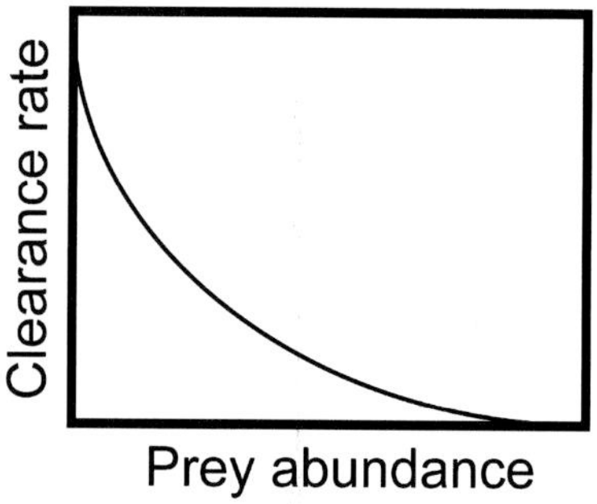

Fig. B4.1 The inappropriate relationship between clearance rate and prey.

The units and dimensions of the axes of the functional response vary between studies. Often, prey numbers are used (prey ingested per predator per unit time versus prey abundance). Alternatively, prey are converted to carbon, through direct measurements or by converting easily measured biovolumes to carbon (e.g., see Montagnes et al. 1994; Menden-Deuer & Lessard 2000). Less frequently, ingestion is presented as specific ingestion, the biomass uptake by the tintinnids divided by its biomass content, but as we will see in the section "Cell size, number, and growth rate" on page 105, tintinnid size may change with prey abundance, so applying a single estimate of tintinnid biomass, would likely be inappropriate in these estimates.

There are several methods used to estimate ingestion rate. Laboratory studies tend to measure either the uptake of prey directly or the loss of prey from the culture container. Particles that have been used for uptake experiments include starch particles, fluorescent beads, heat-killed and stained flagellates, live flagellates that differ from those previously fed to the tintinnids, and radioisotope-labeled prey; for an overview of methods see Båmstedt et al. (2000). Studies that examine the depletion of prey over time use a range of living prey and of course must account for prey growth, which is generally measured in "controls" that lack predators (see Båmstedt et al. 2000). Regardless of the method, the measurement that is made is the removal of particles by tintinnids (prey per predator per unit time), rather than any estimate of the amount of water processed by the tintinnid (volume per predator per unit time), as is often cited (see Box 4.1). The ingestion data that I have examined in this Chapter are primarily laboratory-based (as these provide controlled estimates) and have then been converted to carbon, using literature values. Although not ideal for all purposes, this conversion allows comparison, which in this instance is my main aim; the reader is directed to the literature (Table 4.2) for details on specific responses.

Several of the processes described above will modify the shape of the functional response (Fig. 4.3). For instance, if the tintinnid searching rate were to decrease at low prey concentrations, or if when prey were scarce the area of encounter was to decrease in size, then a type III response ("reduced"; Fig. 4.3) would be expected; see Kiørboe (2008) for more on type II and III responses. Alternatively, tintinnid swimming speed (or at least relative displacement rate) may increase at low prey abundances (Section 4.4), which would alter the shape of the curve by increasing encounter rate and hence making the curve rise earlier ("increased"; Fig. 4.3). In practice, such variations near the origin of the type II response are exceedingly difficult to recognize, and so far no studies have revealed a significant deviation from a type II response for tintinnids.

In contrast, a response that is regularly observed for tintinnids, and other planktonic ciliates, is feeding inhibition at high prey abundance (inhibition; Fig. 4.3). The reason for this has not, to my knowledge, been directly studied, but given that both physical and chemical cues can negatively affect tintinnid behavior (Sections 4.4 and 4.5), it is likely that these play a role in inhibition. Ultimately, it may be useful to assess and parameterize type III and inhibition responses, as they may alter model predictions (e.g., Harrison 1995; Montagnes & Lessard 1999).

However, for now let us assume a type II response. In which case, the parameters (I_{max}, $k_{½}$) will also vary depending on several factors. For instance, the prey and predator size will both alter the encounter area (Fig. 4.2), and the speed at which the predator and prey move will alter the encounter rate (Gerritsen & Strickler 1977). Furthermore, prey and predator size will influence the maximum ingestion rate, as large predators may be capable of processing prey more rapidly, and small prey may be ingested more easily than large prey, although very small prey may be less accessible. A summary of data on tintinnids ingestion rates reveals considerable variation in the responses (Fig. 4.4 and Table 4.2), which is not surprising as these studies were conducted on a range of species, of different sizes, feeding on a range of prey. We can, however, see this variation in the responses by tintinnids as being extremely informative; given their diversity, it may be possible to use tintinnids alone to investigate cross-taxa trends relevant to planktonic ciliates.

In fact, there may be fundamental principles guiding the variation of the responses. For instance, given that feeding is based on the amount of water processed by the ciliate, and this should be proportional to the oral area, we might expect that the maximum clearance rate ($I_{max}/k_{½}$; Box 4.1) is proportional to the oral area, and specifically in tintinnids, the square of the oral lorica diameter; there appears to be some evidence for this (Fig. 4.5a). Also, if, as indicated by Dolan (2010), preferred prey size increases with lorica oral diameter

Table 4.2 Parameters associated with the feeding of tintinnids. Values for I_{max}, $k_{½}$, and maximum clearance are described in the text (see Box 4.1). Data for responses were either obtained directly from the literature or were estimated from data derived from figures presented in the literature (i.e., curves were fit through the obtained data to determine parameters). All values were rounded to two significant figures. Note also that responses are not temperature corrected, and experimental temperatures are presented. The observed maximum ingestion is presented when feeding inhibition occurred at high prey levels and was obtained visually for figures. Note, when the response does not reach an asymptote (e.g., the highest response in Fig. 4.5), then I_{max} is not necessarily the maximum obtained ingestion rate; it is a predicted maximum at an unrealistically high prey concentration and must not be seen as a real maximum. All I_{max} and k values are presented in terms of carbon, which were converted to carbon from prey volume when necessary, by either using estimates provided within the text of the relevant papers or converted using predictions of Menden-Deuer & Lessard (2000). References (Ref): (1) Hansen et al. 1991; (2), Heinbokel 1978a; (3) Jakobsen et al. 2001; (4) Kamiyama et al. 2005; (5) Verity 1985; (6) Verity 1991.

Ciliate	Prey	Prey length (μm)	Prey biomass (pg C)	I_{max} (ng C d^{-1})	Maximum observed I_{max} (ng C d^{-1})	$k_{½}$ (ng C ml^{-1})	Maximum clearance rate (μL h^{-1})	°C	Ref
Amphorides quadrilineata	*Isochrysis galbana*	5	6.82	4.0	3.7	50	3.3	21	(4)
Eutintinnus pectinis	*Isochrysis galbana*	5	–	13	8.2	140	3.9	18	(2)
Favella ehrenbergii	*Heterocapsa triqetra*	20	507	120	100	150	33	18	(1)
Favella tamarense	*Alexandrium tamerense*	30	585	39	36	49	33	15	(4)
Helicostomella subulata	*Isochrysis galbana, Monochrysis lutheri, Dunaliella tertiolecta*	5, 5, 8	–	4.6	4.0	43	4.5	18	(2)
Tintinnopsis acuminate	*Isochrysis galbana, Monochrysis lutheri*	5	–	23	9.2	430	2.2	18	(2)
Tintinnopsis acuminate	*Isochrysis galbana*		3.4	4.0	3.4	57	2.9	20	(5)
Tintinnopsis dadayi	*Katodinium rotundatum* plus other prey	7	40.8	17	13	146	4.8	20	(6)
Tintinnopsis vasculum	*Dicrateria inornata*	5	10.5	18	5.8	71	11	20	(5)

(Fig. 4.5b), and we make the assumption that prey of a larger size (relative to the lorica oral diameter) are more time consuming to process, then maximum ingestion rate might increase with lorica diameter; there seems to be some evidence for this, although the support for the relationship is not great (Fig. 4.5c). Clearly, there will be several factors that alter the functional response. These are briefly examined below, with some examples and an indication of how they might act.

Size selectivity

Given the well accepted notion that planktonic ecosystems are size-structured and that there is a tendency for planktonic organisms to consume prey that are roughly one-tenth their size (Azam et al. 1983; Hansen et al. 1994), it is not surprising that effort has been focused on estimating the size of prey consumed by tintinnids. Rather than considering the cell's oral region as an area of encounter, most studies have focused on the much easier to measure diameter of the lorica as a correlate with prey size (typically prey diameter). This has been most recently reviewed by Dolan (2010), who indicated that, based on ingested material, tintinnids prefer prey that are approximately 20% of their lorica diameter (Fig. 4.5b) and will on average consume prey that are no more than 30% of the lorica diameter. These, improved, estimates deviate from the well-cited values of maximum prey size of 43%

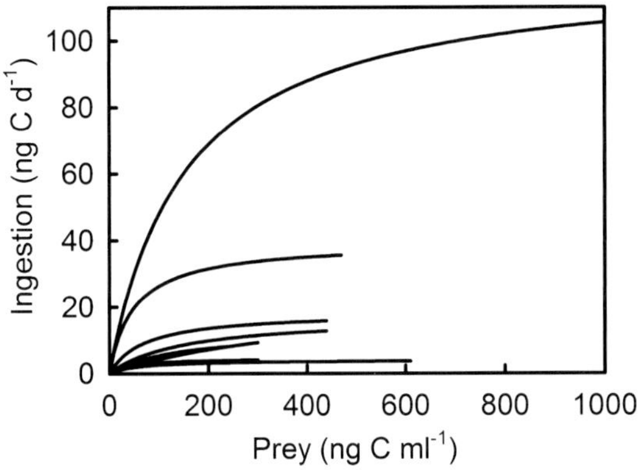

Fig. 4.4 A summary of tintinnid functional responses. Curves are only drawn over ranges where ingestion increased with increasing prey abundance. See Table 4.2 for parameters and literature associated with individual responses.

(Heinbokel 1978b) or 45% (Capriulo 1990), which are based on both prey within tintinnids and prey removed from the environment (presumably by tintinnids). Thus, tintinnids may typically consume prey that are smaller than previously considered, although there are records of tintinnids consuming very large particles, up to 80–100% of their lorical diameter (Capriulo 1982; Kamiyama & Arima 2001).

Of similar consequence to food-web structure is a consideration of the minimum size of prey consumed by tintinnids. Many tintinnids are capable of consuming picoplankton (<2 µm), and for smaller tintinnid species prey that are 2–5 µm may be a dominant dietary component (Bernard & Rassoulzadegan 1993). Furthermore Turner & Tester (1992) indicated that *Favella panamensis* consumes labeled *Escherichia coli* (~2 × 0.5 µm) when exposed to abundances of approximately 5×10^6 ml^{-1} (i.e., five times higher than the typical 10^6 ml^{-1} of most marine systems; see, for example, Azam et al. 1983). Therefore, although they can capture them, it is unlikely that tintinnids are major consumers of bacteria in marine ecosystems; this role is more likely attributed to heterotrophic nanoflagellates (see Azam et al. 1983; Fenchel 1987).

It must be noted though that size, measured as prey radius or length, may be deceptive. Some diatoms (and other prey) possess cellular extensions; specifically diatoms, such as *Thalassiosira*, have β-chitin threads that may be missed by casual observation or broken off by agitation in cultures. These structures inhibit grazing by tintinnids (Verity & Villareal 1986), and as

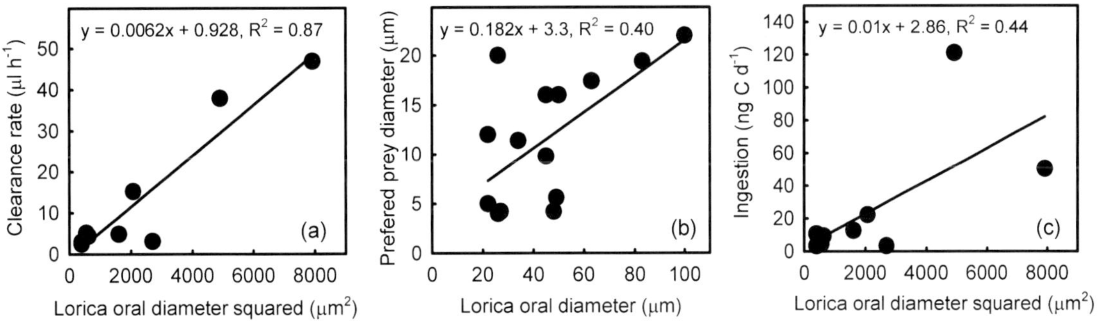

Fig. 4.5 Scaling tintinnid grazing rate with lorica oral diameter (LOD). (a) Maximum clearance rate versus LOD^2. (b) Data from Dolan (2010) plus some additional data, indicating preferred prey diameter versus LOD. (c) Ingestion rate versus LOD^2. Ingestion and clearance data are also presented in Table 4.1. Equations on figures represent the line on the figure. R^2 values are offered to provide an indication of goodness of fit.

will be seen below alter tintinnid swimming behavior and growth rate (Table 4.1). Thus, researchers must be rigorous when estimating the perceived size of prey; this is a lesson clearly illustrated by tintinnids that should apply to work on all microzooplankton, if not all size selective grazers.

Still, there may be some general rules regarding prey size of tintinnids; Kamiyama & Arima (2001) provide an elegant exploration of such a trend. They argue that ingestion by ciliates, and specifically tintinnids, can be described as the number of prey encountering the oral area (e.g., as illustrated in Fig. 4.2). Assuming this is so, ingestion rate (I) of small prey can be expressed as the ratio of total-prey cross-section (A_p) occupying the ciliate's oral area (A_c), over a short time period. Thus, $I = A_p/A_c$, or scaling for prey diameter (D), $I = \alpha D^2$, where α is a constant associated with the tintinnid's affinity to the prey, and thus $D^2 = I/\alpha$. Prey volume can be expressed in terms of the prey diameter: $D^3 = V/\beta$, where $\beta = 1/6\pi$. From here, they obtain $\ln I = -2/3 \ln V + (\ln \alpha + 2/3 \ln \beta)$, suggesting that the slope of $\ln I$ versus $\ln V$ will be −2/3 (−0.67) and that the y-intercept provides information on the tintinnid's ability to capture prey. Kamiyama & Arima (2001) then explore this theoretical value by comparing it to laboratory data on *Favella taraikaensis* (−0.63) and *Favella ehrenbergii* (−0.51), indicating that the relationship seems plausible.

How then do these data on the potential size spectrum inform us about the average tintinnid? The modal oral diameter of tintinnids is approximately 40 μm (sample size ~650, range 5–185 μm; Dolan 2010), and tintinnids typically consume prey that are 8 μm (Dolan 2010). This falls within the range suggested by others of 2–30 μm (Rassoulzadegan et al. 1988). Given that modal tintinnid lorica length is approximately 60 μm, and the predicted predator: prey ratio for planktonic ciliates is 8:1 (Hansen et al. 1994), based on prey size alone, tintinnids seem to be good models of planktonic ciliates. However, size is far from the only character governing capture.

Selectivity other than size

Given that many planktonic taxa in the size range of prey available to tintinnids are toxic to a range of organisms, it is naïve to think this would not also apply to tintinnids, and it would be expected that there are prey preference, even for non-toxic prey. Of course, this is the case. Several organisms that are of an appropriate size are rejected (Taniguchi & Takeda 1988) and even cause mortality (Table 4.1), whereas others are selected for. *Tintinnopsis* sp. exhibited prey preference on similarly sized prey, but this preference was not accounted for by the trophic position of the prey (autotrophic versus heterotrophic), the diet of the heterotrophs, or the gross biochemical composition of the prey (Verity 1991a); instead it was speculated that other factors such as chemoreception play an important role. Similarly, *Favella* sp. exhibits prey selectivity on similarly sized living and inert prey (Stoecker 1988). In fact, tintinnid feeding selectivity can be extended to strains of a single taxon: *Eutintintinnis* sp. exhibits selectivity on strains of *Synechococcus* that are of similar size, elemental composition, and motility; in this case, again, chemical cues were likely at play (Apple et al. 2011).

Surprisingly though, several tintinnids consume taxa that are considered harmful or toxic: *Favella ehrenbergii* ingests the toxic species *Alexandrium ostenfeldii* (Hansen et al. 1992); *Amphorides quadrilineata* ingests the brown tide algal *Aureoumbra lagunesis* (Jakobsen et al. 2001); *Favella taraikaenis* consumes the harmful algae *Heterocapsa circularisquama* (Kamiyama 2000) and *Alexandrium tamarense* (Kamiyama et al. 2005); and *Favella* sp. ingests *Heterosigma akashiwo* and grows when this toxic species occurs in mixed cultures but dies when it is the sole diet (Graham & Strom 2010). In fact, when *Favella taraikaenis* consumes *Alexandrium tamerense* it excretes toxins that are then dissolved in the environment, potentially reducing the toxicity of the algal bloom (Kamiyama & Suzuki 2006). Still, it is difficult to generalize feeding on toxic (or harmful) algal species by tintinnids; the relationship may be population-specific and be influenced by previously exposure to taxa, as has been observed in the relationships between the copepod *Acartia* and *Alexandrium fundyense* (Colin & Dam 2003). Possibly, *Favella taraikaensis* that has been exposed to *Alexandrium tamarense* may too have developed an ability to actively feed on this toxic dinoflagellate (T. Kamiyama, personal communication).

Even though tintinnids may consume harmful algae, this can lead ultimately to their death (Stoecker et al. 2002), so we must not equate grazing with growth. Furthermore, my own observations on non-tintinnid oligotrichs indicates that although one ciliate consumes and grows well on a particular prey species, it may not grow well on another species that is a perfectly suitable food for a different species of ciliate. Although tintinnids such as *Favella* may exhibit prey preference (see, for example, Stoecker et al. 1981), to my knowledge this aspect of prey-specificity being dependent on which tintinnid is examined has not been quantitatively demonstrated for tintinnids or other oligotrichs. Clearly, understanding the impact of predator–prey specificities will enhance our appreciation of both dynamics and diversity in pelagic ecosystems and, therefore, should be encouraged. Given the extensive wealth of information on tintinnids, it seems that they may be ideal candidates to pursue such studies.

4.3 GROWTH

An overview of tintinnid growth

Before assessing the factors that govern tintinnid growth and reproduction, I offer an overview of a generalized tintinnid *cell-* and *life-cycle* (Fig. 4.6), with the assumption that the reader is familiar with the basics of ciliate biology (if not, see Lynn 2008; Hausmann et al. 2003; Hausmann & Bradbury 1996). The *cell-cycle* includes the process of individual cell growth, mitosis (of the micronucleus), amitotic division of the macronucleus, formation of a new oral region, proliferation of somatic cilia, and subsequent cell division; repeated cell divisions lead to exponential population growth (for details on the division of tintinnids and the formation of a new lorica and fate of the old lorica see Chapter 2).

During repeated cell divisions (cell cycle, the left side of Fig. 4.6a), amitotic division of the macronucleus may lead to loss or dilution of essential gene fragments, and the accumulation of deleterious mutations (known as "Muller's ratchet" (Bell 1988)), which may ultimately lead to senescence (inability to conjugate) and death of a clone, owing to lack of vitality (Fig. 4.6). Given that the macronucleus governs cell growth, this may reduce clonal fitness (i.e., the progeny of a single cell); the solution to this decline in fitness is sex (conjugation), which, among other processes, replaces the macronucleus (Fig. 4.6; see Lynn (2008), or for a

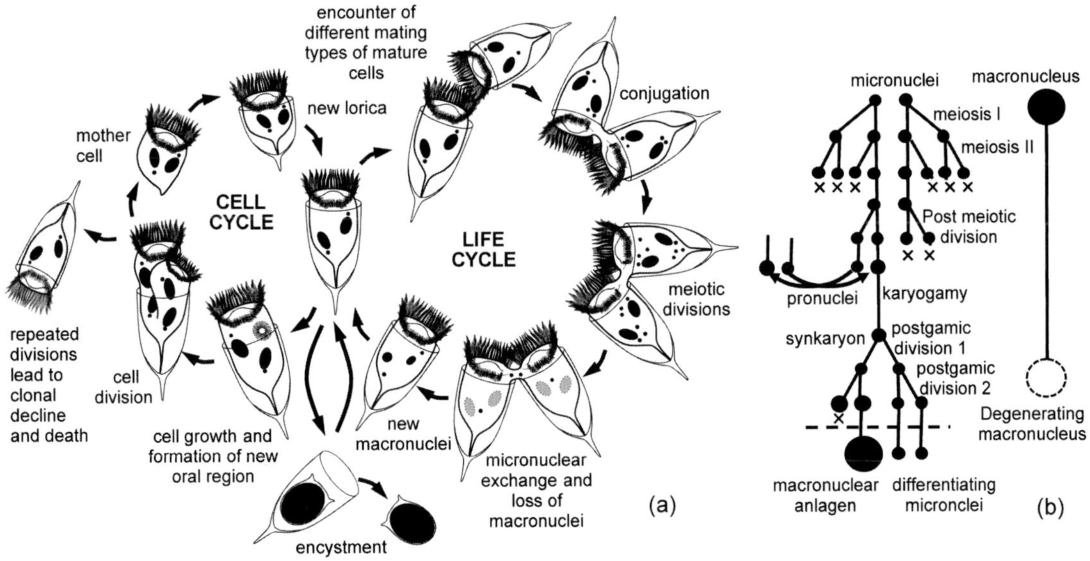

Fig. 4.6 A generalized tintinnid life and cell cycle. (a) This is a schematized version of the process amalgamated from several studies; it should not be seen as an accurate representation of any one species. Note also that it is unclear if encystment is the result of cells in the cell cycle or if it is a post-conjugation event. (b) This is a representation of the nuclear events during and after conjugation. It is based on observations of the non-tintinnid oligotrich *Pelagostrobilidium* (Ota & Taniguchi 2003) and is, therefore, only a prediction of what may occur in tintinnids. ×, degeneration of a micronuclear product. The dashed horizontal line is the predicted time when the two conjugating cells separate.

detailed evaluation of clonal decline see Bell (1988)). In fact, Gold (1971) and Gold & Pollinger (1971) attributed the long-term (over a year) vitality of cultures of *Tintinnopsis beroidea* to the mixing of several isolates (potentially different mating types); in cultures Gold regularly observed conjugation, or at least behavioral events that he attributed to conjugation. This ability of cultures to be rejuvenated through sex is also illustrated by work on a close relative of tintinnids, *Strombidinopsis*, that revealed conjugation within a single clone (selfing) resulting in rejuvenation of the culture (Montagnes et al. 1996). Possibly tintinnids too can "self", but this has not been observed.

The *life-cycle* (the right side of Fig. 4.6a) may include multiple cell divisions, but it begins with conjugation. During conjugation the micronuclei undergo meiosis to form gametic, haploid nuclei, one of which migrates to the adjacent cell and fuses with its complementary, stationary, haploid nucleus; after genetic recombination, the old macronuclei disintegrate and are replaced by newly generated macronuclei, produced by one of the newly formed micronuclei (Fig. 4.6a, b); the details of this process differ between ciliate taxa, and the interested reader is directed to Lynn (2008), Hausmann & Bradbury (1996), and Chapter 3.

There are few recent data on the life and cell cycle of tintinnids, likely because such fundamental research is time-consuming, requires culturing skills, and is perhaps more descriptive than hypothesis-driven and thus is perceived as both costly and unimportant under the present management of science. However, thanks to the efforts of a few dedicated modern studies and several older works, it is possible to assemble a general tintinnid life and cell cycle (Fig. 4.6a, b). This amalgamation is conjectural, and is based primarily on observations of *Favella* (Laval-Peuto 1983; Stoecker et al. 1983) and is augmented by observation on a range of other tintinnids (Apstein 1893; Laackmann 1908 and references therein; Biernacka 1952; Coats & Heinbokel 1982) and recent work on the near relatives of tintinnids (*Pelagostrobilidium*, Ota & Taniguchi 2003; *Halteria grandinella*, Agatha & Foissner 2009).

The factors that stimulate tintinnid conjugation are unknown, but, as for other ciliates, conjugation likely arises in mature cells that are stimulated by changes in environmental conditions such as temperature shifts and reduced prey abundance (but see Hausmann & Bradbury 1996; Lynn 2008). In *Favella*, conjugation has been divided into three phases: pre-conjugation, when cells actively search for each other and then contact at the oral region (this period lasts ~5 min); formation of a cytoplasmic bridge at the anterior end, meiosis of the micronuclei, and exchange of gametic nuclei (lasting hours); and finally separation. During conjugation the cells remain motile, although the adoral membranelles may beat more slowly, and cells do not feed. This summary of the three steps is a highly abridged version of the detailed description provided by Laval-Peuto (1983) and is supported by observations of Ota & Taniguchi (2003) on *Pelagostrobilidium* and other older observations that are picturesque but at times difficult to follow (e.g., Laackmann 1908).

The nuclear reorganization in conjugating tintinnids is poorly documented, although Laackmann (1908) describes and illustrates some of it in detail for *Tintinnopsis*, but we can predict the sequence of events from detailed observations on the closely related *Pelagostrobilidium* (Ota & Taniguchi 2003; Fig. 4.6b); these generally agree with the events that occur for the less closely related oligotrich *Halteria grandinella* (Agatha & Foissner 2009), but there is some variation between observations as would undoubtedly occur between tintinnid species. For instance, Laackmann (1908), among other subtle differences from the process illustrated in Fig. 4.6b, suggests that for some tintinnids two haploid micronuclei migrate to the partner. Finally, there is one, atypical, report on tintinnid conjugation that included "microgametes" (Gold & Pollingher 1971), but as this behavior has only been observed once it must be viewed with caution (see Chapter 6, where it may be suggested that these microgametes are in fact parasites).

Several tintinnid species produce cysts (Biernacka 1952; Paranjape 1980; Stoecker et al. 1983; Kamiyama 1994a, 1997a; Kamiyama et al. 1995); see Chapter 7 for details on encystment and excystement. It is unclear which factors stimulate cyst production, although Paranjape (1980) argued that it was seasonal in temperate waters, possibly with cysts acting as an overwintering stage, and it is unclear if cysts are a post-conjugation product or occur throughout the cell cycle (Chapter 7). Excystment, in contrast, has been well investigated: it occurs when cysts are exposed to warmer temperatures, increased light levels, and when exposed to phytoplankton or their exudates (Paranjape 1980; Kamiyama 1994a, 1997a; Kamiyama et al. 1995). Likely all of these factors relate to improved conditions in the spring, leading credence to Paranjape's (1980) argument that cysts act as an overwintering stage. More detailed work on related species of oligotrichs supports this seasonal relevance (Kim & Taniguchi 1997), although some encystment events of naked oligotrichs (see, for example, Reid 1987; Müller 1996) may follow short-term blooms; thus tintinnids may form cysts in response to both short- and long-term pressures to survive poor biotic and abiotic conditions (again, Chapter 7 deals with this in more detail).

Population growth

As indicated above, individual tintinnids will increase in size, over the cell cycle (Fig. 4.6a). However, generally when growth is referred to in the literature, it is synonymous with *specific growth rate*, the instantaneous rate of increase (the novice reader is referred to any general book on population biology for a more detailed description); this constant is then used to determine *population growth* (see section "Incorporation of tintinnids into models" on page 118). In laboratory cultures, specific growth rate (denoted by convection as r or μ, with dimensions of time^{-1}) is typically determined, following standard population-biology methods, from the change in the natural logarithm of numbers over time (t), but older studies on tintinnids may use $\log_2$ or $\log_{10}$, whereas others report doubling time, all of which are interconvertible.

In actively growing cultures, measurements of abundance (or biomass) can be made over several time points, and linear regression analysis may then be applied to the change in the natural logarithm of numbers (or biomass) over time; this can be reduced to two points: the initial (N_0) and final count (N_t) can then be used to determine growth rate following: $r = \ln(N_0/N_t)/t$. As we shall see below, however, using numbers (or biomass as a bulk measure) alone, rather than examining individual cell biomass, may be misleading, if estimating biomass production is the goal, as cell size can change over time or under different prey regimes; i.e., a large ciliate with the same specific growth rate (determined by biomass or numbers) as a small ciliate will produce more biomass over the same time period. Furthermore, although growth rate estimates always assume exponential growth, it is clear that this cannot be maintained indefinitely. As an example, data on a constantly fed natural assemblage of ciliates, including tintinnids (Fig. 4.7) revealed that tintinnids may either reach a maximum sustainable number (i.e., a carrying capacity), as was the case for *Eutintinnus lususundae*, or they may decline after a maximum is obtained, as was the case for *Favella taraikaensis*; these data illustrate that some caution must be applied when choosing time points to estimate growth rate. Generally, field populations rarely reach more than 10 individuals per milliliter (Chapter 9), and this may be a good rule of thumb to use when examining maximum numbers in cultures. Although, some tintinnids reach far higher numbers in culture; e.g., by increasing the prey abundance, Gold (1971) was able to raise *Tintinnopsis beroidea* to more than 10^3 cells per milliliter. Such data suggest that tintinnids may be "bottom-up controlled" under natural conditions (i.e., their abundance is limited by prey resources rather than grazing, but see Chapter 5).

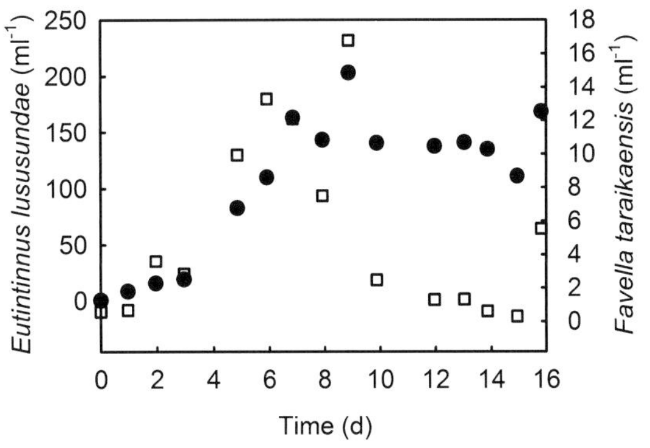

Fig. 4.7 An example of tintinnid population growth over time, illustrating initial exponential growth, a carrying capacity, and decline in late stationary phase. Data from Taniguchi & Kawakami (1983): ●, *Eutintinnus lususundae*; □, *Favella taraikaensis*.

Numerical response

The tintinnid numerical response (i.e., the change in specific growth rate, r, with prey abundance or biomass) is possibly as important as the functional response (see section "Functional response" on page 93). Like the functional response, it tends to follow a rectangular hyperbolic (type II-like) shape, at times with inhibition at high prey abundance. The typical function that is fit to such data differs from Equation 4.2 only in that it has a non-zero prey-intercept

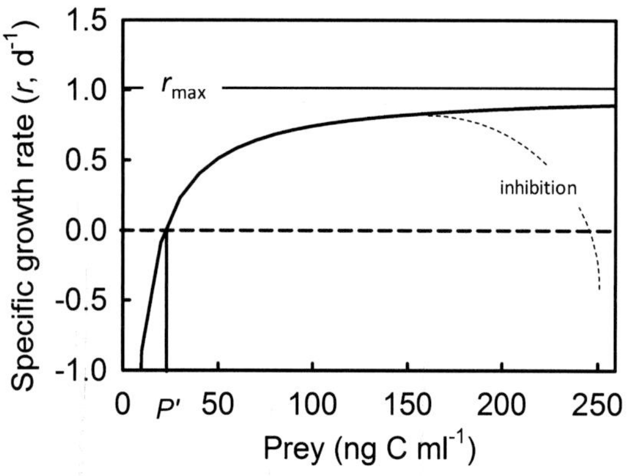

Fig. 4.8 The numerical response. Specific growth rate increases to the asymptote r_{max} in the typical response (solid line), with the response passing through the *x*-axis at a level above zero (P'). However, behavioral changes (see text) with prey abundance may "inhibit" growth.

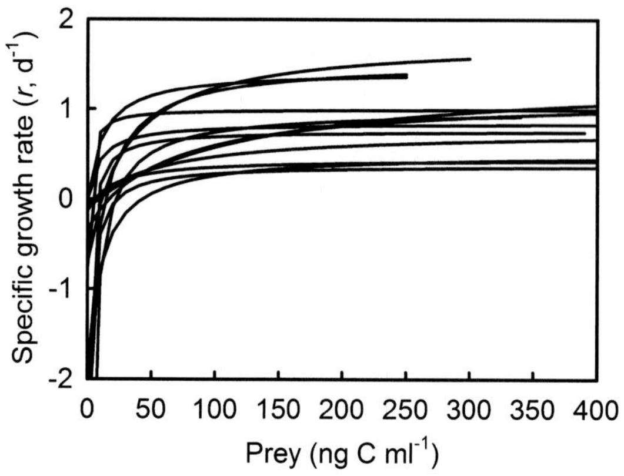

Fig. 4.9 An indication of the diversity of tintinnid numerical responses. Curves are only drawn over ranges where ingestion increased with increasing prey abundance. See Table 4.3 for parameters and literature associated with individual responses.

(Fig. 4.8) that represents the prey abundance required to maintain tintinnids at zero growth (Equation 4.3),

$$r = \frac{r_{max}(P - P')}{k + P - P'} \qquad (4.3)$$

with, r_{max} being the maximum growth rate, P being prey abundance, P' being the prey "threshold level" where $r = 0$, and k being a constant (note, k is not the half-saturation constant as in the functional response $k_{½}$; $k - P'$ is homologous to a half-saturation constant; see Fenton et al. (2010)).

This function (Equation 4.3) has been widely applied to tintinnid numerical responses for over 30 years (e.g., Heinbokel 1978a), and is extremely useful as it reveals mortality rates at sub-threshold levels, and coupled with the functional response may reveal other attributes (see section "Gross growth and assimilation efficiency" on page 105 and Box 4.2). Again, as for the functional responses, I have focused on laboratory studies of the numerical responses and have converted prey to carbon for constancy (Fig. 4.9 and Table 4.3). These growth incubation experiments typically assume exponential growth is constant over a short incubation period in vessels on the order of 10–1000 ml, and determine r, as described in "Population growth" on page 101.

Besides the obvious variation in growth responses (Fig. 4.9), likely because of different predator and prey species, notable similarities exist. First, the threshold concentration (i.e., the level required to sustain tintinnids) ranges from virtually 0 to 50 ng C ml^{-1} and is typically between 10 and 20 ng C ml^{-1}; this range is on the low end of estimates found for naked oligotrichs (reviewed by Gismervik 2005) and may imply a selective advantage for tintinnids at low prey abundances. Second, many of the growth rates reach a maximum at approximately 1 d^{-1} (i.e., ~1.5 divisions d^{-1}), and generally by 50 ng C ml^{-1} most species have achieved their maximum growth rate; this is in the same range as other oligotrichs (see, for example, Gismervik 2005).

The ecological implications of the above findings have been discussed in terms of the naked planktonic ciliates (Montagnes 1996): prey (2–20 μm) biomass levels in many coastal waters range between 5 and 50 ng C ml^{-1}. Although this suggests that at times there are insufficient prey available to support tintinnid growth, it also suggests that tintinnids are capable of achieving their maximum growth rates at (and above) the higher end of typical prey levels; i.e. most, but not all, of the time tintinnids are bottom-up controlled, being limited by prey availability (but see Chapter 5). Furthermore, tintinnids will have maximum grow rates equal to or greater than their prey (see Banse 1982; Montagnes 1996). Consequently, at times, when prey blooms occur, tintinnids may exploit them, and then die due to starvation as prey are driven to extinction. Tintinnids will also have a selective advantage over competing mesozooplankton grazers

Table 4.3 Parameters associated with the growth of tintinnids. Values for r_{max}, k, and P' are described in the text. Data for responses were either obtained directly from the literature or were estimated from data derived from figures presented in the literature (i.e., curves were fit through the obtained data to determine parameters). All values were rounded to two significant figures. Note also that responses are not temperature corrected, and experimental temperatures are presented. The observed maximum growth rate is presented when growth inhibition occurred at high prey levels and was obtained visually for figures. Note, when the response does not reach an asymptote (e.g., some of the highest response in Fig. 4.9), then r_{max} is not necessarily the maximum obtained growth rate; it is a predicted maximum at an unrealistically high prey concentration and must not be seen as a real maximum. All P' and k values are presented in terms of carbon, which were converted from prey volume when necessary, by either using estimates provided within the text of the relevant papers or converted using predictions of Menden-Deuer & Lessard (2000). References (Ref) : (1) Hansen 1995; (2) Heinbokel 1978a; (3) Jakobsen et al. 2001; (4) Kamiyama et al. 2005; (5) Rosetta & McManus 2003; (6) Stoecker et al. 1983; (7) Verity 1985; (8) Verity 1991.

Ciliate	Prey	Prey length (μm)	Prey biomass (pg C)	r_{max} (d^{-1})	Observed maximum, r_{max} (d^{-1})	k (ng C ml^{-1})	P' (ng C ml^{-1})	°C	Ref
Amphorides quadrilineata	*Isochrysis galbana*	5	6.8	0.38	0.36	27	24	21	(3)
Eutintinnus pectinis	*Isochrysis galbana*	5	–	1.5	1.3	24	14	18	(2)
Favella ehrenbergii	*Heterocapsa triqetra*	20	510	0.77	0.72	60	5.1	15	(1)
Favella ehrenbergii	*Prymnesium parvum*	10	45	0.51	0.45	54	43	20	(5)
Favella sp.	mixed diet	–	–	1.4	1.1	130	8.8	17	(6)
Favella sp.	*Heterocapsa triqetra*	20	510	0.85	0.84	9.6	0	15	(6)
Favella tamarense	*Alexandrium tamerense*	30	590	1.0	0.99	3.1	1.3	15	(4)
Helicostomella subulata	*Isochrysis galbana*, *Monochrysis lutheri*, *Dunaliella tertiolecta*	5, 5, 8	–	0.76	0.74	9.6	7.5	18	(2)
Metacylis angulata	*Prymnesium parvum*	5	45	0.45	0.43	26	5.6	20	(5)
Tintinnopsis acuminata	*Isochrysis galbana*, *Monochrysis lutheri*	5	–	1.4	1.4	12	0	18	(2)
Tintinnopsis acuminate	*Isochrysis galbana*	5	3.4	1.8		39	11	20	(7)
Tintinnopsis dadayi	*Katodinium rotundatum* plus other prey	7	41	1.2	0.97	77	17	20	(8)
Tintinnopsis vasculum	*Dicrateria inornata*	5	11	1.0	0.92	26	22	15	(8)

(Chapter 9), as their growth rates are higher. There are instances of such tintinnid-blooms occurring: for example, Admiraal & Venekamp (1986) observed tintinnid blooms associated with an algal bloom of *Phaeocystis pouchetii* and suggested tintinnid grazing pressure was in part responsible for the demise of the algal bloom; similarly Andersen & Sørenson (1986) argued that blooms of *Helicostomella subulata* might control prey populations over short periods in a Danish fjord.

It might also be expected that tintinnids move between local patches of prey and can bloom under conditions where prey are abundant. Such speculations were made by Stoecker et al. (1984) who found a correlation between the abundance of *Favella* and its prey, and this supports other work that indicates that naked oligotrichs may bloom when prey are abundant and graze down their prey (see, for example, Montagnes & Lessard 1999). Of course, this idea is not new; it was posed by Banse (1982) who argued such boom–bust dynamics should occur as ciliates have equally high, or higher, growth rates as their prey. In Section 4.4, we will see that tintinnids may have an ability to find and remain in regions of high prey abundance and thus increase their growth rate.

Scaling tintinnid maximum growth rate to size

Several studies have explored and then used the strong inverse relationship between plankter size and growth rate (e.g., Hansen et al. 1997; Glazier 2006). These functions have both pragmatic application in size-structured analysis of food-web dynamics (e.g., Montagnes et al. 2010) and more theoretical applications, such as in the assessment of metabolic scaling (e.g., Glazier 2006). Tintinnids, because of their loricae, are relatively difficult to place into this context, as most studies consider body (or cell) size as the independent variable; the lorica obscures the cell, preventing accurate measurement. It is, therefore, not surprising that work on tintinnids has focuses on the relationship between lorica size and growth rate; Dolan (2010) indicated that the lorical oral diameter (LOD) was a good metric for predicting growth rate, and here I build on his data set (provided online by Dolan (2010)) and explore the use of this response (Fig. 4.10).

There is a weak but significant relationship between specific growth rate (r) and LOD, indicating that LOD could be used as a predictor. Furthermore, the allom-

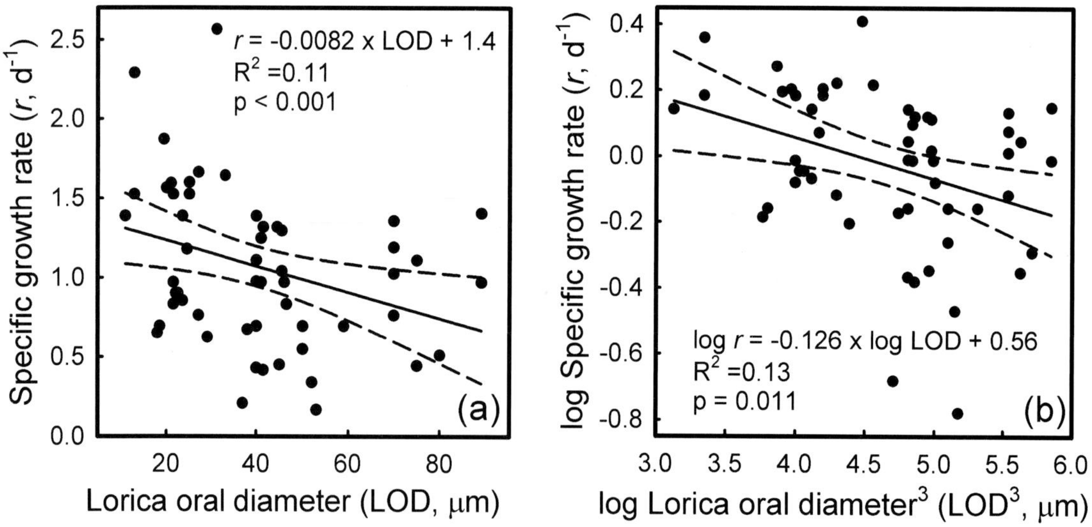

Fig. 4.10 (a) The relationship between specific (r, d^{-1}) growth rate and LOD. (b) The allometric relationship between log$r$ and logLOD^3 (the latter acting as a proxy for volume). Data are from Dolan (2010), augmented by data from Hansen (1995), Heinbokel (1978a), Jakobsen et al. (2001), Kamiyama et al. (2006), Rosetta & McManus (2003), and Verity (1991a, 1985).

etric relationship of log$r$ versus logLOD^3 (the latter acting as a proxy for volume) has a slope of approximately −0.13, which is less than the typically predicted −0.25, based on energetic arguments (see Hansen et al. 1997; Glazier 2006). However, the data presented in Fig. 4.10 are neither temperature corrected nor are they based solely on maximum growth rates. Both of these concerns must be addressed if this relationship for tintinnids is to be pursued, and likely this will require increasing the data set.

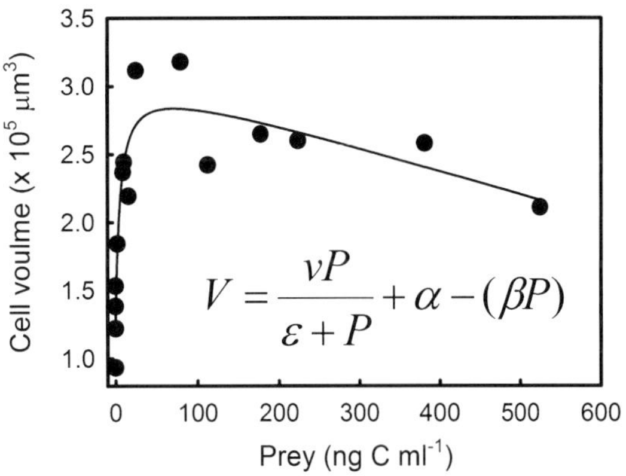

Fig. 4.11 An indication of how tintinnid biovolume may change with increasing prey abundance. Data obtained from Kamiyama et al. (2005). The line through the fit is a phenomenological response, following the equation depicted on the figure, where V is cell volume, P is prey abundance, and α, β, and ε are constants (see Box 4.2 for details on this response).

Cell size, number, and growth rate

As indicated above, most tintinnid growth rate estimates rely on a change in numbers, and from a perspective of population dynamics, this is entirely appropriate, especially as it is individuals that feed, conjugate, and disperse. However, biomass- or bioenergetic-based models often work with bulk estimates (e.g., carbon, see section "Incorporation of tintinnids into models" on page 118). If cell biomass (or its surrogate biovolume; see Menden-Deuer & Lessard 2000) remains constant, then the product of tintinnid abundance and an average cell biomass can be used in such models, and this approach has been used in the past. However, there is good evidence that cell volume of tintinnids (and closely related ciliates to tintinnids) changes with prey abundance (e.g., Fig. 4.11; Montagnes & Lessard 1999). Therefore, applying an average biomass to reflect the numerical response in terms of biomass is inappropriate. Instead, as factors such as prey abundance (Fig. 4.11), temperature (Atkinson et al. 2003), and culture age (personal observations) may influence cell size of ciliates, such biotic and abiotic interactions must be considered.

Gross growth and assimilation efficiency

Gross growth efficiency (GGE, sometimes referred to as yield or gross production efficiency) is the amount of ingested prey biomass that is incorporated into predator biomass, whereas assimilation efficiency (AE or e, see Box 4.2) is the amount of prey biomass that is assimilated (i.e., not egested) by the predator. One or the other of these parameters is often incorporated into population and ecosystem models to estimate the numerical response from functional response data (see section "Incorporation of tintinnids into models" on page 118) or to estimates ingestion when growth is known (Straile 1997); consequently they require careful consideration.

Given the variety of diets that are available to tintinnids and the range of preferences for these (Table 4.1), it should not be surprising that GGE and AE should both be species dependent. In fact, growth may not just depend on prey species but also on prey clones (Table 4.1, Hansen 1989). Furthermore, there is a substantial amount of literature that indicates that efficiency depends very much on the prey size, composition, and palatability (e.g., Kiørboe 2008). To my knowledge, the influence of these factors have not been explored for GGE or AE of tintinnids.

Of the two, GGE is the easier to estimate, by dividing specific ingestion rate (I/predator biomass) by specific growth rate (r); this is equivalent to dividing the biomass ingested over a set period by the biomass incorporated into new cells over the same period. However, there are issues with the application of GGE. First, as tintinnid biomass may vary with prey abundance (see section "Cell size, number, and growth rate" on page 105), a constant predator biomass may be inappropriate to apply in calculations, as several authors have done in the past for tintinnids (see, for example, Verity 1985, 1991a). Second, as the numerical response (Equation 4.3) has a positive x-intercept, whereas the functional response (Equation 4.2) passes through the origin, GGE will approach zero at the threshold prey level (P', Equation 4.3) and will be negative below it; thus GGE cannot be constant, and calculations may result in, non-intuitive, negative values where predator mortality occurs. Finally, as the functional and numerical responses often do not have similar shapes (i.e., the asymptotic response differs, with ingestion tending to increase when growth rate has reach a maximum, see Box 4.2 for an example), as prey abundance changes so does GGE, typically decreasing with increasing prey (Straile 1997). Consequently, determining and applying a single mean GGE value in model calculations, as is often done, can be misleading.

As suggested above, GGE of tintinnids appears to change with prey abundance. For instance, the data presented in Fig. 4.12 suggest some variation with prey, and for *Favella* sp. Stoecker & Evans (1985) indicate a reduction from 0.5 to 0.25 as prey levels increased from approximately 20–100 to 350–400 ng C ml^{-1}. Generally, tintinnid GGE seems to range between 0.3 and 0.7, with a mean of approximately 0.45, when it is measured above the prey threshold level (Equation 4.3, P'). This is above the average value determined for ciliates (0.3) and other planktonic protozoa and animals (0.3 or below), in general (Straile 1997). However, these observations (Fig. 4.12) might be regarded with caution when applying them to tintinnids, given that they are based on only four studies, by three separate research groups, on four species (three of which are in the same genus). Furthermore, as we will see in Section 4.5 and as Verity (1985) indicates, temperature will alter GGE of tintinnids.

Assimilation efficiency is pragmatically and computationally more challenging to study on protozoa, as it either requires measuring egestion rate (i.e., the production of fecal material), respiration rate (e.g. oxygen uptake; for tintinnids see Verity (1985)), or relying on biomass based functional and numerical responses and a rather complex set of calculations (Fenton et al. 2010; Box 4.2). Still, there are some estimates of tintinnid assimilation efficiency, under single prey abundances: 71% for *Favella* sp. and 67% for *Favella ehrenbergii* (Stoecker 1984; Rassoulzadegan 1978, respectively).

However, like GGE, AE should vary with prey abundance. Fenton et al. (2010), reanalyzing the data of Kamiyama et al. (2005), indicate that at least for *Favella taraikaensis*, assimilation efficiency varies with prey abundance; this follows the trend support by Fenton et al. (2010) that AE generally varies with prey abundance. So far, there has been no application of the biomass-based methods offered by Fenton et al. (2010) to tintinnid data. Here I use one set of data to illustrate the method's potential use and issues arising from problematic data (Box 4.2); the example should stimulate more rigorous research in this area rather than discouraging the reader.

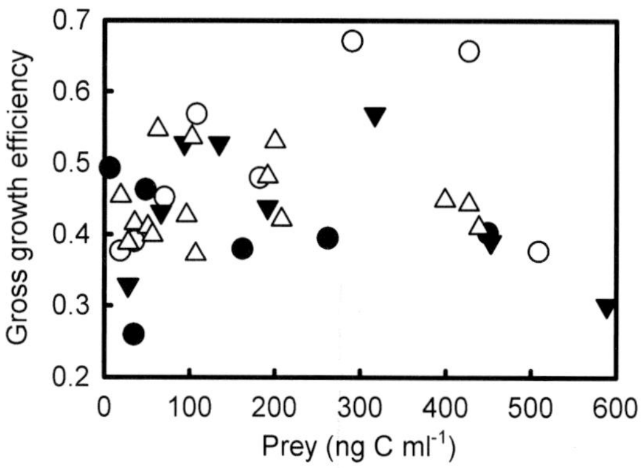

Fig. 4.12 Some examples of how GGE may vary with prey abundance. Data from Kamiyama (2005) on *Favella taraikaensis* (●) at 15 °C, Verity (1985) on *Tintinopsis acuminata* (○) and *Tintinopsis vasculum* (▼) at 15 °C, and Verity (1991a) on *Tintinnopsis dadayi* (△) at 20 °C.

4.4 SWIMMING BEHAVIOR

The basics of swimming

Before describing tintinnid swimming, we need to contextualize briefly the environment in which they live. Others have done a more thorough job than I, and the interested reader is directed to Vogel (1996) for an overview, to Purcell (1977) for microorganisms in general, to Kiørboe (2008) for plankton, and to Fenchel (1987) for protozoa specifically. However, to summarize these works briefly, tintinnids live in an environment of low Reynolds numbers, one where water does not act as we know it. For tintinnids, water is sticky and viscous, akin to us swimming through glycerol or syrup. There are also virtually no inertial forces; when a tintinnid stops propelling itself, it stops moving, and there will be a portion of the water that

Box 4.2 Assessing assimilation efficiency

It is pragmatically difficult to determine assimilation efficiency (e) by measuring protozoan, and for our purposes tintinnid, egestion rate, as collecting protozoan fecal pellets is challenging, although possible (Stoecker 1984). Instead, we can adopt a different method to assess how e varies with prey abundance. Fenton et al. (2010) prove that e can be determined from biomass-based functional and numerical responses (Equations 4.2 and 4.3, respectively) and derive Equation B4.3, with parameters as defined in the main text.

$$e = \frac{kr_{max}(k_{1/2} + P)}{I_{max}(k - P')(k - P - P')} \qquad \text{(B4.3)}$$

However, Equation B4.3 assumes that predator biomass is constant, with respect to prey abundance, and as we have seen (section "Cell size, number, and growth rate" on page 105), this is undoubtedly not so (e.g., Figs 4.11 and B4.2c). Consequently, Fenton et al. (2010) modified their analysis to account for changes in predator ingestion rate, growth rate, and biomass with prey abundance. Here I use literature data on *Favella taraikaensis* (from Kamiyama et al. 2005) to illustrate: 1) how this numerical procedure might be applied and 2) some concerns regarding the analysis of data by this procedure.

The data of Kamiyama et al. (2005) provide Fig. B4.2a–c, with the associated equations. Using the

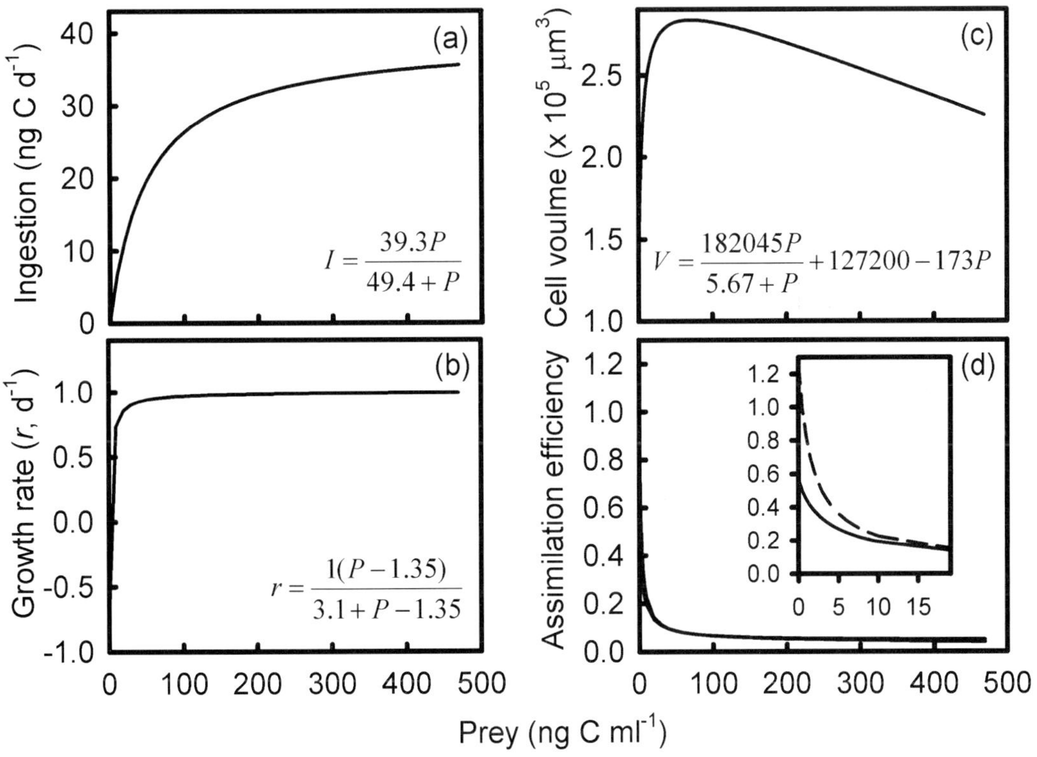

Fig. B4.2 Reponses used to determine the relationship between assimilation efficiency (e) and prey abundance: the functional response (a); the numerical response (b); the volume response to prey abundance (c). The response of assimilation efficiency to prey abundance (d) was obtained from Equation B4.3 and procedures described in this box. Data are from Kamiyama et al. (2005); equations in panels a and b are parameterized from values presented in the source, whereas parameters in the equation in panel c are derived from fitting the phenomenological equation to data presented in the source. The inset on d is the initial portion of the main response in this panel.

Continued

functional and numerical response parameters and by assuming a constant predator volume we can then determine how assimilation (*e*) might vary with prey abundance (Fig. B4.2d, dashed line). However, there are issues with this response in that it indicates $e > 1$ as prey abundance decreases (Fig. B4.2d, insert). We can then use the volume response to correct for changes in biomass with prey abundance by multiplying *e* (from Equation B4.3) by the ratio of predator volume at a particular prey abundance (Fig. B4.2c) and the maximum predator volume (obtained from Fig. B4.2c = $2.84 \times 10^4 \mu m^3$). This provides a predator biomass-corrected estimate of *e* (Fig. B4.2d, solid line). Although this conversion solves one issue (*e* becomes <1), it still reveals a response that convention, and possibly logic, would question: it is unlikely that $e < 0.05$ over much of the observed range.

The underlying reasons for this unlikely result are unclear, but they undoubtedly result from a mismatch between the functional and numerical responses. As can be seen in Fig. B4.2b, growth rate reaches an asymptote far earlier than ingestion. There may be methodological issues here; it is possible that the one-day acclimation period, at each prey abundance, applied by Kamiyama et al. (2005) was insufficiently long. For instance, if the tintinnid had a surplus of food before acclimation, then growth rate at low prey levels may have been higher than would be determined if the ciliate had been acclimated for several days at low prey levels. In fact, the very low threshold level of $1.3 ng C ml^{-1}$ (see Table 4.3) suggests that this may have been an issue. These data thus potentially reveal an important issue. Acclimation time and carefully controlled experiments are essential if we plan to use responses to model ecological and ecophysiological processes.

Alternatively, the unexpectedly low estimate of *e* may result from the choice of prey, *Alexandrium tamarense*, which is relatively large for tintinnid food and is toxic. Such factors may impose energetic costs on the predator (T. Kamiyama, personal communication) and could result in both a slower digestion time and a reduced cell size of the tintinnid at high prey abundances. Clearly, these factors too must be considered.

is effectively deflected away from the flow, as illustrated (Fig. 4.13) by Fauré-Fremiet (1908). Finally, because water is effectively sticky around the ciliate, a tintinnid will move with a "coat" of water around it, and the ciliate will rely on diffusion to attain signals. Thus, unlike many invertebrates that passively filter-feed allowing water to flow to them (e.g., feather worms), tintinnids must constantly move themselves or the water around them to obtain food. As indicated above, for tintinnids, swimming and food acquisition are inextricably linked.

Most of the study of swimming has focused on a single genus: *Favella* (Taniguchi & Takeda 1988, Buskey & Stoecker 1988, 1989, Hansen 1989, Strom et al. 2007, Harvey & Menden-Deuer 2011), so the following description relies heavily on these works. Typically, tintinnids swim in a helix (which may vary in direction rotation, even for individuals: S. Strom, personal communication, S. Menden-Deuer, personal communication) punctuated by turns, during which the ciliate swims backward for a short distance, and reorients itself (Fig. 4.14; Hansen 1989). Tintinnids do not "jump", as do some of their near relatives, the naked oligotrichs (see, for example, Gilbert 1994; Jakobsen 2001); possibly the lorica prevents this movement, which in fact may be a disadvantage, as, relatively, they can be more susceptible to predators (Broglio et al. 2001 and Chapter 5). One wonders then why they have a lorica (but see Chapter 1).

Swimming speeds of *Favella* range from 0.4 to $1.2 mm s^{-1}$, and the only study on another tintinnid (*Metacylis*, Broglio et al. 2001) supports this general speed ($0.8 mm s^{-1}$). This is similar to rates of up to about $1 mm s^{-1}$ for the aloricate oligotrich *Strombidium sulcatum* (Fenchel & Jonsson 1988) but is faster than that of *Strobilidim spiralis*, (= *Pelagostrobilidium spirale*) which moves on average approximately $0.3 mm s^{-1}$, punctuated by about six jumps per minute, which are about 0.4 mm long at a speed of approximately $4 mm s^{-1}$ (Broglio et al. 2001). Apparently, from these limited data, tintinnids do not swim in exactly the same manner as their near relatives.

Besides speed, tintinnid turning behavior also varies, with reorientation, expressed in the literature as degrees per second, ranging around 140 (Buskey & Stoecker 1988, 1989). Decreased swimming speed and increased reorientation occur at higher food concentration, resulting in tintinnids remaining in a prey patch (Buskey & Stoecker 1988). Changes in swimming behavior (speed and number of turns) may also occur when prey produce toxins or have fine spines that may physically disturb the ciliates (Fig. 4.14;

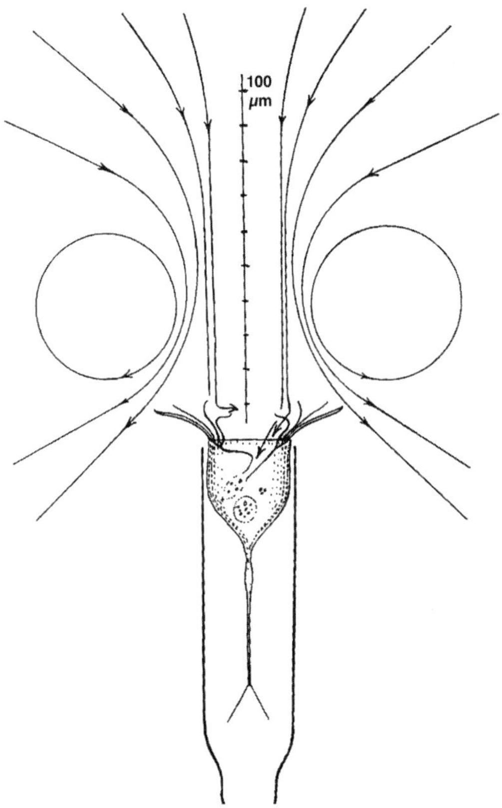

Fig. 4.13 From Fauré-Fremiet (1908). Water flow generated by *Tintinnidium inquilinum* (*Eutintinnus iniquilinus*).

Hansen 1989, Buskey & Stoecker 1988, 1989; Harvey & Menden-Deuer 2011).

An assessment of swimming motion

Recognizing the general swimming behavior, I now provide a speculative analysis of its underlying mechanisms. This description is based primarily on the study of the non-tintinnid oligotrich *Strombidium sulcatum* (Fenchel & Jonsson 1988) and is augmented by work on tintinnids. Fenchel & Jonnson (1988) proposed that swimming is composed of two rotational motions: movement of the ciliate in circles, caused by asymmetry of the oral ciliature, and the rotation of the ciliate around its long axis (Fig. 4.15). The transportation (or displacement rate, v) of the ciliate is then defined by Equation 4.4,

$$v = \frac{\omega_0 r_0}{\left[1+(\omega_0 / \omega_1)^2\right]^{0.5}} \tag{4.4}$$

where, ω_0 is the rotational rate of the ciliate in a circle, r_0 is the radius of the circle when there is no displacement, and ω_1 is the rotation rate of the ciliate around its long axis (Fig. 4.15). When ω_1 is 0, the ciliate swims in a circle, whereas as ω_0 approaches infinity the ciliate swims increasingly in a straight line. Thus, by altering these two attributes, the ciliate may change its dispersal rate, without changing its speed and, therefore,

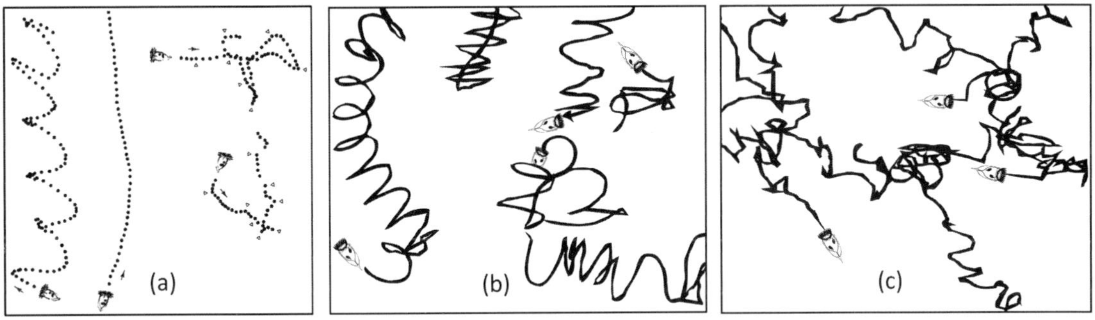

Fig. 4.14 Swimming patterns of *Favella*, redrawn from (a) Hansen (1989) and (b, c) Buskey, E.J. & Stoecker, D.K. (1989) Behavioral responses of the marine tintinnid Favella sp. to phytoplankton: influence of chemical, mechanical and photic stimuli. Journal of Experimental Marine Biology and Ecology, 132, 1–16, with permission from Elsevier. The two tracks on the left of panel a illustrate *Favella* swimming in a relatively tight spiral and a spiral that appears almost to be a straight line; on the right of panel a are two illustrations of the ciliate reversing in the presence of the toxic dinoflagellate *Alexandrium tamarense* (the triangles represent periods of up to 1 s, when the ciliate was immobile). In panel b, *Favella* is swimming in the presence of the prey species *Heterocapsa triquetra*, whereas in panel c it is swimming in the presence of *Thalassiosira* sp. that possesses chitinous threads, which inhibit swimming.

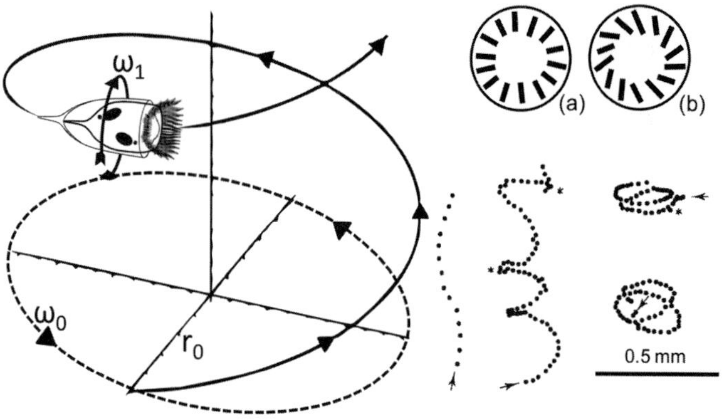

Fig. 4.15 A proposed generalized tintinnid swimming pattern, based on the predicted behavior of *Strombidium sulcatum* by Fenchel & Jonnson (1988). Symbols in the spiral swimming diagram are as described in the text (Equation 4.4). Panels a and b are the theoretical orientations of the oral membranelles when the cell is not rotating around its own axis (a) and when it is (b). The dotted images are traces of swimming patterns for *Strombidium sulcatum*, indicating displacement while swimming in a spiral, similar to that observed for tintinnids (Fig. 4.14).

not necessarily changing the processing of water or feeding rate.

For both *S. sulcatum* (Fenchel & Jonnson 1988) and *Favella taraikaensis* (Taniguchi & Takeda 1988) the adoral membranelles beat in a metachronal fashion around the oral end with a frequency of approximately 10–30 Hz, although for tintinnids held in place, a reduced rate occurs when less favored prey are present (Taniguchi & Takeda 1988), suggesting a reduced swimming speed; through this process, water is forced between the membranes (acting analogously to a peristaltic pump; see Fenchel (1987) for a review of this process), from the inner area (the oral region of the tintinnid) outward (Fig. 4.2). Fenchel & Jonsson (1988) suggest an intriguing explanation: that by changing the angle of the membranelles from a radial arrangement to a more tangential angle (Fig. 4.15a, b), the ciliate will increase rotation around its own long axis (ω_1) and thus change its trajectory following arguments made above (Equation 4.4).

We can also make simple estimates of the maximum clearance rate (see Box 4.1) from predictions of swimming speed and the size of the oral region: assuming the oral region of *Favella* is approximately 100 μm in diameter, the oral area and thus the area of encounter would be about 7900 μm^2, and the volume encountered by a ciliate swimming 1000 $\mu m\, s^{-1}$ would be approximately 30 $\mu l\, h^{-1}$. This is within the same range as measurements of maximum clearance presented in Fig. 4.6, and is quite similar to the estimate of 38 $\mu l\, h^{-1}$ that I have made, above, from data presented in Hansen (1991), but it is lower than the estimate of 131 $\mu l\, h^{-1}$ made by Buskey & Stoecker (1988). Possibly the method of extrapolating the maximum rate made by Buskey & Stoecker (1988) overestimated this value, or the oral encounter area of *Favella* is, in fact, larger than I estimate (e.g., if the diameter was 150 μm, maximum clearance would be more than twice my calculation). Still, this simple prediction suggests a means to support the trend illustrated in Fig. 4.5, offers a further research avenue to pursue, and provides us with a mechanism that may apply to other planktonic ciliates. Furthermore, if we could estimate the handling time of a tintinnid (likely limited by the time to recycle food vacuole membrane, see above), then it might be possible to discard the tedious process of estimating the functional response through measuring grazing rates across an extensive number of prey concentrations and simply predict the functional response from maximum ingestion (I_{max} = 1/handling time) and maximum clearance rate (= I_{max}/k); see Box 4.1.

Changes in swimming behavior

Like many protozoa, tintinnids tend to accumulate in regions of high prey abundance, likely influenced by chemical or physical cues produced by potential prey

(Buskey & Stocker 1988, Verity 1988). The response to accumulate is undoubtedly not through movement toward a recognized source (taxis), as their small size and lack of appropriately spaced sensory structures will reduce the ability of tintinnids to perceive a gradient. Instead, tintinnids will rely on kinesis, a combination of "runs" (spiral swimming) and "tumbles" (reversal and reorientation) leading to the apparent recognition of a gradient. By modifying the length and speed of the run and the frequency of tumbles, tintinnids will alter their dispersal and remain in favorable conditions; for a detailed account of this process in protozoa see Fenchel (1987); Fenchel & Blackburn (1999); Blackburn & Fenchel (1999). Possibly, following logic described above (Fig. 4.15), tintinnids could refrain from tumbles and instead simply swim in circles to reduce dispersal, and thus their options seem to be greater than for some other protozoa. There is good evidence that as conditions become favorable (increased prey), *Favella* exhibits increased turning behavior and reduced speed, and thus reduces its displacement (Buskey & Stoecker 1988).

Reversal of swimming may also occur to avoid conditions, and using this behavior tintinnids have been used to assess the effect of toxic algae on protozoa. When *Favella ehrenbergii* is exposed to the toxic dinoflagellate *Alexandrium tamarense*, it increased the number of "tumbles" and the period of swimming backward increased (Fig. 4.14; Hansen 1989); similar behavior was observed for *Favella taraikaensis* exposed to *Heterocapsa circularisquama* (Kamiyama & Arima 1997). Fenchel (1987) refers to such behavior, a series of quick reversals that does not persist but is followed by resumed swimming, as transient. This behavior allows escape from an environment that has been entered, as if tumbles were to persist the protozoan would remain in the source and never escape! Complementing this behavior, another kinetic response has evolved for some protists: long runs, once a non-favorable environment is experienced; whether tintinnids exhibit this is unknown.

A final directional swimming of tintinnids is negatively geotactic in nature (i.e., they tend to swim upward), which appears to be hydrodynamically driven. Few ciliates possess any structure associated with sensing gravity, although some species have statocyst-like organelles and other use the entire cell "protoplast" as a gravity sensor (Hemmersbach & Braun 2006; Hemmersbach-Krause et al. 1994; Hausmann et al. 2003; Hausmann & Bradbury 1996). However, it is possible that planktonic ciliates reorient when they sink, so that their anterior faces upward (Fenchel 1987). This appears to be the case for oligotrichs in general and for tintinnids specifically; not surprisingly, the mass of the lorica may aid in reorientation (Jonsson 1989). This advantage appears to be reflected in the distribution of tintinnid in field samples, where they were more strongly associated with surface waters than their "naked" close relatives (Jonsson 1989). Possibly a selective advantage of a lorica is that it aids in directional swimming (Kofoid & Campbell 1939)?

Swimming synthesis

Swimming behavior of tintinnids is evidently a complex combination of responses to prey, chemical cues, and gravity. These changes in behavior will have direct and indirect effects on their success. For instance, increased swimming rate will increase contact with prey, and upward migration may bring tintinnids into sunlit regions, where autotrophic prey are more abundant. Ideally, such behaviors should be incorporated into the assessment of their survival and their impact in models (see section "Incorporation of tintinnids into models" on page 118). However, given that most of the research on swimming has focused on relatively few of the myriad of tintinnid species, there is a need to explore taxonomic variation in behaviors across this diverse group. In doing so, we may find scalable traits, such as the relationship between swimming speed and organism size.

4.5 RESPONSE TO ABIOTIC FACTORS AND INTERACTIONS WITH BIOTIC FACTORS

So far in this chapter, the focus has been on how biological factors, primarily prey abundance and type, influence tintinnid behavior and ecophysiology. Now I turn to the few abiotic factors that have been investigated for tintinnids and suggest which others require attention. In doing so, I briefly raise issues associated with interaction between abiotic and biotic factors, and again suggest directions.

Temperature

As tintinnids are globally distributed from the poles to the equator (see Chapter 10), and local waters will change in temperature seasonally and with depth, one of the most important abiotic influences on tintinnids will be temperature. However, we must also recognize that species (and possibly strains) undoubtedly are adapted (i.e., evolved) to local environments, and measurements from one region may not be applicable to others. For instance, as will be seen below, *Favella* sp. obtained from temperate waters grows poorly at 8 °C, at not more than 0.2 d^{-1} (Aelion & Chisholm 1985), whereas *Tintinnopsis* spp. and *Parafavella gigantea*, examined in polar waters, grow well at 5 °C (1.7 and 0.38 d^{-1}, respectively) (Hansen & Jensen 2000).

With this caveat in mind, how do we account for temperature changes? Based on theoretical and empirical chemistry, rate reactions should increase exponentially with temperature, following the Arrhenius function, which is similar, but not identical, to the Q_{10} function (see Schmidt-Nielsen 1990). As I will illustrate below, there is good evidence that neither of these exponential functions are very appropriate, but given that many studies on tintinnids (and other protozoa) rely on Q_{10}, a brief review of the concept is provided below.

The historically used, biology-based temperature quotient is Q_{10}, the change in a rate over 10 °C (Equation 4.5), where R_1, and R_2 are the rates at two temperatures (T_1 and T_2, respectively).

$$Q_{10} = \left(\frac{R_2}{R_1}\right)^{10/(T_2-T_1)} \qquad (4.5)$$

Below a critical temperature reactions will slow to the point where rates decrease rapidly, and above a critical temperature reactions will be inhibited as proteins begin to deform and denature (Schmidt-Nielsen 1990). Generally, however, it is appreciated that there is a region where rates (e.g., growth, ingestion, swimming) increase with temperature (Fig. 4.16a).

Stoecker & Guillard (1982) found no obvious effect of temperature (8, 15, and 20 °C) on the ingestion rate of *Favella* sp. In contrast, work on *Favella* sp. by Aelion & Chisholm (1985) both supports the trend that growth and ingestion rates increase with temperature (Fig. 4.16b, c) and unintentionally illustrates the inappropriateness of determining Q_{10} values from such data, as there is no apparent exponential increase. Like many other studies, these authors still considered estimates of Q_{10} (an exponential function, Equation 4.5 and Fig. 4.16a) between several temperatures (Aelion & Chisholm 1985); this illustrates a common problem: we often, unquestioningly, accept well-established approaches that are blatantly wrong. More recently, the acceptance of Q_{10} was taken to task, and it appears that protozoan growth rate increases linearly over the range where growth increases, with a slope of 0.07 ± 0.005 $d^{-1}\,°C^{-1}$ (Montagnes et al. 2003). Fitting a linear regression though the appropriate portion of data from Aelion & Chisholm (1985) provides a slope of 0.0563 $d^{-1}\,°C^{-1}$ (Fig. 4.16b), well within the range observed by Montagnes et al. (2003), but lower than the mean. It would be inappropriate to speculate that

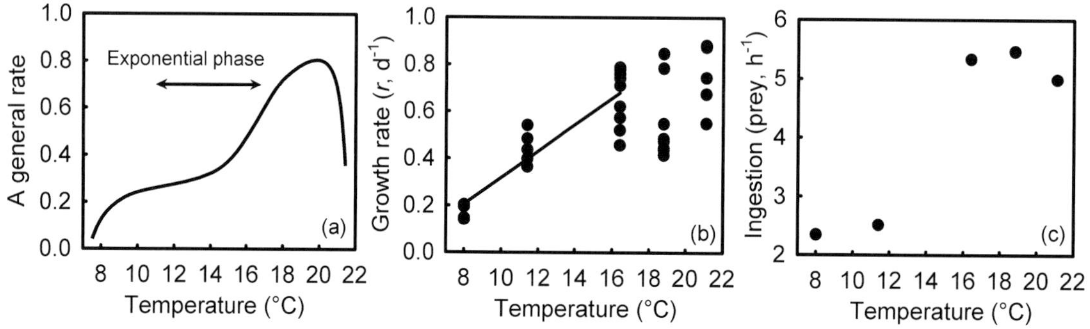

Fig. 4.16 An indication of the influence of temperature on tintinnid growth and ingestion rates. (a) An indication of how general rates are classically expected to increase with temperature, indicating that at high and low temperatures rates decrease, and in a mid-phase (arrow) they increase exponentially, following Equation 4.5. (b, c) The effect of temperature on the growth and ingestion rates of *Favella* sp., based on data from Aelion & Chisholm (1985). The line through the growth data is the best linear fit through the portion where there is an increase in growth rate with temperature (see the text for details on this line).

this one species represents tintinnids. However, given that neither the growth nor ingestion data are shown to increase exponentially (Fig. 4.16), this seems to be yet another area for tintinnid-based study, especially as temperature corrections such as Q_{10} are regularly applied both in meta-analyses (e.g., Hansen et al. 1997) and in ecosystem models (e.g., Blackford et al. 2004).

One other measurement that may vary with temperature is tintinnid cell size. For, protists, in general, Atkinson et al. (2003) have shown that for a 1 °C increase there is a cell-size reduction of 2.5% of the volume observed at 15 °C. To my knowledge, there are no data on tintinnids (neither lorica nor cell size) to indicate that they too follow this trend. Given the arguments above that volume (and hence biomass) changes can be important in ecological calculations (see section "Cell size, number, and growth rate" on page 105), and oral lorica diameter relates to feeding (Section 4.2), this also is an area that requires study. For instance, we might ask if both cell and lorica size are influenced by temperature.

I now turn attention to the interaction between temperature and other factors. There is an interactive affect of temperature and food abundance on a growth and ingestion rates of other heterotrophic protists (see Kimmance et al. 2006; Montagnes et al. 2008a). We might then expect the same to occur for tintinnids. By replotting data from Verity (1985), we can see that changes in both ingestion and growth rates do suggest interactions between temperature and prey abundance on these processes (Fig. 4.17). By parameterizing responses for such data, Montagnes et al. (2008a) were able to predict the influence of temperature shifts on a non-tintinnid ciliate. Possibly the data presented in Fig. 4.17 could be used to the same end.

Other abiotic factors

pH

The influence of several other abiotic factors on tintinnids has been explored, but not to the extent that has been placed on temperature. Twenty years ago we would not have thought to consider the affect of pH on pelagic organisms, but now with CO_2-driven ocean acidification being recognized as a serious threat, low pH has become topical in planktonic studies (e.g., Joint et al. 2011). It is rather ironic, then, that the only pH-based work on tintinnids focuses on raised levels, potentially as a result of phytoplankton blooms removing CO_2 in localized areas. Pedersen & Hansen (2003) indicate that the growth rate of *Favella ehrenbergii* is, relative to several other protists, very sensitive to pH: above 8.8 mortality occurred, and this may be driven by autotrophs in cultures. By extension, in localized bays and fjords, where phytoplankton bloom, this tintinnid may be driven to extinction. Similarly, in cultures, where pH may increase, tintinnids may suffer; this may explain the decline in growth rate and demise of tintinnids observed in cultures by Taniguchi & Kawakami (1983) (e.g., Fig. 4.7). Clearly though, these data do not address our present crisis of decreasing pH. There is thus great scope for the use of tintinnids here, especially as ocean acidification could alter lorica formation, notably of species whose loricae incorporate calcareous matter, such as coccoliths (see Chapters 1 and 2).

Light

The effect of light on tintinnids has also been assessed, to some extent. *Favella* sp. had higher ingestion rates in

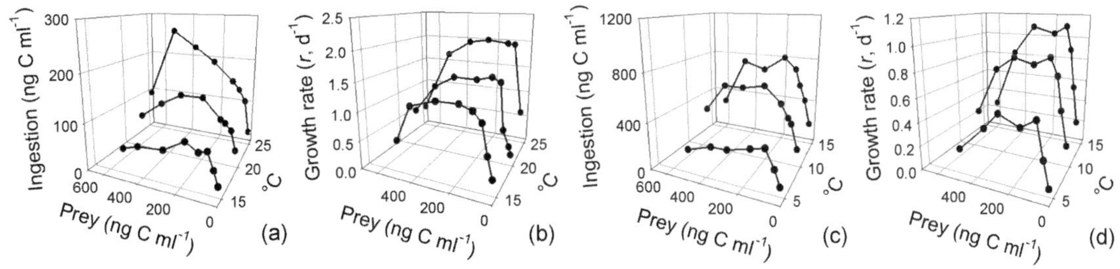

Fig. 4.17 An indication of the interactive affect of temperature and food concentration on the growth and grazing rates of two tintinnids (data replotted from Verity (1985)). (a, b) *Tintinnopsis acuminate*; (c, d) *Tintinnopsis vasculum*.

the light, which Stoecker & Guillard (1982) attributed to the ciliate exploiting prey patches (formed by the positively phototactic prey; see section "Changes in swimming behaviour" on page 110). *Coxliella* sp. also exhibited increased growth and ingestion rates when grown in the light (Strom 2001). Two explanations for this were that increased processing of prey at saturating food levels would reduce residence time, effectively then, as we have seen in Section 4.2, handling time might be reduced. Alternatively, behavioral changes in swimming, such as increased swimming rates in the light, as has been shown for the naked oligotrich *Strombidium reticulatum* (Jonsson 1989), might increase contact with prey, again as indicated in the section on "Changes in swimming behaviour". Increased growth rate might simply be a function of increased ingestion, but Strom (2001) proposes several light-mediated physiological changes that could enhance digestion and thus also improve growth. Possibly, as individual tintinnids may be immobilized for study (Taniguchi & Takeda 1988; Stoecker et al. 1995), it would be feasible to follow the rate of food vacuole processing under various light levels, to test such hypotheses.

Trace metals and salinity

Trace metals have long been recognized to affect oligotrich growth rate, with low levels being essential and higher levels causing inhibition (e.g., Gifford 1985). It is, therefore, not surprising that levels of copper and zinc above those typically found in open waters had deleterious effects on the swimming and growth rate of *Favella* sp. (Stoecker et al. 1986). However, it is also noteworthy that there was interaction, as these two metals will likely compete in binding at cell sites, with increased zinc levels resulting (under some conditions) in a reduction of copper toxicity on *Favella* sp. (Stoecker et al. 1986). These data clearly illustrate the importance of multivariate analysis, when appropriate and possible.

Salinity is the single, major abiotic influence on tintinnids that seems to have been generally overlooked, with some notable exceptions: for example, in a laboratory study, Jonsson (1989) found that several tintinnids were not limited in their distribution by a 4–6 practical salinity unit (PSU) halocline; Stoecker et al. (1984) found no correlation between salinity and the distribution of *Favella* sp. in a small estuary. However, given that tintinnids reside in estuaries (Chapter 10), the little attention that has been paid to salinity appears to be a glaring oversight. Ironically, it may be the, apparent, euryhaline distribution of tintinnids (see, for example, Dolan & Gallegos 2001) that has deterred ecologists from pursuing work in this direction. Still, I contest that it is the tintinnids' apparent ability to survive over a great range of salinities that should stimulate behavioral research across their observable ranges. Such data could be instrumental in assessing competitive advantages. In fact, a recent study (Harvey & Menden-Deuer 2011) proposed that strong haloclines could prevent *Favella ehrenbergii* from occurring where the toxic flagellate *Heterosigma akashiwo* occurs; this would then benefit the flagellate, potentially allowing it to bloom. Clearly, as I seem to reiterate, there is much scope here.

4.6 TINTINNIDS AS MODELS AND IN MODELS

General models

At the start of this chapter, I indicate that tintinnids are not the major group of planktonic ciliates. Furthermore, as we have seen, their lorica makes tintinnids functionally and behaviorally atypical of most planktonic ciliates (Chapter 1). Why, then, is their study useful, in an ecophysiological and behavioral context? Of course, understanding tintinnid biology is useful, as they do represent a component of the plankton (Chapters 1 and 9), but here I raise the question: can they act as a model for all planktonic ciliates?

To answer this, it might be instructive to assess first the role of model organisms. Montagnes et al. (2012) propose several criteria that make good models; by using a subset of these traits, we can consider if tintinnids are useful in this context (Table 4.4). In general, based on these criteria, I suggest that tintinnids have substantial advantages over their naked relatives as models, with the caveat that their loricae make them distinct and, therefore, limits some aspects of study. Thus, recognizing the guidance of Box & Draper (1987), "Remember that all models are wrong; the practical question is how wrong do they have to be to not be useful.", we can conclude that tintinnids are good models of planktonic ciliates.

Table 4.4 Tintinnids as model planktonic ciliates?

A model species should be:	Level to which tintinnids possesses this trait	
	Positive	**Negative**
Easier to study than the target it is modelling	• Generally this seems to be so, as tintinnids are as easy or easier to culture than their naked relatives.	
Relevant • ecologically • behaviorally	• They are globally distributed and at times are abundant, even forming blooms (Chapter 9). • The behavior of tintinnids is relatively easy to study, as the lorica can be held in place	• They generally represent a minority of the oligotrichs, the major group of ciliates (see Chapter 9) • Their behavior may differ from other ciliates, when the lorica acts as a confounding factor
Similar in size to the target	• They cover the same size range as the naked oligotrichs	
Inexpensive to obtain, maintain, and practical to use	• They can be collected all over the world, and cultures are relatively easily maintained	
Not rare or threatened	• Given the wide distribution and abundance, tintinnids are generally not thought to be rare or threatened	
Easy to identify	• They have distinctive morphology based on lorica, and this taxonomy is well established	• The lorica provides an inadequate means of identification at times (see Chapter 3)
Easy to culture	• Methods to culture them have been well established (Gold 1970; Gifford 1985; Kemp et al. 1993)	
Easy to manipulate in experiments (e.g., robust to manipulation)	• Generally they are easy to manipulate, as the lorica means that they move slowly and makes them more robust than naked olgotrichs	
Able to survive storage (e.g., cryopreservation)	• They produce cysts, which may in the future be used to store cultures	• The use of cysts for storage has not yet be developed and they cannot be cryopreserved
Fecund, with rapid generation times	• They have up to about two divisions per day, which is reasonably rapid for protozoa of its size	• Their life cycles are poorly understood

Continued

Table 4.4 *Continued*

A model species should be:	Level to which tintinnids possesses this trait	
	Positive	**Negative**
Able to produce clonal lines (to separate environmental from genetic effects)	• Their ability to do this is well established	
Stable in characteristics over many generations		• This has yet to be rigorously tested
Useful for interdisciplinary (e.g., molecular–ecological–numeric modelling) use	• The vast study of them over the past approximately 50 years, coupled with modern methods, suggests that they could be used for interdisciplinary work	
Long-standing, with a history of use and substantial background	• There is over 100 years of research on them	
Unemotive (e.g., not being a target for animal rights)	• They do not tend to elicit emotive responses commensurate with acts of animal cruelty	

Favella

Often, rather than a group of organism being the focus of study, one taxon emerges as a model of robust use. Classic examples of organisms used in an interdisciplinary manner include the white rat (mammals), *Daphnia* (freshwater zooplankton), and *Escherichia coli* (bacteria); we then might hope to find one such tintinnid. Undoubtedly, the reader has recognized that one taxon has arisen repeatedly in studies cited in this chapter: *Favella* (Figs 4.18 and 4.19). I now turn attention to *Favella* and assess the extent to which it represents a useful model of tintinnids and possibly for planktonic ciliates in general.

Again, we can rely on Table 4.4 as a guide to indicate the use of *Favella*. This genus ranges in length from approximately 30 to 300 μm, covering much of the range of tintinnids and other naked ciliates. Its lorica is hyaline, possibly limiting some aspects of study to this sub-group of tintinnids but allowing the cell to be seen when it is retracted; this makes observations such as food capture and food vacuole processing tractable when it is held in position (Taniguchi & Takeda 1988; Stoecker et al. 1995). Furthermore, species of *Favella* are widely distributed in neritic waters almost from pole to pole (Chapter 10), and it is relatively easy to isolate, manipulate, and culture using established methods (see papers cited in the legend to Fig. 4.19), making its study pragmatic.

Favella may be one of the dominant planktonic ciliates, at times (Stoecker et al. 1984), making it ecologically relevant. The relatively simple lorica shape (Fig. 4.18), its well-defined oral and somatic ciliature (Kim et al. 2010), and its general cell shape also make it useful to examine behavioral traits. Furthermore, studies have revealed that *Favella* undergoes conjugation and forms cysts (Laval-Peuto 1983; Stoecker et al. 1983; Chapter 7), allowing these processes to be studied. In fact, possibly the most detailed studies on tintinnid encycstment and excycystment have focused on *Favella* (Chapter 7). Similarly, *Favella* has been a focal species for the study of tintinnid parasitism (Chapter 6); again the hyaline lorica allows observations.

Identification of *Favella* can be problematic as lorica shape is not necessarily reliable and morphological identification is taxing (Chapter 3). Nevertheless, there are both historic and recent studies identifying key species (see Kim et al. 2010), and the National Center for Biotechnology Information (NCBI) GenBank data

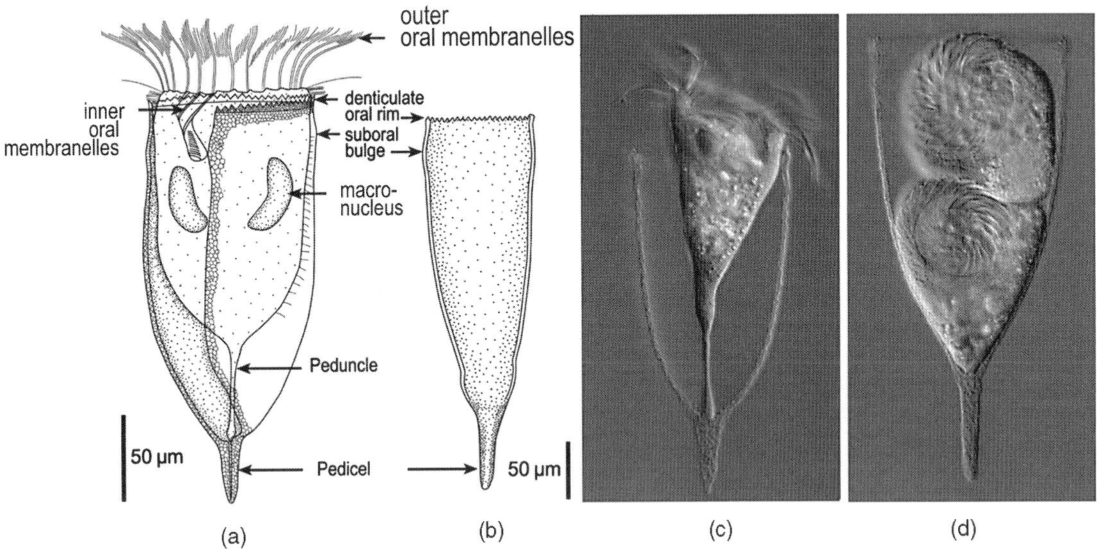

Fig. 4.18 An indication of *Favella*, a model tintinnid, based on images of *Favella serrata*. (a, b) from http://www.liv.ac.uk/ciliate/Datasheets/f_serr.htm; (c, d) with permission from F. Neidl.

set (September 2011) provides 37 sequences: 17 for *Favella ehrenbergii*; 17 for *Favella* sp. and one for each for *Favella taraikaensis*, *Favella campanula*, and *Favella panamensis*. However, it should be noted that the genus might be polyphyletic (see Chapter 3). Furthermore, it is becoming apparent that many protozoan species exhibit strong differences between clones (e.g., this is the case for the model protozoan *Oxyrrhis marina* (Montagnes et al. 2011)); similarly, unpublished work by D. Stoecker (personal communication) revealed that *Favella* clones isolated from the same sample exhibited different thermal optima for growth. Consequently, for this genus (and other tintinnids), researchers must recognize clonal variability.

Storage of *Favella* culture is also problematic, as it is with most planktonic ciliates. Cultures tend to last for months, rather than years, and then require replacement as they lose vitality (D. Stoecker, personal communication), presumably because of clonal decline (see section "An overview of tintinnid growth" on page 99). However, for those researchers located by the coast, new cultures are readily accessible, by enriching natural samples (collected with a fine-mesh net) with appropriate prey (see Table 4.1). Finally, *Favella* has been used in a range of studies over the past 35–40 years, indicating its long history of use, the extensive data set on it, and its potential interdisciplinary role in the study of microzooplankton (Fig. 4.19).

How well does *Favella* relate to naked planktonic ciliates in general? Immediately, the issue of possessing a lorica makes any tintinnid, including *Favella*, inappropriate for studying all aspects of naked ciliates. For instance many naked non-loricate ciliates are capable of jumping (or at least moving rapidly for short distances), and this is especially so for the "Strobilidiid-like" ciliates. As this swimming behavior allows naked ciliates to escape predation (Jakobsen 2001), *Favella* would clearly be a poor model to investigate escape responses of this sort. In contrast, *Favella* would be an ideal "control", to compare with jumping ciliates. The oral ciliature of the tintinnids is similar to the choreotrichs but differs from that of the other major planktonic group, the oligotrichs (the former has a closed circle of adoral membranelles, whereas the latter has a C-shaped arrangement, see Chapter 3). Thus, *Favella*, and in fact all tintinnids, would probably be inappropriate to provide a detailed model of water flow associated with the oral region of oligotrichs. Finally, *Favella* cysts are similar in morphology to the few naked planktonic ciliate cysts that have been studied (Chapter 7).

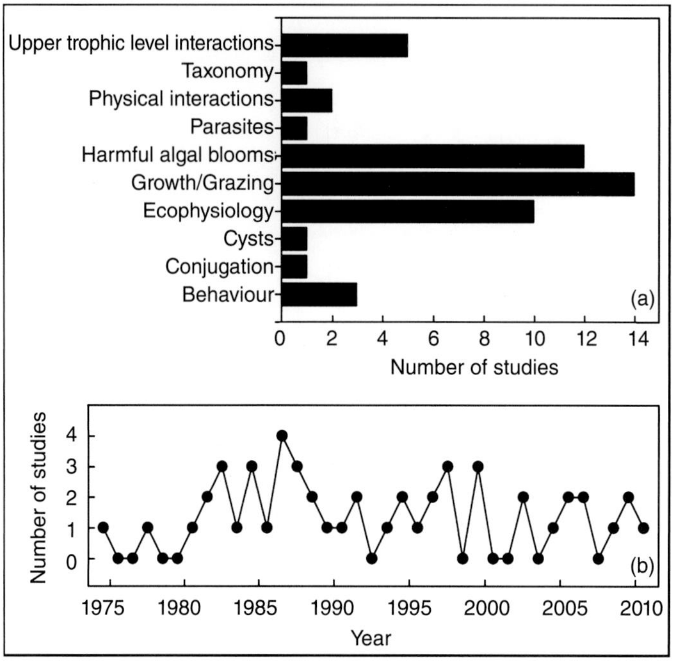

Fig. 4.19 An indication of the breadth and depth of study on the model genus *Favella*. (a) Fifty major studies on *Favella* were each placed in a single general category that best represented the work, although there may be overlap with other categories. (b) Number of studies per year were plotted over the past three and a half decades, to indicate the constant, but relatively low-level, effort of work on *Favella*. Data are from Aelion & Chisholm (1985); Ayukai (1987); Buskey & Stoecker (1988, 1989); Carlsson et al. (1990); Coats et al. (1994); Fredrickson & Strom (2009); Fulco (2007); Graham & Strom (2010); Hansen (1989, 1995); Hansen et al. 1991); Harvey & Menden-Deuer (2011); Kamiyama (1997, 2000); Kamiyama & Suzuki (2006); Kamiyama et al. (2006); Kim et al. (2010); Kuwahara et al. (1975); Laval-Peuto (1983); Maniero et al. (2000); Nagano et al. (2000); Pedersen & Hansen (2003); Rassoulzadegan (1978); Rosetta & McManus (2003); Shimeta et al. (1995); Stoecker (1984, 1988); Stoecker & Egloff (1987); Stoecker & Evans (1985); Stoecker & Guillard (1982); Stoecker et al. (1981, 1983, 1986, 1987a, b, 1995); Strom & Loukos (1998); Strom et al. (1998, 2007); Suzuki & Taniguchi (1995); Taniguchi & Takeda (1988); Taniguchi & Kawakami (1983, 1985); Turner & Tester (1992).

It, therefore, seems that with some caveats, *Favella* could be a useful model to assess general traits. We can also determine some of the areas that may require further study by examining the breadth and depth of work on this genus over the last three and a half decades (Fig. 4.19); this should provide guidance and encouragement for future studies.

Incorporation of tintinnids into models

Given the long history of study and relatively extensive data collected on tintinnids, it is not surprising that they have been used to inform pelagic ecosystem and population models (e.g., Buitenhuis et al. 2010). Typically microzooplankton (organisms between 20 and 200 μm; Chapter 9) have been represented by a single box (see, for example, Fig. 4.20), although more detailed exceptions exist, where the microzooplankton are recognized as too complex to be adequately modeled by a single component (e.g., Stickney et al. 2000). Here, I explore the extent to which we can use tintinnid data to parameterize the single microzooplankton box models.

In models similar to that depicted in Fig. 4.20, the microzooplankton, and for our purposes the tintinnids (T) and their prey (P), can be represented by the following differential equations:

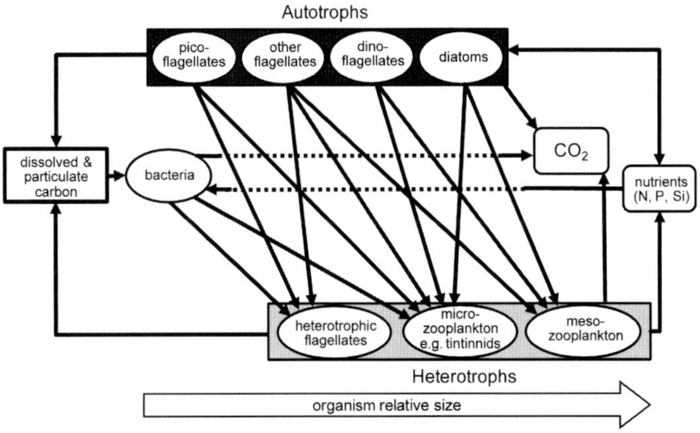

Fig. 4.20 An example of a schematic model of the pelagic ecosystem: the ERSEM model, redrawn from Blackford, J.C., Allen, J.I & Gilbert, F.J. (2004) Ecosystem dynamics at six contrasting sites: a generic modelling study. Journal of Marine Systems, 52, 191–215, with permission from Elsevier.

$$\frac{dP}{dt} = \text{prey growth} - \text{tintinnid ingestion}$$

$$\frac{dT}{dt} = \text{tintinnid growth} - (\text{mortality} + \text{respiration})$$

These two equations are often parameterized based on the Rosenzweig–MacArthur model (Turchin 2003), as:

$$\frac{dP}{dt} = \mu P\left(1 - \frac{P}{K}\right) - \frac{I_{\max}P}{k_{1/2} + P}T \tag{4.6}$$

$$\frac{dT}{dt} = e\frac{I_{MAX}P}{k_{1/2} + P}T - \delta T \tag{4.7}$$

In this simple couplet of differential equations, prey growth is represented by logistic growth (μ is the maximum prey growth and K is the prey carrying capacity); clearly in more complex models (e.g., Fig. 4.20, but for an overview also see Miller (2004)) prey growth will be governed by other factors, such as nutrient availability, and the recycling of nutrients by tintinnids may play a part in this. Prey loss in Equation 4.6 is a function of tintinnid ingestion, which the astute reader will recognize as the product of a type II functional response (Equation 4.2) and the tintinnid population (T). Tintinnid population growth (Equation 4.7) is dictated by the fraction of ingested prey that is assimilated (i.e., the product of assimilation efficiency, e, and the type II functional response) minus the proportion of the population that dies, owing to a constant mortality rate (δ). In some models this loss of tintinnids may be divided into two components, mortality and respiration, whereas in larger models an added loss term would be imposed on the tintinnids by "top-down control" by grazing of mesozooplankton (see, for example, Blackford et al. 2004; Chapter 5) or potentially by parasites (see Montagnes et al. 2008b; Chapter 6). Typically, in such plankton models immigration and emigration (the other two drivers of population abundance (Turchin 2003)) are ignored.

As can be seen in the previous sections of this chapter, there are good estimates of the type II functional response, for a range of tintinnids. There are also some estimates for tintinnid assimilation efficiency (e), although these are few (see section "Gross growth and assimilation efficiency" on page 105), and as indicated in Box 4.2, e may vary with prey abundance. In contrast, estimates for tintinnid mortality are non-existent (and potentially will vary with prey abundance, as has been observed for other another ciliate (Minter et al. 2011)), and there are few estimates of their respiration (see Verity 1985). Consequently, Equation 4.7 will be poorly parameterized.

Recently, Fenton et al. (2010) have formally recognized another option, which admittedly has been applied to planktonic ciliates for over a decade (see, for example, Montagnes & Lessard 1999). Rather than relying on estimates of e and δ, it might be far simpler to directly measure growth and mortality of tintinnids and establish a numerical response (Equation 4.3). Equation 4.7 could then be replaced by Equation 4.8, which incorporates Equation 4.3.

$$\frac{dT}{dt} = T\frac{r_{max}(P-P')}{k+P-P'} \tag{4.8}$$

There are issues associated with using these two equations. Possibly the most disconcerting of these is the parameterization of the functional and numerical responses. If the two are not carefully determined, it is possible to develop a set of equations (Equations 4.6 and 4.8) that fail to adhere to mass conservation (e.g., more biomass could be made than is possible, given consumption); this parallels issues associated with estimating assimilation efficiency (Box 4.2). Consequently, there is a need to carefully determine functional and numerical responses under similar, well-acclimated conditions, to prevent the generation of unrealistic data.

Above, I have emphasized the potential for using tintinnids to parameterize large-scale ecosystem models such as that depicted in Fig. 4.20. However, protozoa, and specifically tintinnids may also be used to parameterize relatively simple models that test fundamental principles, associated with plankton dynamics and general food-web theory. For instance, by solving differential equations, based on a three-compartment food web that was initially studied in the laboratory (Fig. 4.21), Stoecker & Evans (1985) indicated that increasing the rate of tintinnid predation on the intermediate predator (*Balanion*) may increase stability (persistence) of the system. In another study, Strom & Loukos (1998) observed through laboratory work that *Favella* altered its preference for prey, in response to changing ratios of prey in incubation chambers. They then extended the empirical study by using modelling methods, akin to those described above (Equations 4.6 and 4.7) but with modifications, including terms to account for prey preference. Through these efforts Strom & Loukos (1998) revealed that switching of prey preference based on selectivity (see, for example, section "Selectivity other than size" on page 98) led toward stability (i.e., providing a refuge for preferred prey at low abundances). A different example of a model, derived from laboratory studies on a tintinnid supports arguments made above (see section "Numerical response" on page 101) that ciliates, and specifically tintinnids may control prey blooms of the harmful alga (Kamiyama et al. 2005). Using their functional and numerical responses (also presented in Box 4.2) and assuming a constant growth rate of the harmful dinoflagellate *Alexandrium tamarense*, these authors modeled bloom dynamics and indicated, among other issues, that, if *A. tamarense* concentration exceeds allowable limits for harvesting shellfish, and if there are sufficient *Favella taraikaensis* (within realistic levels found in the field), then tintinnid grazing may reduce toxicity of bivalves that consume *A. tamarense* influences.

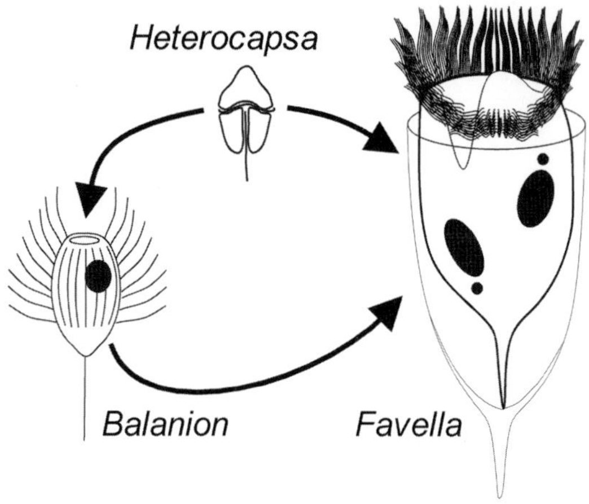

Fig. 4.21 The three-compartment food web studied by Stoecker & Evan (1985), consisting of the tintinnid *Favella* sp., the non-tintinnid ciliate *Balanion* sp., and the autotrophuic dinoflagellate *Heterocapsa triquetra*. Arrows represent the trophic integrations.

Thus, tintinnid-based models offer substantive scope in their ability to address a range of ecological issues. Given the wealth of data on tintinnids and their relative tractability as laboratory organisms, they should lend themselves to further investigation to this end. Specific issues that we might consider addressing are as follows: (1) simultaneously establishing coupled functional and numerical responses, to apply Equations 4.6 and 4.8 in models; (2) establishing the influence of biological factors such as prey switching, multiple trophic levels (e.g., Fig. 4.21), and changes in swimming/searching behavior, and incorporate these into models; (3) establishing the influence of abiotic factors, such as temperature, pH, and salinity on responses; and, possibly most difficult, (4) assessing interactive effects of multiple factors on tintinnid responses. In doing so, tintinnids, and specifically key taxa such as *Favella*, could provide us with useful insights in to the role of microzooplankton in planktonic ecosystems.

4.7 KEY POINTS

1. Feeding is a complex process in tintinnids, composed of several steps that include behavioral traits such as handling and prey rejection.
2. Although it is complex, tintinnid feeding can be characterized by the functional response (ingestion rate versus food abundance) that follows a typical type II response, which provides useful information about their biology and can be incorporated into food-web models.
3. Attributes associated with tintinnid feeding can be correlated to the oral diameter of the lorica, potentially providing a metric for scaling responses.
4. Tintinnid feeding selectivity is based both on size and other characteristics such as prey shape, movement, and composition.
5. It is important to recognize that increase in numbers (reproduction) and sex are not linked in tintinnids (as is the case for most protozoa).
6. The tintinnid growth cycle is potentially complex, including cyst formation and clonal decline.
7. The numerical response (growth rate versus food abundance) can be established for tintinnids and provides useful information about their biology and can be incorporated into food-web models.
8. Like feeding attributes, some growth attributes of tintinnids may be scaled to the oral lorica diameter.
9. Not only do growth rate and feeding rate change with food abundance, tintinnid cell size changes with food abundance; this can complicate measurements of tintinnid biomass production.
10. Both gross growth efficiency and assimilation efficiency may change with food abundance, and these relationships need to be recognized when examining tintinnids.
11. Tintinnid swimming behavior changes owing to a variety of factors (e.g., food abundance, type, toxicity), and these behavioral changes may alter their feeding and thus their growth rates.
12. A range of abiotic factors affect tintinnid feeding and growth; the best studied of these is temperature, but none (pH, salinity, toxins) have been sufficiently examined.
13. Interaction between biotic and abiotic factors influences tintinnid feeding and growth and undoubtedly affects other processes.
14. Tintinnids may make good model organisms for representing planktonic ciliates in general.
15. One genus, *Favella*, stands out as a model tintinnid.
16. Tintinnids can be used to parameterize large- and small-scale food-web models, and there is a need to improve estimates of several of the processes and behaviors (outlined above) to allow them to be better used in these models.

ACKNOWLEDGMENTS

I thank the co-authors of this book for their comments. I also thank Klaus Hausmann, Denis Lynn, Tom Fenchel, and others for their comments and support.

Chapter 5

PREDATORS OF TINTINNIDS

Diane K. Stoecker

University of Maryland Center for Environmental Science, Horn Point Laboratory, Cambridge, MD, USA

The Biology and Ecology of Tintinnid Ciliates: Models for Marine Plankton, First Edition. Edited by John R. Dolan, David J.S. Montagnes, Sabine Agatha, D. Wayne Coats, and Diane K. Stoecker.

5.1 INTRODUCTION

Tintinnids, as members of the nanoplankton and microplankton (Chapter 9), are prey of suspension-feeding zooplankton and benthic invertebrates. Most investigations of feeding by zooplankton and benthos have focused on grazing on phytoplankton, but laboratory and field data show that protistan nano- and microzooplankton can be an important component of the diet suspension-feeders (reviewed in Stoecker & McDowell Capuzzo 1990; Gifford 1991; Pierce & Turner 1992; Sanders & Wickham 1993; Calbet & Saiz 2005). Small raptorial predators, such as fish larvae, also consume microzooplankton (Montagnes et al. 2010). In the past 30 years, the contribution of microzooplankton, including ciliates, to planktonic biomass in surface waters has been quantified. Ciliates are an important component of the prey field available to zooplankton, fish larvae, and benthic invertebrates, particularly when phytoplankton biomass is low or dominated by small size (<10 μm) cells (Chapter 9). However, information on the role of protistan microzooplankton, and in particular tintinnids and other ciliates, in the diet of most zooplankton, icthyoplankton, and benthic invertebrate species is either lacking or semi-quantitative.

Experimental data on predation on tintinnids are rare. There are limited data on ingestion and clearance rates of tintinnids by copepods (mostly *Acartia* spp.), ctenophores (comb jellies), and some cnidarians (jellyfish) from experiments using natural assemblages containing tintinnids or cultured tintinnids. There are few or no experimental data on predation by most types of protistan and metazoan zooplankton (Tables 5.1–5.5), most species of fish larvae (Table 5.6), and most suspension-feeding benthic invertebrates (Table 5.7) on tintinnids.

The evidence for ingestion of planktonic ciliates by most predators and grazers is the observation of intact tintinnids or tintinnid loricae in gut or fecal pellet contents. However, lack of tintinnids in gut and fecal pellet contents does not always indicate lack of predation on tintinnids. Some protistan and invertebrate predators extract the tintinnid cell from the lorica before ingestion (Stoecker & Sanders 1985; Tackx et al. 1995; Uchida et al. 1997; Smalley et al. 1999). Whether or not ingestion of the lorica occurs probably depends on the size and species of zooplankton as well as size and type of lorica. On the other hand, presence of empty loricae in gut contents or fecal pellet contents does not always indicate predation, especially in detritivores that ingest organic aggregates, which may contain empty loricae. This may particularly be a problem in evaluating gut or fecal pellet contents of deep-sea zooplankton and benthic invertebrates that feed on sinking organic material (Harding 1974; Gowing & Wishner 1986, 1992; Steinberg 1995).

To estimate the contribution of tintinnids to the diet based on gut contents, it is critical to have information on digestion times, which vary widely for different prey types. For example, ctenophores (comb jellies) digest aloricate ciliates in a minute, take about 30 minutes to digest tintinnids and up to 2.2 hours to digest adult copepods (Sullivan 2010; Granhag et al. 2011). As a result, gut contents can under-represent the contribution of tintinnids to diets compared with larger or more robust prey such as crustacean zooplankton and diatoms but over-represent their contribution compared with aloricate ciliates and other "naked" microzooplankton. The problems in determining diet from fecal pellet contents are similar. For example, pieces of copepod carapace, diatom frustules, and loricae are often preserved, but aloricate ciliates and athecate dinoflagellates are almost never recognizable in fecal pellet contents. Thus, our knowledge of the diets of marine organisms based on gut or fecal pellet contents is biased.

The application of molecular methods to detection of tintinnids and other ciliates in the diet of zooplankton and other suspension feeders is in its infancy. Ohman et al. (1991) showed that immunochemical techniques could detect predation by fish larvae on an aloricate ciliate in the laboratory. More recently, quantitative polymerase chain reaction (PCR) has been applied to detection or estimation of copepod feeding on specific phytoplankton species (Nejstgaard et al. 2008; Simonelli et al. 2009; Haley et al. 2011). Sequences are now available for several tintinnids (Chapter 3) as well as for aloricate ciliates common in the marine plankton (Doherty et al. 2010; Tamura et al. 2011) and thus it should be possible to apply quantitative PCR to estimation of predation on particular ciliate taxa. However, although differential digestion can still be a problem with molecular techniques just as it is with traditional gut contents analyses (Troedsson et al. 2009), application of molecular methods for prey detection should yield new insights into predation on tintinnids and specifically on the relative importance of tintinnids and aloricate ciliates in the diets of many predators.

Table 5.1 Protistan predators of tintinnids.

Predator	Tintinnid prey	Evidence	Reference
Dinoflagellates			
Gyrodinium instriatum	*Favella azorica*, *F. taraikaensis*, *Eutintinnus tubulosus*	Observations of feeding	Uchida et al. 1997
Ceratium furca	*Unidentified tintinnids*	Food vacuole contents	Smalley et al. 1999
Noctiluca scintillans	*Tintinnopsis* spp., *Codonellopsis* spp., *Tintinnidium* spp., *Tintinnidium muciocola*, *Stenosemella nivalis*	Food vacuole contents	Andersen & Sørensen 1986; Fonda-Umani et al. 2004; Sato et al. 2010
Other ciliates			
Favella panamensis	*Tintinnopsis tubulosa*	Laboratory feeding experiment	Robertson 1983
Favella serrata	*T. beroidea*	Food vacuole contents	Lebour 1922
Favella sp.	*T. minuta*, *T. acuminata*	Food vacuole contents	Verity 1986b
Radiolarians			
Collozoum caudatum	Unidentified tintinnids	Loricae in fecal strands	Swanberg & Anderson 1981
Collozoum longiforme	Unidentified tintinnids	Loricae in fecal strands	Swanberg & Harbison 1980; Swanberg & Caron 1991
Diplosphaera sp.	Unidentified tintinnids	Tintinnids in pseudopodial network	Anderson 1983
Hexastylus sp.	Unidentified tintinnids	Tintinnids in pseudopodial network	Anderson 1983
Spongodrymus sp.	Unidentified tintinnids	Tintinnids in pseudopodial network	Swanberg et al. 1986, Anderson 1983
Physematium muelleri	Unidentified tintinnids	Tintinnids in pseudopodial network	Swanberg et al. 1986, Swanberg & Anderson 1985
Phaeodarian radiolarians	Unidentified tintinnids	Empty loricae but not cellular remains in food vacuoles	Gowing & Béntham 1994
Planktonic Foraminifera			
Globigerinoides sacculifer	Unidentified tintinnids	Tintinnids in pseudopodial network	Caron & Bé 1984, Anderson 1993

Table 5.2 Copepod predators of tintinnids.

Predator	Tintinnid prey	Evidence	Reference
Calanoid			
Acartia clausi	*Favella tarakaensis*, *Helicostomella fusiformis*, *Metacylis* sp.	Feeding experiment	Ayukai 1987; Broglio et al. 2001
Acartia hongi	*Eutintinnus sp.*	Feeding experiment	Yang et al. 2010
Acartia hudsonica	*Eutintinnus pectinis*	Feeding experiment	Turner & Anderson 1983
Acartia tonsa	*Favella panamensis*, *Favella sp.; Eutintinnus* sp. *Tintinnopsis* sp., *T. beroidea*, *T. brasiliensis*, *T. parva*, *T. tubulosa*	Feeding experiment	Robertson 1983; Stoecker & Sanders 1985; Stoecker & Egloff 1987; Gifford & Dagg 1988, 1991; Tackx et al. 1995; Vargas & González 2004; Diodato & Hoffmeyer 2008
Acartia spp.	*Eutintinnus sp.* , *Codonellopsis sp.*	Feeding experiment	Rollwagen-Bollens & Penry 2003
Aetideopsis antarctica	*Cymatocylis antarctica* *Laackmanniella naviculaefera*, *Codonellopsis gaussi*,	Gut contents	Hopkins 1987
Calanus australis	Unidentified tintinnids	Feeding experiment	Sanchez et al. 2011
Calanus finmarchicus	*Tintinnus subulatus*, *Stenosemella ventricosa* *Helicostomella subulata*	Gut contents	Marshall 1924
Calanus cristatus, *C. plumchrus*	Unidentified tintinnids	Gut contents	Arashkevich 1969
Calanus pacificus	Unidentified tintinnids	Feeding experiment	Leising et al. 2005a
Calanus tenuicornis	Unidentified tintinnids	Fecal pellet	Turner 1984a
Calanus propinquus	*Codonellopsis gaussi*, *Laackmanniella naviculaefera*, *Cymatocylis antarctica*	Gut contents	Hopkins 1987
Centropages typicus	*Helicostomella subulata*	Gut contents	Lebour 1922
Eucalanus bungii	Unidentified tintinnids	Gut contents	Arashkevich 1969
Eucalanus pileatus	*Salpingella* sp.	Fecal pellet	Turner 1984b
Euchaeta antarctica	*Codonellopsis gaussi*, *Laackmanniella naviculaefera*, *Cymatocylis antarctica*	Gut contents	Hopkins 1987
Euchaeta erebi	*Codonellopsis gaussi*, *Cymatocylis antarctica*	Gut contents	Hopkins 1987
Euchaeta similis	*Cymatocylis antarctica*	Gut contents	Hopkins 1987
Gaedius pungens	Unidentified tintinnids	Gut contents	Arashkevich 1969
Gaetanus pileatus			

Continued

Table 5.2 *Continued*

Predator	Tintinnid prey	Evidence	Reference
Metridia spp.	Unidentified tintinnids	Feeding experiment	Lonsdale et al. 2000
Metridia gerlachei	*Codonellopsis gaussi, Laackmanniella naviculaefera, Cymatocylis antarctica*	Gut contents	Hopkins 1987
Metridia ornata, Metridia ochotensis	Unidentified tintinnids	Gut contents	Arashkevich 1969
Oncaea curvata	*Codonellopsis gaussi*	Gut contents	Hopkins 1987
Paracalanus quasimodo	Unidentified tintinnids	Fecal pellet	Turner 1984b
Pleuromamma xiphas	Unidentified tintinnids	Gut contents	Arashkevich 1969
Pseudocalanus newmani	Unidentified tintinnids	Feeding experiment	Leising et al. 2005b
Pseudochirella polyspina	Unidentified tintinnids	Gut contents	Arashkevich 1969
Tortanus setacaudatus	*Favella panamensis*	Feeding experiment	Robertson 1983
Cyclopoid			
Limnoithona tetraspina	Unidentified tintinnid	Feeding experiment	Bouley & Kimmerer 2006
Oithona similis	*Codonellopsis gaussi*; Unidentified tintinnids	Gut contents Feeding experiment	Hopkins 1987; Nishibe et al. 2010
Oithona spp.	Unidentified tintinnids	Feeding experiment	Lonsdale et al. 2000

Table 5.3 Euphausiid and other large crustacean zooplankton predators of tintinnids (all evidence from gut contents).

Predator	Tintinnid prey	Reference
Euphausiids		
Bentheuphausia amblyops	Unidentified tintinnids	Mauchline 1980
Euphausia diomedeae	Unidentified tintinnids	Mauchline 1980
Euphausia crystallorophias	*Codonellopsis gaussi, Laackmanniella naviculaefera, Cymatocylis antarctica*	Hopkins 1987
Euphausia pacifica	Unidentified tintinnids	Mauchline & Fisher 1969; Nakagawa et al. 2001
Euphausia similis	Unidentified tintinnids	Mauchline 1980
Euphausia superba (Antarctic krill)	Unidentified tintinnids	Mauchline 1980
Meganyctiphanes norwegica	*Helicostomella spp., Paravella* spp.	Hedin 1975b
Nematoscelis gracilis, N. mirops	Unidentified tintinnids	Mauchline 1980
Thysanoessa inermis T. longicaudata, T.longipes, T.raschii, T. tricuspidata	Unidentified tintinnids	Mauchline & Fisher 1969
Thysanoessa macrura	*Codonellopsis gaussi, Laackmanniella naviculaefera, Cymatocylis antarctica*	Hopkins 1987

Table 5.3 *Continued*

Predator	Tintinnid prey	Reference
Mysids		
Antarctomysis ohlinii	*Codonellopsis gaussi*, *Laackmanniella naviculaefera*, *Cymatocylis antarctica*	Hopkins 1987
Mysis mixta	Unidentified tintinnids	Mauchline 1980
Neomysis mixta	Unidentified tintinnids	Rudstam et al. 1989
Penaeids		
Bentheogennema borealis	Unidentified tintinnids	Omori 1974
Sergestes similis	Unidentified tintinnids	Omori 1974
Caridea		
Hymenodora frontalis	Unidentified tintinnids	Omori 1974
Physetocaris microphthalma	Unidentified tintinnids	Omori 1974
Amphipods		
Epimeriella macronyx	*Codonellopsis gaussi*, *Laackmanniella naviculaefera*, *Cymatocylis antarctica*	Hopkins 1987
Eusirus tridentatus	*Codonellopsis gaussi*, *Laackmanniella naviculaefera*, *Cymatocylis antarctica*	Hopkins 1987
Hyperiella dilatata	*Codonellopsis gaussi*, *Laackmanniella naviculaefera*, *Cymatocylis antarctica*	Hopkins 1987
Orchomene plebs	*Codonellopsis gaussi*, *Laackmanniella naviculaefera*	Hopkins 1987
Orchomene rossi	*Cymatocylis antarctica*	Hopkins 1987

Table 5.4 Gelatinous zooplankton predators of tintinnids.

Predator	Tintinnid prey	Evidence	Reference
Cnidarians			
Hydromedusae			
Aglantha digitale (pink helmet)	Unidentified tintinnids	Gut contents	Larson (1987)
Mitrocoma cellularia (cross jellyfish)	Unidentified tintinnids	Gut contents	Larson (1987)
Aglaura hemistoma	Unidentified tintinnids	Gut contents (live specimens)	Colin et al. (2005)
Mitrocomella polydiademata	Unidentified tintinnids	Gut contents	Larson (1987)
Phialidium spp.	Unidentified tintinnids	Gut contents	Larson (1987)
Proboscidactyla flavicirrata	Unidentified tintinnids	Gut contents	Larson (1987)
Sarsia princeps	Unidentified tintinnids	Gut contents	Larson (1987)

Continued

Table 5.4 *Continued*

Predator	Tintinnid prey	Evidence	Reference
Scyphomedusae			
Aurelia aurita (moon jelly)	*Tintinnopsis* sp., *Favella* sp.	Feeding exp	Stoecker et al. 1987b
Chrysaora quinquecirrha (sea nettle) ephyrae	Unidentified tintinnids	Feeding exp	Olesen et al. 1996
Cotylorhiza tuberculata	Unidentified tintinnids	Gut contents	Pérez-Ruzafa et al. 2002
Rhizostoma pulmo	Unidentified tintinnids	Gut contents	Pérez-Ruzafa et al. 2002
Stomolophus meleagris (cannonball jelly)	Unidentified tintinnids	Gut contents	Larson 1991
Ctenophores			
Bolinopsis infundibulum	Unidentified tintinnids	Gut contents	Costello & Coverdale 1998
Pleurobrachia bachei	Unidentified tintinnids	Gut contents	Larson (1987)
Pleurobrachia pileus	Unidentified tintinnids	Gut contents	Costello & Coverdale 1998
Mnemiopsis leidyi, larvae and post larvae	*Favella* sp., *Tintinnopsis* sp., unidentified tintinnids	Feeding exp; gut contents	Stoecker et al. 1987a; Sullivan & Gifford 2007; Sullivan 2010; Granhag et al. 2011
Planktonic tunicates			
Salps	*Favella philippinensis*, *Tintinnopsis* spp., *Leprotintinnus* spp., *Eutintinus tenuis*	Gut contents	Naidu 1983
Doliolids	*Favella philippinensis*, *Tintinnopsis* spp., *Leprotintinnus* spp., *Eutintinus tenuis*	Gut contents	Naidu 1983
Oikopleura spp./ *O. vanhoeffeni*	*Favella philippinensis*, *Tintinnopsis spp.*, *Leprotintinnus spp.*, *Eutintinus tenuis; Stenosemela steini*	Gut contents; Fecal pellet contents	Naidu 1983; Urban et al. 1992
Pteropods			
Limacina helicina	Unidentified tintinnids	Gut contents	Paranjape 1968 cited in Sanders & Wickham 1993; Gilmer & Harbison 1991
Chaetognaths			
Sagitta elegans, chaetognath (young, stage I)	*Helicostomella*, *Parafavella*	Gut contents	Pearre 1973, 1981, Alvarez-Cadena 1993

Table 5.5 Other holoplanktonic predators of tintinnids.

Predator	Tintinnid prey	Evidence	Reference
Rotifers			
Synchaeta vorax	*Tintinnopsis* sp.	Gut contents	Lam-Hoai et al. 1997
Ostracods			
Conchoecia belgicae	*Codonellopsis gaussi*, *Laackmanniella naviculaefera*, *Cymatocylis antarctica*	Gut contents	Hopkins 1987
Conchoecia isocheira	*Cymatocylis antarctica*	Gut contents	Hopkins 1987
Cladocerans			
Evadne nordmanni *Penillia* sp.	*Favella philippinensis*, *Tintinnopsis* spp., *Leprotintinnus* spp., *Eutintinus tenuis*	Observed holding prey after preservation; gut contents	Bainbridge 1958 Naidu 1983

Table 5.6 Larval fish predators of tintinnids (all evidence solely from gut contents unless otherwise noted).

Predator	Tintinnid prey	Reference
Family Acanthuridae		
Paracanthurus hepatus (common surgeonfish)	*Amphorellopsis acuta*	Nagano et al. 2000
Family Ammodytidae		
Ammodytes personatus (Japanese sand lance)	*Codonellopsis morchella*, *Helicostomella fusiformis*, *Stenosemella nivalis*	Nagano et al. 2001
Ammodytes tobianus (sand lance)	Unidentified tintinnids	LeBour 1918
Family Atnerinidae		
Hypoatherina tropicalis (silverside)	Unidentified tintinnids	Schmitt 1986
Family Atherinopsidae		
Atherinopsis californiensis (jacksmelt)	*Stenosemella* sp.	Watson & Davis 1989
Leuresthes tenuis (California grunion)	*Stenosemella* sp.	Watson & Davis 1989
Larval atherinids	Unidentified tintinnids	Kauffman et al. 1981
Family Carangidae		
Trachurus symmetricus (jack mackerel)	Unidentified tintinnids	Arthur 1976
Family Centropomidae		
Centropomus undecimalis (common snook)	Unidentified tintinnids	E-Wittenrich et al. 2009

Continued

Table 5.6 *Continued*

Predator	Tintinnid prey	Reference
Family Clinidae		
Heterostichus rostratus (giant kelpfish)	Unidentified tintinnids	Stepien 1986
Family Clupeidae		
Anchoa mitchilli (Bay anchovy)	Unidentified tintinnids	Detwyler & Houde 1970
Brevoortia patronus (Gulf menhaden)	*Favella* sp.	Govoni et al. 1983; Stoecker & Govoni 1984
Clupea harengus (Atlantic herring)	*Stenosemella ventricosa* *Tintinnopsis* spp.	Hentschel 1950; Sherman & Honey 1971; Lebour 1919b; 1923; Courtois & Dodson 1986; Sanders & Wickham 1993
Sardina pilchardus (sardine or European pilchard)	Unidentified tintinnids	Morote et al. 2010
Sprattus sprattus (synonym: *Clupea sprattus*) (sprat)	*Tintinnopsis ventricosa* (= *S. ventricosa*)	Lebour 1919b; Fortier & Harris 1989
Sardinops sagax (Pacific sardine)	Unidentified tintinnids	Arthur 1976
Family Engraulidae		
Engraulis mordax (Californian anchovy)	Unidentified tintinnids	Brewer & Kleppel 1986; Arthur 1976
Engraulis ringens (anchoveta)	*Codonellopsis pusilla*, *Craterella urceolata*, *Helicostomella longa*	De Mendiola 1974
Mallotus villosus (capelin)	*Tintinnopsis* sp.	Courtois & Dodson 1986
Encrasicholina purpurea (synonym: *Stolephorus* purpureus) (*nehu*)	Unidentified tintinnids	Schmitt 1986
Family Gadidae		
Gadus morhua (cod)	Unidentified tintinnids	Bainbridge & McKay 1968 cited in Last 1978b; Last 1978b; Van Der Meeren & Naess 1993; Fossum & Ellertsen 1994
Melanogrammus aeglefinus (haddock)	*Stenosemella nucula*, *S. ventricosa*, *Acanthostemeela norvegica*, *Favella denticulata*	Ogilvie 1938
Merlangius merlangus (whiting)	Unidentified tintinnids	Last 1978b
Micromesistius poutassou (blue Whiting)	Unidentified tintinnids	Conway 1980
Trisopterus luscus (bib)	Unidentified tintinnids	Last 1978b
Family Gobioididae		
Gobiid larvae	Unidentified tintinnids	Houde & Lovdal 1984
Family Haemulidae		
Orthopristis chrysoptera (pigfish)	Unidentified tintinnids	Houde & Lovdal 1984

Table 5.6 *Continued*

Predator	Tintinnid prey	Reference
Family Labridae		
Labrus (*bergylta* type) wrasse	*Tintinnopsis beroidea*, *Cittarocyclis denticulate*	Lebour 1919c
Family Merlucciidae		
Macruronus novaezelandiae (blue grenadier)	*Codonella elongate*, *Petalotricha pacifica*, *Parundella caudate*, *Proplectella fastigata*, *Dictyocysta fenestrata*, *D. reticulata*, *Amphorella brandtii*	Murdoch 1990
Family Ophichthidae		
Myrophis sp. (worm eel)	*Codonellopsis* spp.	Govoni 2010
Family Paralepididae		
Paralepis sp.	Unidentified tintinnids	Duka & Gordina 1973
Family Paralichthyidae		
Paralichthys californicus (California halibut)	*Stenosomella* sp.	Watson & Davis 1989
Paralichthys dentatus (Summer flounder)	Unidentified tintinnids	Grover 1998
Family Pleuronectidae		
Limanda limanda (dab)	Unidentified tintinnids	Last 1978a
Platichthys flesus (European flounder)	Unidentified tintinnids	Last 1978a
Parophrys vetulus (*syn = Isopsetta isolepis*) (English sole)	Unidentified tintinnids	Gadomski & Boehlert 1984
Pseudopleuronectes americanus (Winter flounder)	Unidentified tintinnids	Shaheen et al. 2004 Pearcy 1962
Family Rhombosoleidae		
Ammotretis rostratus (longsnout flounder)	*Tintinnopsis sp.*, *Favella sp.*	Jenkins 1987, 1988
Rhombosolea tapirina (Greenback flounder)	*Favella sp*, *Tintinnopsis sp.*	Jenkins 1987, 1988
Family Sciaenidae		
Genyonemus lineatus (white croaker)	*Stenostomella* sp., *Codonaria* sp.	Brewer & Kleppel 1986; Watson & Davis 1989
Leiostomus xanthurus (spot)	*Tintinnopsis* sp., *Stenosemella* sp., *Codonellopsis* sp., *Dictyocysta* sp.	Govoni et al. 1983; Govoni & Chester 1990

Continued

Table 5.6 *Continued*

Predator	Tintinnid prey	Reference
Micropogonias undulatus (Atlantic croaker)	*Stenosemella* sp.	Govoni et al. 1983
Seriphus politus (queenfish)	*Stenosemella* sp., *Codonaria* sp.	Watson & Davis 1989
Family Scophthalmidae		
Zeugopterus norvegicus (synonyms: *Phrynorhombus norvegicus*, *Scophthalmus norvegicus*) Norwegian topknot	*T. ventricosa* (= *S. ventricosa*)	Lebour 1919c
Family Scorpaenidae		
Sebastiscus marmoratus (firefish)	Unidentified tintinnids	Nagano et al. 2001
Family Serranidae		
Epinephelus septemfasciatus (convict grouper)	*Favella taraikaensis*	Nagano et al. 2000
Paralabrax sp.	*Stenosemella* sp.	Watson & Davis 1989
Family Soleidae		
Microchirus variegates (thickback sole)	Unidentified tintinnids	Lebour 1918
Solea solea (sole)	Unidentified tintinnids	Last 1978a
Solea ovata (= *Solea variegata*)	*Tintinnopsis ventricosa* (= *S. ventricosa*)	Lebour 1919
Family Sparidae		
Archosargus rhomboidalis (Western Atlantic seabream)	Unidentified tintinnids	Stepien 1976
Family Sternoptychidae		
Maurolicus sp. (deep sea hatchetfish)	Unidentified tintinnids	Duka & Gordina 1973
Family Syngnathidae		
Syngnathus sp. (pipefish)	*Favella ehrenbergii*	Capriulo & Ninivaggi 1982

E, feeding or rearing experiment.

Table 5.7 Benthic invertebrate and meroplanktonic larval predators of tintinnids.

Predator	Tintinnid prey	Evidence	Ref
Mollusks			
Bivalve and gastropod veligers	*Favella philippinensis*, *Tintinnopsis* spp.	Gut contents	Naidu 1983
Mytilus edulis (blue Mussel)	*Tintinnopsis* spp.	Enclosure experiments	Horsted et al. 1988
Ostrea edulis (oyster)	Unidentified tintinnids	Ingested	Krsinic 1987 cited in Sanders & Wickham 1993

Table 5.7 *Continued*

Predator	Tintinnid prey	Evidence	Ref
Crinoids			
Lamprometra klunzingeri, *Heterometra savignii*, *Capillaster multiradiatus* (feather-stars)	*Cyttarocylis eucecryphalus*, *Codonellopsis morchella*, *Proplectella* spp., *Epiplocylis undella* and other spp.	Gut contents	Rutman & Fishelson 1969
Antedon bifida (feather star)	Unidentified tintinnids	Gut contents	La Touche & West 1980
Cnidaria			
Aurelia aurita polyps	*Favella ehrenbergii*	Feeding exp	Kamiyama 2011
Crustacea			
Aratus pisonii zoeae (grapsid crab larvae)	*Favella ehrenbergi*	Feeding exp	Schwamborn et al. 2006
Uca sp. Stage 1 zoea (fiddler crab larvae)	*Favella panamensis*, *Tintinnopsis tubulosa*	Feeding exp	Robertson 1983
Ectoprocta (bryozoans)			
Cyphonautes (bryozoan larvae)	*Tintinnopsis* sp.	Gut contents	Lebour 1922
Phoronida (horseshoe worms)			
Actinotrocha (larvae of worm, *Phoronis* sp.)	*Helicostomella subulata*, *Tintinnoposis beroidea*	Gut contents	Lebour 1922

Tintinnid abundance and species composition is variable both seasonally and spatially (Chapter 10). In most plankton communities most of the time, tintinnids are less abundant than aloricate ciliates and other microzooplankton; however, episodically tintinnids can dominate microzooplankton assemblages, especially in coastal waters (Chapter 9). Because most field studies provide "snapshots" of the diet, tintinnids are often reported as a dietary component of a grazer or predator in one investigation but not another, even from the same location or season. The presence of tintinnids in the diet is often an indicator that planktonic ciliates in general are an important, and probably overlooked, dietary component.

5.2 PREDATORS OF TINTINNIDS

Other protists

The microbial food web is complex, with protists preying on other protists, including tintinnids (Table 5.1). Tintinnids can prey on smaller tintinnids. *Favella* spp. prey on smaller *Tintinnopsis* spp. as well as on dinoflagellates and aloricate ciliates (LeBour 1922; Robertson 1983; Verity 1986). Predation on other tintinnids is probably more common in *Favella* than in other many other tintinnid species because of *Favella*'s comparatively large size, but it is also more obvious than it would be in many other species because of *Favella*'s hyaline lorica, which allows easy microscopic observation of ingested prey (Fig. 5.1).

Both heterotrophic and mixotrophic dinoflagellates prey on tintinnids (Table 5.1). The bloom-forming heterotrophic dinoflagellate, *Noctiluca scintillans*, preys on microplankton, including tintinnids (Fonda Umani et al. 2004; Sato et al. 2010). Some mixotrophic dinoflagellates, including *Ceratium furca* and *Gyrodinium instriatum*, ingest tintinnid cells without the lorica (Uchida et al. 1997; Smalley et al. 1999). *G. instriatum* has been observed to extract large tintinnids from their lorica and then engulf the cell (Fig. 5.2). Because many dinoflagellates do not ingest the lorica, their predation on tintinnids can be difficult to detect. Many mixotrophic

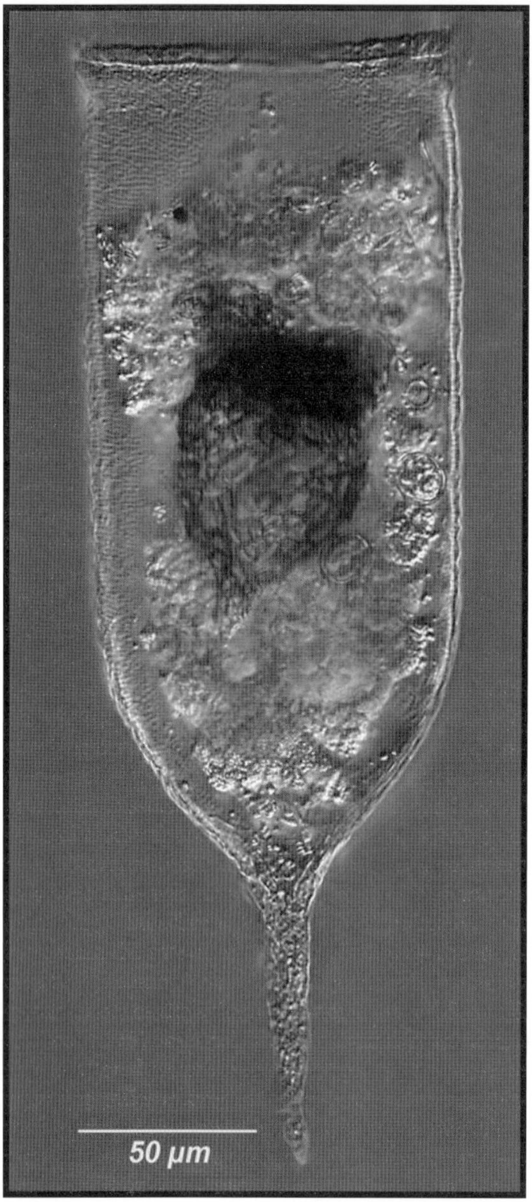

Fig. 5.1 *Favella* from an Indian mangrove estuary with an ingested *Tintinnopsis*, one of several with ingested tintinnid remains from the same sample (J. Dolan, personal communication).

and heterotrophic dinoflagellates prey on a wide size range of protists using a peduncle (Jeong et al. 2010) and it is likely that they include tintinnids in their diet. In addition to traditional predators, tintinnids also have dinoflagellate ectoparasites, *Duboscquella* and *Tintinnophagus*; these could be considered predators as they devour the host from the outside (see Chapter 6).

In tropical and subtropical oceans, the planktonic sarcodines (Radiolaria, Acantharia and Foraminifera) are important in biogeochemical cycles as well as common predators on tintinnids (Caron & Swanberg 1990) (Table 5.1). Tintinnids and their empty loricae are abundant in the fecal strands of large colonial radiolaria (*Collozoum* spp.) and are common in the pseudopodial networks of solitary polycystine radiolarian (Table 5.1). Swanberg & Caron (1991) investigated sarcodine diets in surface waters of the subtropical Atlantic. Tintinnids were observed in the pseudopodial networks of 17% of the polycystine radiodiolaria, 19% of Acantharia, and 3% of the Foraminifera but accounted for only 3, 7, and <1%, respectively, of the estimated biomass ingested. Tintinnids appear to be a preferred prey of Acantharia and polycystine Radiolaria but not Foraminifera (Caron & Be 1984; Swanberg & Caron 1991). Low numbers of aloricate ciliates are found in the pseudopodial networks of the sarcodines, but because aloricates are digested rapidly, their contribution to the diet is probably severely underestimated relative to the tintinnids. Below the euphotic zone, phaeodarian Radiolaria are more abundant than polycystine Radiolaria. Gowing & Bentham (1994) observed empty tintinnid loricae, but not loricae with cellular content, in the food vacuoles of phaeodaria from sediment traps in the North Pacific. The phaeodarians may have ingested sinking, empty loricae along with detritus.

Copepods

Copepods are the dominant metazoan zooplankton in food webs that support fisheries. Based on gut or fecal pellet contents (e.g., Fig. 5.3), predation on tintinnids is common among copepods including members of the calanoid copepod families Acartidae, Aetideidae, Calanidae, Clausocalanidae, Eucalananidae, Euchaetidae, Metridinidae, Paracalanidae, and Tortanidae (Table 5.2). This includes most of the common suspension-feeding copepods regarded as herbivorous

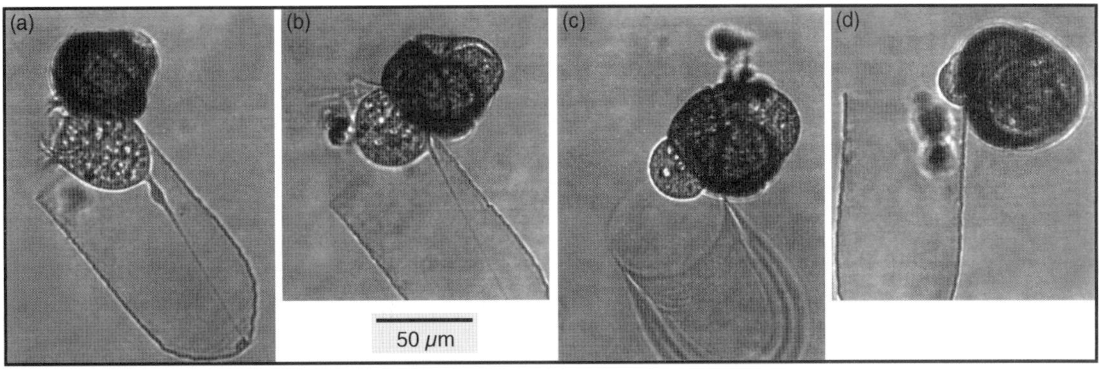

Fig. 5.2 Time course (a–d) of the feeding of *Gyrodinium instriatum*, a mixotrophic dinoflagellate, on the tintinnid *Favella azorica*. Figure from Uchida, T., Kamiyama, T. & Matsuyama, Y., Predation by a photosynthetic dinoflagellate *Gyrodinium instriatum* on loricated ciliates. Journal of Plankton Research, 1997, 19, 603–608, by permission of Oxford University Press.

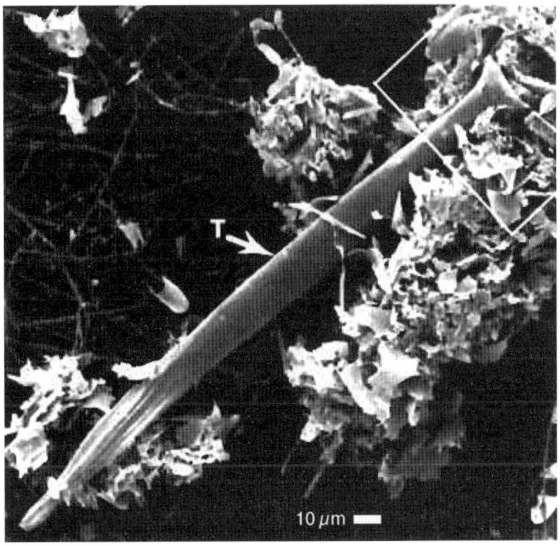

Fig. 5.3 Tintinnid lorica (*Salpingella* sp.) in a fragmented fecal pellet of the copepod *Eucalanus pileatus* from the Gulf of Mexico. Figure from Turner (1984b) with permission of InterResearch.

or omnivorous (for example *Acartia*, *Calanus*, *Eucalanus*, *Paracalanus*) as well as predatory calanoids such as *Euchaeta* and *Tortanus* (Mauchline 1998). Tintinnids are also prey of the planktonic ambush-feeding cyclopoid copepods, *Oithona* and *Limnoithona* (Table 5.2). In most cases, tintinnids, even when they are a frequent component of the gut or fecal pellet contents, appear to be a minor component of the diet in terms of biomass. An exception to this is during tintinnid "blooms" when they may account for most of the fecal pellet contents (Turner & Anderson 1983).

Compared with gut and fecal content studies, there have been relatively few measurements of feeding by copepods on tintinnids (Table 5.2). Laboratory feeding experiments have been conducted with mixtures of cultured ciliates and phytoplankton that are within the range of densities observed in coastal environments for these prey types. The common estuarine and coastal copepod *Acartia tonsa* has higher clearance for tintinnids and other ciliates than for phytoplankton in these prey mixtures (Stoecker & Sanders 1985; Stoecker & Egloff 1987). Inclusion of tintinnids in a phytoplankton-based diet can enhance egg production by calanoid copepods (Stoecker & Egloff 1987; Kleppel 1993). Feeding experiments with natural prey assemblages and the calanoid copepods *Acartia*, *Calanus*, *Metridia*, and *Pseudocalanus* spp. are consistent with the results using cultures. Copepods usually have higher clearance rates for tintinnids (if abundant) and aloricate ciliates than for phytoplankton (Turner & Anderson 1983; Gifford & Dagg 1988, 1993; Lonsdale et al. 2000; Rollwagen Bollens & Penry 2003; Leising et al. 2005b; Diodato & Hoffmeyer 2008). This suggests that copepod predation on tintinnids as well as on other microzooplankters can lead to trophic cascades, releasing phytoplankton from microzooplankton grazing pressure (Stoecker & Sanders 1985; Irigoien et al. 2005; Reaugh et al. 2007; Calbet & Saiz 2005).

An interesting question is whether tintinnids are more or less susceptible to copepod predation than other microzooplankton. In laboratory feeding experiments, adult *Acartia tonsa* have higher clearance rates for the aloricate ciliate *Strobilidium* spp. than the larger tintinnid *Favella* sp. but have higher clearance for *Favella* sp. than a smaller *Tintinnopsis* sp. However, copepod nauplii appear to prefer *Tintinnopsis* species over *Favella* species (Stoecker & Egloff 1987). In experiments with *A. clausi*, clearance of the tintinnid *Metacylis* species was higher than of the similar size aloricate ciliate, *Strobilidium spiralis*. Broglio et al. (2001) suggest that this was because *S. spiralis* had jumping behavior that reduces capture. However, not all aloricate ciliates "jump". In fact, in experiments with natural assemblages of phytoplankton and microzooplankton, clearance by *Acartia* spp. of tintinnids is usually lower than of aloricate ciliates (Rollwagen Bollens & Penry 2003). In feeding experiments with *Calanus australis* and natural prey assemblages, the copepods showed preference for tintinnids over non-thecate heterotrophic dinoflagellates (Sanchez et al. 2011).

Cyclopoid copepods are ambush predators on motile prey and show strong prey preferences among microplankton. *Oithonia* spp. have higher clearance for tintinnids and other ciliates than for heterotrophic dinoflagellates, but higher clearance for aloricate ciliates than co-occurring tintinnids in natural prey assemblages (Lonsdale et al. 2000; Nishibe et al. 2010). *Limnoithona tetraspina*, another cyclopoid copepod, feeds readily on aloricate ciliates such as *Mesodinium* and *Strombidium* spp. but has relatively low feeding rates on tintinnids (Bouley & Kimmerer 2006).

Presumably the generally higher clearing rates of calanoid and cyclopoid copepods for tintinnids and other ciliates than for dinoflagllates is because the ciliates generate a greater hydrodynamic signal than the dinoflagellates, and thus are detected more readily (Jakobsen et al. 2005). The generally greater clearance by copepods of aloricate ciliates than of tintinnids may be due to differences in their hydrodynamic signal strength, differences in handling time and predator defenses or a combination of these factors.

Euphausiids and other large crustacean zooplankton

Another important group of marine crustacean zooplankton is the euphausiids (krill); because of their large size and swarming behavior, euphausiids are important in the diet of many fish, seabirds and marine mammals. Tintinnids are frequently found in the gut contents of euphausiids including *Euphausia pacifica*, *E. superba* (Antarctic krill), *E. crystallorophias* (ice krill), and *Meganyctiphanes norwegica* (northern krill) (Mauchline & Fisher 1969) (Table 5.3). Krill are thought to feed primarily on large diatoms, but tintinnids can constitute over 50% of the recognizable stomach contents in *Euphausia pacifica* in the northern northwest Pacific (Nakagawa et al. 2001). However, tintinnids may not be preferred prey. In experiments in a Chilean fjord, *E. vallentini* fed mostly on phytoplankton, and clearance of tintinnids and other microzooplankton was not significant (Sanchez et al. 2011). Overall, the significance of tintinnids in the diet of krill is unknown, but it is possible that tintinnids and other microzooplankton, although not major components of the diet, are nutritionally important high-quality food items. Because of their abundance in the environment, it is likely that aloricate ciliates are generally more important than tintinnids as prey for euphausiids, but because almost all data on euphausiid diets are from gut contents, it is not possible to evaluate the relative importance.

Among other large crustacean zooplankton, tintinnids or their loricae occur in the gut contents of mysids, penaeid and carid shrimp, and amphipods (Table 5.3). The overall significance of tintinnids and other microzooplankton in their diets is unknown.

Gelatinous zooplankton including cnidarians, ctenophores, chaetognaths, pteropods and pelagic tunicates

Tintinnid ciliates are consumed by gelatinous predators including cnidarians (medusae), ctenophores (comb jellies), and chaetognaths (arrow worms) as well as gelatinous suspension feeders such as the thecostome pteropods and pelagic tunicates (Table 5.4).

Medusae are primarily carnivorous, and most of the larger medusa consume crustacean zooplankton, fish larvae and fish eggs. However, some medusae are suspension feeders on microplankton, including tintinnids and other ciliates. Suspension feeding on microplankton may be particularly common among the smaller hydromedusae (Colin et al. 2005). Tintinnids were 25% of the recognizable prey items, but only 1% of the estimated prey biomass, in the gut of the hydromedusa *Proboscidactyla flavicirrata* (Larson 1987). In the other five hydromedusae investigated by Larson, tintinnids accounted for less than 1% of the recognizable prey (Table 5.4). Colin et al. (2005)

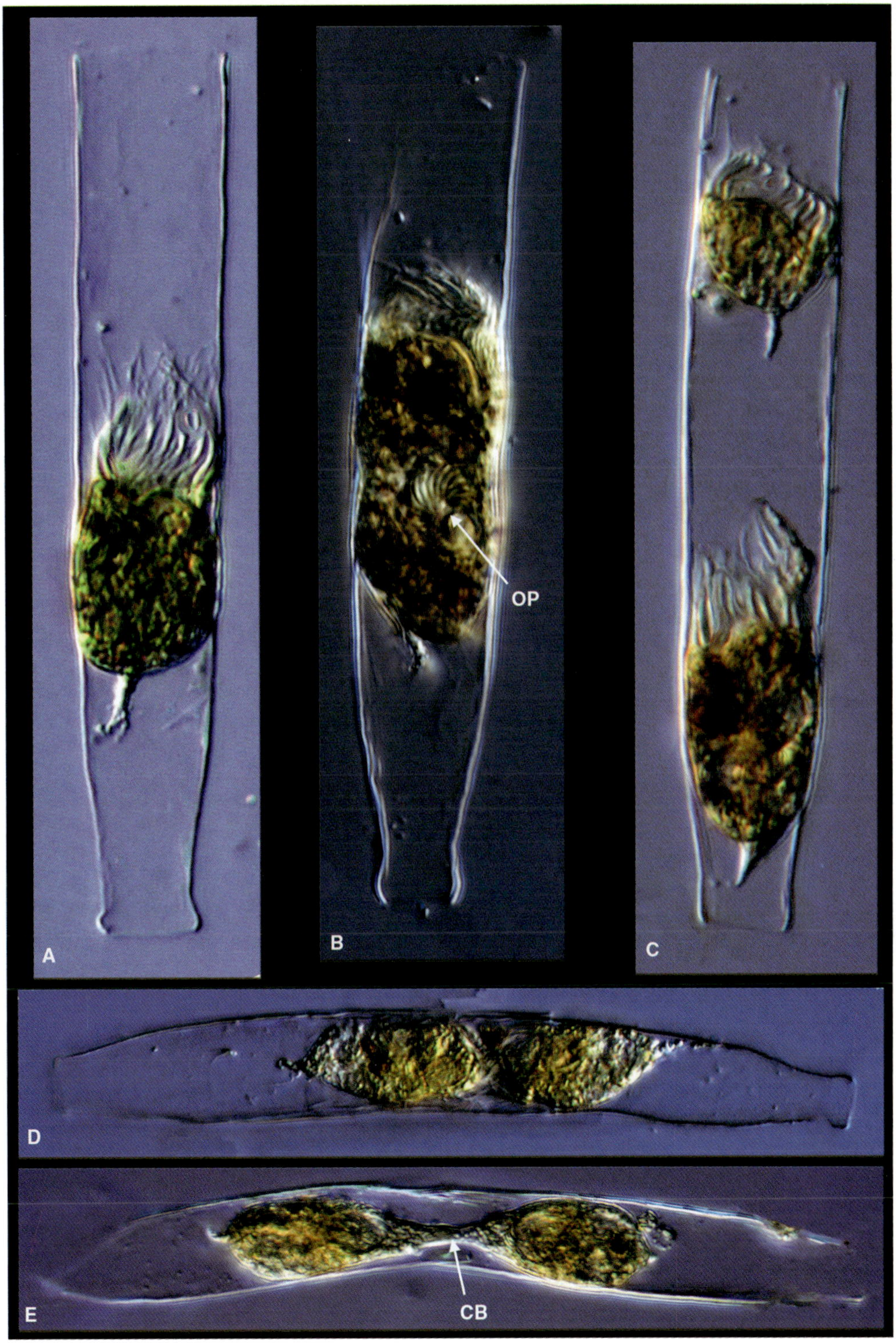

Plate 1.1 Life-cycle stages in *Eutintinnus inquilinus*. (a) A trophont (feeding stage) cell. (b) An early stage of binary fission showing the oral primordium (OP), which will develop into a new mouth. (c) The final stage of cell division. (d) An early stage of conjugation. (e) A conjugating pair showing a well-developed cytoplasmic bridge (CB) through which nuclear material will be exchanged. Loricae are about 100 μm long. Images are of Lugol's-fixed cells; sample from a mixed culture provided by F. Lombard.

The Biology and Ecology of Tintinnid Ciliates: Models for Marine Plankton, First Edition. Edited by John R. Dolan, David J.S. Montagnes, Sabine Agatha, D. Wayne Coats, and Diane K. Stoecker.

Plate 1.4 Some tintinnid species with agglutinated loricae: *Tintinnopsis radix* (a), *T. campanula* (b), *T. dadayi* (c), *T. spiralis* (d), *Codonella elongata* covered with coccoliths (e), *Tintinnidium* sp. (f), *Stenosemella ventricosa* (g), *Leprotintinnus pellucidus* (h), and *Codonellopsis schabi* (i). Species names are attributed based on lorica morphology. All the specimens are Lugol's-fixed cells except for *Codonella elongata* which was a live specimen.

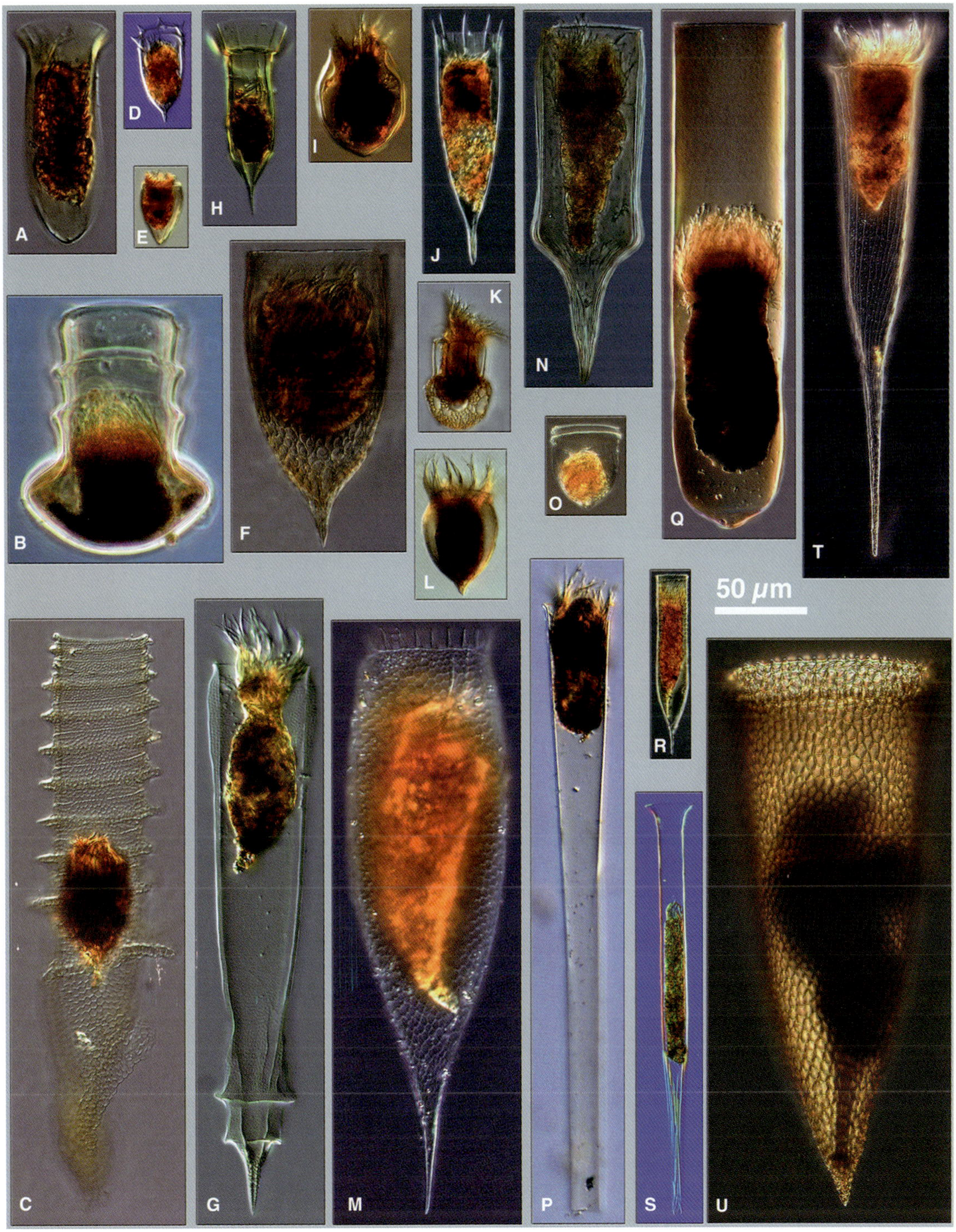

Plate 1.5 Examples of tintinnid species with hyaline loricae: *Amphorides quadrilineata* (a), *Amplectella collaria* (b), *Climacocylis scalaria* (c), *Acanthostomella conicoides* (d), *Protorhabdonella simplex* (e), *Epiplocylis blanda* (f), *Xystonellopsis paradoxa* (g), *Ormosella trachelium* (h), *Proplectella elipisoida* (i), *Dadayiella ganymedes* (j), *Dictyocysta lepida* (k), *Metacylis mediterranea* (l), *Parafavella parumdentata* (m), *Parundella messinensis* (n), *Ascampbellia tortulata* (o), *Eutintinnus stramentus* (p), *Undella hyalina* (q), *Helicostomella subulata* (r), *Salpingella acuminata* (s), *Rhabdonella spiralis* (t), and *Cyttarocylis cassis* (u). All the specimens are Lugol's-fixed cells.

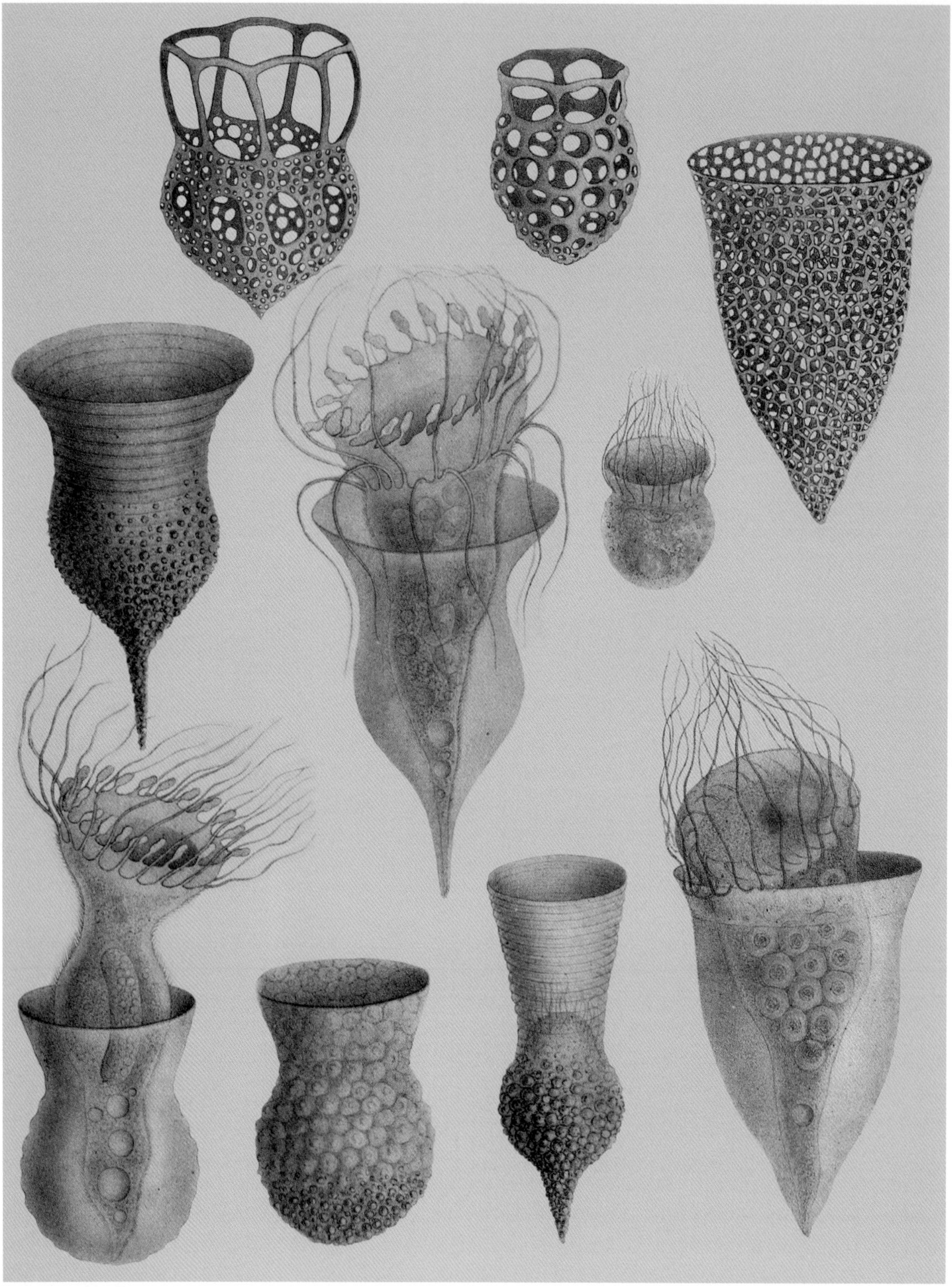

Plate 1.14 Tintinnids of Ernst Haeckel. Restored drawings from Ernst Haeckel's 1873 paper, the first scientific article devoted to tintinnid ciliates. Image by J.R. Dolan.

Plate 1.15 "La Boheme: A Portrait of Today's Oceans in Peril", the 2012 sculpture by Mara G. Haseltine depicts two tintinnids ensnarled in plastic. Dimensions: 91 cm × 122 cm × 185 cm (length × width × height). Materials: uranium-infused blown glass and plastic. By permission of Mara Heseltine.

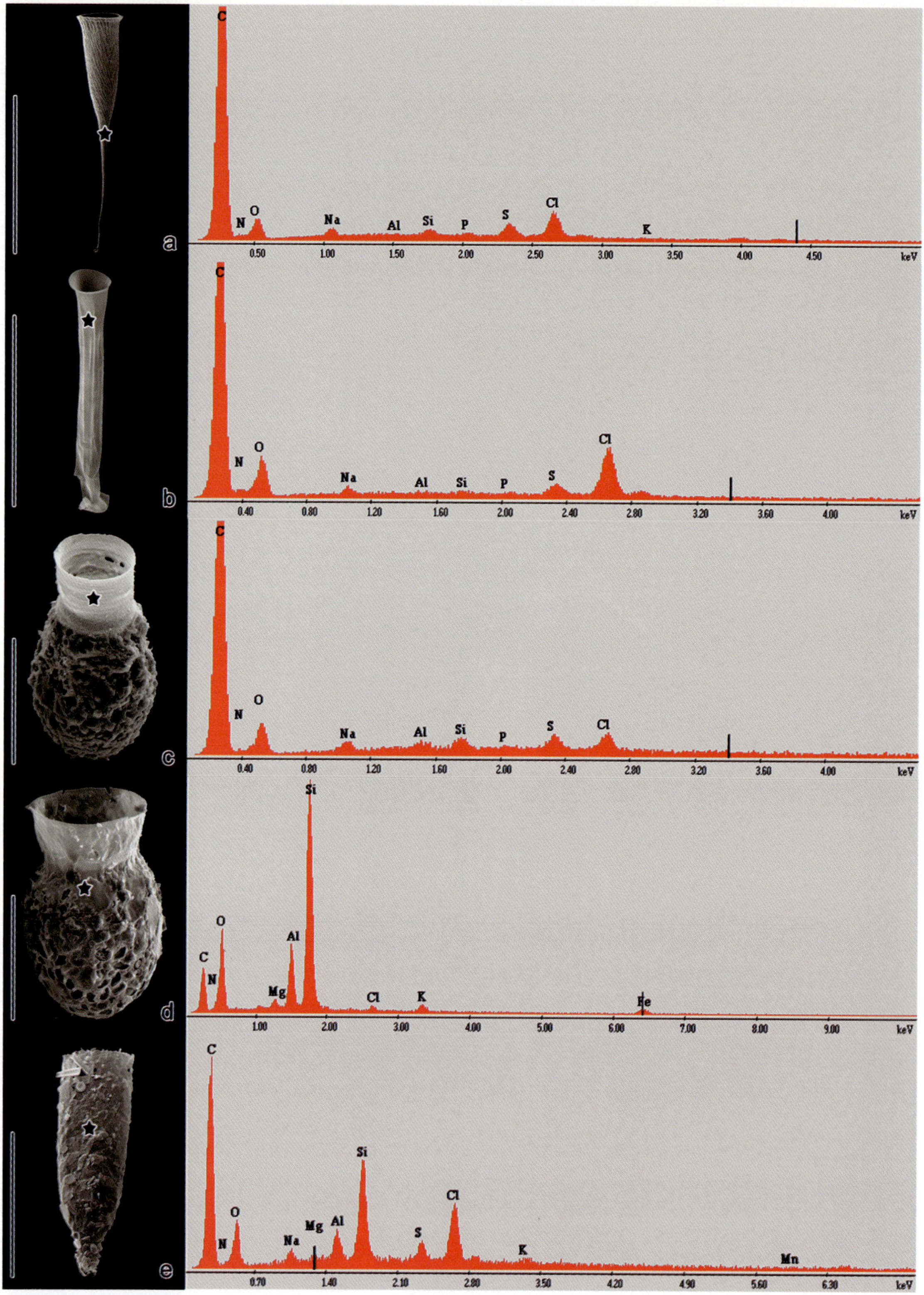

Plate 2.1 Scanning electron micrographs (left column) and EDX spectrograms (right column; adapted from Agatha & Simon 2012). After drying, the preserved loricae were coated with carbonate. The regions analyzed are marked by asterisks. (a) *Rhabdonella spiralis*. (b) *Eutintinnus brandti*. (c) *Codonellopsis schabi*. (d) *Codonella aspera*. (e) Tintinnopsis *cf. cylindrica* – Note the presence of nitrogen in all loricae. Scale bars 200 μm (a, b) and 50 μm (c–e).

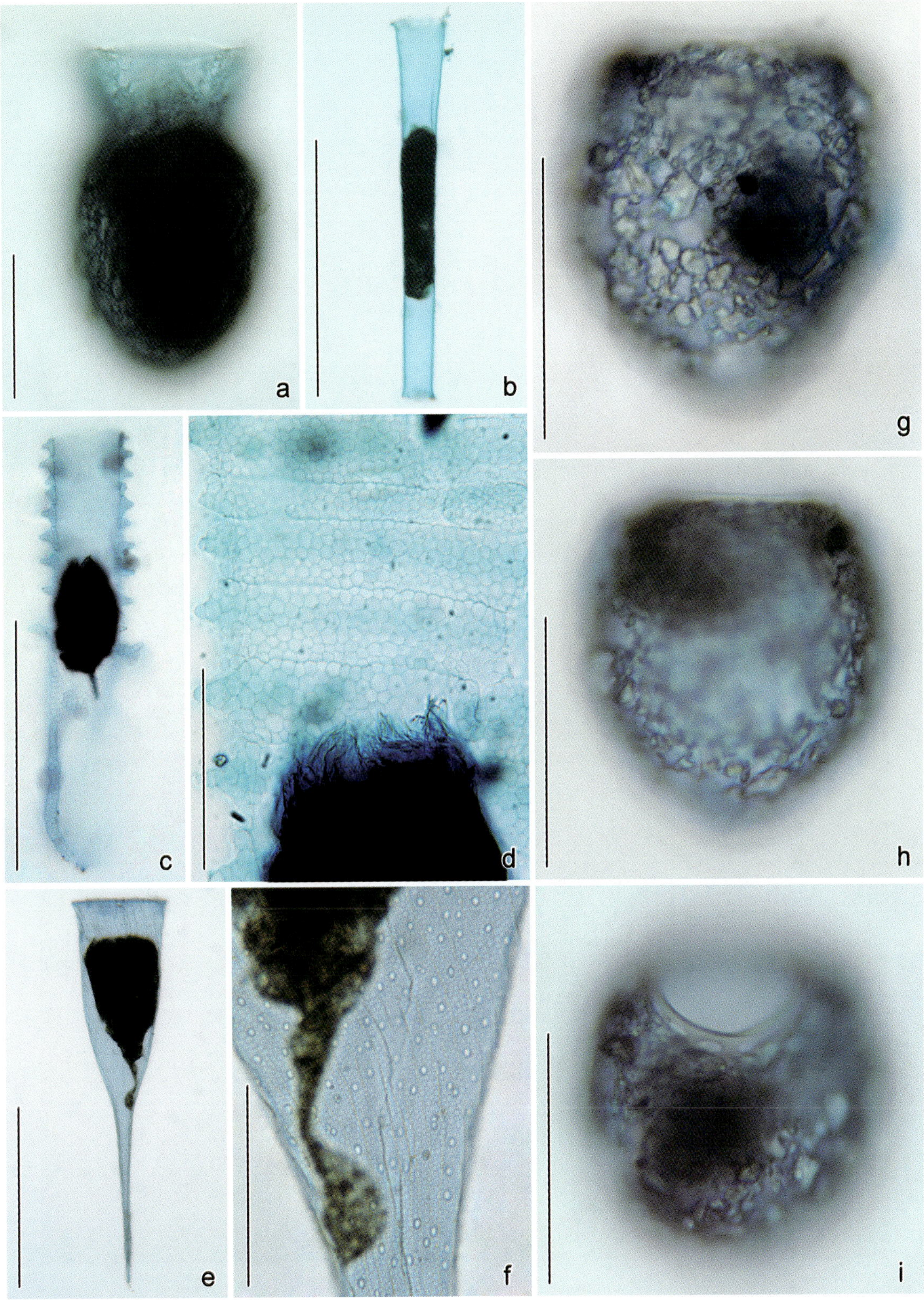

Plate 2.2 Loricae after mercuric bromophenol blue (a–f) and alcian blue stain (g–i); adapted from Agatha & Simon (2012). (a) *Codonella aspera*; the staining is restricted to the lorica matrix, namely the embedded mineral particles are not colored. (b) *Eutintinnus brandti*; the lorica with its monolaminar and compact texture is uniformly stained. (c, d) *Climacocylis* species; the alveolar texture of the lorica wall is well recognizable. (e, f) *Rhabdonella spiralis* showing the alveolar texture, the pores, and the spiraled surface ridges. (g–i) *Stenosemella ventricosa*; the lateral (g, h) and the oblique top (i) views show that the staining is restricted to the bowl matrix, namely the mineral particles embedded in the bowl and the hyaline collar are not colored. Scale bars 50 μm (a, g–i), 200 μm (b, c), 40 μm (d), 100 μm (e), and 20 μm (f).

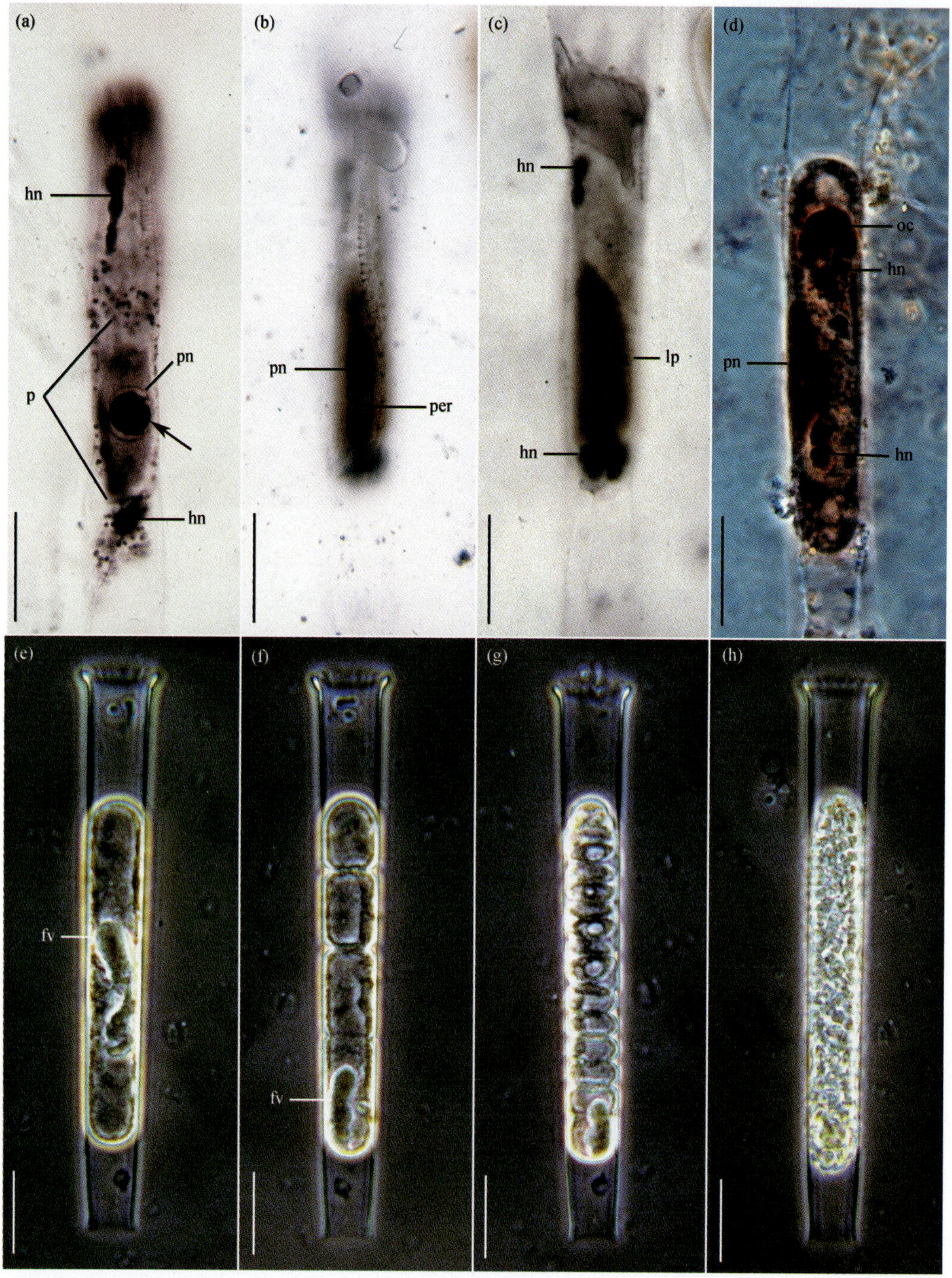

Plate 6.1 *Euduboscquella cachoni* from *Eutintinnus pectinis* (adapted from Coats 1988). (a) Mid-infection showing trophont nucleus containing a single large nucleolus (arrow). (b, c) A late trophont with well-developed perinema and lamina pharyngea. (d) A recently formed tomont with ingested host cell relatively undigested. (e–h) Successive stages in sporogenesis. Host nuclei (hn); food vacuole (fv); oral cilia of host (oc); parasite (p); parasite nucleus (pn); perinema (per); lamina pharyngea (lp). Scale bar, 20 μm.

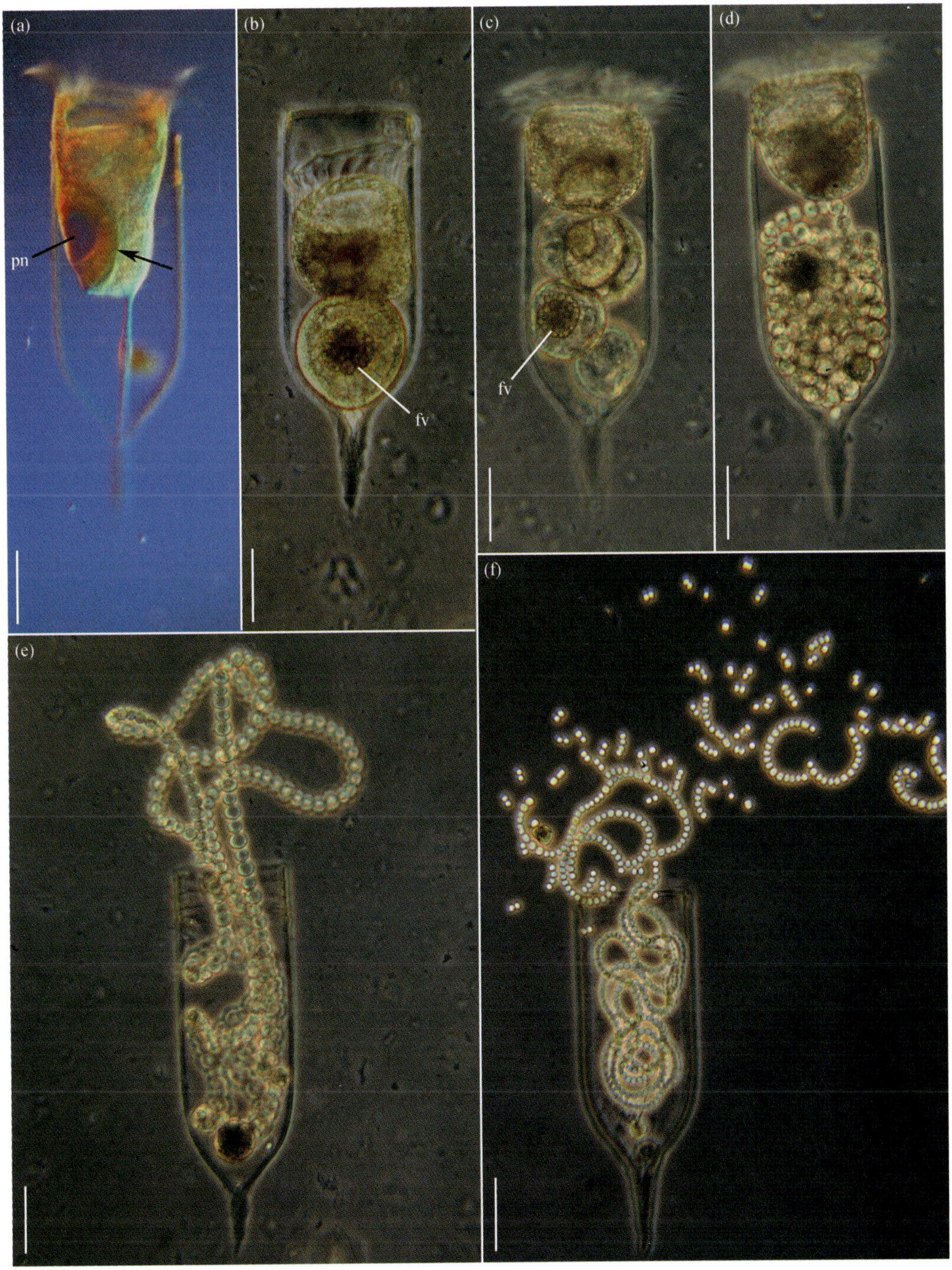

Plate 6.2 Stages *in vivo* during the life cycle of *Euduboscquella aspida* from *Favella panamensis* (adapted from Coats et al. 1994). (a) Late trophont (arrow). (b) Tomont located posterior to host cell. (c) Four-cell state of sporogenesis. (d–f) Progressively later stages in the development of beaded sporocytes. Food vacuole (fv); parasite nucleus (pn). Scale bar, 50 μm).

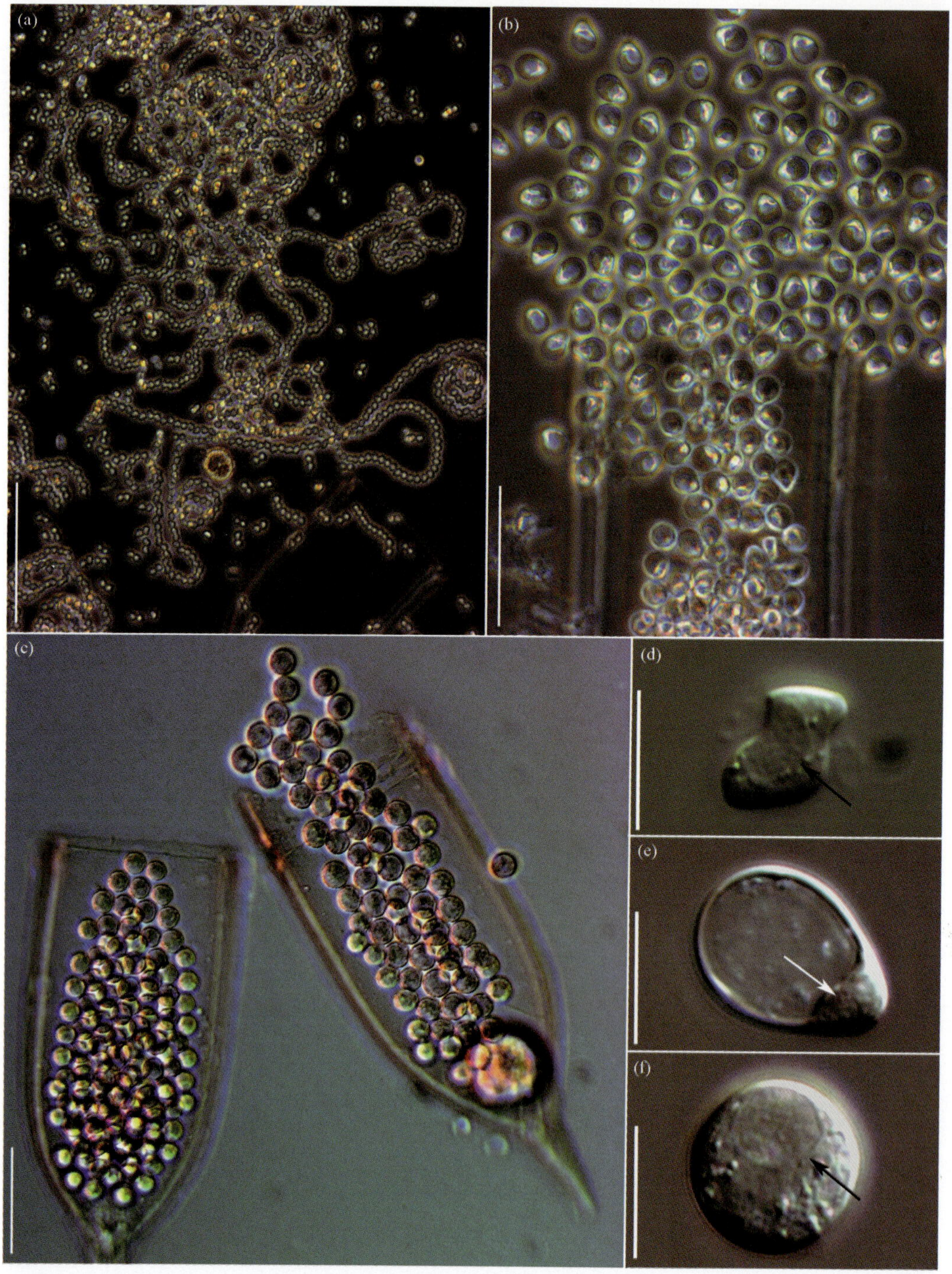

Plate 6.3 Spores produced by *Euduboscquella crenulata* from *Favella panamensis* (adapted from Coats, D.W., Bachvaroff, T.R. & Delwiche, C.F. (2012) Revision of the family Duboscquellidae with description of *Euduboscquella crenulata* n. gen., n. sp. (Dinoflagellata, Syndinea), an intracellular parasite of the ciliate *Favella panamensis* Kofoid and Campbell, 1929. Journal of Eukaryotic Microbiology, 85, in press). (a) Sporocytes emerging from host lorica as beaded strings (scale bar, 50 μm). (b) Non-motile egg-shaped spores dispersing from host lorica (scale bar, 50 μm). (c) Non-motile spherical spores in host loricae (scale bar, 50 mm). (d–f) Dinokont, non-motile egg-shaped, and non-motile spherical spore, respectively Scale bars, 10 μm; arrows indicate spore nuclei.

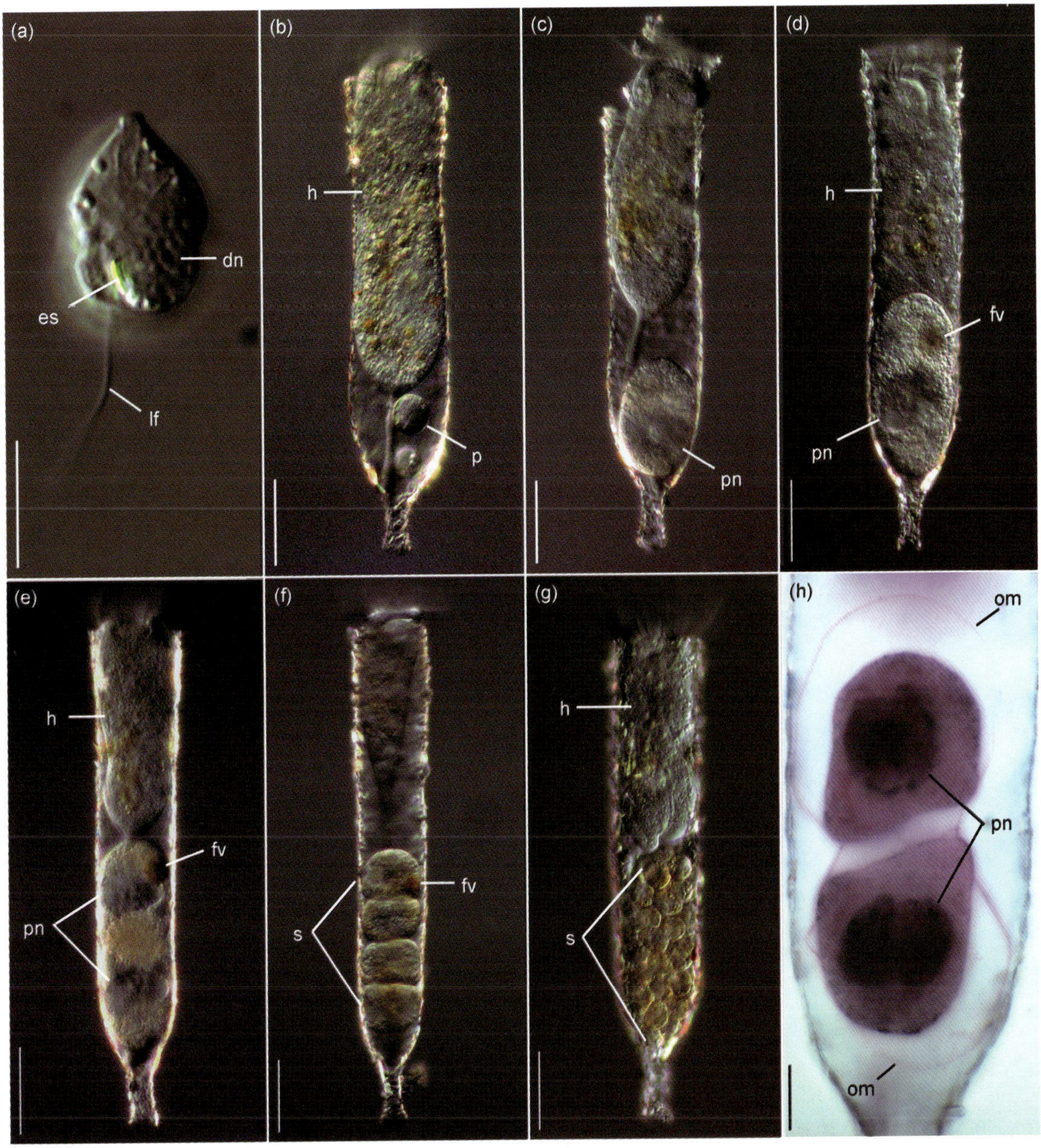

Plate 6.4 Stages in the infection cycle of *Tintinnophagus acutus* from *Tintinnopsis cylindrica* (adapted from Coats, D.W., Kim, S., Bachvaroff, T.R., Handy, S.M. & Delwiche, C.F. (2010) *Tintinnophagus acutus* n. g., n. sp. (Phylum Dinoflagellata), an ectoparasite of the ciliate *Tintinnopsis cylindrica* Daday 1887, and its relationship to *Duboscquodinium collini* Grassé 1952. Journal of Eukaryotic Microbiology, 57, 468–482). (a) Infective dinospore. (b–d) Growth of the trophont. (e–g) Sporogenesis. (h) Outer membrane and nuclei of protargol-stained specimen entering second division. Dinospore nucleus with condensed chromosomes (dn); eyespot (es); food vacuole (fv); host cell (h); outer membrane (om); parasite (p); parasite nucleus (pn); sporocytes (s). Scale bars, 10 μm (a, h), 40 μm (b–g).

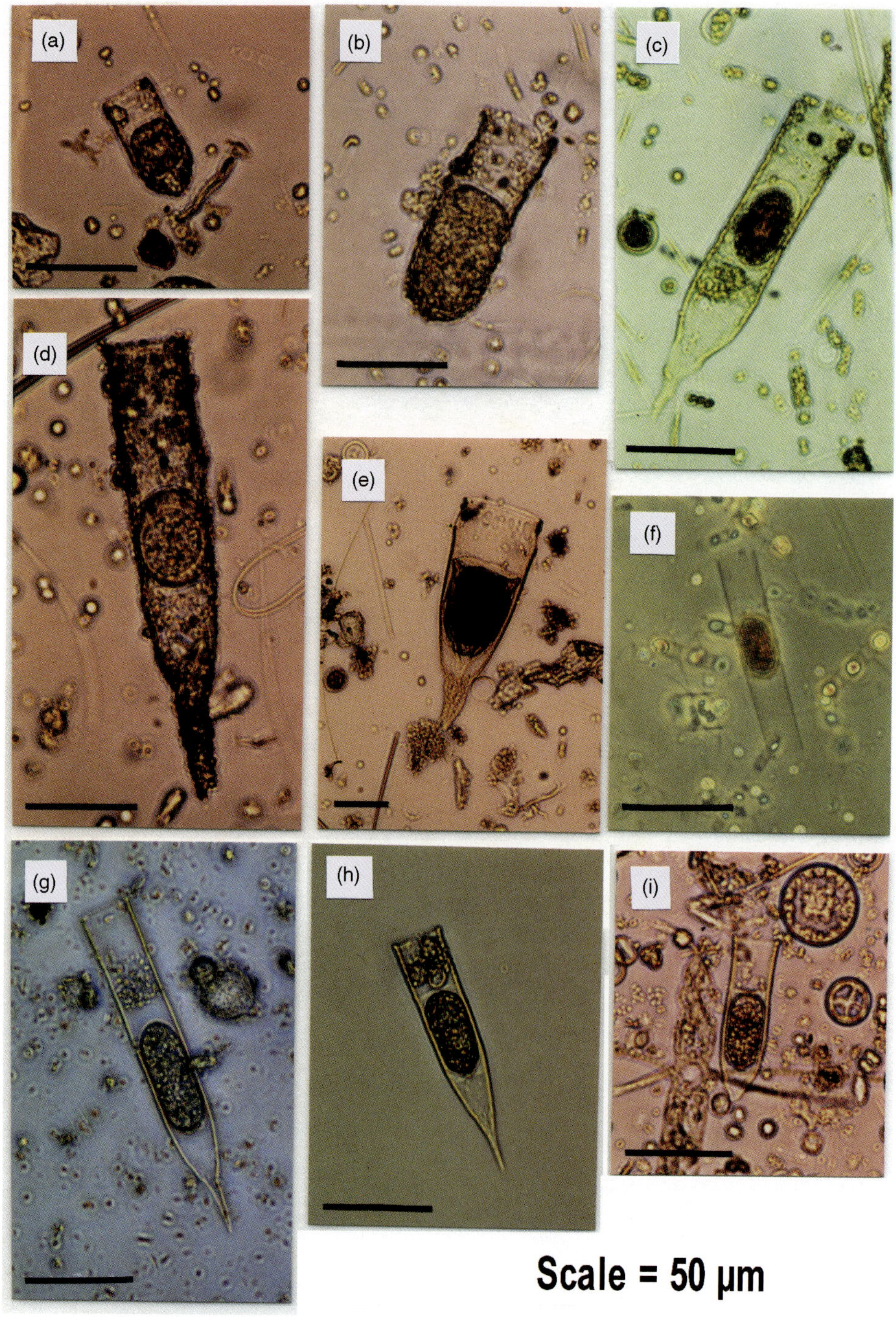

Plate 7.1 Photographs of tintinnid cysts from coastal waters of central and western Japan (from Kamiyama 1994b). (a) *Tintinnopsis beriodea*. (b) *Tintinnopsis* sp. (c) *Tintinnopsis corniger*. (d) *Tintinnopsis kofoidi*. (e) *Favella taraikaensis*. (f) *Eutintinnus tubulosus*. (g) *Helicostomella subulata*. (h) *Helicostomella fusiformis*. (i) *Helicostomella longa*. Scale bar, 50 µm. Note that *T. kofoidi* may be a synonym of *T. cylindrica* (Agatha & Reidel-Lorjé 2006).

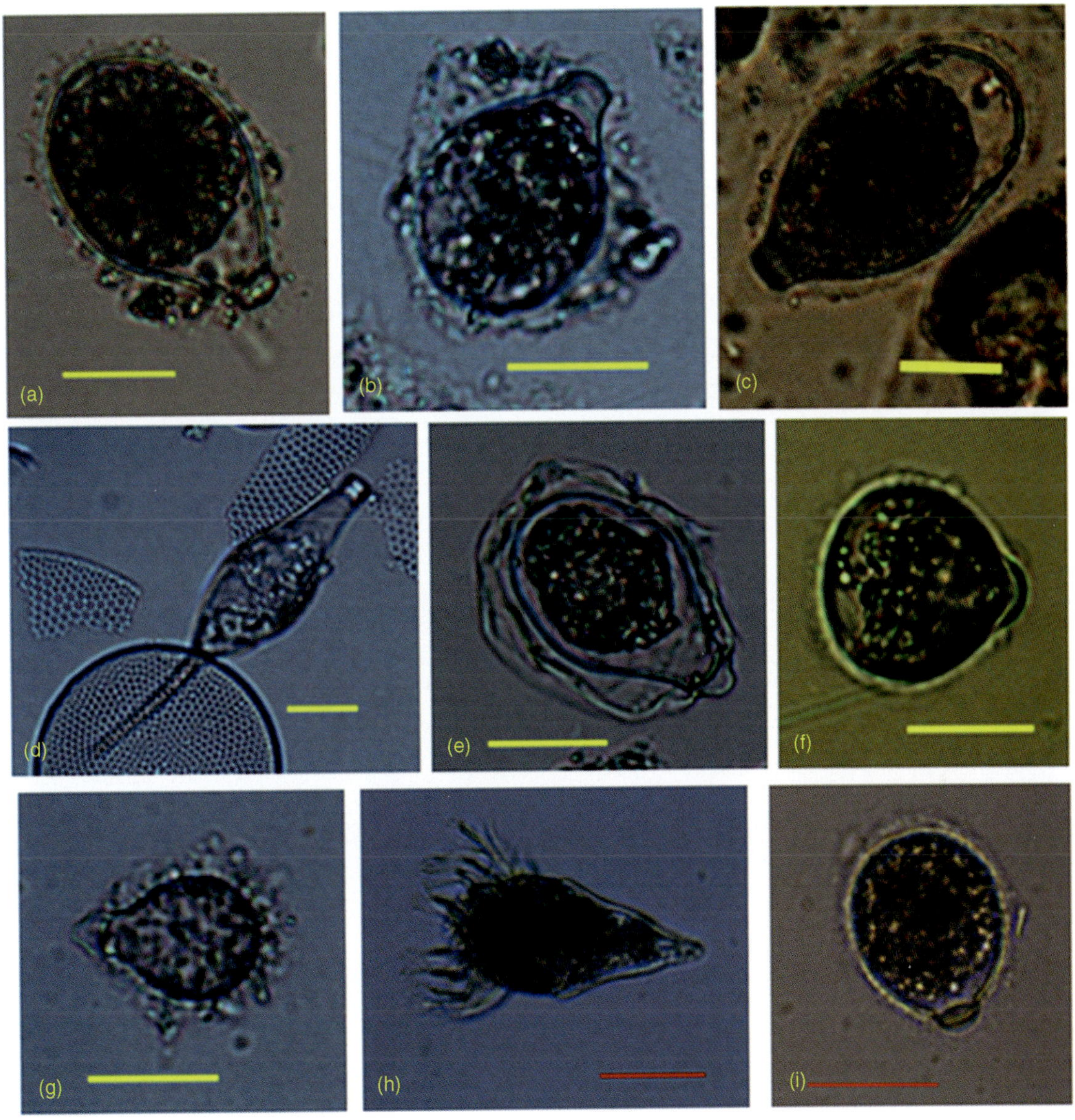

Plate 7.2 Photographs of various flask-shaped cysts of marine ciliates (a–g) observed in sediments of northern Japan, and a vegetative cell (h) of the oligotrich ciliate *Strombidium chlorophilum* identified from protargol-stained cells and the cyst (i) produced in a laboratory culture. Scale bars, 20 μm (a–g), 50 μm (h, i).

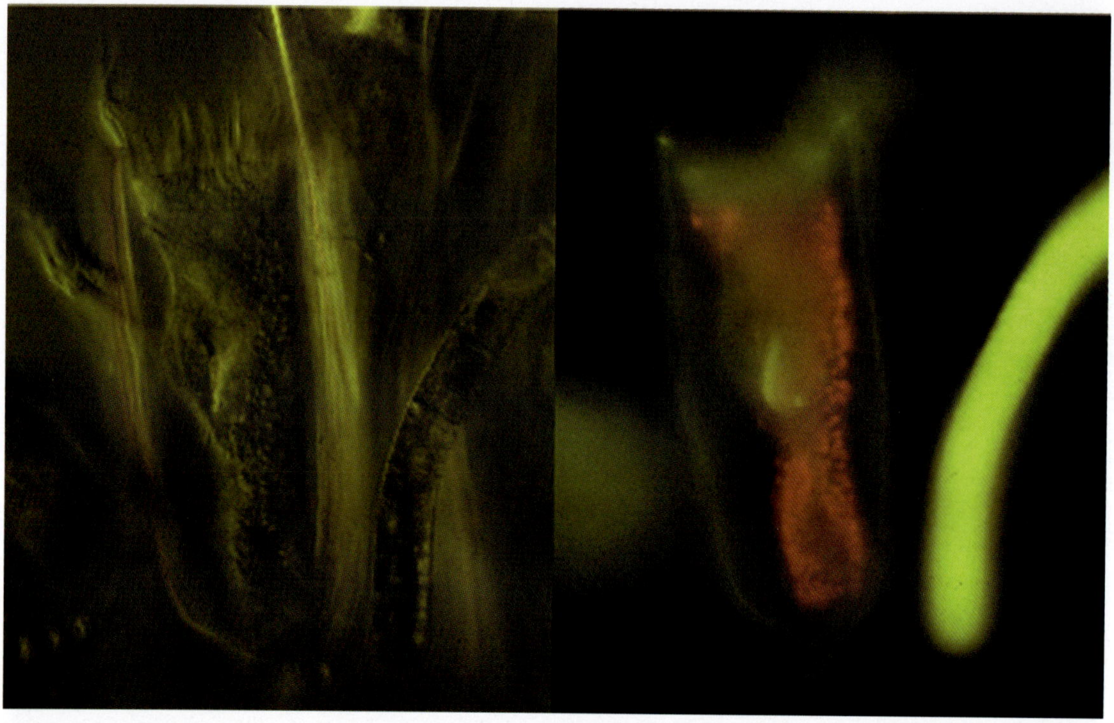

Plate 9.5 Differential interference contrast (left) and fluorescence (right) images of a large tintinnid collected 40 km north of Hawaii in 1983 by Mary Silver of the Unversity of California at Santa Cruz. Bright red autofluorescent bodies (white in the right panel), indicative of chlorophyll, were seen inside the tintinnid cell, and their distribution within the cytoplasm indicates that they are not food vacuole contents. The shape of the lorica suggests that this is a member of the genus *Favella* or *Metacylis*. For scale, the filament to the right of the tintinnid is a trichome of the cyanobacterium *Trichodesmium* sp., about 10 μm in width. Image courtesy of Dr Mary Silver.

Plate 10.1 Examples of species with distributions restricted to high latitudes. *Ptychocylis acuta* (a) and *Parafavella parumdendata* (b) are found only in Arctic and Subarctic waters while *Laackmanniella prolongata* (c) and *Cymatocylisdrygalskii* (d) are restricted to Antarctic waters.

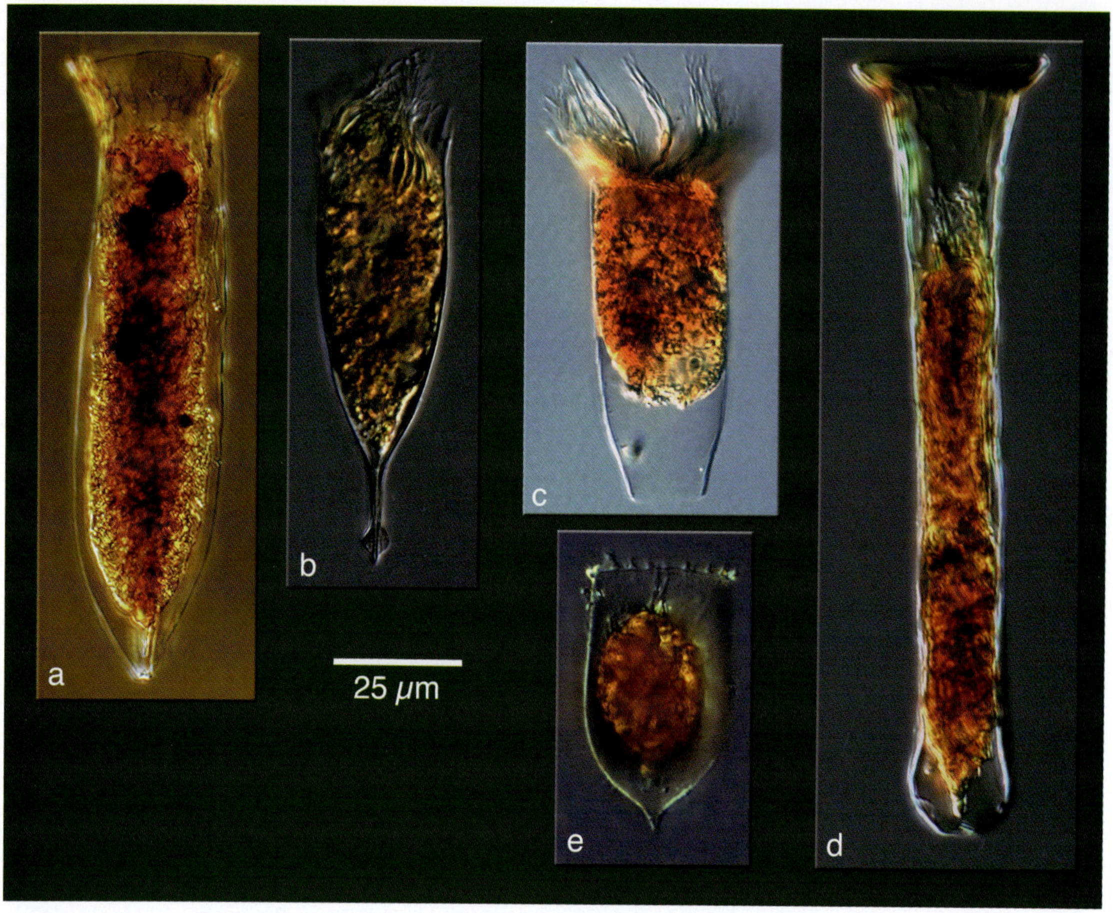

Plate 10.2 The most commonly reported and widely distributed tintinnid species. *Amphorides quadrilineata* (a) specimen from the Scripps Canyon area in the Pacific Ocean; *Dadayiella ganymedes* (b) from the Indian Ocean collected during the Tara Oceans Expedition; *Eutintinnus apertus* (c) and *Steenstrupiella steenstrupii* (d) both from the Bay of Villefranche in the N.W. Mediterranean Sea; *Acanthostomella norvegica* (e) from the Bering Sea. All specimens were preserved with Lugol's solution.

observed suspension feeding on microplankton by the small, cosmopolitan hydromedusa *Aglaura hemistoma*. Examination of preserved specimens revealed few items in the gut, but examination of live specimens showed that 100% contained unrecognizable protistan remains and about 20% contained tintinnids.

Among the larger scyphomedusae, at least five species, including bloom forming species such as the moon jelly (*Aurita aurita*), sea nettle (*Chrysaora quinquequecirrha*), and cannonball jelly (*Stomolophus melagris*) consume tintinnids (Table 5.4). The cannonball jelly is unusual among scyphomedusae in that it is a small particle feeder. In natural prey assemblages, clearance by the cannonball jelly of tintinnids is higher than of copepods but lower than of veliger larvae of mollusks. Tintinnids constituted 9% of the recognizable items in the gut (Larson 1991). In feeding experiments with the moon jelly, *Aurita aurita*, aloricate ciliates were selected over tintinnids (Stoecker et al. 1987). Olesen et al. (1996) observed that ephyrae of the sea nettle, *Chrysaora quinquecirrha*, had higher clearance for an aloricate ciliate, *Strobilidium*, than for tintinnids in natural assemblages.

The cannonball jelly digests tintinnids more rapidly than other recognizable prey (Larson 1991). This observation, along with the feeding experiments showing that the feeding rate of jellyfish on aloricate ciliates can be higher than on tintinnids, although aloricate ciliates are not recognizable in gut contents, suggests that the role of ciliates in the diet of jellyfish has been severely underestimated. It is possible tintinnids, along with other microzooplankton, are important in the diets of both small and large medusae. At high jellyfish densities, when jellyfish predation can deplete copepod and other metazoan zooplankton populations, rapidly growing microzooplankton populations may be a "maintenance" diet prolonging jellyfish blooms.

Ctenophores (comb jellies) are another ecologically important group of gelatinous zooplankton, particularly in estuarine, shelf and slope waters. Both cydipid and lobate ctenophores consume tintinnids (Table 5.4). Like the cnidarians, ctenophores are primarily predators on copepods but ingestion of tintinnids and other microplankton may be important for newly hatched tentaculate larvae, which are too delicate to capture and ingest adult copepods without getting damaged (Stoecker et al. 1987; Sullivan & Gifford 2007). However, Sullivan & Gifford (2007) noted that newly hatched *Mmemiopsis leidyi* larvae selected against ciliates and diatoms relative to thecate dinoflagellates and euglenoid flagellates. As in the case of cnidarians, limited data suggest that aloricate ciliates are preferred over similar-sized tintinnids (Stoecker et al. 1987).

Chaetognaths (arrow worms) are aggressive predators on copepods and fish larvae in shelf, slope, and oceanic waters. The adults do not feed on microzooplankton, but young (stage I) of the chaetognath *Sagitta elegans* prey on motile microplankton, including tintinnids, as well as on copepods (Pearre 1973, 1981; Alvarez-Cadena 1993). Young chaeotognaths probably also prey on aloricate ciliates, but these would be unrecognizable in gut contents. The importance of predation on tintinnids and other ciliates to survival and growth of young chaetognaths is unknown.

Pelagic tunicates (salps, doliolids, and appendicularians) are generalist ciliary mucous-net feeders with high clearance rates for a wide range of particle sizes. The appendicularians (larvaceans) have an incurrent filter on their house that limits ingestion of larger microplankton, which although cleared, are not ingested (Alldredge & Madin 1982; Vargas & Gonzalez 2004). Thus, with the exception of the appendicularians, which do not ingest large microplankton, the pelagic tunicates probably prey nonselectively on nano- and microplankton, including tintinnids. Tintinnids have been observed in the guts contents of salps, doliolids, and the appendicularian *Oikopleura* sp. (Table 5.4).

The mucous-net feeding thecosome pteropod *Limacina helicina* is an important component of polar and subpolar planktonic food webs and is prey for zooplanktivorous fish. In Arctic seas, it selectively feeds on tintinnids and other motile zooplankton (Gilmer & Harbison 1991). During midsummer, tintinnids can constitute 19% of the carbon in the gut of adult *L. helicina*. However, in Antarctica, although both *L. helicina* and tintinnids are present, Hopkins (1987) did not observe tintinnids in the guts of net caught specimens.

Other holoplankton

The importance of tintinnids and other microzooplankton in the diets of the less studied zooplankton taxa is largely unknown. Rotifers, ostracods, and cladocera ingest tintinnids (Table 5.5). The estuarine cladoceran *Evadne nordmanni* captures and holds, and presumably ingests, tintinnids (Bainbridge 1958). However, Sanchez et al. (2011), in feeding experiments with plankton from a Chilean fjord, found that the cladoceran *Podon leuckarti* did not ingest tintinnids and other microplankton.

Ichthyoplankton

The diets of fish larvae are a topic of interest to fisheries biologists because food limitation can be a significant source of mortality to fish larvae in the first few weeks of their lives (Hjort 1914; Lasker 1981; Turner 1984; von Herbing & Gallager 2000; Nagano et al. 2001; Montagnes et al. 2010). Tintinnids, usually along with thecate dinoflagellates, have long been recognized as components of the gut contents of larval fish, particularly first-feeding or yolk-sac larvae of clupeids (anchovies, menhaden, herring, sardine), gadids (cod, haddock, whiting, bib), flat fish (halibut, flounders, dab, sole), sciaenids (spot, croaker), acanthurids (surgeonfish), and the Ammondytidae (sand lances) (Table 5.6). Ingestion of tintinnids and other microplankton is important in initiating feeding. Once larval fish are few days old, they start to capture and ingest copepods successfully. The occurrence of tintinnids in the gut usually drops drastically once the larvae start eating copepods (Last 1978a, b; Jenkins 1987; Nagano et al. 2001; Stoecker & Govoni 1984; Van Der Meeren & Naess 1993; Fossum & Ellertsen 1994; Morote et al. 2008a; 2010; Grover 1998; Gadomski & Boehlert 1984; Shaheen et al. 2004; Wittenrich et al. 2009). Observations of feeding on tintinnids based on gut contents have been confirmed in laboratory experiments with first-feeding gulf menhaden (Stoecker & Govoni 1984), convict grouper (Nagano et al. 2000), in mesocosm experiments with cod (Van Der Meeren & Naess 1993) and in rearing experiments with common snook (Wittenrich et al. 2009).

The frequency of occurrence of tintinnids in the guts of first-feeding larvae is often very high, and numerically, tintinnids can constitute a significant proportion of the recognizable prey items. In a mesocosm experiment, Van Der Meeren & Naess (1993) found that tintinnids accounted for over 75% of the prey items in the guts of cod larvae younger than 17 days old. Nagano et al. (2001) reported that in first-feeding Japanese sand lance, tintinnids were 78% of the prey items in the gut. In both these instances tintinnid densities in the water were unusually high. More commonly, tintinnids account for less than 25% of the recognizable gut contents of larval fish (Govoni et al. 1983; Govoni & Chester 1990; Ogilvie 1938; Last 1978a, b; Jenkins 1987; Nagano et al. 2001; Stepien 1976; Figueiredo et al. 2007; Sherman & Honey 1971; Morote et al. 2008b; 2010; Watson & Davis 1989; Arthur 1976; Llopiz & Cowen 2009). Because of the relatively small individual biomass of tintinnids, the estimated biomass contribution of tintinnids to the diet is usually much lower than the numerical contribution. For example, Grover (1998) estimated that tintinnids contributed 29% of the numerical composition but only 3% of the volumetric composition of the diet of preflexion summer flounder larvae. In a mesocosm experiment in which tintinnids numerically accounted for most of the prey in the guts of cod larvae, they accounted for less than 25% of the ingested biomass (van der Meeren & Naess 1993).

Analyses of fish gut contents have focused on prey with hard parts that are resistant to digestion, such as copepods and tintinnids; "soft" prey have generally been ignored. A large proportion of the gut contents usually are not recognizable, consisting of mushy material (van der Meeren & Naess 1993; Govoni et al. 1983; Ohman et al. 1991). Aloricate protists, including choreotrich ciliates, turn to "mush" almost instantly in the gut (Ohman et al. 1991). In most studies of larval gut contents, fish are preserved with fixatives that do not preserve ciliates well and the gut contents are examined at relatively low magnification; this has probably led to a severe underestimation of ciliates in larval guts (Stoecker & Govoni 1984; Pierce & Turner 1992; Figueiredo et al. 2005; Montagnes et al. 2010). Fukami et al. (1999), using a combination of staining and epifluorescence microscopy, observed numerous aloricate ciliates and ciliate-like cells in various stages of digestion in the gut contents of fish larvae collected in Japanese coastal waters. Protists were abundant or moderately abundant in all but 9 out of 52 taxa of fish larvae, suggesting that they are a component of the diet of most larval fish. Figueiredo et al. (2007), using fixatives appropriate for ciliates, found aloricate ciliates in the guts of a variety fish larvae including spat, sand lance, dab, gobies, and cod collected in the Irish Sea. With application of a correction factor to account for the rapid digestion of protists, they estimated that protists constituted at least 25% of the diets of six species of larval fish (Figueiredo et al. 2005; Montagnes et al. 2010). Laboratory feeding experiments support these findings. Larvae of the northern anchovy (*Engraulis mordax*) ingest cultured *Strombidium* sp. in laboratory experiments; at high ciliate densities, the larvae can ingest ciliate carbon equivalent to 85–95% of their body carbon per day (Ohman et al. 1991). First-feeding cod larvae preferentially ingested an aloricate ciliate, *Balanion* sp., over faster swimming, larger copepod nauplii in laboratory experiments (von Herbing & Gallager 2000). Flounder larvae (*Platichthys flesus*) can consume hundreds of cultured *Euplotes* per hour (Figueiredo et al. 2007).

In contrast to most zooplankton, fish larvae are visual predators. Detection, capture, and ingestion of prey by larval fish depend on prey size, visibility contrast, prey swimming patterns and escape responses (Buskey et al. 1993). Although fish larvae will ingest non-motile particles such as copepod eggs and large diatoms, they prefer motile prey. Abundance in larval gut contents suggests they select motile protists (ciliates and dinoflagellates) over non-motile protists (diatoms) (Figueiredo et al. 2007). In laboratory experiments, menhaden larvae select the large tintinnid *Favella* sp. over smaller dinoflagellates (Stoecker & Govoni 1984). In the few field studies in which aloricate ciliates were included as well as tintinnids, data suggest that most fish larvae ingest more aloricate ciliates (which are usually more abundant) but that tintinnids are observed in the guts despite their low abundances in the environment (Figueiredo et al. 2007). Among the aloricate ciliates, fish larvae seem to prefer pigmented, mixotrophic ciliates to transparent, heterotrophic ciliates, probably due to the greater visibility of the pigmented species (Figueiredo et al. 2007; Van Der Meeren & Naess 1993). Although fish larvae often ingest tintinnids with a hyaline lorica, such as *Favella* spp., they may preferentially prey on tintinnids with agluttinated loricae, such as *Stenosemella* spp., because loricae formed with mineral particles may increase the visibility of the tintinnids to larvae (Govoni & Chester 1990). The preferences of fish larvae for microplankton prey is complicated and depends on prey size, visibility, movement, and prey escape responses as well as the experience of the fish larvae (Govoni et al. 1983; von Herbing & Gallager 2000).

Presence of tintinnids in the gut contents is an indicator of the important role of ciliates, in general, in the diets of fish-feeding fish larvae. Protists, including tintinnids, appear to be a largely overlooked trophic link, and sometimes the primary food source for young larvae (Montagnes et al. 2010). Nagano et al. (2001) associated feeding of Japanese sand lance larvae on aggregations of tintinnids with enhanced survival. In laboratory experiments, presence of tintinnid prey was associated with enhanced survival of first-feeding grouper, *Epinephelus septemfasciatus* (Nagano et al. 2000).

It is worth noting that tintinnids and other ciliates are likely a significant component of fish food webs, beyond the stage of the first-feeding larvae. One of the first attempts to depict a marine planktonic food web was Alistair Hardy's food web of the herring (Hardy 1924). The web showed the wide variety of organisms eaten by the different life-history stages of herring as well as the prey of the herring food items. Tintinnids were included as prey of small larvae. We now know that tintinnids and other ciliates are probably of importance to herring throughout its life directly or indirectly (Fig. 5.4).

Larvae and adults of benthic invertebrates

The meroplankton includes the larvae of many benthic invertebrates. Diet and nutrition affect larval survival, which in turn can affect benthic recruitment. Tintinnids are part of the diet of planktonic larvae of crabs, bryozoans and some worms (Table 5.7). Naidu (1983) saw tintinnids in the guts of veliger larvae of mollusks. However, most tintinnids are too big for veligers to ingest. For example, veligers of the eastern oyster, *Crassostrea virginica*, can only ingest particles up to about 30 μm (Baldwin & Newell 1991) and thus only smaller tintinnids are probably consumed by most larval mollusks.

Suspension-feeding adult bivalves can clear tintinnids from the water column and then either ingest them or deposit them in pseudo-feces (Table 5.7). Feeding by dense beds of blue mussels (*Mytilus edulis*) and zebra mussels (*Dreissena polymorpha*) can significantly reduce populations of tintinnids in shallow waters (Horstad et al. 1988; Pace et al. 1998). Unfortunately, because of the sampling methods used in these studies, there are no comparable data on removal of aloricate ciliates. However, in feeding experiments with green shell mussels (*Perna canaliculus*), high clearance of ciliates (primarily oligotrich ciliates) has been measured (Zeldis et al. 2004). Direct comparisons of bivalve grazing on tintinnids and aloricate ciliates are not available.

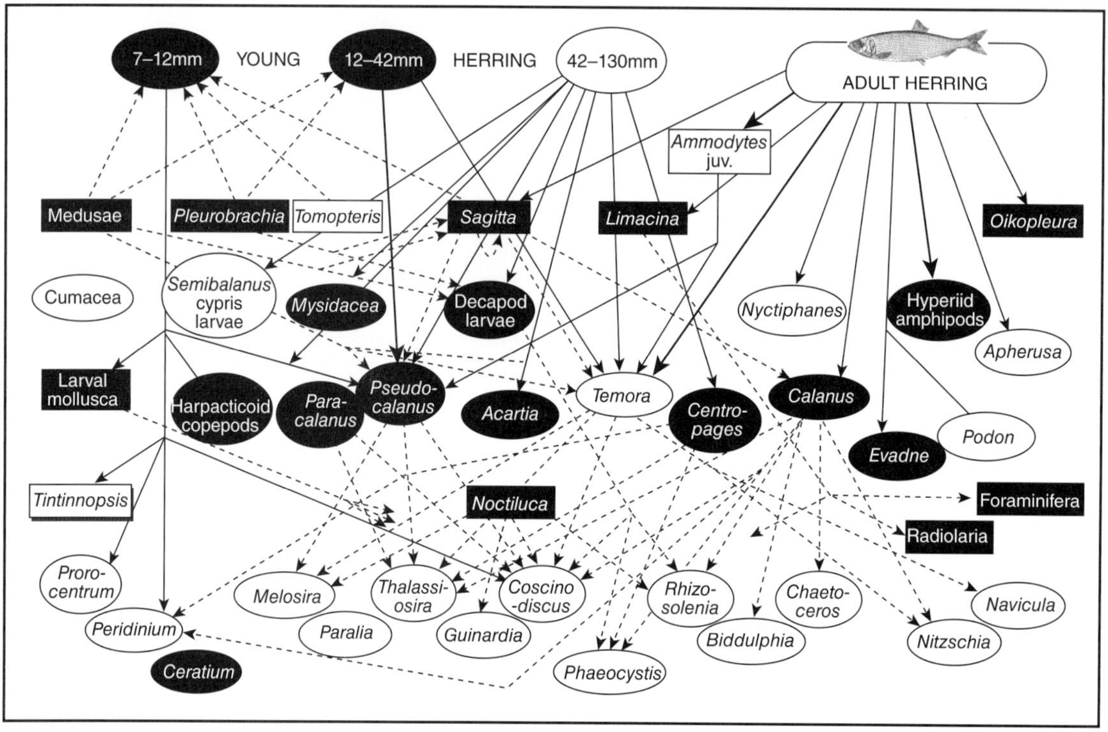

Fig. 5.4 The herring food web of Hardy (1924) showed the tintinnid *Tintinnopsis* (shadowed box in lower left corner) as part of the diet of small larval herring. Here the food web is re-drawn to show the organisms depicted by Hardy that are now known to feed on tintinnids and probably other planktonic ciliates (black backgrounds). In the original figure the line thickness denoted the importance of the link; the dashed lines denoted links based on the work from laboratories other than that of Hardy, primarily that of LeBour.

Tintinnids are a component of the diet of microphagous suspension feeders such as crinoids (feather stars) in shallow waters (Rutman & Fishelson 1969; La Touche & West 1980). Rutman & Fishelson (1969) found loricae of over 29 species of tintinnids in the gut contents of crinoids from the Red Sea and 49% of the recognizable gut contents were tintinnids or their loricae. However, crinoids feed on detritus as well as plankton, and empty loricae can persist in the detritus. Much of the ingestion might have been of empty loricae rather than intact tintinnids. The preservation techniques used were not suitable for preservation of intact ciliate cells.

Benthic polyp stages of scyphomedusae ingest ciliates and other protists. Kamiyama (2011) observed feeding by *Aurelia aurita* (moon jelly) polyps on ciliates, including the tintinnid *Favella ehrenbergii*. Clearance by polyps of *F. ehrenbergii* and of aloricate ciliates was similar. Stable isotope (N_{15}) labeling of the prey showed that the polyps assimilated nitrogen derived from the ciliates.

Many suspension-feeding benthic invertebrates probably consume tintinnids, but for most species there is little information on their rates of predation on microzooplankton. Most feeding experiments with benthic suspension feeders have been with phytoplankton. However, similarly to the planktonic copepods that have been studied, microzooplankton may be qualitatively and quantitatively an important component of the diet of benthic suspension feeders. There are some suggestions that tintinnids with heavy, agglutinated loricae occur closer to the bottom than aloricate choreotrichs, and thus may be relatively more available as prey to benthic suspension feeders (Rutman & Fishelson 1969).

5.3 ANTI-PREDATOR DEFENSES OF TINTINNIDS

The lorica may partly protect tintinnids against some types of predation. Many protistan predators consume tintinnids but do not ingest the lorica; perhaps the presence of a lorica provides protection or increases handling time, reducing predation efficiency on tintinnids. Tintinnids do not have escape "jumping" responses such as found in some aloricate ciliates such as *Mesodinium*, *Halteria*, and *Strobilidium* spp. However, tintinnids often become motionless for a short period in response to disturbances, which may allow them to escape detection by the predator or sink out of the encounter zone of the predator (Broglio et al. 2001). Capriulo et al. (1982) calculated that the dense lorica of *Stenosemella* sp., agluttinated with mineral particles, could allow the species to sink rapidly after disturbance, decreasing predation by copepods.

Some tintinnids, including *Codonaria*, *Codonella*, *Codonellopsis*, *Dityocysta*, *Salpingacantha*, and *Salpingella*, possess a membranous apparatus that can close the lorica opening in response to disturbance (Agatha 2010a). It is possible that closure is protective (Biedermann 1893 and Daday 1887, cited in Agatha 2010a) and although not complete, might decrease release of chemical signals from the tintinnid that elicit responses from the predator and thus might reduce successful attacks (Agatha 2010a). Unfortunately, there are no behavioral observations of predator–prey interactions involving tintinnids with a lorica closing mechanism.

The lorica may also be somewhat effective as a post-capture defense against predation by crustacean zooplankton. Copepods can, in some cases, ingest the tintinnid cell without ingesting the lorica, although in other cases tintinnids appear to be ingested whole (Stoecker & Sanders 1985; Tackx et al. 1995; Robertson 1983). In experiments with the tintinnid *Metacylis* sp. and the copepod *Acartia clausi*, Broglio et al. (2001) observed that once captured by the copepod, the tintinnid was ingested 96% of the time, suggesting that for this predator–prey pair, the lorica is not an effective post capture deterrent. Regardless of the precise mechanisms involved, there is some evidence that tintinnids, compared with similar sized aloricate ciliates, may be less subject to predation by copepods. Nishibe et al. (2010) recorded lower clearance rates for *Oithona simulis* feeding on tintinnids than either large or smaller aloricate ciliates in shipboard experiments with natural communities (Fig. 5.5).

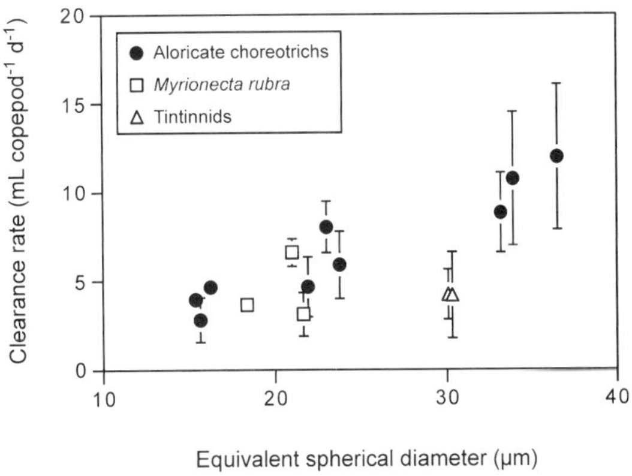

Fig. 5.5 Tintinnids appeared to be less subject to predation than aloricate ciliates such as aloricate choreotrichs and *Myrionecta rubra* (= *Mesodinium rubrum*) by the cyclopoid copepod *Oithona simulus*, as shown in the results of experiments with natural communities of ciliates. Error bars represent standard deviation. Figure from Nishibe et al. (2010) with permission from the Plankton Society of Japan.

Defense against gelatinous predators, such as cnidarians and ctenophores, may be quite different from defense against crustacean predators. Both medusae and ctenophores appear to have much higher clearance rates for aloricate ciliates than for tintinnids (Stoecker et al. 1987a, b; Olesen et al. 1996). Medusae and ctenophores sort prey from inert materials based on mechanical and chemical clues, with amino acids and proteins stimulating feeding and particle retention (Southward 1955). The non-living lorica, particularly one covered with mineral particles, may allow a proportion of the tintinnids to avoid triggering capture responses or may result in post-capture rejection. The ctenophore *Mnemiopsis* tends to reject spiny polycheate larvae as prey; the acute shape of the posterior end of the lorica and spiny protrusions on some loricae may discourage predation as well (Stoecker et al. 1987). Behavioral responses of tintinnids to disturbance, cessation of swimming, withdrawal into the lorica, and, in some species, closure of the oral opening (Agatha 2010a) may help to disguise tintinnids as inert particle to some gelatinous predators. Tintinnids are ingested and digested by cnidarians and ctenophores, so these potential defenses are, at best, only partly effective.

Attachment between the lorica and diatoms or other surfaces may reduce predation risk. The association of *Eutintinnus apertus* and *E. pinguis* with spiny diatoms

(*Chaetoceros* spp.) is well known, but other tintinnid taxa have also been observed physically associated with large diatoms (Kofoid & Campbell 1939; Taylor 1982; Gomez 2007). For example, *Salpingella subconica* forms associations with the ribbon-forming diatom *Fragilariopsis doliolus* (Froneman et al. 1998). The tintinnid may benefit from attachment to the diatom in several ways, including decreased susceptibility to predation (Froneman et al. 1998; Gomez 2007). Attachment to diatoms may increase the effective size of the tintinnid and decrease its vulnerability to small size predators but perhaps increase risk from slightly larger predators. Predation risk would depend on the predators present and their preferred prey size. The coastal tintinnid *Eutintinnus inquilinus* attaches its lorica to solid surfaces, including suspended particles. *E. inquilinus* is less susceptible to predation by copepods than is a morphologically similar tintinnid that is entirely free swimming (Jonsson et al. 2004). It seems likely that attachment to diatoms or inert particles would also decrease susceptibility of tintinnids to protistan predators and some gelatinous predators.

Some, but not all, species of tintinnids form cysts (Chapter 7). Encystment may be both a strategy to survive unsuitable growth conditions and a way to avoid mortality due to predation (Reid 1987). Predation and parasitism (Chapter 6) are thought to be the main causes of mortality in microzooplankton. Populations cannot withstand high rates of mortality when the growth rate is low owing to unsuitable conditions. Cyst formation may be a way to maintain stable populations during unfavorable conditions.

5.4 TOP-DOWN CONTROL OF TINTINNIDS

There have been several studies of top-down control of planktonic ciliate populations, but the focus has been on aloricate planktonic ciliates, because tintinnids rarely dominate the ciliate assemblage. Maximum intrinsic growth rates (r_{max}) for tintinnids range from approximately 0.5 to 1.8 d^{-1} (0.7 to 2.5 doublings d^{-1}) (Chapter 4; also reviewed in Pierce & Turner 1992), but unless conditions are ideal, lower intrinsic growth can be expected. Mortality (m) due to predation, parasitism, and other factors can reduce net growth rate (K), with $K = r - m$. In most planktonic ecosystems, predation is a major source of mortality for tintinnids and other ciliates, although, particularly when tintinnid populations are dense, parasitism can also be important (Chapter 6). The dominant predators on planktonic ciliates in most marine ecosystems are copepods with predation pressure highest on the larger ciliates. In temperate, coastal waters, Nielsen & Kiørboe (1994) estimated that the copepod community could check growth of ciliates larger than 50 µm year round but that copepod predation had only a moderate effect on smaller ciliates because of their rapid growth. Calbet & Saiz (2005) estimated, based on global averages of ciliate abundance and copepod ingestion rates, that adult copepod densities of approximately 7 l^{-1} would be required to reduce the ciliate standing stock by 50% (mortality due to predation, or g, ~0.7 d^{-1}). Such high copepod densities are mostly restricted to copepod swarms in small areas and to highly productive coastal waters (Calbet & Saiz 2005). Assuming that clearance of copepods for tintinnids is similar to those for ciliates in general, in most waters copepod predation alone should not prevent the net growth of tintinnid populations. Predation pressure by copepods on ciliates is often highest near the end of phytoplankton blooms when copepods are abundant, algae are scarce but microzooplankton grazers are abundant, at these times predation may prevent net growth of ciliates (Stoecker & Sanders 1985; Stoecker and Egloff 1987). Predation by copepods has a minor influence on ciliate standing stocks in many waters (Lonsdale et al. 2000; Nishibe et al. 2010; Broglio et al. 2004). However, under other circumstances copepod predation on ciliates can lead to trophic cascades, releasing phytoplankton from microzooplankton grazing pressure (Stoecker & Sanders 1985; Irigoien et al. 2005; Calbet & Saiz 2005; Reaugh et al. 2007; Nejstgaard et al. 2001).

Gelatinous zooplankton might exert top down control on ciliates as many gelatinous zooplankton can occur in dense swarms (Alldredge & Madin 1982; Olesen et al. 1996). Although medusae and ctenophores consume tintinnids, they also exert top down control on copepods, the net result may be that they reduce predation on tintinnids and other microplankton due to trophic cascading (Stoecker et al. 1987a, b). Planktonic sarcodines sometimes occur at high densities in surface waters of tropical and subtropical oceans; they may be significant predators on tintinnids and other ciliates in the open ocean. Most other planktonic predators, including fish larvae (Jenkins 1987; Montagnes et al. 2010), probably have little impact on tintinnid populations. However, in shallow waters, dense beds of benthic bivalves, owing to their high clearance rates, can exert significant top down control on tintinnids (Horstad et al. 1988; Pace et al. 1998).

Although predation may seldom prevent the net growth of tintinnid populations, it may be a major factor in the evolution of tintinnids by selecting for lorica shape and size (Chapter 2) and in maintaining tintinnid diversity (Chapter 10). In natural assemblages, copepods often have the highest clearance rates for the most abundant tintinnid species (Dolan & Gallegos 2001) or on medium-sized rather than small or large tintinnid species (Cariou et al. 1999). Copepod predation can decrease the diversity of slow-growing tintinnid assemblages, but increase the diversity of tintinnids in assemblages dominated by fast growing species (Dolan and Gallegos 2001). Dolan et al. (2002) observed that in the Mediterranean Sea, tintinnid species richness, diversity of lorica diameters, and lorica lengths were not correlated with copepod abundance, suggesting that tintinnid diversity is more influenced by resource diversity than predation.

5.5 IMPORTANCE OF TINTINNIDS AS PREY FOR THE PREDATOR

Although tintinnids occasionally "bloom" in coastal waters or frontal areas, their abundance and biomass are much less than those of oligotrichs and aloricate choreotrichs and thus they are probably less important as prey than the aloricate ciliates (Chapter 9) (Pierce & Turner 1992; Figueiredo et al. 2009). An exception to this may be predators that appear to specialize on tintinnids such as some protistan predators, young chaetognaths, and first-feeding fish larvae (Tables 5.1–5.6). However, in many cases apparent preference for tintinnids appears due to underestimation of predation on aloricate ciliates and other soft-bodied microplankton.

Clearance rates of most zooplankton for ciliates, including tintinnids, are usually higher than for co-occurring phytoplankton. This is partly because most ciliates are greater than 10 μm in size, and thus in the preferred prey size range of most suspension feeders, whereas in many environments the phytoplankton is dominated by cells less than 5 μm in size, which are too small to be efficiently cleared by crustacean grazers. Ciliates, including tintinnids, are quantitatively important in the diet of copepods, especially when abundances of large-cell-size phytoplankton are low (Ohman & Runge 1994; Gifford & Dagg 1991). Globally, the contribution of ciliates to the diet of copepods is estimated to be approximately 30% of the daily ration (Calbet & Saiz 2005). Less is known about the role of ciliates in the diet of other zooplankton and almost nothing about their contribution to the diets of benthic suspension feeders. In addition to their quantitative significance in the diet (carbon contribution), ciliates, in general, may be qualitatively significant as high-quality food items. Inclusion of *Favella* sp. in the diets of copepods boosts egg production (Stoecker & Egloff 1987). Ciliates often have lower C:N ratios and are a better source of amino acids and proteins than phytoplankton and detritus (Stoecker & Capuzzo 1990; Gifford 1991). Metazoan zooplankton, benthic invertebrates, and fish all require specific essential polyunsaturated fatty acids, amino acids, and sterols in their diet. Tintinnids and other microzooplankton are probably an important source of these essential nutrients for suspension feeders (reviewed in Stoecker & Capuzzo 1990; Sanders & Wickham 1993; Wickham 1995; Broglio et al. 2003). However, fatty-acid profiles of ciliates vary with that of the prey they ingest. Broglio et al. (2003) found that in the copepod *Acartia tonsa* a mono-specific ciliate diet did not support higher egg production efficiency than mono-specific algal diets.

5.6 TINTINNIDS AS VECTORS FOR ALGAL TOXINS

Large tintinnids, particularly *Favella* spp., are often associated with dinoflagellate blooms. They can ingest toxic dinoflagellates including *Alexandrium* spp., which can contain paralytic shellfish poisoning (PSP) toxins, *Dinophysis* spp., which can contain okadaic acid (responsible for diarrheic shellfish poisoning), and small dinoflagellates that produce azaspiracids, which are polyether toxins implicated in shellfish poisoning in humans (see Maneiro et al. 2000; Kamiyama & Suzuki 2006; Krock et al. 2009). Tintinnid predation may help regulate toxic blooms, but tintinnids may be vectors of toxin transfer to higher trophic levels (White 1981). The PSP toxin content per unit biomass of *Favella taraikaensis* cultured on *A. tamarense* is lower than that of their dinoflagellate prey, indicating that the tintinnids do not bio-magnify PSP (Kamiyama & Suzuki 2006). *F. taraikaensis* rapidly excreted the ingested PSP, lowering the particulate toxin content of the water.

However, tintinnids can be vectors for toxin transfer because one tintinnid can contain several ingested toxic cells. By repackaging the toxin into a larger, although less concentrated package, grazers that might not ingest the dinoflagellate could be exposed to the dinoflagellate toxin. PSP toxins can be found in plankton size fractions containing tintinnids and heterotrophic dinoflagellates, but few un-ingested *Alexandrium* cells (Turner et al. 2005). This could be important in transfer of PSP toxins to fish larvae and copepods.

Grazing experiments with *Favella serrata* during blooms of *Dinophysis acuminata*, a dinoflagellate producing okadaic acid, have shown that the tintinnids ingest the toxic dinoflagellates and continue to grow (Maneiro et al. 2000). Okadaic acid content in seston size fractions correlated positively with *F. serrata*, indicating that the tintinnids could be a vector for toxin transfer. As in the case of *Favella* and *Alexandrium*, there was no indication that the tintinnids biomagnified the toxins.

In Scottish coastal waters, concentrations of azapiracids have been associated with plankton size fractions containing *Favella ehrenbergii*. However, tintinnids isolated from the tows and cultured on non-toxic dinoflagellates in the laboratory rapidly lost their azapiracid content, indicating that the tintinnids were temporary vectors for the toxin (Krock et al. 2009).

5.7 KEY POINTS

1. Tintinnids are prey for many zooplankton, suspension-feeding benthic invertebrates, and fish larvae. Most predators on tintinnids are omnivores.

2. Presence or absence of loricae in guts or fecal pellet contents is not always a reliable indicator of predation or lack of predation on tintinnids.

3. One of the functions of the lorica appears to be as a partial defense against predation. Predation on tintinnids appears to be lower than on similar-sized aloricate ciliates. However, conclusive data are lacking.

4. Under most circumstances, predation does not appear to prevent net growth of tintinnids. The exceptions to this include top-down control in coastal waters by high copepod populations and by dense beds of bivalves in shallow waters.

5. The presence of tintinnids in food vacuoles of protists, in gut contents of metazoa, and in fecal pellets is an indicator of the general importance of planktonic ciliates in the diets of these organisms.

6. Tintinnids, along with aloricate ciliates, may be significant components in the diet of many suspension feeders usually categorized as grazers on phytoplankton and in the diet of first-feeding larval fish.

ACKNOWLEDGMENTS

I thank Dr R.R.L. Guillard for introducing me to tintinnids by providing a sample of water containing live *Favella*, and then providing me with microalgae from his wonderful collection of marine phytoplankton so that I could try to culture them. His mentorship when I was a postdoctoral scholar many years ago fostered my interest in the ecology and physiology of tintinnids and other ciliates and provided me with the tools to study them experimentally.

Chapter 6

PARASITES OF TINTINNIDS

D. Wayne Coats and Tsvetan R. Bachvaroff
Smithsonian Environmental Research Center, Edgewater, MD, USA

The Biology and Ecology of Tintinnid Ciliates: Models for Marine Plankton, First Edition. Edited by John R. Dolan, David J.S. Montagnes, Sabine Agatha, D. Wayne Coats, and Diane K. Stoecker.

6.1 BACKGROUND

Heterotrophic symbionts of ciliates

Freshwater and marine ciliates host a diverse array of heterotrophic microbes including viruses, bacteria, fungi, and various protists (see Ball (1969) and Lauckner (1980) for review of the early literature). Relationships between ciliates and their endo- and ectosymbionts range from passive associations to mutualism and parasitism. True viral infections of ciliates are apparently unknown (Munn 2006), but ciliates can acquire viruses directly from the water via macropinocytosis (Pinheiro et al. 2007) or through ingestion of food items (Kucera 1992; Lobo-da-Cunha & Azevedo 1992). Ingested viruses can be deactivated by digestion (Pinheiro et al. 2008) or can persist, even multiply, within the ciliate food vacuole to be released in fecal pellets (Lauckner 1980). Thus, ciliates may act as passive vectors for the dispersal and transmission of viral disease.

About 60 types of bacteria, many of which are likely mutualistic (Vannini et al. 2003), are known to occur in the cytoplasm or nuclei of ciliates (Fokin et al. 2003). A striking example from the perspective of human health is the ability of pneumonia-causing *Legionella* bacteria to live and reproduce in the cytoplasm of *Tetrahymena*, without apparent harm to the ciliate (Fields et al. 1984; Barbaree et al. 1986; Steele & McLennan 1996). When ingested by *Tetrahymena*, fecal pellets of virulent *L. pneumophila* are released from the ciliate and may contribute to the transmission of Legionnaires' disease (Berk et al. 2008). Some endosymbiotic bacteria, however, clearly have negative effects on their ciliate hosts. *Holospora* species that invade the micronucleus of *Paramecium* are an interesting example, as they destroy the ciliate's generative nucleus, thereby resulting in genetic castration (Görtz 1988). Other bacterial symbionts have a more direct effect on ciliate survival, as they release toxins into the medium that kill sensitive individuals of the same ciliate species or are transferred to sensitive cells during the sexual cycle of the host organisms (Verni et al. 1977; Heckmann 1983).

The wide assortment of protists that parasitize ciliates encompasses fungi (Foissner & Foissner 1986, 1995; Canter & Dick 1994), euglenoids (Schönfeld 1959), kinetoplastids (Wille et al. 1981; Görtz & Dieckmann 1987), dinoflagellates (Cachon & Cachon 1987), and even other ciliates (Canter et al. 1990). Infected ciliates typically succumb to their parasite and often appear unable to reproduce; however, there are some notable exceptions. *Euplotespora binucleata*, a microsporidial parasite of *Euplotes woodruffi*, appears to have little effect on its host, as infected and uninfected cultures of the ciliate have comparable growth rates (Fokin et al. 2008). In some host–parasite systems, complete recovery from parasitism seems possible. During stationary growth, the macronucleus of *Paramecium trichium* is overtaken by *Leptomonas karyophilus*, resulting in death of the host. Under food-replete conditions, however, reproduction of the ciliate exceeded that of the parasite, resulting in some host cells devoid of parasites (Gillies & Hanson 1963). Resistance to parasitism can also differ across host strains. *Leptomonas ciliatorum*, a macronuclear parasite of the hypotrich ciliate *Paraholosticha sterkii*, had no discernable effect on growth rate of its original host isolate, but killed a second host strain obtained from a different location (Görtz & Dieckmann 1987).

Lifestyles of ciliate pathogens represent obligate intracellular parasites, typically free-living species that are opportunistically endoparasitic, and extracellular forms that could easily be considered predators, as they often feed gregariously, devouring the host from the outside. Indeed, some species once classified as ectoparasites (e.g., *Spiromonas gonderi*; Foissner & Foissner 1984) are now considered to be free-living, micropredators of ciliates (Simpson & Patterson 1996). The distinction between predatory and ectoparasitic relationships among protists is often difficult to assess, particularly when dealing with species that kill and consume their host/prey on time scales of minutes to hours. For well-studied groups like the heterotrophic dinoflagellates, species fall along a predator-parasite continuum, with trophic classification of intermediate forms being rather arbitrary (Coats 1999). Gaines & Elbrächter (1987) distinguished parasitic dinoflagellates as species that have morphologically different feeding and reproductive stages and that produce more than two daughter cells after each feeding event. Because all but one parasite reported to infect tintinnids is a dinoflagellate, Gaines and Elbrächter's definition is adopted here.

Discovery of tintinnid parasites

In samples from Messina, Italy, and the Canary Islands, Haeckel (1873) found small, spherical cells in the cytoplasm of two tintinnid species, *Cyttarocylis cassis* (reported as *Dictyocysta cassis*) and *Tintinnopsis campanula* (reported as *Codonella campanella*), interpreting them to be ciliate reproductive propagules or spores (note that tintinnid taxonomy used in this chapter follows Kofoid & Campbell (1929, 1939), with exceptions as adopted in Chapter 3). Haeckel's spherical spores (Fig. 6.1a) are now considered the first report of dinoflagellate parasitism in ciliates. Some years later, Laackmann (1906, 1908) mistook parasitism of *Tintinnopsis campanula* and *Coxliella helix* (reported as *Cyttarocylis helix*; basionym: *Tintinnus helix* Claparède & Lachmann 1858) as sexual reproduction, reporting infections as "sporocysts." His "sporocysts" of *T. campanula* were certainly an intracellular parasite (Fig. 6.1b) that emerged from the host to produce either spherical microspores or gymnodinoid macrospores. By contrast, "sporocysts" of *T. helix* were clearly ectoparasites, as small individuals attached to the surface of the ciliate increased in size as the host became gradually smaller (Fig 6.1c, d). The ectoparasite of *T. helix* eventually divided to produce gymnodinoid macrospores (Fig. 6.1e) similar in appearance and behavior to those Laackmann observed in *T. campanula*. Lohmann (1908), a mentor of Laackmann, described colorless, gymnodinoid macrospores that emerged from the lorica of tintinnids he identified as *Tintinnopsis nucula*, but that Kofoid & Campbell (1939) considered to be *Stenosemella nivalis*. Lohmann recognized the parasitic nature of the spores arising from *S. nivalis* and provided for them the name *Gymnodinium tintinnicola*. Thus, Lohmann provided the first formal description of a dinoflagellate parasitic on tintinnids. The macrospores of *G. tintinnicola* (Fig. 6.1f) were clearly dinophycean, having a deep cingulum that separated the anterior portion of the cell, the episome, from the posterior portion, the hyposome, that was creased by a well-defined sulcus. The two flagella had a dinokont arrangement, as they arose laterally on the cell, with one transverse, beating in an undulating fashion within the cingulum, and the other longitudinal, extending posteriad from the juncture of the cingulum and sulcus. For detailed consideration of dinoflagellate morphology and structural terminology, the interested reader is referred to Taylor (1987).

Entz (1909b) encountered spores of a dinoflagellate that he considered to be a parasite of *Favella ehrenbergii* (reported as *Cyttarocylis ehrenbergii*), but did not

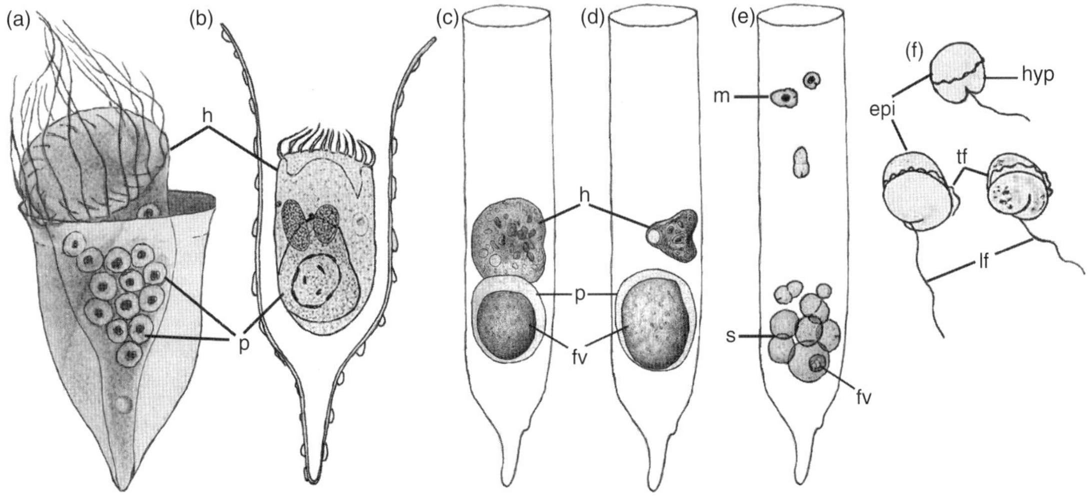

Fig. 6.1 *Duboscquella tintinnicola* and unnamed parasites mistaken as tintinnid reproductive stages. (a) Parasites in the cytoplasm of *Cyttarocylis cassis* (adapted from Haeckel 1873). (b) Intracellular parasite of *Tintinnopsis campanula* (adapted from Laackmann 1908). (c–e) Extracellular parasite of *Tintinnus helix* (adapted from Laackmann 1908). (f) Macrospores of *Duboscquella* (*Gymnodinium*) *tintinnicola* (adapted from Lohmann 1908). Episome (epi); food vacuole (fv); host (h); hyposome (hyp); longitudinal flagellum (lf); mature macrospore (m); parasite (p); sporocytes, sporogenic division products preceding spore formation, (s); transverse flagellum (tf).

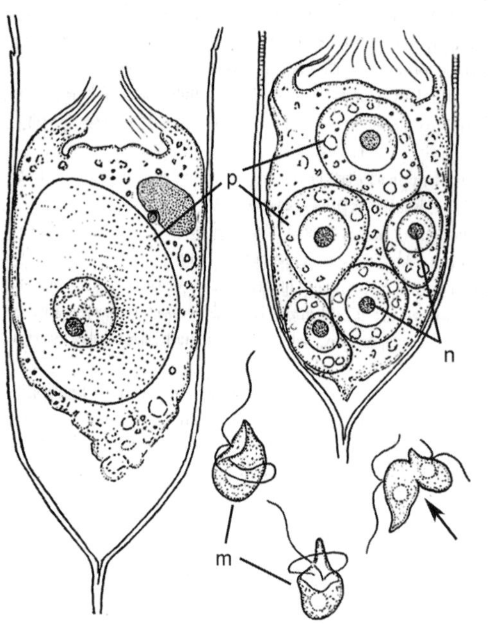

Fig. 6.2 Intracellular parasite of *Favella ehrenbergii* (adapted from Chatton 1920). Mature macrospore (m); nucleolus (n); parasite (p); last sporogenic division (arrow) mistaken by Duboscq & Collin (1910) as gamete fusion.

provide illustrations or a species name. A year later, Duboscq & Collin (1910) described what they interpreted as sexual reproduction in an unnamed parasite infecting *Favella ehrenbergii*. Although their account of the sexual cycle is likely erroneous (Cachon 1964), their observations and illustrations subsequently published by Chatton (1920) provided important details on the morphology of the parasite and its spores (Fig. 6.2). The actively growing parasite, the trophont, was clearly intracellular and, except for the large central nucleolus, closely resembled Laackmann's (1908) sporocyst of *T. campanula* (Fig. 6.1b). Some specimens contained several parasites similar to the report of Haeckel (1873), leading Duboscq and Collin to believe that sporogenesis began inside the host, with further divisions outside the host resulting in spores. Because cell division of the parasite inside the host cytoplasm was not observed, it is probable that Duboscq and Collin actually saw multiple infections of the host (Cachon 1964; Coats et al. 1994), rather than intracellular sporogenesis. Unlike the dinophycean macrospores observed by Laackmann (1908) and Lohmann (1908), the parasite studied by Duboscq & Collin produced *Oxyrrhis*-like macrospores that had a pointed anterior end, a rounded posterior, and a pair of laterally placed flagella with one extending anteriad and the other wrapping loosely around the cell.

Chatton (1920) created the genus *Duboscquella* to include all the tintinnid parasites mentioned above. He designated *Duboscquella tintinnicola* (Lohmann, 1908) Chatton, 1920 as the type species and accepted the report of Duboscq & Collin (1910) as a redescription of *Gymnodinium tintinnicola*, even though spore morphology was not consistent with that reported by Lohmann (1908). Over the past 90 years, parasites resembling the organism studied by Duboscq & Collin (1910) have been reported as *Duboscquella tintinni* (Campbell 1927, *lapsus pennae* for *Duboscquella tintinnicola*), *Duboscquella tintinnicola* (Hofker 1931), and *Duboscquella* sp. (Stoecker et al. 1983; Akselman & Santinelli 1989; Agatha & Riedel-Lorjé 2006; Harada et al. 2007; Konovalova 2007). In addition, seven other species of *Duboscquella* have been described (Chatton 1952; Cachon 1964; Coats 1988), four of which are intracellular parasites of tintinnids. Based on lifestyle and lack of permanently condensed chromosomes in these species, the genus *Duboscquella* has been placed in the subphylum Syndinea. Recently, Coats et al. (2012) used spore morphology and apparent ectoparasitic lifestyle of *Duboscquella tintinnicola* to separate the species from other members of the genus *Duboscquella*, placing the latter in a new genus, *Euduboscquella*. Thus, *Duboscquella tintinnicola* now belongs to the subphylum Dinokaryota as an ectoparasite of tintinnids, whereas intracellular parasites previously included in *Duboscquella* remain in the subphylum Syndinea as species of *Euduboscquella*.

Three other genera of parasitic dinoflagellates, *Amoebophrya*, *Duboscquodinium*, and *Tintinnophagus*, contain species that infect tintinnids: *Amoebophrya tintinni* is a syndinean parasite of *Xystonella lohmanni* (Cachon 1964), whereas *Duboscquodinium collini*, *Duboscquodinium kofoidi*, and *Tintinnophagus acutus* are dinophycean parasites of *Eutintinnus fraknoii*, *Tintinnopsis campanula*, and *Tintinnopsis cylindrica*, respectively (Chatton 1952; Coats et al. 2010). Lastly, Campbell (1926) described *Karyoclastis tintinni* (Fig. 6.3) as a small ovoid parasite (~1 μm diameter) that reproduces by binary fission, filling and eventually rupturing the macronuclei of its host, *Stenosemella nivalis* (reported as *Tintinnopsis nucula*). Subsequent accounts

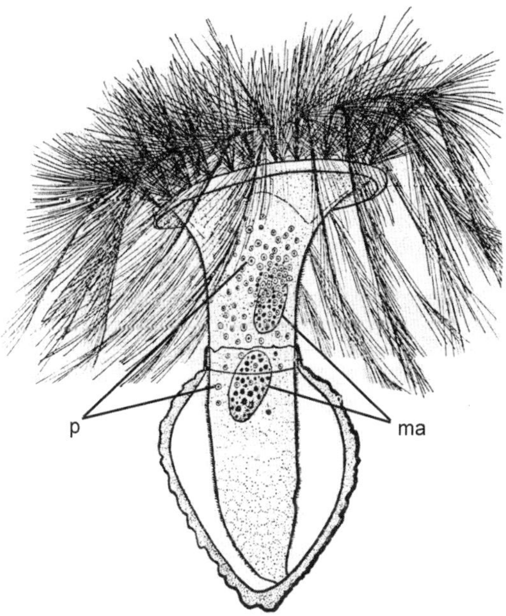

Fig. 6.3 *Karyoclastis tintinni* infecting *Stenosemella nivalis* (from Campbell 1926). Parasites (p) are emerging from the anterior of the two host macronuclei (ma).

of *K. tintinni*, or organisms resembling it, are very limited (Campbell 1927; Hada 1932a), with Cachon (1964) concluding that Campbell (1926) mistook "enclaves resulting from the metabolism of the ciliate" (translated from the original French) as parasitism.

6.2 DINOFLAGELLATE PARASITES: MORPHOLOGY AND INFECTION CYCLE

Syndinean parasites

Six species representing two genera of syndinean dinoflagellates (subphylum Syndinea Corliss, 1984) are intracellular parasites of tintinnids: *Amoebophrya tintinni*, *Euduboscquella anisospora*, *E. aspida*, *E. cachoni*, *E. cnemata*, and *E. crenulata* (Chatton 1920; Cachon 1964; Coats 1988; Coats et al. 2012). The two genera have similar life cycles, with infections directly transmitted to new hosts by spores, but show marked divergence in trophont development, timing of nuclear division relative to the infection cycle, and morphogenetic events associated with emergence of the parasite from the host cell. Differences also exist in spore morphology and diversity. Although both genera form motile dinokont spores that have a poorly defined girdle and sulcus, some species of *Euduboscquella* also produce non-motile or weakly motile spores that apparently lack flagella or have a single flagellum (see below).

Amoebophrya infections are established when the dinospore actively penetrates the host cell membrane (Fig. 6.4). The dinospore loses its flagella when entering the host cytoplasm, but retains and replicates the flagellar basal bodies. As the basal bodies proliferate, they form a spiraled row on the surface of the growing trophont. With continued growth, the trophont takes on a "beehive" appearance, with an elaborate cavity, the mastigocoel, housing numerous flagella that arise from the spirally arranged basal bodies. At the end of the infection cycle, the flagella of the mature parasite begin to move and the opening to the mastigocoel expands outward, with the margin of the hyposome moving backward away from the apex of the episome (Fig. 6.4). As a result, the episome presses against and ruptures the host cell membrane, releasing the parasite vermiform into the surrounding water. The motile vermiform is an ephemeral multinucleate, multiflagellate stage that quickly increases in length and soon fragments to form dinospores. The duration of the infection cycle is unknown for *A. tintinni*, but ranges from two to four days in other species (Park et al. 2004).

Cachon (1964) and Cachon & Cachon (1987) describe formation of the "beehive" stage by expansion of the trophont hyposome to enclose the episome, except for the narrow opening leading to the mastigocoel (Fig. 6.4). That pattern, however, does not appear to be consistent across all species, as a recent ultrastructural study of *Amoebophrya* sp. from the dinoflagellate *Akashiwo sanguinea* (Miller et al. 2012) shows the mastigocoel forming as a closed cavity within the trophont cytoplasm. The mastigocoel of *Amoebophrya* sp. from *A. sanguinea* does not open to the host cytoplasm until late in the infection cycle. Nuclear development also differs among *Amoebophrya* species, with some undergoing nuclear division along with growth of the trophont, polyenergide development, whereas others show nuclear enlargement and delayed fission, synenergide development (Fig. 6.5a,b). Cachon (1964) reported both forms of nuclear development in *A. tintinni*, with nuclear fission along with trophont growth being most common. Whether synenergide or polyenergide in development, the mature "beehive" of *A. tintinni* is

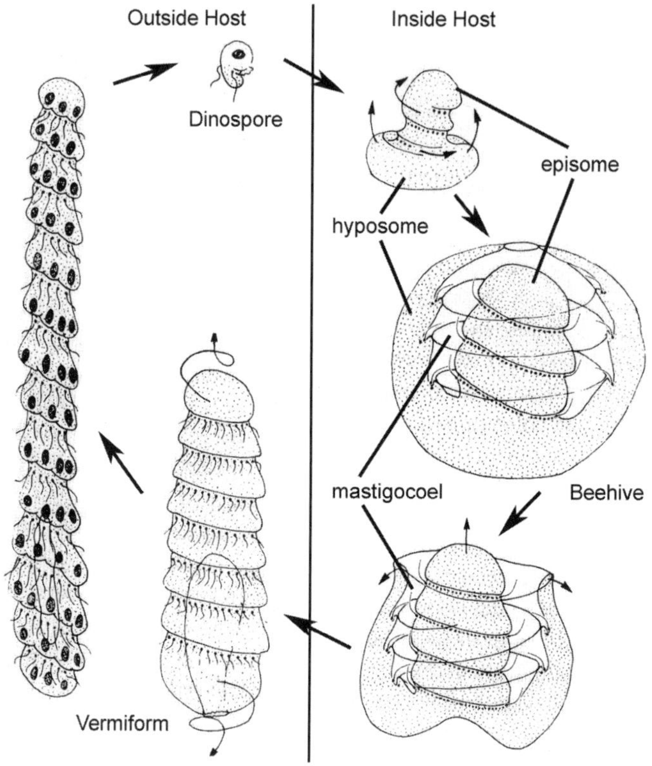

Fig. 6.4 Generalized infection cycle for *Amoebophrya* (adapted from Cachon, J. & Cachon, M. (1987) Parasitic dinoflagellates. The Biology of Dinoflagellates (ed. F.J.R. Taylor), pp. 571–610. Blackwell, Oxford). Dinospore and vermiform stages are free swimming in the aqueous environment. Trophont stages, including the "beehive", reside in the host cytoplasm or nucleus.

multinucleate, with the nuclei underlying the helically arranged basal bodies (Fig. 6.5c). Nutrition of the trophont likely involves osmotrophy, but ingestion of host material through a microtubule-lined cytopharynx is also probable (Cachon 1964; Miller et al. 2012).

Infection of tintinnids by *Euduboscquella* species appears to be a passive process. Cachon (1964) inoculated natural populations of tintinnids with spores of *E. aspida*, maintaining non-inoculated samples as controls. After incubation for 24 hours, the infection rate was three times higher in the inoculated samples, and recently established trophonts were located near the host cytostome, leading to the conclusion that infections are transmitted when the host ingests, but fails to digest spores of *Euduboscquella*. Parasitism has little influence on tintinnid swimming and feeding behavior, but hinders reproduction of some host species. When infected by *E. cachoni*, *Eutintinnus pectinis* appears unable to divide and succumbs to parasitism, while *Favella panamensis* parasitized by *E. aspida* reproduces normally and sometimes survives the infection (Cachon 1964; Coats 1988; Coats et al. 1994). Estimates for the duration of the infection cycle range from 0.75 to 4 days. Sporogenesis lasts another 0.25–3 days, depending on environmental conditions and the host–parasite system under study (Cachon 1964; Coats 1988; Coats et al. 1994).

Trophonts of *Euduboscquella* species are subspherical to irregular in shape, have a synenergide nucleus, and grow into a large, often slightly yellow mass that fills much of the host cytoplasm. During growth, the episome becomes defined on the surface of the parasite

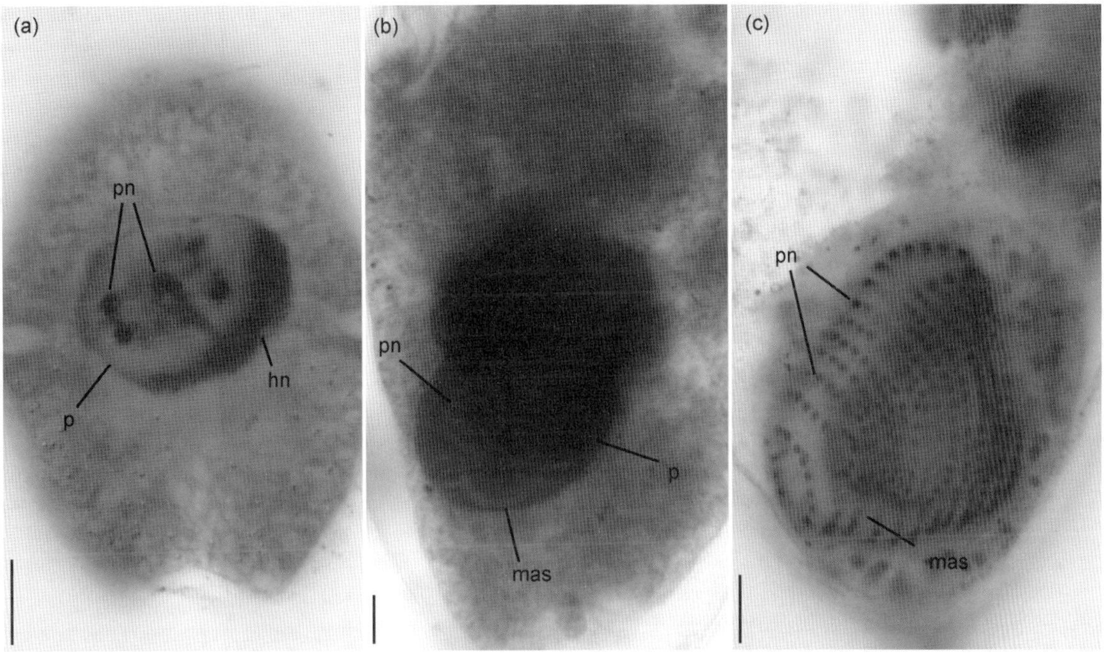

Fig. 6.5 Nuclear development in *Amoebophrya* species. (a) *Akashiwo sanguinea* infected by a polyenergide species of *Amoebophrya*. (b) Synenergide trophont of *Amoebophrya tintinni* in *Favella panamensis*. (c) Beehive stage of *Amoebophrya tintinni* in *Favella panamensis*. Host nucleus (hn); mastigocoel (mas); parasite (p); parasite nucleus (pn). Scale bars, 10 µm. Previously unpublished micrographs provided by D.W. Coats.

as a circular region having a rigid appearance and a flattened to slightly convex profile (Fig. 6.6 and Plate 6.1). In mid- to late infections, the rigid, episomal surface, or shield, of maturing parasite is creased by one or more grooves, the arrangement of which is often species-specific (Table 6.1). Bordering the shield is a fibrous, contractile annulus, the perinema, from which arises a microtubular tube, the lamina pharyngea (Fig. 6.7). The trophont appears to feed by osmotrophy (Cachon 1964), but ingestion of host material through the lamina pharyngea, as noted in *Amoebophrya*, is a distinct possibility. Emergence of the mature trophont from the host cell involves dramatic invagination of the hyposome, expansion of the episome, and contraction of the perinema to form a food vacuole enclosing all (Fig. 6.6) or part (Fig. 6.8) of the host cell. Although the function of the lamina pharyngea is uncertain, Cachon (1964) believed that it facilitated the ingestion of host remains during the emergence process. The post-feeding cell, or tomont, partly digests host material contained in the food vacuole before undergoing palintomy, serial nuclear and cytoplasmic divisions without interruption until spores are formed.

Sporogenesis of *Euduboscquella* species follows one of three patterns (Cachon 1964; Coats 1988; Coats et al. 2012). In *E. cachoni*, palintomy produces tightly packed daughter cells, sporocytes, (pattern 1; Fig. 6.6 and Plate 6.1) that separate only after the final sporogenic division to produce either dinokont or non-dinokont spores. In *E. aspida*, *E. crenulata*, and *E. cnemata*, the first four to five divisions yield sporocytes that are linked together like a string of beads, with subsequent division elaborating very long strings of sporocytes that mature as dinokont spores (pattern 2; Fig. 6.8 and Plates 6.2 and 6.3) or producing loosely clustered sporocytes that mature as non-dinokont spores (pattern 3; Plate 6.3). Sporogenesis of *E. anisospora* is not well described, but figures provided by Grassé (1952, in

Table 6.1 Host taxa and morphological attributes of *Euduboscquella* species that infect tintinnids (see text for consideration of sporogenic patterns).

Parasite	Host species	Trophont	Sporogenesis	Spores
E. aspida	*Favella* spp. *Coxliella laciniosa* *Tintinnopsis campanula* *Eutintinnus fraknoii*	Shield with sagittal groove, underlain by crisscrossing fibers; single large nucleolus	Pattern 2 and 3	Two types: reniform, highly motile dinokont spores, 4–7 μm long, numbering ~1000; weakly motile non-dinokont spores with rounded episome and pointed hyposome, 2–3 μm long, numbering over 50 000
E. anisospora	*Favella* sp.	Shield with 6–7 nearly parallel grooves; single large nucleolus	Appears to follow pattern 3	Two types: highly motile dinokont spores with bulbous episome and narrow hyposome; two distinct sizes: 20 μm and 12.5 μm long
E. cachoni	*Eutintinnus pectinis* *Eutintinnus tenuis*	Shield with sagittal groove; single large nucleolus	Pattern 1	Three types: highly motile dinokont spores with bulbous episome and narrow hyposome, 5–8 μm long, number ~50; spherical, weakly motile mono-flagellated spores, 2–3 μm diam., numbering ~500; non-motile spindle-shaped spores, 4–8 μm long, numbering ~600
E. cnemata	*Favella ehrenbergii*	20–30 grooves curving out from the center of the shield; single large nucleolus	Appears to follow Pattern 2	One type reported: reniform dinokont spores, 3–4 μm long
E. crenulata	*Favella panamensis*	Numerous intertwining grooves, multiple nucleoli distributed peripherally	Pattern 2 and 3	Three types: highly motile sigmoidal dinokont spores, 7–10 μm long, several hundred in number; non-motile spherical spores, 10–12 μm diameter, numbering 100–200; non-motile egg-shaped spores, 10–16 μm long; 100–200 in number

Chatton 1952) appear to follow pattern 2. Although a given parasite follows only one of these patterns, a host organism with more than one infection may have parasites that display different patterns.

Euduboscquella species, other than *E. cnemata*, produce at least two types of spore that differ in size, shape, and sometimes motility (Table 6.1). Historically, highly motile dinokont spores were called macrospores, whereas smaller, often weakly motile spores were termed microspores (Chatton 1952; Cachon 1964; Coats 1988). That terminology has become problematic, as the highly motile dinokont spores of *E. crenulata* are smaller than the parasite's non-motile spores (Coats et al. 2012). Thus, the macrospore–microspore dichotomy of the older literature is abandoned in subsequent discussion.

Cachon (1964) reported only dinokont spores for *E. cnemata*, but recognized that his observations were insufficient to exclude the possibility of other spore morphologies. *Euduboscquella anisospora* and *E. aspida* form large and small flagellated spores (Fig. 6.9). In *E. anisospora*, both sizes are highly motile, clearly

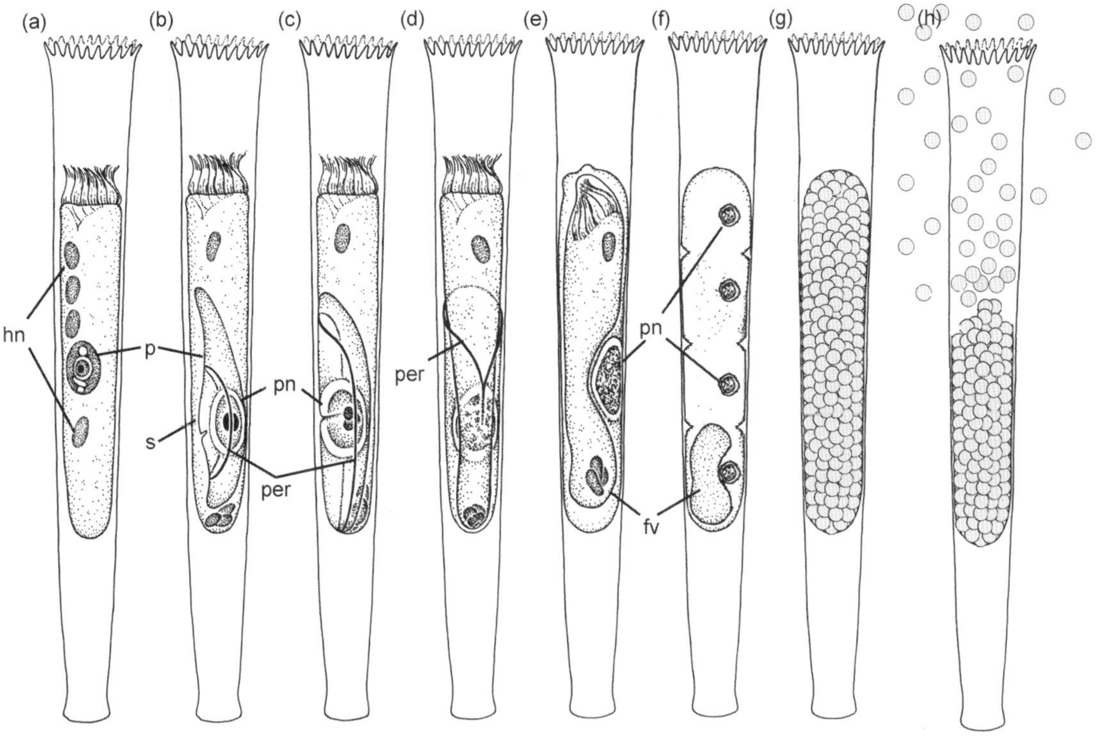

Fig. 6.6 Stages in the infection cycle of *Euduboscquella cachoni* from *Eutintinnus pectinis* (adapted from Coats 1988). (a–c) Growth of parasite trophont in host cytoplasm. (d) Emergence of parasite from host cell. (e) Parasite tomont with remains of host cell enclosed in a food vacuole. (f–h) Palintomic sporogenesis leading to tightly packed sporocytes and release of spores. Host nucleus (hn); parasite (p); parasite nucleus (pn); perinema (per); food vacuole (fv); shield (s).

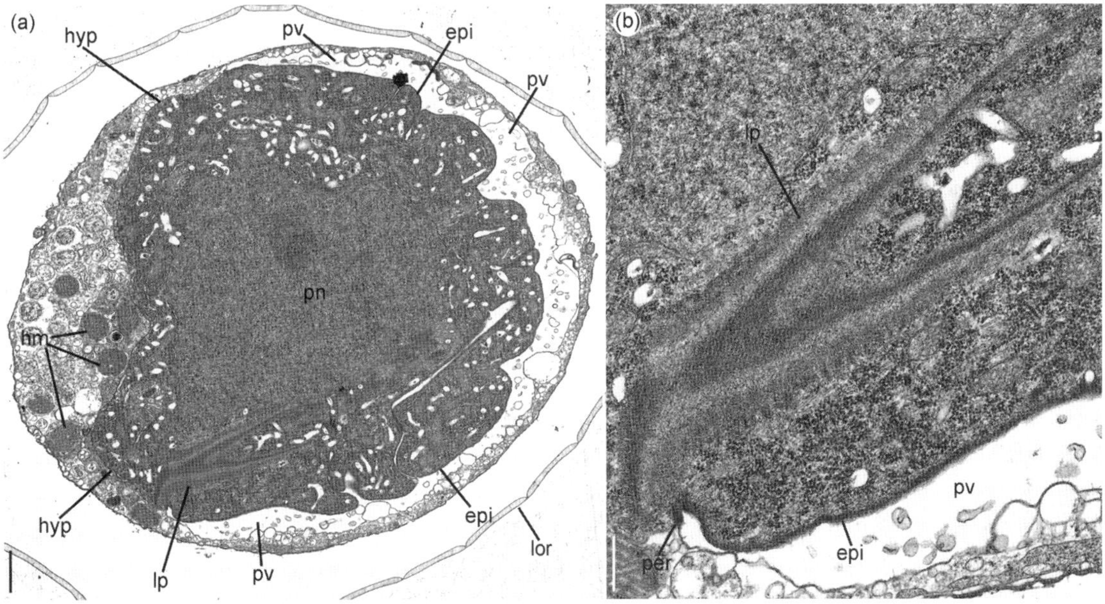

Fig. 6.7 Transmission electron micrographs of *Euduboscquella cachoni* from *Eutintinnus pectinis* showing position and structure of the lamina pharyngea. (a) Cross-section of parasite at level of the lamina pharyngea; scale bar, 1 μm. (b) Enlargement showing microtubular structure of lamina pharyngea; scale bar, 200 nm. Host mitochondria (hm); lamina pharyngea (lp); host lorica (lor); parasite episome (epi); parasite hyposome (hyp); parasite nucleus (pn); parasitophorous vacuole (pv); perinema (per). Previously unpublished micrographs provided by D.W. Coats.

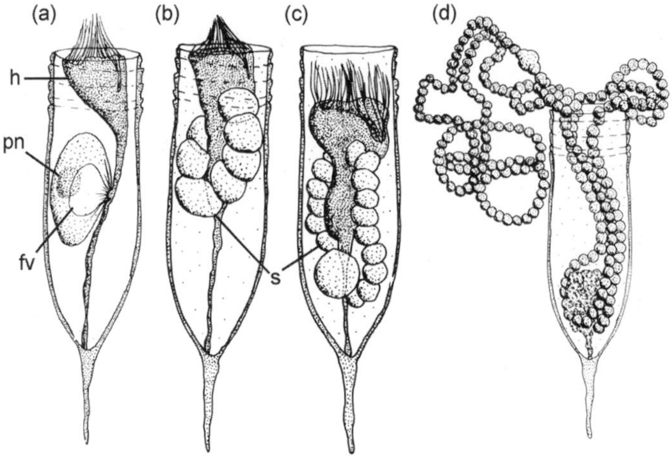

Fig. 6.8 Stages in the infection cycle of *Euduboscquella aspida* from *Favella ehrenbergii* (adapted from Cachon, J. & Cachon, M. (1987) Parasitic dinoflagellates. The Biology of Dinoflagellates (ed. F.J.R. Taylor), pp. 571–610. Blackwell, Oxford). (a) Parasite emerging from host cell. (b–d) Palintomy leading to beaded sporocytes. Host cell (h); parasite nucleus (pn); food vacuole (fv); sporocytes (s).

dinokont, and have a bulbous episome above a narrower, cylindrical hyposome. The large spores of *E. aspida* are reniform and clearly dinokont, whereas its smaller spores have a rounded episome, pointed hyposome, and do not appear to be dinokont. Indeed, Cachon (1964) was not certain that the small spores of *E. aspida* were biflagellated, remarking that "the abrupt movements of the protist can only be explained by the action of a second flagellum that remains pressed against the cell body at rest, but we weren't able to see it in the living specimen; however, Mann's staining allowed us to see, near the blepharoplast, a thin transverse strand that could be a flagellum" (translated from original French). *Euduboscquella cachoni* and *E. crenulata* each form three types of spores, but spore morphologies differ considerably between the two species (Coats 1988; Coats et al. 2012). Spores of *E. cachoni* include highly motile dinokont cells resembling those of *E. anisospora*, small spherical cells with a single flagellum, and non-motile, spindle-shaped cells that may represent cysts (Fig. 6.10). By contrast, *E. crenulata* forms sigmoid dinokont spores, non-motile spherical spores, and non-motile egg-shaped spores (Plate 6.3). *Euduboscquella* species, other than *E. anisospora*, release only one type of spore from each infection, but more than one kind of spore may be produced

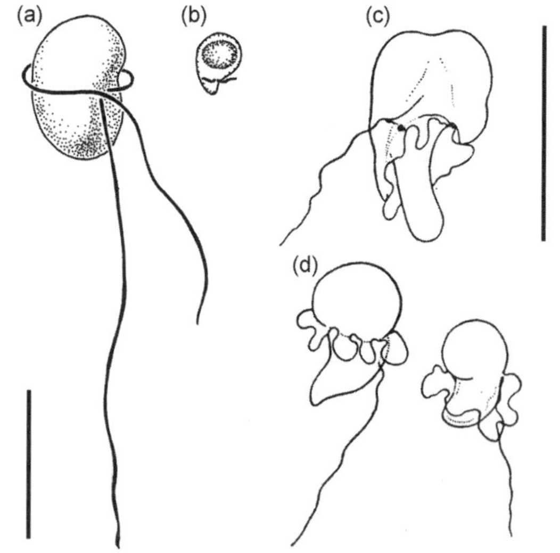

Fig. 6.9 Large and small flagellated spores of *Euduboscquella aspida* (a, b, respectively) and *Euduboscquella anisospora* (c, d, respectively). Left scale bar for *E. aspida* spores, 5 μm. Right scale bar for *E. anisospora* spores, 20 μm. (a, b) Redrawn from Cachon (1964). (c, d) Adapted from Chatton (1952).

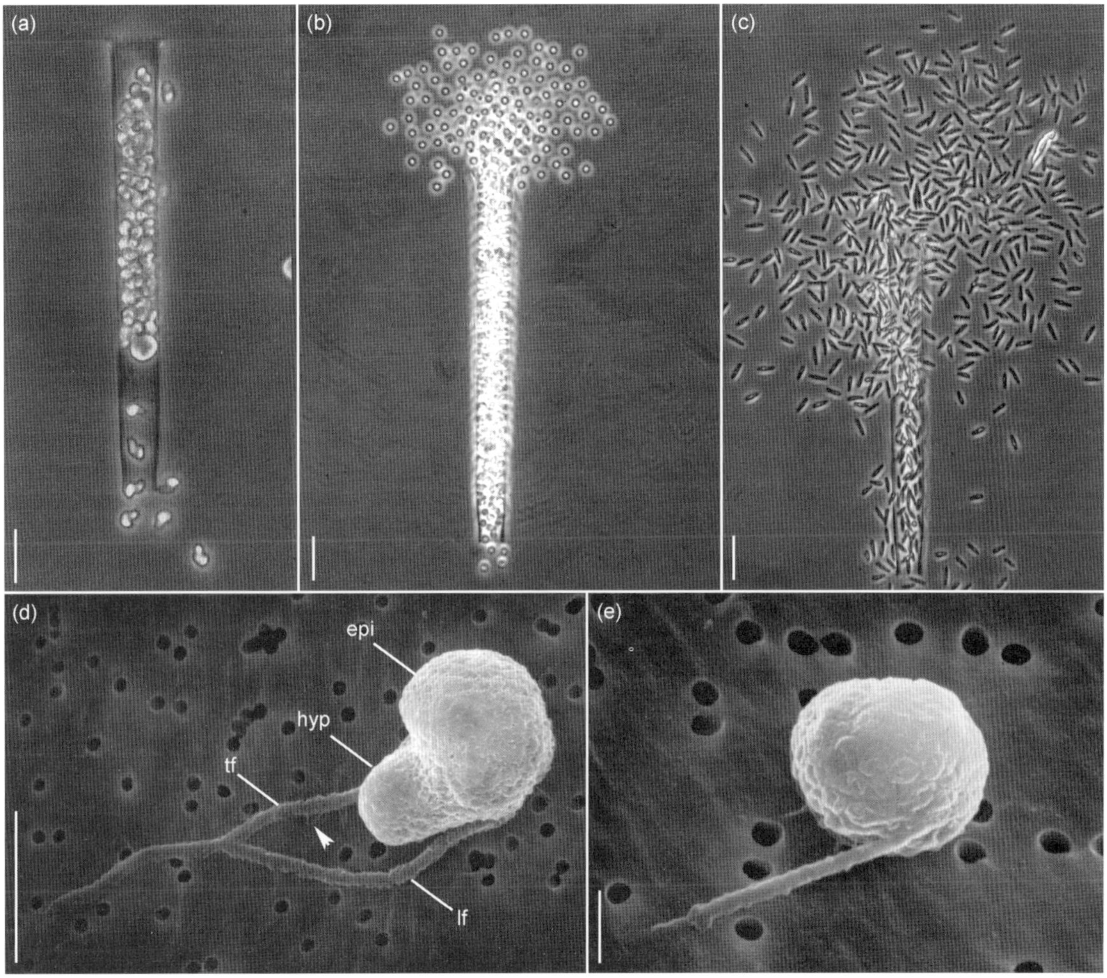

Fig. 6.10 Spore produced by *Euduboscquella cachoni* from *Eutintinnus pectinis* (adapted from Coats 1988). (a–c) Highly motile dinokont spores, weakly motile non-dinokont spores, and cyst-like spores, respectively, emerging from host lorica. (d, e) Scanning electron micrographs of dinokont and non-dinokont spores, respectively. Episome (epi); hyposome (hyp); longitudinal flagellum (lf); transverse flagellum (tf); mastigonemes (arrowhead). Scale bar, 20 m for (a–c); 5 μm for (d); 1 μm for (e).

from a host having multiple infections. *E. anisospora* differs as it apparently forms large and small dinokont spores from the same infection (Chatton 1952).

Dinophycean parasites

Four species representing three dinophycean genera (subphylum Dinokaryota Fensome et al., 1993; class Dinophyceae Pascher, 1914) are parasites of tintinnids (Chatton 1952; Coats et al. 2010, 2012). *Duboscquella tintinnicola*, *Duboscquodinium collini*, and *Tintinnophagus acutus* are, or appear to be, ectoparasitic, whereas the lifestyle of *Duboscquodinium kofoidi* is unknown. *T. acutus* infections are actively transmitted to new hosts by highly motile, predator-like dinospores, with infections of the other species presumably propagated in the same manner. The dinospores of *Duboscquodinium*

Table 6.2 Host taxa and morphological attributes of dinophycean species that infect tintinnids.

Parasite	Host species	Tomont	Sporogenesis	Spores
Duboscquella tintinnicola	*Stenosemella nivalis*	Without double outer wall or membrane; food vacuole filling most of cell	First division not differential; sporocytes without double outer wall or membrane	Athecate (?) with rounded episome, 17–20 μm long by 10–12 μm wide; lacking conspicuous eyespot; numbering 10–24
Duboscquodinium collini	*Eutintinnus fraknoii*	With double outer wall or membrane; dinokaryotic	First division differential, forming trophocyte and gametocyte; sporocytes leave outer wall or membrane	Gymnodinoid; 16 in number
Duboscquodinium kofoidi	*Tintinnopsis campanula*	Without double outer wall or membrane	First division not differential; sporocytes forming rose-shaped aggregate without double outer wall or membrane	Gymnodinoid
Tintinnophagus acutus	*Tintinnopsis cylindrica*	Without double outer wall or membrane; dinokaryotic; small food vacuole	First division not differential; sporocytes with double outer wall or membrane	Weakly thecate with pointed episome and conspicuous eyespot;13–16 μm long by 9–12 μm wide; numbering 18–46

collini and *Duboscquodinium kofoidi* are not well characterized, but those of *Duboscquella tintinnicola* and *T. acutus* are unmistakably dinophycean, having a clearly defined girdle that cradles the transverse flagellum and a distinct sulcus with longitudinal flagellum. All four species are presumably dinokaryotic, but that condition has only been confirmed for *Duboscquodinium collini* and *T. acutus*. Except for *T. acutus*, trophont morphology is poorly characterized or unknown. Thus, dinophycean parasites of tintinnids are largely distinguished by features of their post-feeding stages (Table 6.2).

Duboscquella tintinnicola and *Duboscquodinium kofoidi* have not been reported since their original descriptions, whereas *Duboscquodinium collini* was recently rediscovered, but not fully characterized (Coats et al. 2010; J. Dolan, unpublished observations). In reviewing the early literature, Coats et al. (2012) concluded that *Duboscquella tintinnicola* is ectoparasitic, lacks a second outer membrane or cyst wall around the tomont, and forms dinospores having a rounded episome, but no conspicuous eyespot. The last two attributes distinguish *Duboscquella tintinnicola* from *T. acutus*. In addition, *Duboscquella tintinnicola* completes sporogenesis without forming a degenerative trophocyte or a "rosace" configuration of sporocytes, features characteristic of *Duboscquodinium* species.

As described by Grassé (in Chatton 1952), the tomont of *Duboscquodinium collini* has a double outer membrane and dinokaryotic nucleus, with the first division forming a gametocyte and a trophocyte (Fig. 6.11a, b). The gametocyte continues to divide, producing sporocytes that emerge from the outer membrane to become gymnodinoid dinospores, while the trophocyte receives the partly digested food vacuole, fails to divide, and eventually degenerates. Recent examination of *Duboscquodinium collini* from the Mediterranean Sea (Coats et al. 2010) confirmed morphological fea-

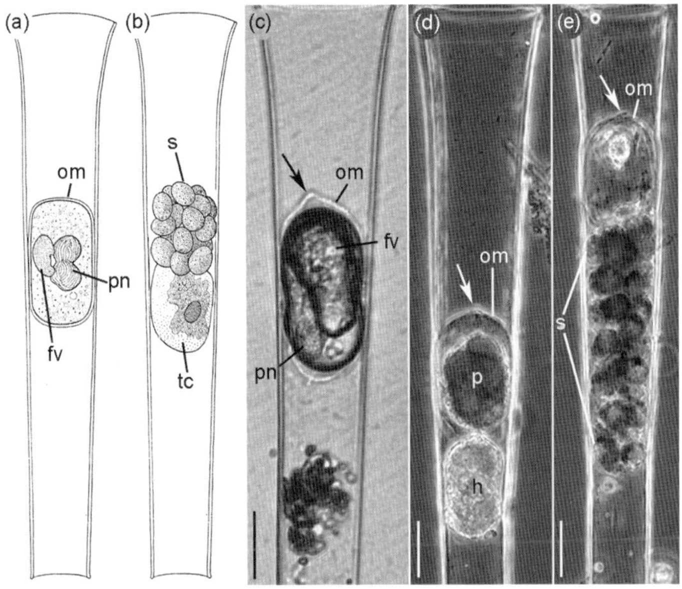

Figue 6.11 *Duboscquodinium collini* from *Eutintinnus fraknoii*. (a, b) Tomont and sporogenic stages, respectively (adapted from Chatton 1952). (c) Tomont from the Mediterranean Sea showing noticeable protuberance (arrow) in outer membrane; from Coats, D.W., Kim, S., Bachvaroff, T.R., Handy, S.M. & Delwiche, C.F. (2010) *Tintinnophagus acutus* n. g., n. sp. (Phylum Dinoflagellata), an ectoparasite of the ciliate *Tintinnopsis cylindrica* Daday 1887, and its relationship to *Duboscquodinium collini* Grassé 1952. Journal of Eukaryotic Microbiology, 57, 468–482. (d, e) Trophont feeding on host cell and sporocytes emerging from outer membrane, respectively. Note protuberance in outer membrane (arrows). Previously unpublished images of Lugol's preserved specimens from 50 m at a station (32° 41′ S, 84° 04′ W) in the southeastern Pacific Ocean; images courtesy of J. Dolan. Food vacuole (fv); outer membrane (om); parasite nucleus (pn); sporocytes (s); trophocyte (tc). Scale bars, 25 μm.

tures described by Grassé, but showed a distinctive protuberance in the outer membrane of the trophont that was consistently directed toward the oral opening of the host lorica (Fig 6.11c). Specimens from the southeast Pacific (J. Dolan, unpublished observations) also had a distinctive protuberance in the outer membrane and appeared to be feeding on the host cell (Fig. 6.11d), supporting the notion that *Duboscquodinium collini* is an ectoparasite. Sporocytes from Dolan's samples appeared to emerge from the outer membrane of the tomont before maturing into dinospores (Fig. 6.11e), further supporting Grassé's observations.

Grassé (in Chatton 1952) tentatively assigned a second species to the genus *Duboscquodinium*, providing the name *Duboscquodinium (?) kofoidi* (Fig. 6.12). Grassé appears to have aligned *Duboscquodinium kofoidi* with *Duboscquodinium collini*, as both species form gymnodinoid dinospores. *Duboscquodinium kofoidi*, however, differs from *Duboscquodinium collini* by the absence of a double outer membrane and the formation of an unusual "rosace" stage during sporogenesis. Were it

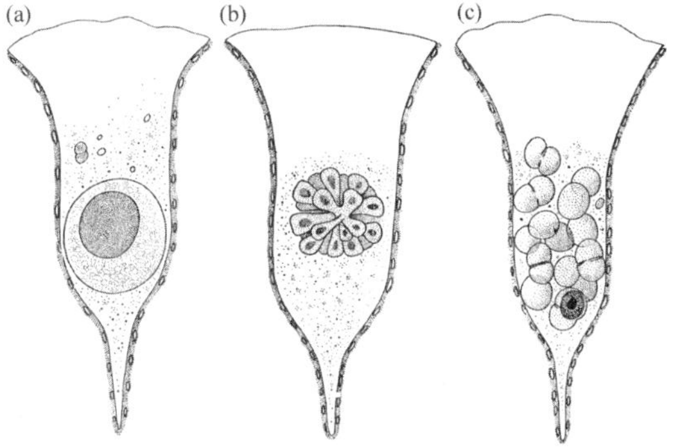

Fig. 6.12 *Duboscquodinium kofoidi* from *Tintinnopsis campanula*. (a) Tomont without double outer membrane. (b) "Rosace-stage" of sporogenesis. (c) Sporocytes differentiating into gymnodinoid dinospores. Adapted from Chatton (1952).

not for this "rosace" stage, *Duboscquodinium kofoidi* would be indistinguishable from the sporocysts found by Laackmann (1908) in the loricae of *Tintinnus helix* and now considered to be *Duboscquella tintinnicola* (Coats et al. 2012). Should future studies establish synonymy of *Duboscquella tintinnicola* and *Duboscquodinium kofoidi*, then the latter would fall as a junior synonym.

Tintinnophagus acutus is by far the most completely characterized dinophycean parasite of tintinnids. Its dinospores appear weakly thecate, have a sharply pointed episome, large nucleus with condensed chromosomes, and a conspicuous eyespot (Plate 6.4). After entering the host lorica, *T. acutus* attaches posteriorly on the tintinnid zooid via a feeding tube and forms a spherical trophont without flagella, girdle, and sulcus. The trophont ingests host material through a feeding tube and grows slowly, with more than a 50-fold increase in volume over 4 days. During that time, the nucleus becomes quite large, presumably synenergide, and a small, typically reddish food vacuole forms at the anterior end of the parasite. Sporogenesis is palintomic, but prolonged, requiring 2 days to produce mature dinospores. Sporocytes, but not the tomont, are encased in a second outer membrane or cyst wall. Host cells generally survive infection, but show aberrant nuclear morphology and appear unable to reproduce (Coats et al. 2010).

Sexuality

A sexual cycle, often encompassing formation of resting cysts, is well documented among members of the Dinokaryota, but has yet to be reported for dinophycean species that parasitize tintinnids (Pfiester & Anderson 1987; Anderson et al. 1998). By contrast, sexuality has been implicated in only three syndinean species, two of which are parasites of tintinnids. As mentioned earlier, Duboscq & Collin (1910) observed paired flagellated spores in an unidentified parasite of *Favella ehrenbergii*, interpreting them to be fusing gametes (Fig. 6.2). The paired spores were of equal size and shape (isospores), lost their flagella, and rounded into a mass, presumably a zygote, but subsequent development was not followed. Cachon (1964) was unable to repeat the observations of Duboscq and Collin, finding instead that paired isospores resulted from sporogenic fission and always separated when maintained in a dish. When placed on a microscope slide, however, the paired cells rounded into a mass and degenerated. Thus, it seems likely that Duboscq & Collin (1910) mistook the final division of flagellated spores as gamete fusion.

Sexuality involving fusion of motile anisospores has been reported for the tintinnid parasite *Euduboscquella anisospora* by Grassé (in Chatton 1952) and for *Coccidinium mesnili*, a syndinean parasite of dinoflagellates (Chatton & Biecheler 1936). According to Grassé, spores of *E. anisospora* produced inside the same host lorica paired in various orientations, lost their flagella, and fused to form a zygote. Illustrations were not provided, and the fate of the presumptive zygote was not determined, leaving uncertain whether Grassé had actually observed part of a sexual cycle. Chatton & Biecheler's (1936) account of sexuality in *C. mesnili* is more convincing, as the authors provided illustrations (Fig. 6.13) and a time-frame for gamete fusion. Small capped (c-type) and rostrate (r-type) spores coupled perpendicularly at their flagellar bases and became motionless as fusion progressed. Fusion lasted about 5 min, occurring first in the hyposome region, with flagella of both spores being retained. The resulting quadriflagellate zygote regained motility, but subsequent development was not determined. *Coccidinium* is a problematic genus known only from the original species descriptions. Early developmental stages closely resemble *Amoebophrya* and *Duboscquella*, with characterization of the genus *Coccidinium* possibly based on the inadvertent combination of features from several unrelated organisms (Loeblich III 1982; Fensome et al. 1993). Thus, the relevance of the sexual cycle reported for *C. mesnili* is difficult to assess. Interestingly, however, several strains of *Amoebophrya* have been maintained in culture, apparently without undergoing a sexual cycle (Park et al. 2004).

Most recently, Coats et al. (2012) described sexuality involving two of the three spore types formed by *Eudu-*

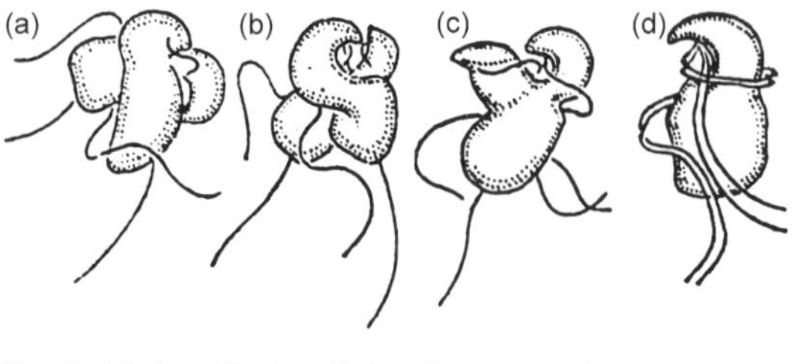

Fig. 6.13 (a–c) Fusion of *Coccidinium mesnili* spores to form (d) a motile, quadriflagellate zygote (adapted from Chatton & Biecheler 1936).

boscquella crenulata (Fig. 6.14). Small dinokont spores paired with non-motile spherical spores to form non-motile zygotes, but did not respond to non-motile egg-shaped spores. Fusion of the spores, or more appropriately the gametes, required about 20 min and was accompanied by nuclear fusion and flagellar resorption. The non-motile zygote remained quiescent for about 2 hours and then divided twice, presumably undergoing meiosis, to produce four daughter cells similar in appearance to the non-motile egg-shaped spores of *E. crenulata*.

Taken together, these observations strongly support the existence of a sexual cycle in syndinean dinoflagellates in general and particularly in species that infect tintinnids. What remains unclear is the ability of the various extracellular life-history stages to initiate new

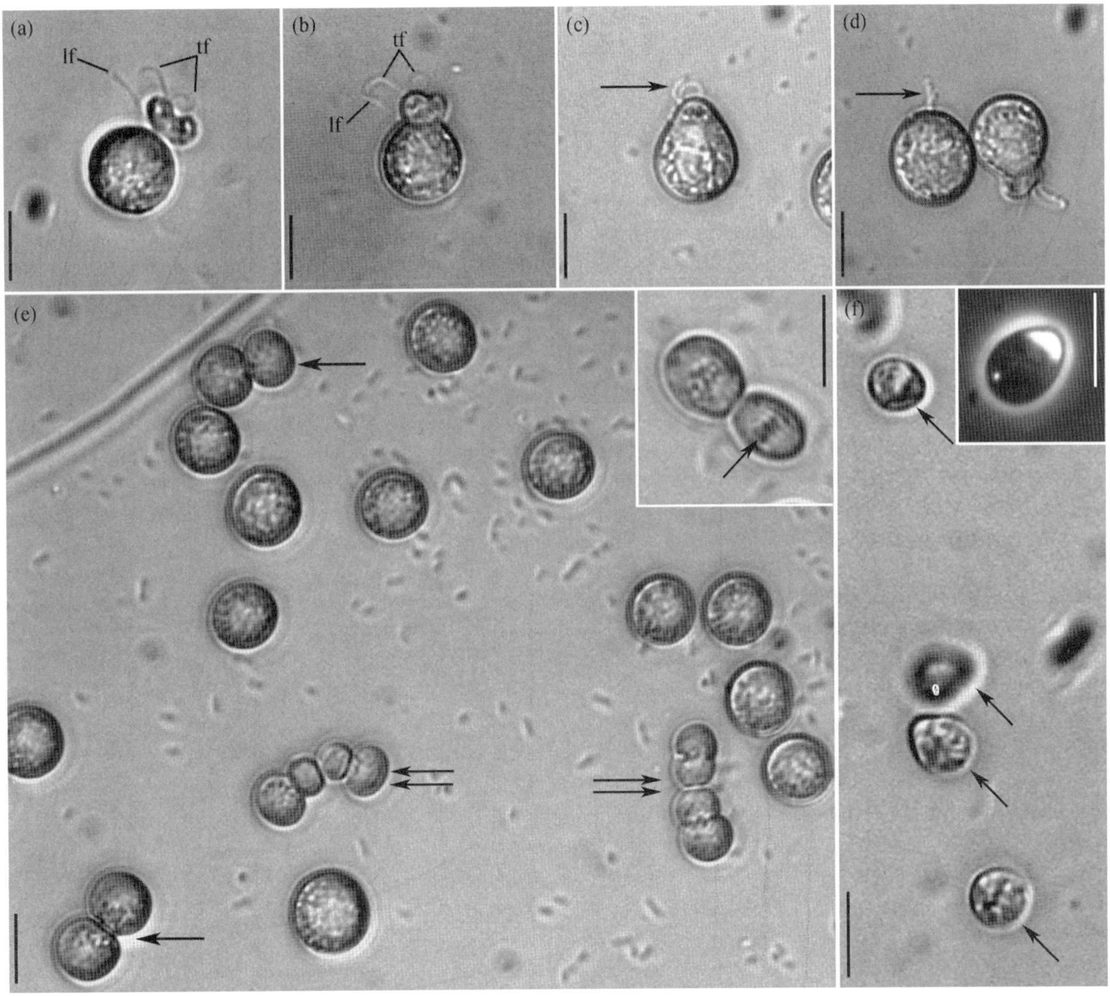

Fig. 6.14 Sexual processes in the life cycle of *Euduboscquella crenulata* from *Favella panamensis* (adapted from Coats, D.W., Bachvaroff, T.R. & Delwiche, C.F. (2012) Revision of the family Duboscquellidae with description of *Euduboscquella crenulata* n. gen., n. sp. (Dinoflagellata, Syndinea), an intracellular parasite of the ciliate *Favella panamensis* Kofoid and Campbell, 1929. Journal of Eukaryotic Microbiology, 85, in press). (a–d) Fusion of motile dinokont and non-motile spherical spores. Arrows indicate partly resorbed flagella. (c) Dividing zygotes. Single arrows indicate first and double arrows second zygotic division. Insert shows metaphase configuration during the first division. (f) Egg-shaped daughter cells (arrows) resulting from one zygote, with insert showing a daughter cell in phase contrast. Longitudinal flagellum (lf); transverse flagellum (tf). Scale bars, 10 μm.

infections or to form dormant stages. Some spore types may serve only as gametes, lacking the ability to infect host cells, while others may transmit infections or produce resting cysts. Cysts or other free-living life-history stages provide a refuge for parasites when host organisms are unavailable and are critical for the survival of host-specific parasites that must kill their host to reproduce (i.e., parasitoids). Thus, it is likely that at least some parasites of tintinnids form cysts. So far, resting cysts have not been unambiguously documented for syndinean dinoflagellates; however, dormancy of *Amoebophrya* sp. within the cyst of its dinoflagellate host has been demonstrated (Chambouvet et al. 2011).

Key to the dinoflagellate parasites of tintinnids

Given the extensive diversity of syndinean dinoflagellates revealed by environmental sequencing (Guillou et al. 2008) and recent recognition that ciliates are also parasitized by dinophycean species (Coats et al. 2010), it seems reasonable to expect discovery of additional tintinnid parasites. Indeed, single cell sequencing of tintinnids infected by syndinean dinoflagellates has revealed greater sequence diversity than can be accounted for by currently described species (Bachvaroff et al. 2012). Thus, the following key is provided as a tentative guide in identifying dinoflagellates that parasitize tintinnids, with revision certainly to be required as new species are described and as currently valid, but poorly characterized, species are redescribed.

Key to the dinoflagellate parasites of tintinnids

1.	Extracellular parasite; dinokont spores with well-developed cingulum and sulcus	2
1′.	Intracellular parasite; multiple spore types possible, but dinokont spores lack strongly developed cingulum and sulcus	5
2.	Tomont with double outer membrane	*Duboscquodinium collini*
2′.	Tomont without double outer membrane	3
3.	Sporocytes forming "rosace" stage	*Duboscquodinium kofoidi*
3′.	Sporogenesis without "rosace" stage"	4
4.	Dinospores with pointed episome and conspicuous eyespot	*Tintinnophagus acutus*
4′.	Dinospores with rounded episome; lacking conspicuous eyespot	*Duboscquella tintinnicola*
5.	Nuclear division in trophont forming "beehive" stage	*Amoebophrya tintinni*
5′.	Nuclear division occurring in tomont	6
6.	Sporocytes as tightly packed cluster	*Euduboscquella cachoni*
6′.	Sporocytes not tightly packed	7
7.	Shield with single sagittal groove	*Euduboscquella aspida*
7′.	Shield with six or seven nearly parallel grooves	*Euduboscquella anisospora*
7″.	Shield with 20–30 grooves spiraling out from center	*Euduboscquella cnemata*
7‴.	Shield with numerous intertwining grooves	*Euduboscquella crenulata*

6.3 MOLECULAR EVOLUTION OF DINOFLAGELLATES PARASITIZING TINTINNIDS

Placement of tintinnid parasites within the dinoflagellates

The broadest, most inclusive clade of dinoflagellates has been placed at various taxonomic levels (e.g., Dinozoa (Adl et al. 2005) or Dinoflagellata (Dodge & Lee 2000)) and can be very roughly divided into two large bins: the syndineans (subphylum Syndinea Corliss, 1984) and the dinokaryotes (subphylum Dinokaryota Fensome et al., 1993) (Fig. 6.15a). The blastodiniphyceans (subphylum Dinokaryota; class Blastodiniphyceae Fensome et al., 1993) were once considered a distinct parasitic lineage within the dinokaryotes, but have since been shown to be grossly polyphyletic and scattered across molecular phylogenies (Litaker et al. 1999; Saldarriaga et al. 2001; Levy et al. 2007; Skovgaard et al. 2007; Coats et al. 2010). Several troublesome genera including *Oxyrrhis*, *Perkin-*

sus, *Parvilucifera*, and *Thallassomyces* are affiliated with the dinoflagellates on molecular trees, but do not fit well in either the dinokaryote or syndinean bin (Noren et al. 1999; Saldarriaga et al. 2003; Silberman et al. 2004). None of these troublesome genera are parasites of tintinnids, so they can be safely ignored in this treatment.

Dinokaryotes, as implied by the name, have several unique, presumably derived characters of the nucleus including the following: (1) condensed chromosomes during interphase, (2) unusual closed mitosis with extranuclear spindle, (3) low protein:DNA ratios, (4) DNA not packaged with histones into nucleosomes, (5) high DNA content, (6) modified DNA bases, and finally (7) many highly duplicated protein-coding genes (Spector 1984; Bachvaroff & Place 2008). Syndineans have closed mitosis with an extranuclear spindle (Ris & Kubai 1974), but differ from the dinokaryotes by their apparent lack of condensed chromosomes during interphase, small number of chromosomes indicating low DNA content, and a sufficient quantity of basic proteins associated with chromosomes to be detected histochemically, although organization of histones as nucleosomes has not been demonstrated (Cachon 1964; Loeblich III 1976).

Dinoflagellate parasites of tintinnids appear to have evolved independently several times. Historically, all dinoflagellates that use tintinnids as hosts were considered to be syndineans (Fensome et al. 1993). Recent investigations combining molecular and morphological approaches, however, have placed two of those genera, *Duboscquella* and *Duboscquodinium*, among the dinokaryotes and revealed a previously undescribed dinokaryote genus, *Tintinnophagus*, that infects tintinnids (Coats et al. 2010, 2012). Thus, the syndinean and dinokaryote lineages now each contain at least two different genera that parasitize tintinnids. Despite recent advances in assessing the diversity of dinoflagellate parasites of tintinnids, the number of described genera and species will likely continue to expand in the future.

Dinokaryote parasites of tintinnids

The recently described tintinnid parasite *Tintinnophagus acutus* was placed without strong support near a "*Pfiesteria*-group" composed of heterotrophic free-living genera (*Luciella*, *Pfiesteria*, *Pseudopfiesteria*, *Stoeckeria*, and *Cryptoperidinopsis*), as well as the parasitic species *Amyloodinium ocellatum* and *Paulsenella vostonochi* (Fig. 6.15b). It is worth noting that small-subunit ribosomal RNA (SSU rRNA)-based molecular phylogenies for dinokaryotes are notoriously poorly supported, with only a few well-supported clades consistently found (Saunders et al. 1997; Saldarriaga et al. 2001; Murray et al. 2005; Shalchian-Tabrizi et al. 2006). Despite the vagueness of the molecular phylogeny, *T. acutus* and related heterotrophs can be seen as sharing extracellular feeding by a peduncle, followed by a non-motile division stage. In some species (*A. ocellatum*, *Pseudopfiesteria shumwayae*, *Paulsenella*, and *T. acutus*) the division cyst produces multiple dinospores, whereas the other species produce only two daughter cells (Parrow & Burkholder 2003; Calado et al. 2009)

Another dinokaryote parasite of tintinnids, *Duboscquodinium collini*, has only been reported twice in the literature (Chatton 1952; Coats et al. 2010). The SSU rRNA from *Duboscquodinium collini* is scarcely distinguishable from the homologous sequences from several *Scrippsiella* spp. and *Peridinium polonicum*, with only four base positions differing between *D. collini* and *S. trochoidea*, 0.2% difference (Coats et al. 2010). On SSU rRNA trees, the genera *Peridinium* and *Scrippsiella* are polyphyletic (Logares et al. 2007a; Coats et al. 2010). The very small amount of SSU sequence difference among *D. collini*, *Scrippsiella* spp., and *Peridinium polonicum* may reflect either recent divergence or high sequence conservation within this group of dinokaryotes. *Scrippsiella hangoei* and *Peridinium aciculiferum* are even a more extreme example. The two species have identical rRNA loci despite clear differences in mitochondrial genes, plate tabulation, and salinity preferences (Logares et al. 2007b). If a similar pattern occurs within other related lineages, SSU rRNA gene trees may not effectively capture species or genus level relationships.

In reference to *Duboscquodinium collini*, details of life cycle, development, and host preference or specificity are not completely understood. Although *Duboscquodinium collini* does not appear to be photosynthetic, *Scrippsiella* would be considered a photosynthetic genus. In dinoflagellates, photosynthesis appears to be an unreliable character, with the genera *Blastodinium* and *Polykrikos* containing both photosynthetic and non-photosynthetic species (Chatton 1920; Hoppenrath et al. 2010). Alternately, *Scrippsiella* and *Duboscquodinium*, like many photosynthetic dinoflagellates, may be mixotrophic, and the chloroplasts may be reduced or inconspicuous while the dinoflagellate is feeding (Hansen 2011).

Despite the difficulty in defining dinokaryote relationships based on SSU gene trees, *Tintinnophagus*

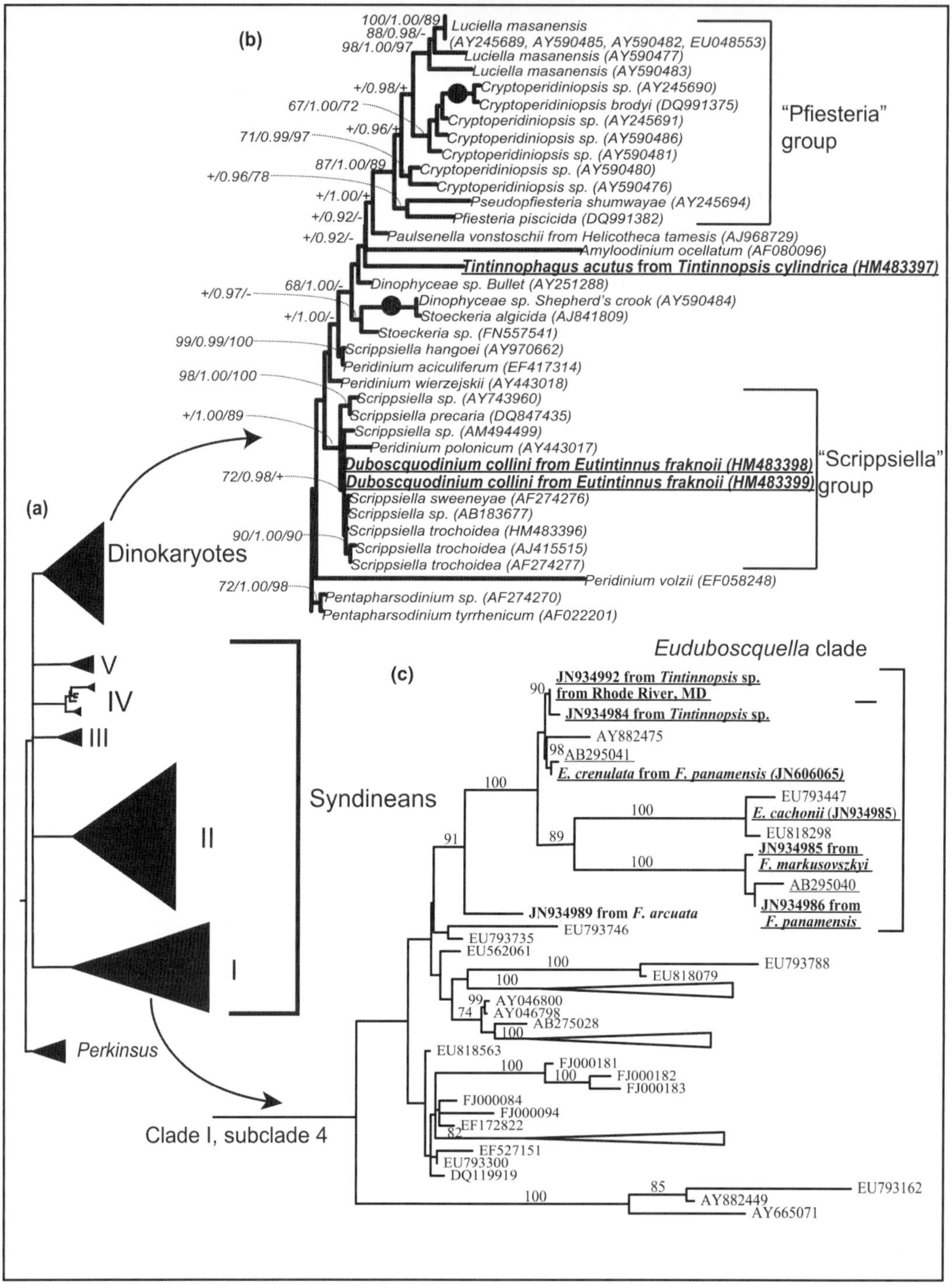

Fig. 6.15 Phylogeny of dinoflagellates using SSU rRNA. (a) A schematic phylogeny of the dinoflagellate lineage showing the different syndinean clades and the dinokaryotes as a polytomy (adapted and revised from Guillou, L., Viprey, M., Chambouvet, A., Welsh, R.M., Kirkham, A.R., Massana, R., Scanlan, D.J. & Worden A.Z. (2008) Widespread occurrence and genetic diversity of marine parasitoids belonging to Syndiniales (Alveolata). Environmental Microbiology, 10, 3349–3365). (b) Partial phylogeny of the dinokaryotes including two species parasitic on tintinnids. Values above nodes are ML/MrBayes/ ML-distance bootstrap or p values (adapted from Coats, D.W., Kim, S., Bachvaroff, T.R., Handy, S.M. & Delwiche, C.F. (2010) *Tintinnophagus acutus* n. g., n. sp. (Phylum Dinoflagellata), an ectoparasite of the ciliate *Tintinnopsis cylindrica* Daday 1887, and its relationship to *Duboscquodinium collini* Grassé 1952. Journal of Eukaryotic Microbiology, 57, 468–482). Underlined species are parasites of tintinnids. (c) Partial phylogeny of syndinean dinoflagellates from clade I showing only subclade 4. Underlined sequences are from tintinnid hosts, all other sequences are from environmental clone libraries. Maximum likelihood bootstrap values are shown above the branches (adapted from Bachvaroff, T.R., Kim, S., Guillou, L., Delwiche, C.F. & Coats, D.W. (2012) Diversity of the syndinian genus *Euduboscquella* based on single cell PCR. Applied and Environmental Microbiology, in press. Copyright © American Society for Microbiology).

acutus and *Duboscquodinium collini* are clearly dinokaryotes based both on molecular trees and morphology. In addition, the two genera have little affinity on SSU gene trees, suggesting they have evolved independently. At present, based exclusively on molecular data, two distinct dinokaryote species parasitize tintinnids, with a third, *Duboscquella tintinnicola*, awaiting molecular characterization. In addition, two other completely uncharacterized tintinnid parasites have been found within the dinokaryotes (T.R. Bachvaroff, unpublished observations).

Syndinean parasites of tintinnids

The first view of the broad molecular diversity of syndinean dinoflagellates was based on environmental clone libraries. These sequences were directly acquired from whole or size-fractionated water samples, without knowing the identity of the organisms contributing the sequences. Phylogenies from such datasets produced two broad and diverse "Marine Alveolate" clades, simply labeled clades I and II (Moon-van der Staay et al. 2001). Both clades are here considered syndinean dinoflagellates. Clade II contained sequences attributed to the parasitic genus *Amoebophrya*, while clade I contained no sequences attributed to described species. The most striking features of these trees were the diversity and divergence of the sequences. Subsequent increases in the number of environmental clone library sequences and novel sequences for described syndineans have developed a broad but not fully resolved picture. Within syndinean dinoflagellates, the original clades I and II have been expanded to five (Guillou et al. 2008) or eight (Richards & Bass 2005) deeply diverging clades. The most commonly found sequences are still those associated with the original clades I and II, and these are the two clades containing parasites of tintinnids.

Over the past several years, rRNA sequences from representatives for five of the 14 syndinean genera (excluding *Duboscquodinium*) recognized by Fensome et al. (1993) have been placed in GenBank: *Amoebophrya* (Janson et al. 2000; Gunderson et al. 2002; Salomon et al. 2003; Kim 2006; Kim et al. 2008), *Euduboscquella* species previously placed in *Duboscquella* (Harada et al. 2007; Bachvaroff et al. 2012; Coats et al. 2012), *Hematodinium* (Small et al. 2006, 2007; Ryazanova et al. 2010), *Ichthyodinium* (Mori et al. 2007; Yuasa et al. 2007; Skovgaard et al. 2009), and *Syndinium* (Skovgaard et al. 2005). Sequences available for the several isolates of *Amoebophrya*, *Hematodinium*, and *Euduboscquella* are consistent with multiple species within each genus. Overall, syndinean environmental sequences present in GenBank continue to dwarf those from described species and genera.

In SSU phylogenies of syndineans and dinokaryotes, there are several well-supported and consistent relationships interspersed among poorly supported and inconsistent relationships (Saunders et al. 1997; Saldarriaga et al. 2001; Murray et al. 2005; Shalchian-Tabrizi et al. 2006). Starting from the most expansive view of the dinoflagellate lineage, the relationships among the genera *Perkinsus* and *Oxyrrhis*, the dinokaryotes, and the numerous syndinean clades are ambiguous, even when using multiple protein-coding genes for phylogeny (Saldarriaga et al. 2003; Bachvaroff et al. 2011). Monophyly of the dinoflagellate lineage, as defined above, is consistent, but rarely well supported (Litaker et al. 1999; Leander et al. 2003; Richards & Bass 2005; Skovgaard et al. 2005, 2009; Harada et al. 2007; Guillou et al. 2008), except when protein-coding genes are used (Bachvaroff et al. 2011). Bootstrap support for syndinean clades I through V as distinct lineages within the Syndinea is consistently recovered, while support for monophyly of the dinokaryotes is often poor when large numbers of syndinean sequences are included (compare Coats et al. (2010) to Guillou et al. (2008)).

Constructing good phylogenies can be difficult because of the sheer number of sequences, and the SSU alignment can be uncertain and dependent on taxon sampling. Several difficult long-branch genera, such as *Oxyrrhis*, *Noctiluca*, and *Thallassomyces*, and the specter of chimeric environmental clone library sequences are other concerns. Chimeric sequences can result from artificially ligating partial sequences from two or more organisms into a single linear sequence and can be difficult to uncover in the absence of a well-sampled set of reference sequences obtained from individuals (see Berney et al. 2004). These factors can limit the analytic methods. However, even with robust likelihood methods and subsampling of the data, the phylogenies strongly support the clades labeled with roman numerals, but not the relationships among them, and the bootstrap support for the dinokaryotes is still poor even after specific long-branch taxa are removed (Fig 6.15a). From a strict perspective, the dinokaryotes can be viewed as an uncertain paraphyletic group intermingled with the syndineans, but these rRNA phylogenies contrast with both morphologic features and poorly sampled protein-coding gene trees that together support monophyly of the dinokaryotes.

The deep branching syndinean clades I and II diverge as deeply as the dinokaryotes or *Perkinsus* on SSU trees, but are here treated as a single lineage, the syndineans (Fig. 6.15a). Although the clade label nomenclature anticipates major higher-level taxonomic revision, these phylogenies were exclusively based on rRNA, and few of the syndinean clades contain sequences attributed to described genera or species. Using the dinokaryotes as a cautionary example, rRNA evolution in the syndineans may be discontinuous with equal rates of evolution in different lineages. The combination of molecular data, morphology, and good sampling across the different lineages will be required to describe these relationships accurately.

The genus *Euduboscquella* is embedded in syndinean clade I with good support (Harada et al. 2007; Guillou et al. 2008; Skovgaard et al. 2009;Bachvaroff et al. 2012) (Fig. 6.15c). Within clade I, a total of eight different subclades have been defined using Arabic numerals (Guillou et al. 2008), a nomenclature that will be followed here. *Euduboscquella* forms a portion of subclade 4, creating a well-supported *Euduboscquella* clade containing mostly sequences attributed to the genus, with just a few environmental clone library sequences. The only other subclade containing sequences attributed to a described species or genus is subclade 3 with *Ichthyodinium* (Skovgaard et al. 2009). Subclade 3 is the only subclade within syndinean clade I that was poorly supported by bootstrapping (Bachvaroff et al. 2012; Guillou et al. 2008). In a maximum likelihood analysis of syndinean clade I, the vast majority of total sequences are found in subclades 1 and 5 (Guillou et al. 2008; Bachvaroff et al. 2012). Subclades 1, 2, 5, and 7 form one well-supported lineage, while subclades 4, 6, 8, and 3 form a second well-supported lineage (data not shown). The relationships among subclades 4, 6, 8, and 3 or among subclades 1, 2, 5, and 7 are not well resolved.

What is clear, based on SSU phylogenies and confirmed by ITS2 datasets, is that *Euduboscquella* probably contains more species than have been formally described. Although five species have been formally described, only two of seven distinct *Euduboscquella* ITS2 ribotypes could be attributed to described species (Coats et al. 2012). This leaves five different parasite ribotypes that have not yet been clearly attributed to described species. The boundaries of the genus *Euduboscquella* on phylogenies remains very uncertain, perhaps reflecting the lack of sequences for species that infect non-tintinnid hosts. Two species, *Euduboscquella melo* and *Euduboscquella nucleocola*, are known to infect dinoflagellates (Cachon 1964), but rRNA sequences are not available for either. In addition, some parasites of radiolarians have features of *Euduboscquella* (Suzuki et al. 2009), and some rRNA sequences from radiolarian associates fall within clade I (Dolven et al. 2007). However, sequences clearly attributable to *Euduboscquella* are not yet available for radiolarian parasites.

A similar situation exists in syndinean clade II, where there are many environmental clone sequences and a handful of sequences attributed to the genus *Amoebophrya*. This syndinean genus is most commonly found as an intracellular parasite of photosynthetic and heterotrophic dinokaryotes; however, one species, *A. grassei*, infects syndineans of the genus *Oodinium*, two species parasitize ciliates (*A. tintinni* from tintinnids and *A. rosei* from aloricate ciliates living in siphonophores and chaetognaths), and one species is known from each of the rhizarians *Acanthometra pellucida* and *Sticholonche zanclea* (Cachon 1964). Unfortunately, molecular data for *Amoebophrya* species from hosts other than photosynthetic dinokaryotes are not yet available.

The phylogenetic structure of syndinean clade II differs strongly from clade I, with up to 44 different subclades recognized, but with distance-based trees providing poor support for relationships between and within the different subclades (Guillou et al. 2008). Sequences for *Amoebophrya* from photosynthetic dinokaryote hosts are not monophyletic in these trees and do not form simple monophyletic lineages, even with much smaller sampling of environmental clone sequences (Kim et al. 2008). Based on culture experiments, some *Amoebophrya* strains that infect photosynthetic dinokaryotes have clear host specificity, while others have a broader host range, suggesting species level differences among strains. Sequence diversity for these strains also suggests that more *Amoebophrya* species are present than just the single species *A. ceratii* formally described from photosynthetic dinoflagellate hosts. Such results suggest that connecting molecular data with *Amoebophrya* from tintinnid and other non-dinokaryote hosts may reveal yet another complicated pattern of diversity.

6.4 ECOLOGY OF TINTINNID PARASITES

Distribution, seasonality, and host range

Parasitism of tintinnids by dinoflagellates occurs circumglobally in temperate to subtropical waters (Fig. 6.16); however, data are insufficient to assess the distribution of particular parasite species. Given the recent recognition of dinophycean infections in tintinnids (Coats et al. 2010), it is not surprising that most accounts involve parasitism by syndinean dinoflagellates, particularly *Euduboscquella* spp. Nonetheless, the limited data available for dinophycean parasites of tintinnids suggest that they are as broadly distributed as their syndinean counterparts. Also understandable is that most observations are from coastal waters where tintinnid abundance is generally higher (Chapter 9) and frequent sampling is possible. Infections of *Eutintinnus fraknoii* off the coast of Chile (J. Dolan, unpublished observations) and *Parafavella* spp. in the Arctic Ocean (Meunier 1910; reported as spores) suggest that high host abundance is not a prerequisite for tintinnid parasitism, with tintinnid parasites likely to occur in other pelagic settings.

Parasitism of tintinnids occurs throughout the year, but seasonal patterns differ among parasite taxa. The syndinean parasites *Amoebophrya tintinni* and *Euduboscquella* spp. are most common in summer months when phytoplankton biomass is sufficient to support elevated host densities (Cachon 1964; Coats & Heisler 1989; Coats et al. 1994), although tintinnids infected by these species have been reported from March to

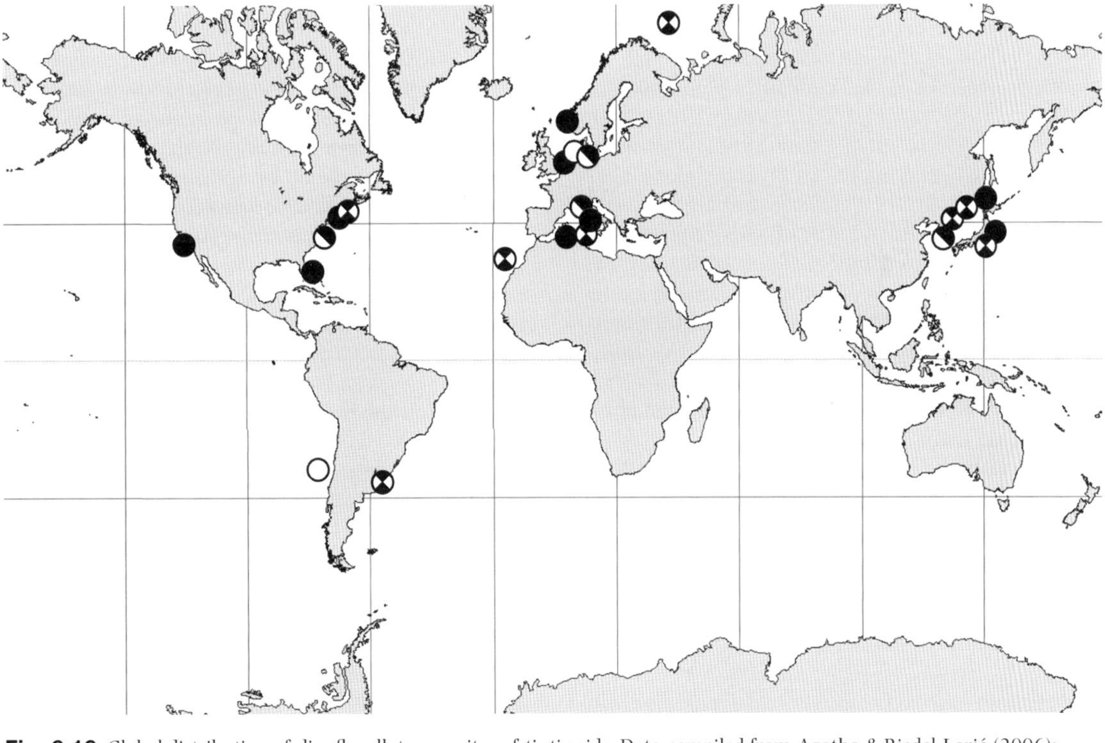

Fig. 6.16 Global distribution of dinoflagellate parasites of tintinnids. Data compiled from Agatha & Riedel-Lorjé (2006); Akselman & Santinelli (1989); Bachvaroff et al. (2012); Cachon (1964); Campbell (1927); Chatton (1952); Coats & Heisler (1989); Coats et al. (1994; 2010; 2012); J. Dolan (unpublished observations); Duboscq & Collin (1910); Entz (1909); Haeckel (1873); Hada (1932a); Harada et al. (2007); Hofker (1931); Konovalova (2007); Laackmann (1908); Lohmann (1908); Meunier (1910); Pierce & Turner (1994); Stoecker et al. (1983). Syndinean dinoflagellates (filled circle); dinophycean parasites (open circle); both syndinean and dinophycean parasites (half-filled circles); unknown parasites (circle with filled quarters).

October (Stoecker et al. 1983; Cachon 1964; Coats & Heisler 1989; Bachvaroff et al. 2012). Dinophycean parasites apparently have a broader temperature range, as they have been reported during summer and winter, with *Duboscquella tintinnicola* parasitizing *Stenosemella nivalis* from July to September (Lohmann 1908), *Duboscquodinium collini* found in *Eutintinnus fraknoii* during September and December (Coats et al. 2010; J. Dolan, unpublished observations), and *Tintinnophagus acutus* occurring in *Tintinnopsis cylindrica* from July through February (Agatha & Riedel-Lorjé 2006; Coats et al. 2010). Infection of tintinnids (*Codonella* sp.) in winter was also reported by Pierce & Turner (1994), but the taxonomic affinity of the parasite is uncertain.

Parasitic dinoflagellates, or spores most likely derived from parasitic infections, have been reported from over 30 tintinnid species representing the genera *Amphorellopsis, Codonella, Coxliella, Cyttarocylis, Eutintinnus, Favella, Helicostomella, Parafavella, Rhabdonella, Salpingella, Steenstrupiella, Stenosemella, Tintinnopsis*, and *Xystonella* (Haeckel 1873; Laackmann 1908; Lohmann 1908; Entz 1909; Meunier 1910; Hofker 1931; Campbell 1927; Cachon 1964; Coats 1988; Coats et al. 1994, 2010; Pierce & Turner 1994; Konovalova 2007; Bachvaroff et al. 2012). Parasitism has been observed on only a single occasion in two-thirds of the tintinnid species, with an equal number of species hosting unidentified parasites. Thus, it is impossible to assess the host range of tintinnid parasites reliably. Of the 10 dinoflagellate species described as parasites of tintinnids, three are known to infect more than one host species: *Duboscquella tintinnicola* in *Stenosemella nivalis* and *Coxliella helix*; *Euduboscquella aspida* in *Coxliella laciniosa, Eutintinnus fraknoii, Favella ehrenbergii, F. panamensis*, and *Tintinnopsis campanula*; *Euduboscquella cachoni* in *Eutintinnus pectinis* and *E. tenuis* (Cachon 1964; Coats 1988; Coats et al. 1994, 2012; Bachvaroff et al. 2012). *Amoebophrya tintinni* is known only from *Xystonella lohmanni, Duboscquodinium collini* from *Eutintinnus fraknoii, Duboscquodinium kofoidi* from *Tintinnopsis campanula, Euduboscquella anisospora* and *E. cnemata* from *Favella ehrenbergii, Euduboscquella crenulata* from *Favella panamensis*, and *Tintinnophagus acutus* from *Tintinnopsis cylindrica* (Chatton 1952; Cachon 1964; Coats et al. 2010, 2012).

A few tintinnids serve as hosts for at least two different parasites, with *Duboscquodinium collini* and *Euduboscquella aspida* parasitizing *Eutintinnus fraknoii, Euduboscquella anisospora* and *E. aspida* infecting *Favella ehrenbergii, Euduboscquella crenulata* and *E. aspida* reported from *Favella panamensis*, and *Duboscquodinium kofoidi* and *Euduboscquella aspida* occurring in *Tintinnopsis campanula* (Chatton 1952; Cachon 1964; Coats et al. 1994, 2012). In addition, the same host cell can be simultaneously infected by more than one parasite species (D.W. Coats, unpublished observations).

Parasite prevalence and impact on host populations

Cachon (1964) provided the first insight into the ecological importance of parasitism in tintinnids when he noted that populations of *Favella ehrenbergii* in Algiers harbor showed dramatic fluctuations during summer, with most cells being infected by *Euduboscquella* during abrupt declines in tintinnid abundance. Cachon (1964) also noted that, of the several co-occurring tintinnids, the species infected by *Euduboscquella* differed from day to day. Unable to explain that observation, he speculated that morphologically indistinguishable host-specific parasites might be parasitizing different tintinnid species over time. Although sequential infection of tintinnid species by different parasites seems likely, it is also possible for multiple host species to be simultaneously infected by one or more parasite species. That is clearly the case in Chesapeake Bay, where *Stylicauda platensis, Tintinnopsis major, Tintinnopsis panamensis, Tintinnopsis* cf. *subacuta* and *Tintinnopsis radix* showed co-occurring infections (Fig. 6.17). *S. platensis, T. major, T. panamensis*, and *T.* cf. *subacuta* were parasitized by *Euduboscquella* spp., whereas *T. radix* was infected by an unidentified dinophycean parasite. All five hosts were present in the lower Bay, but only four were parasitized in that region. The failure of parasites to spread to *T.* cf. *subacuta* in the lower Bay, along with infection of that host species in the upper Bay, suggests that at least two species of *Euduboscquella* were infecting the tintinnids. As mentioned above, specificity and host range of tintinnid parasites remains unresolved; however, recent molecular data support the notion that tintinnids are parasitized by cryptic species of parasitic dinoflagellates (Bachvaroff et al. 2012).

Stoecker et al. (1983) observed outbreaks of *Euduboscquella* sp. near the end of *Favella* sp. blooms in Perch Pond (Woods Hole, Massachusetts) and suggested that mortality due to parasitism may have been an important factor in determining tintinnid net growth rates.

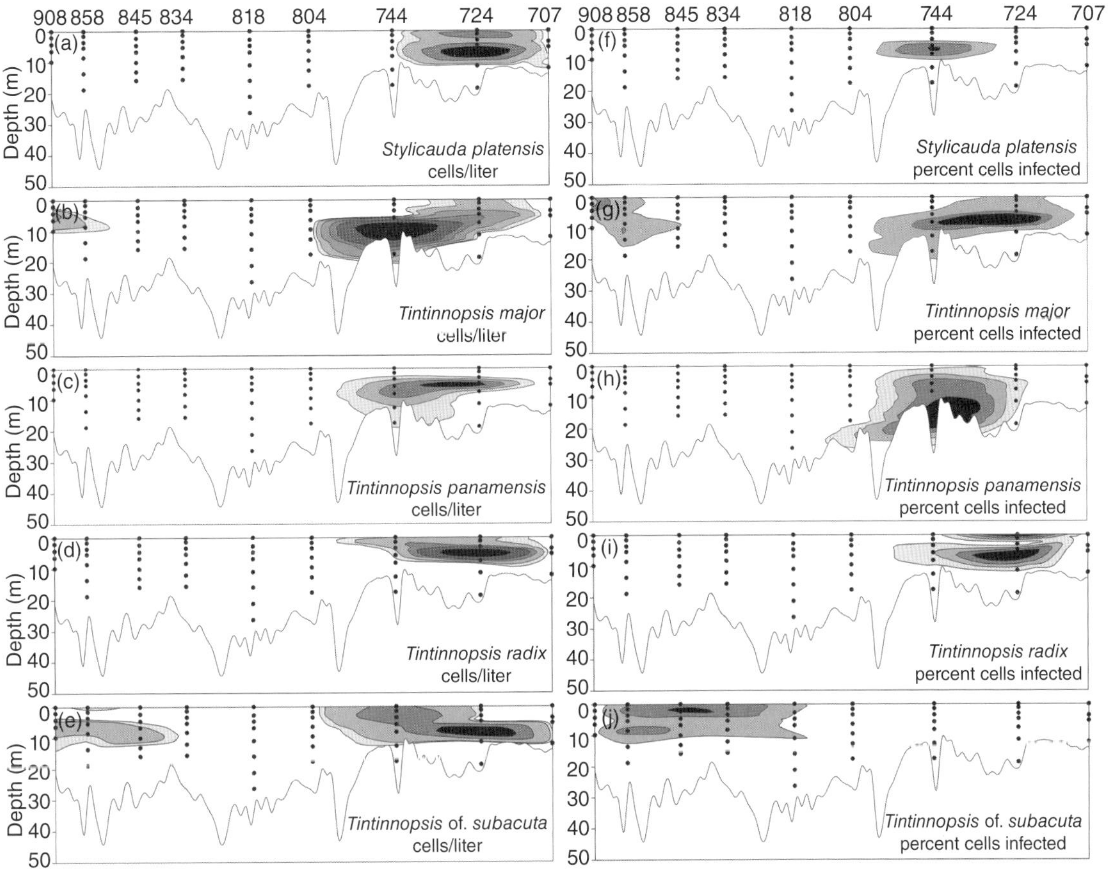

Fig. 6.17 Abundance and parasite prevalence for five tintinnid species along the longitudinal axis of Chesapeake Bay on July 23–24, 1990. Station designations located on upper graphs indicate minute and degree north latitude (e.g., 908 = 39° 08′; 707 = 37° 07′). Filled circles show sample depths and the solid undulating lines represent the bottom contour. (a–e) Host abundance with increasing density of shading indicating >50–100, >100–300, >300–500, and >500 cells/liter for (a) *Stylicauda platensis*; >15–25, >25–50, >50–100, >100–150, >150–300, and >300 cells/liter for (b) *Tintinnopsis major*; >40–80, >80–120, >120–160, and >160 cells/liter for (c) *Tintinnopsis panamensis*; >25–50, >50–100, >100–150, >150–300, and >300 cells/liter for (d) *Tintinnopsis radix*; >50–100, >100–500, >500–1000, and >1000 cells/liter for (e) *Tintinnopsis* cf. *subacuta*. (f–j) Percentage of host cells infected with increasing density of shading indicating >5–10, >10–15, and >15% for (f) *S. platensis*; >5–15, >15–20, and >20% for (g) *T. major*; >20–40, >40–60, >60–80, and >80% for (h) *T. panamensis*; >1–3, >3–5, >5–7, and >7% for (i) *T. radix*; >10–25, >25–50, and >50% for (j) *T.* cf. *subacuta*. Data obtained following methods of Coats et al. (1994), except that parasite prevalence was determined from specimens stained using the quantitative protargol technique (Montagnes & Lynn 1987). Previously unpublished data provided by D.W. Coats.

In studies of two temperate estuaries, Coats & Heisler (1989) and Coats et al. (1994) also argued that parasitism had a significant impact on tintinnid populations. In Chesapeake Bay, kilometer-scale patches of *Eutintinnus pectinis* infected by *Euduboscquella cachoni* persisted from July to September, with epizootics infecting 20–50% of host cells over broad areas. Bay-wide, parasite prevalence averaged 5–18%, with parasite-induced mortality removing 7–24% of host standing stock per day, a loss rate roughly equivalent to grazing pressure

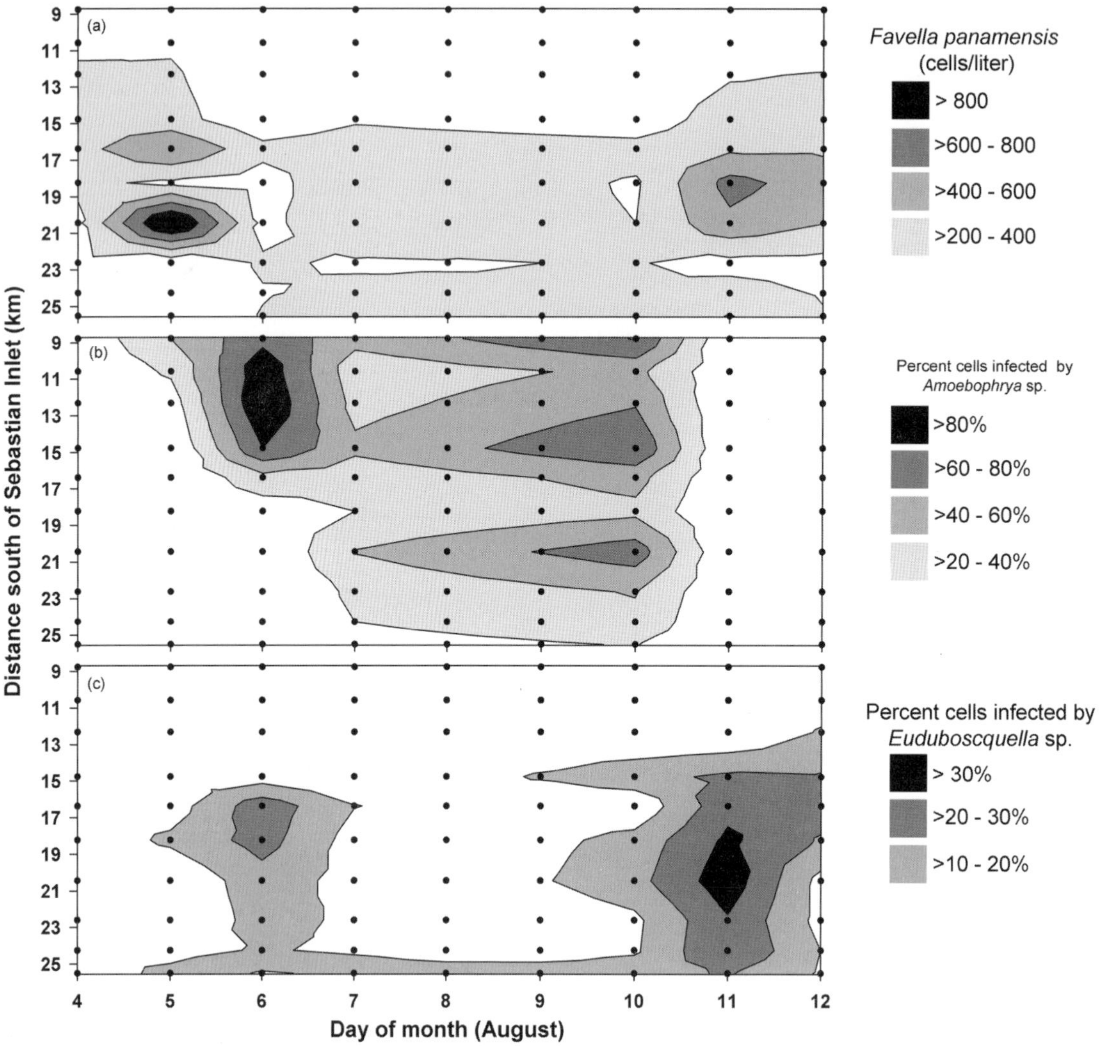

Fig. 6.18 Parasitism of *Favella panamensis* in the Indian River Lagoon during August 1992. (a) Abundance of *F. panamensis* in surface water samples collected from Sebastian Inlet south to Vero Beach. (b) Percentage of cells infected by *Amoebophrya* sp. (c) Percentage of cells infected by *Euduboscquella* sp. Host abundance and parasite prevalence were determined following methods of Coats et al. 1994. Filled circles indicate sample locations. Previously unpublished data provided by D.W. Coats.

from the copepod *Acartia tonsa*. Similar results were obtained for *Favella panamensis* in Chesapeake Bay and the Indian River Lagoon, Florida, where parasitism removed approximately 10% of host biomass per day. In the latter system, *F. panamensis* was infected by two different parasites, *Euduboscquella* sp. and *Amoebophrya* sp. (Fig. 6.18). Epizootic infection by *Amoebophrya* sp. in the northern lagoon and moderate infection by *Euduboscquella* in the southern lagoon were associated with a marked decline in host abundance in early August. Host density recovered over the following week, only to experience epizootic infection by *Euduboscquella* sp. in the southern lagoon. Thus, the two parasites appear to partition host resources in time and space.

6.5 SUMMARY AND FUTURE DIRECTIONS

Although parasitism of tintinnids has been known for over a century, we are still in the discovery phase with respect to parasite diversity, morphology, life history, cell biology, and ecology. Indeed, our understanding of such topics is currently limited to parasitic dinoflagellates, with other parasite taxa known to infect ciliates being unexplored among the tintinnids. Recent studies of tintinnid parasites have revealed new higher taxa of dinoflagellates, previously unrecognized lifestyles, and cryptic diversity. Those discoveries suggested the need for continued investigations and re-examination of the older, often poorly described species. From a molecular perspective, properly ordinating the evolutionary changes between syndineans and dinokaryotes into derived and primitive features can be challenging. The poorly supported topologies on molecular trees combined with difficulty in determining even basic morphology, cell biology, or genomic characters for syndinean dinoflagellates means that the syndineans can alternately be viewed as primitive in comparison to dinokaryotes, or as having lost dinokaryotic features. Resolution of such problems will require combined morphological, developmental, and molecular approaches.

Research on the ecological function of parasites in terrestrial and aquatic ecosystems has shown parasitism to be the most popular lifestyle on Earth, with about half of all biodiversity derived from parasites (Dobson et al. 2008). Parasitism represents 75% of the trophic links in ecosystems, with parasite biomass and production rivaling that of top predators (Dobson et al. 2008; Kuris et al. 2008). Parasites shape host populations, influence the flow of energy and matter in food webs, and promote ecosystem health by sustaining biodiversity (Horwitz & Wilcox 2005; Hudson et al. 2006; Lefèvre et al. 2008). Far less is known about planktonic ecosystems, but information on the ecological significance of parasitism among planktonic organisms is slowly growing. Tintinnid parasites occur in all seasons and in coastal to pelagic systems. Epizootic infections of tintinnids are not uncommon and appear to drive the decline of host species; however, the influence of such events on tintinnid species succession is unclear. From the perspective of the marine pelagic food web, tintinnid parasites compete with larger consumers (metazoan zooplankton, fish larvae, and benthic invertebrates; Chapter 5) for an important and potentially high-quality food source. By cropping tintinnid biomass, parasitic dinoflagellates short circuit the microbial loop, retaining resources at the microbial level by transforming host biomass into nano-sized, typically short-lived, parasite progeny and by recycling host biomass as dissolved and particulate organic matter released upon death of infected cells.

As Pomeroy & Wiebe (1988) indicated over two decades ago, our ability to model and quantify microbial food webs is limited by our understanding of the life histories and feeding strategies of parasitic and predatory microorganisms. With that realization, the study of planktonic parasitism promises to be a fertile area for future research. From the perspective of tintinnids as model organisms, it is informative to recognize that sporogenic stages of parasitic dinoflagellates remain in the lorica for some time after death of the host, making assessment of parasite abundance and host identification easier and more reliable. By contrast, parasites that infect aloricate ciliates and most other protists are immediately dispersed into the water after death of the host, making identification of the host species and quantification of sporogenic stages problematic. Thus, tintinnids may well serve as an ideal model for field studies assessing the ecological importance of parasitism in planktonic protists.

6.6 KEY POINTS

1. Of the various prokaryote and eukaryote organisms known to infect ciliates, only dinoflagellate parasites are well documented for tintinnids.

2. Dinoflagellate parasites of tintinnids encompass endoparasitic syndinean species that grow in the cytoplasm or nucleus of their host and ectoparasitic dinophycean species that attach to the outside of the host cell and act much like predators.

3. Life histories of syndinean and dinophycean parasites of tintinnids are complex, appear not to require an intermediate host, and are often incompletely known, with a sexual cycle clearly documented for only one species.

4. As with many protistan groups, the taxonomy and phylogeny of dinoflagellate parasites is in a state of flux, with undiscovered taxa and cryptic species likely to emerge with new morphological, developmental, and molecular data.

5. Parasitism of tintinnids is widespread, occurring in all seasons, in coastal to pelagic settings, and in arctic to subtropical environments.

6. Parasitic dinoflagellates typically prevent reproduction of their tintinnid hosts and are often lethal, with epizootics contributing to the decline of host populations.
7. Tintinnid parasites compete with metazoan consumers and facilitate recycling of materials at the microbial level.
8. Developmental stages of parasitic dinoflagellates are retained within the host lorica, facilitating identification and enumerations of infected species. Thus, tintinnids provide a unique opportunity for studying the ecological significance of parasitism among planktonic protists.

ACKNOWLEDGMENTS

This material is based in part on work supported by the National Science Foundation under Grant No. 8911316 and 0629624. Any opinions, findings, and conclusions or recommendations expressed in this material are those of the authors and do not necessarily reflect the views of the National Science Foundation. Contribution No. 870 of the Smithsonian Marine Station at Fort Pierce, Florida. We thank Lois Reid for illustrations and Paula Villanueva for assistance in processing Chesapeake Bay samples.

Chapter 7

COMPARATIVE BIOLOGY OF TINTINNID CYSTS

Takashi Kamiyama

Tohoku National Fisheries Research Institute, Fisheries Research Agency, Shiogama, Miyagi, Japan

The Biology and Ecology of Tintinnid Ciliates: Models for Marine Plankton, First Edition. Edited by John R. Dolan, David J.S. Montagnes, Sabine Agatha, D. Wayne Coats, and Diane K. Stoecker.

7.1 INTRODUCTION

What is a cyst?

In the life cycle of a microorganism, a "cyst" is a resting or dormant stage. Often the special morphology of a cyst allows it to withstand harsh environmental conditions. The cyst is commonly thought of as a state of suspended animation in which the metabolic processes of the cell are slowed down and the cell ceases activities such as feeding and locomotion (e.g., Wikipedia). Besides protists and bacteria, some sort of a resting stage or cyst is common to many taxa of planktonic organisms of variety of trophic levels (Belmonte et al. 1997). Cysts of planktonic organisms are often likened to "seeds" with the benthos as a "seedbank" and although they allow survival under unsuitable environmental conditions, they can, like seeds, also play a role in dispersal. In some taxa, cysts represent an obligatory stage, the product of genetic recombination resulting from conjugation (compare with section "Factors inducing encystment" on pages 177–178). Cysts can influence population dynamics, serving as a "source" when individuals emerge from cysts (excystment) or as a "sink" when part of the population undergoes cyst formation (encystment).

Cysts of tintinnid ciliates are relatively conspicuous because they are usually formed within loricae, and have thus often been observed in natural plankton assemblages. In many coastal species of tintinnids, the occurrence of the planktonic stage (= vegetative cells) is distinctly seasonal with cells found in the water column for only a relatively short period of time (see Chapter 10). The regular seasonality of coastal plankton species, with sudden appearance and disappearance, has long been thought to be associated with the existence of resting stages of some sort (Smetacek 1985).

The role of tintinnid cysts in dispersal is unknown. The cysts in marine species which sink to the bottom sediments may have few chances to be dispersed. However, tintinnid cysts in shallow waters are possibly transported by natural dispersal mechanisms (e.g., water circulation, transport on feathers or feet of birds) and by artificial means such as ballast waters (Pierce et al. 1997). Tintinnids also occur in freshwater systems (e.g., Foissner & Wilbert 1979). Although there is no information on cysts in freshwater tintinnids, it is possible that the life cycles of freshwater tintinnids include a cyst stage like those known for oligotrich ciliates in lake waters (Müller & Wünsch 1999).

Historical overview of tintinnid and aloricate ciliate cysts

The extreme changes in physiology and morphology between the vegetative and cyst stages of ciliates have attracted the attention of many biologists over the years. For instance, since the early 1900s, physiological, morphological, and ecological studies have been conducted on cysts of soil and pond ciliates (e.g., the genera *Colpoda*, *Didinium*, *Dileptus*, *Amphileptus*, *Zoothamnium*). Several general reviews on ciliate cysts are available (e.g., van Wagtendonk 1955; Corliss & Esser 1974; Bussers 1984; Bradbury 1987; Matsusaka 2006; Lynn 2008; Verni & Rosati 2011). However, these reviews reveal that information is limited largely to ciliate species that are relatively easy to culture. Nonetheless, even among this small number of ciliate species, the factors inducing encystment and excystment appear diverse with few consistent phenomena. Thus, reviews on ciliate cysts have repeatedly pointed out that most aspects of the physiological and ecological characteristics of ciliate cysts are still unknown (e.g., Corliss & Esser 1974; Bussers 1984; Matsusaka 2006; Foissner et al. 2007; Verni & Rosati 2011).

Tintinnid cysts were first reviewed by Reid & John (1978) who focused mainly on morphological characteristics. Although confusion of tintinnid cysts and parasites was mentioned in their review, they stated that many types of tintinnid cyst have been observed in seawater and sediment samples, suggesting that the cyst stage is likely an important process in the life history of coastal tintinnid species. Since their review, many studies on the occurrence and ecology of a variety of ciliate cysts in both freshwater and seawater systems have been conducted. Unfortunately, our knowledge of tintinnid encystment and excystment remains very incomplete. The physiological and ecological characteristics of cysts of tintinnid and oligotrichs, the main ciliate components in marine plankton communities, appear to be considerably different from those of terrestrial (soil or pond) species. In this chapter, information on occurrence and ecological function of marine tintinnid cysts will be reviewed and compared with findings on cysts of the other common protists of the marine plankton, oligotrichs, and dinoflagellates.

7.2 MORPHOLOGY

Characteristics of tintinnid cysts observed in seawater samples

Tintinnid cysts have been often observed in natural plankton assemblages (e.g., Paranjape 1980) as well as in sediment trap samples (Price & Pospelova 2011). Although the cysts themselves have relatively few morphological characters, because they are in the lorica, presumably we know which species formed the cyst (Fig. and Plate 7.1). Presumptive cysts of more than 20 species were described from seawater or sediment samples (Reid & John 1978). However, it should be noted that distinguishing cysts from a developmental stage of a parasite infection can be difficult (see Chapter 6). Early studies of tintinnids often described cyst (or a "spore") formation that was quite likely a parasite stage (e.g., Haeckel 1873; Laackmann 1908; Meunier 1910). Complete cycles of both encystment and excystment have been observed only in species of *Favella* and *Helicostomella*. Excystment alone has been observed in species of *Eutintinnus* and *Tintinnopsis*.

Typically the cysts are flask-shaped as in *Acanthostomella*, *Eutintinnus*, *Favella*, *Helicostomella*, *Leprotintinnus*, *Metacylis*, *Parundella*, and *Tintinnopsis*, and cylindrical as in *Helicostomella* (Reid & John 1983). In the case of *Favella*, the cyst has been described as granular protoplasm within a flask-shaped container of a thick hyaline wall terminated with a disk-shaped operculum (Reid & John 1978). The droplet-shaped cyst described as "Fusopsis" by Meunier (1910) and depicted as a possible tintinnid cyst by Reid & John (1978, fig. 1h, g.) is now known to very closely resemble the cyst of the oligotrich *Cyrtostrombidium* (Kim et al. 2002b).

Characteristics of oligotrich and dinoflagellate cysts

Oligotrichs and dinoflagellates, many of which also form cysts, occur with tintinnids in the plankton (see Chapter 9). Consequently, in seawater and surface sediment samples, cysts of oligotrichs and dinoflagellates may be found along with tintinnid cysts. There is a very large literature on dinoflagellate cysts by both plankton researchers and paleontologists. In this chapter, dinoflagellate cysts will only be briefly covered; interested readers are directed to general reviews of methods (e.g., Matsuoka & Fukuyo 2000) and distributions (e.g., Marret & Zonneveld 2003). Compared with dinoflagellates, there have been relatively few studies of oligotrich cysts.

The cysts of oligotrich ciliates are typically flask-shaped like a *Favella* cyst and usually have a clear plug-like structure (papula), which is the exit point of the ciliate during the excystment (Fig. and Plate 7.2). The cysts of tintinnids and heterotrich ciliates, which have some common morphological characteristics (flask-shape and papula), were once described as members of "papuliferes" (Meunier 1910). These characteristics of papuliferes led to the suggestion of a common phylogenetic origin different from hypotrich ciliates that produce simple spherical cysts (Reid & John 1983).

From the examples of the cysts of oligotrich shown in Fig. and Plate 7.2, it is clear that the gross morphology is quite variable and the fine structure is quite variable as well. The cyst wall can consist of one thick layer in some species (Kim & Taniguchi 1995; Müller 1996; Kim et al. 2002b) whereas others have an additional outer-membrane such as *Strombidium crassulum* (species name identified provisionally; Reid 1987). In *Strombidium conicum* there is a smooth exterior membrane that can be dissolved with 1 N hydrochloride (Kim & Taniguchi 1995). Still others have no membrane outside of the cysts such as *Cyrtostrombidium boreale* (Kim et al. 2002b).

The cyst walls of *Strombidium capitatum* and *Strombidium biarmatum* are covered with spines (Kim et al. 2002b, 2008; Agatha et al. 2005). In the case of *Meseres corlissi*, a conspicuous coat of extracellular organic scales (termed "lepidosomes") occurs on the cyst wall. The structure of the cyst wall was examined in detail using light and electron microscopy and found to be composed of glycoproteins and chitin (Foissner 2005; Foissner et al. 2005). In *Strombidium crassulum*, detailed qualitative information on the cyst wall obtained using energy-dispersive X-ray spectroscopy and cytochemical analysis suggested it was made of calcium phosphate or calcium carbonate incorporated in an organic matrix (Reid 1987).

A comparative study of the cysts of two oligotrich species, *Halteria grandinella* and *Pelagostrombidium fallax*, and three hypotrich species *Laurentiella strenua*, *Steinia sphagnicola* and *Oxytricha granulifera*, showed that each species had distinct, very likely non-homologous, cyst surface characteristics (spine, thorns, fibrous and tubular lepidosomes, perilemma) (Foissner

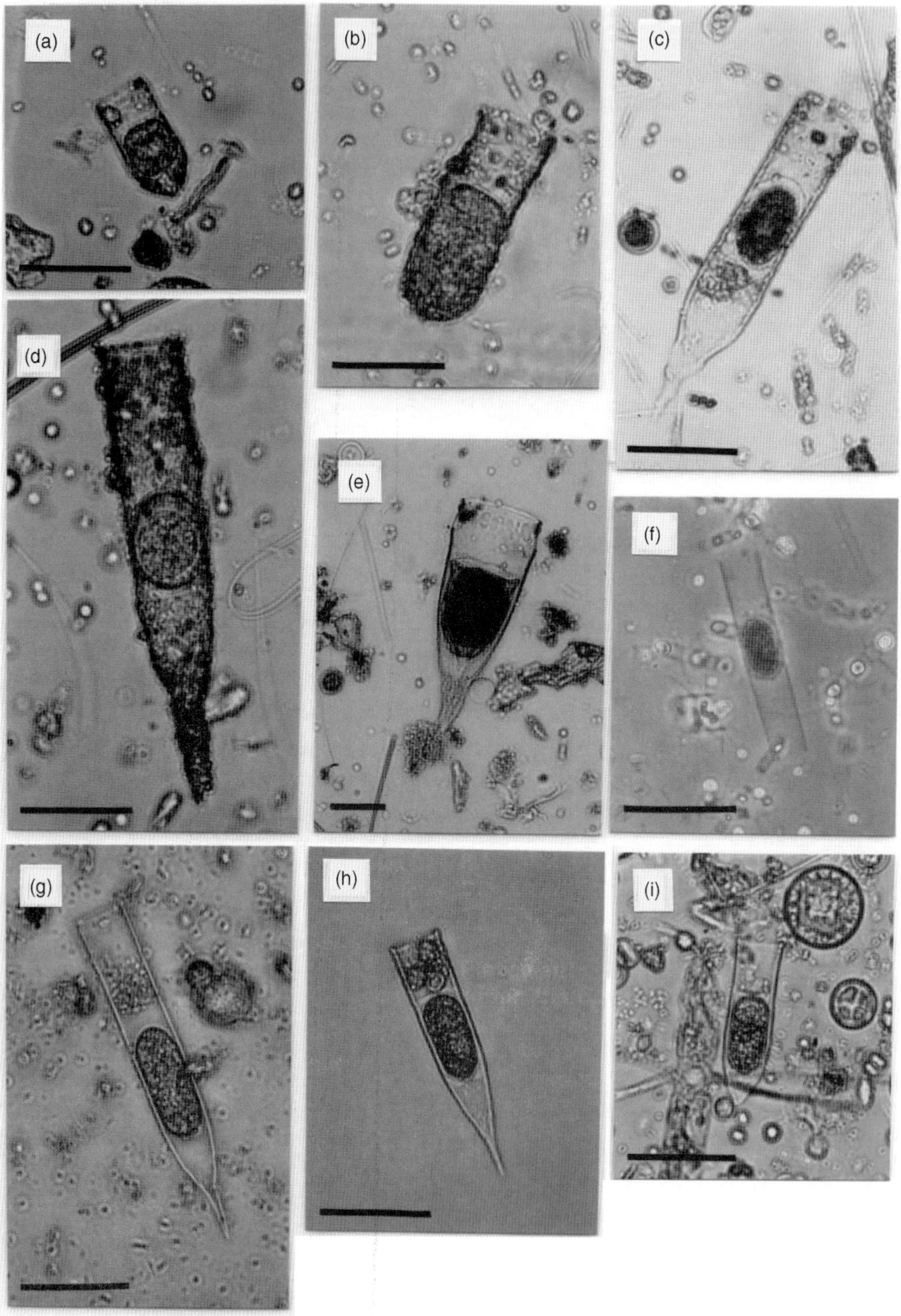

Fig. 7.1 Photographs of tintinnid cysts from coastal waters of central and western Japan (from Kamiyama 1994c). (a) *Tintinnopsis beriodea*. (b) *Tintinnopsis* sp. (c) *Tintinnopsis corniger*. (d) *Tintinnopsis kofoidi*. (e) *Favella taraikaensis*. (f) *Eutintinnus tubulosus*. (g) *Helicostomella subulata*. (h) *Helicostomella fusiformis*. (i) *Helicostomella longa*. Scale bar, 50 μm. Note that *T. kofoidi* may be a synonym of *T. cylindrica* (Agatha & Reidel-Lorjé 2006). For color version, see Plate 7.1.

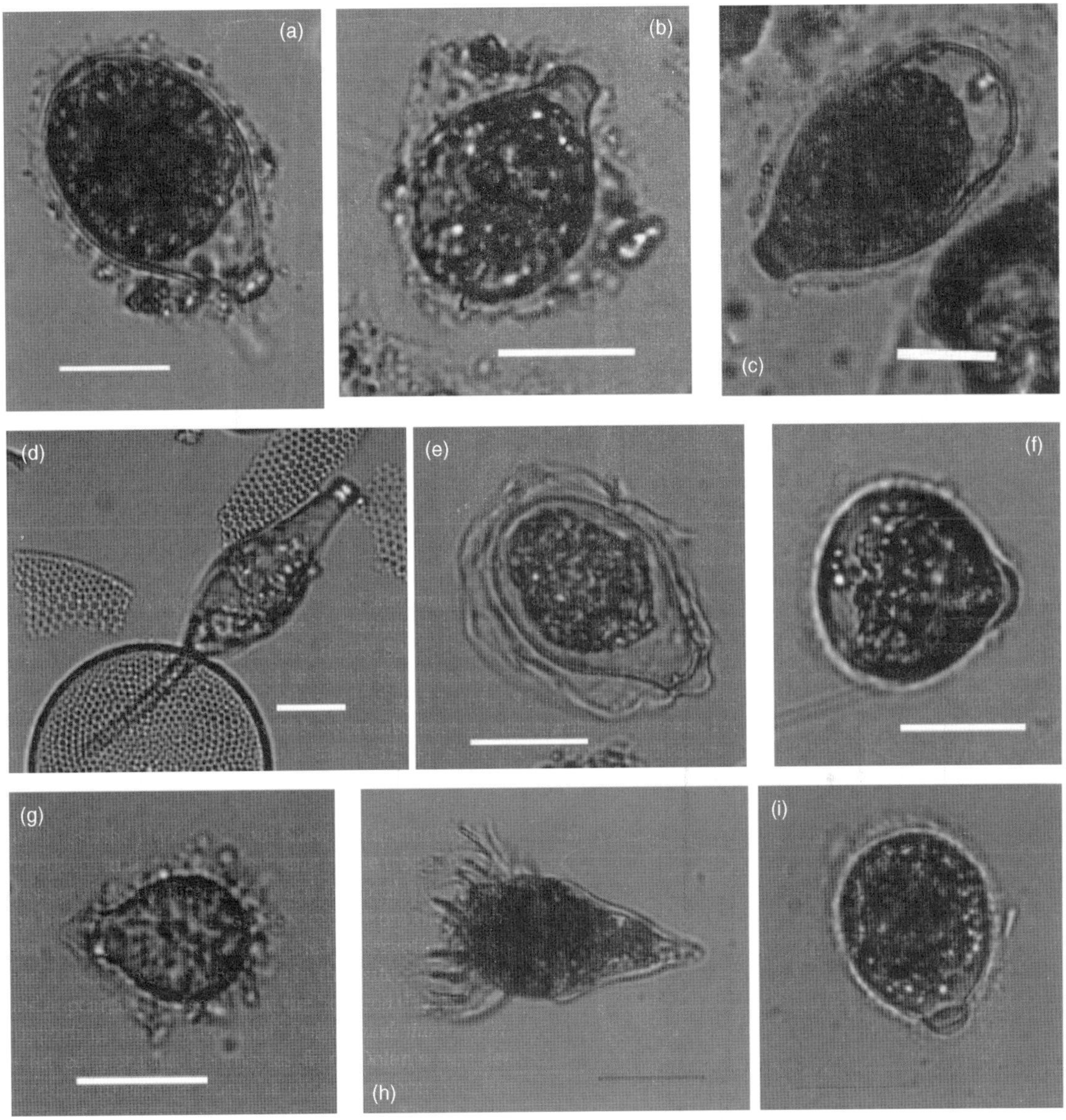

Fig. 7.2 Photographs of various flask-shaped cysts of marine ciliates (a–g) observed in sediments of northern Japan, and a vegetative cell (h) of the oligotrich ciliate *Strombidium chlorophilum* identified from protargol-stained cells and the cyst (i) produced in a laboratory culture. Scale bars, 20 μm (a–g), 50 μm (h, i). For color version, see Plate 7.2.

et al. 2007). Thus, the gross morphology and composition, even among oligotrichs, varies considerably among different species.

Some ciliate cysts have many spine or ridges evident using scanning electron microscopy (Müller and Wünsch 1999; Reid 1987). Cyst walls of some stichotrich and peritrich ciliates comprise three layers termed ectocyst, mesocyst, endocyst (ordered from outside to inside) and show a granular material layer located between the endocyst and the pellicle (e.g., Matsusaka

2006; Calvo et al. 2003). Little is known of the chemical composition of the cyst wall, in general, the main macromolecular cyst wall components are thought to be proteins, glycoproteins, and carbohydrates based on terrestrial ciliate cysts (reviewed in Gutiérrez et al. 1990, 2003; Foissner 2005).

In dinoflagellates, the morphology of cysts is also very diverse with few features in common. As in ciliates, cyst morphology is usually very different from that of the vegetative cells. Among the dinoflagellates, cysts of some species are well known because of interest in the role of excystment in producing harmful algal blooms that can influence fisheries and human health. For example, protocols for the enumeration of *Alexandrium tamarense/catenella* cysts have been established, and distribution of the cysts has been investigated in coastal waters (Anderson et al. 1982; Yamaguchi et al. 1995, 2002).

Specific changes in morphology of most ciliates during the course of cyst formation are commonly a decrease in cell volume, retention of pigments, evacuation of storage products, resorbtion of cilia, and formation of a cyst wall (Gutiérrez et al. 1990, 2001). In almost all species belonging to oligotrichs, stichotrichs, heterotrichs, peritrichs, colpodids, and haptorids, the cyst volume of each species was less than that of the vegetative cell, although the cyst volume of *Strombidium oculatum* was exceptionally 1.6 times larger than the volume of vegetative stage (Foissner et al. 2006). Among marine dinoflagellates, decreases in cell volume do not always occur. In *Alexandrium tamarense*, the cyst is generally larger than the normal vegetative cells, 54 µm × 32 µm (Itakura & Yamaguchi 2005) compared with 26–38 µm × 27–44 µm (Fukuyo et al. 1990) (compare with section "Factors inducing encystment" on pages 177–178).

The retention of pigments in the cyst, hypothesized as perhaps serving as protection from ultraviolet-induced damage (Gutiérrez et al. 2001), may be uncommon in marine ciliates and dinoflagellates. However, some oligotrich and dinoflagellate cysts show adhesion of particles to the surface of the cysts (Yamaguchi et al. 1995; Reid 1987; Kim & Taniguchi 1995), which could serve the same purpose. Adhesion of mineral particles may also promote sedimentation of cysts and make them less susceptible to vertical transport. Cysts can be very resistant. Morphological characteristics of marine ciliate cysts persist for at least a week in hydrochloric acid (Reid & John 1978; Reid 1987; Kim & Taniguchi 1995), but this ability to resist strong acids is much lower than that of dinoflagellate cysts or ciliate cysts in soils and freshwaters (Corliss & Esser 1974), indicating that the chemical characteristics of marine tintinnids and oligotrichs may not correspond to those of other ciliates.

7.3 ENCYSTMENT AND EXCYSTMENT

Encystment

Cyst formation involves drastic changes in the outer and inner structure of ciliate cells. Encystment is sometimes observed in laboratory cultures. Kamiyama (2011) recorded most of the encystment process of a tintinnid ciliate *Favella taraikaensis* in laboratory cultures with time-lapse video connected to an inverted microscope (http://feis.fra.affrc.go.jp/HABD/TPS/Takashi_Kamiyama_video_file.html). Recording was started when a circular movement of *F. taraikaensis* was observed within the lorica. A membranous structure around the cell, which became a cyst wall, was produced after 30 minutes (Fig. 7.3). Then the circular movement of cytoplasm gradually slowed and the membranous structure changed to a cyst wall. When the cyst wall was formed, the cytoplasm darkened and a granular structure of the cell became visible. After 6 hours of observation, movement of cytoplasm stopped and the morphological change into a cyst was apparently finished. As very few reports exist, further work is needed to determine if encystment varies in different tintinnid taxa.

Encystment of the oligotrich *Strombidium conicum* has been observed in laboratory cultures (Endo 2005), and it is likely that the encystment also starts from circular movement of cytoplasm (Fig. 7.4). The vegetative cell, which is conical in shape, changed into a sphere at the initiation of encystment. A flask-shaped cyst wall was produced when the circular movement of cytoplasm stopped. Then, some materials, possibly contents of the food vacuoles and/or materials originating from pigments, were excreted through the opened papula. Finally, the encystment ended with the formation of the papula. In oligotrichs, the encystment process can be rapid, taking just 1 minute in *Strombidium oculatum* (Montagnes et al. 2002a), or a few minutes in *Meseres corlissi* (Foissner et al. 2006).

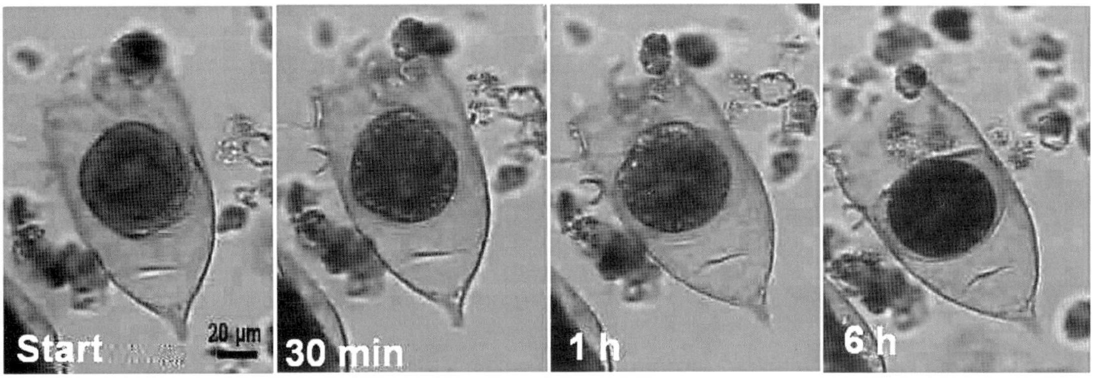

Fig. 7.3 Photographs of the encystment process of the tintinnid ciliate *Favella taraikaensis* observed in a laboratory culture. Observation was started when the circular movement of the cell was observed within the lorica. A membranous structure around the cell was produced after 30 minutes and then gradually changed into a cyst wall. The movement of the cell stopped after 6 hours when the cyst formation was completed. To see the original video, go to http://feis.fra.affrc.go.jp/HABD/TPS/Takashi_Kamiyama_video_file.html (from Kamiyama 2011, with permission of the Plankton Society of Japan).

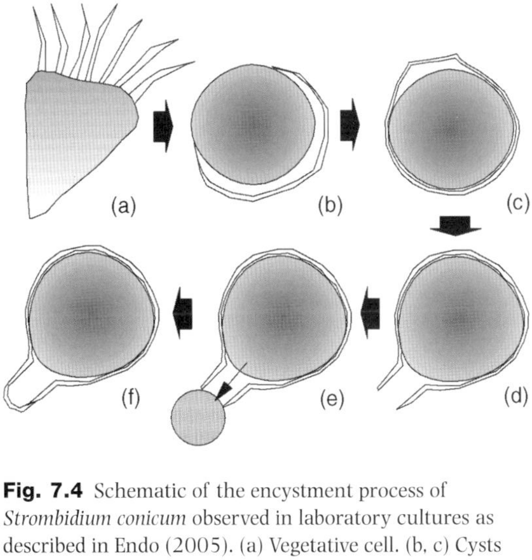

Fig. 7.4 Schematic of the encystment process of *Strombidium conicum* observed in laboratory cultures as described in Endo (2005). (a) Vegetative cell. (b, c) Cysts during wall formation. (d) Completion of cyst wall except for papula. (e) Excretion of unnecessary cell contents. (f) Formation of papula.

Factors inducing encystment

For terrestrial ciliates, the factors inducing encystment have been examined and reviewed in several papers (Corliss & Esser 1974; Gutiérrez et al. 1990, 2001; Matsusaka 2006). In laboratory studies, various factors have been identified: lack of prey, excess prey, over-crowding, accumulation of waste products from the cells, and desiccation (evaporation of culture fluid). The most common exogenous trigger for terrestrial ciliates is probably lack of prey (Matsusaka 2006). However, factors acting in natural environments are still largely unknown (Verni & Rosati 2011). Information on the conditions or stimuli which induce encystment in planktonic marine and freshwater ciliates (including tintinnids) is sparse.

Encystment of the oligotrich *Meseres corlissi* increased dramatically if temperature was below 20 °C, while rarely occurring at more than 20 °C (Foissner et al. 2005). However, in cultures of tintinnids, encystment was observed with no apparent change in culture conditions (e.g., Paranjape 1980). Working with marine oligotrichs, Endo & Taniguchi (2006) reported that environmental conditions such as prey concentration, temperature, and salinity were not critical factors for encystment of the marine coastal oligotrich *Strombidium conicum*. These findings are not inconsistent with the factors for encystment of terrestrial ciliates as described above. Also, they found that the ciliates in mixtures of several different strains formed cysts after conjugation occurred in the culture. Furthermore, encystment was recorded in the mixed clonal cultures after 15–17 binary fissions (Endo & Taniguchi 2006), suggesting that maturation or aging of the population is necessary for encystment. Interestingly, excystment of some cells was also observed after a few days of encystment in the same culture, implying that encystment of *S. conicum* is not uniquely a reaction to unfavorable environmental conditions. Encystment for *S. conicum* may be only a consequence of conjugation and a phase for rejuvenation of the population (Endo 2005).

At present, it is unknown if encystment in tintinnids is related to conjugation or not. Paranjape (1980) noted encystment of *Helicostomella subulata* after the occurrence of conjugation in laboratory cultures and also observed excystment under the same constant conditions after 8–9 weeks from first encystment. In strombidiid ciliates, Müller (2002) observed that all strains derived from excysted individuals readily formed cysts, whereas few cyst forming strains were among those isolated from lake water, and speculated that the genetic variability within the population may create variable responses to external triggers for encystment. The relationship between conjugation and encystment, as well as possible genetic differences among cyst-forming and non-cyst-forming strains, merits further study.

We should note that cyst formation can be unrelated to conjugation. This is evident in perhaps the best-studied marine ciliate with regard to encystment and excystment: the tide-pool oligotrich *Strombidium oculatum* (Fauré-Fremiet 1948; Jonsson, 1994; Montagnes et al. 2002a, b). In this species, encystment and excystment follow a circa-tidal rhythm with encystment during high tides allowing the ciliate to avoid displacment from the tide pool. The timing of the cyst cycle continues in laboratory populations, indicating an endogenous trigger

Factors inducing encystment of dinoflagellates have been clarified mainly for harmful algal bloom species, which display a variety of conjugation patterns and processes. In the case of *Alexandrium tamarense*, after asexual growth of vegetative cells, two types of gamete (female and male) are produced and the conjugation of both gametes takes place, anisogamous conjugation (Anderson 1998). The conjugated gametes become the swimming zygote (planozygote). The planozygote transforms into a dormant resting cyst. However, planozygotes of all dinoflagellates do not always encyst. Planozygotes of *Scrippsiella trochoidea* (Watanabe et al. 1982) and *Gyrodinium instriatum* (Uchida et al. 1996) occasionally develop directly into vegetative cells without going through a cyst stage. In dinoflagellates, a variety of factors have been examined such as temperature, light intensity, photoperiod, nutrient stress, etc., with nutrient stress appearing to be the main factor inducing cyst formation (Pfiester and Anderson 1987). For other phytoplankton taxa, nutrient depletion (especially nitrogen) is also a trigger for encystment or formation of resting spores in raphidophycean flagellates and diatoms (McQuoid & Hobson 1996).

Factors inducing excystment

Excystment of ciliates including tintinnids may be immediately triggered by environmental factors, as well as influenced by two types of endogenous "trigger": inherent, such as a fixed duration of the resting stage; or triggered, for example effects of temperature that cysts have experienced.

Environmental factors

Like encystment, excystment of protists seems to be influenced by a variety of environmental factors. Among terrestrial ciliates, information on factors inducing excystment is sparse compared with that on factors inducing encystment. At present, it appears that dissolved substances influence excystment of terrestrial ciliates (Matsusaka 2006). In contrast, marine and freshwater dinoflagellates, oligotrichs, and tintinnids have been relatively well studied for factors inducing excystment. Temperature is probably the most common factor inducing excystment of both dinoflagellates and ciliates. For example, excystment of dinoflagellates occurs within a species-specific temperature range termed the "temperature window" (Pfiester & Anderson 1987), allowing prediction of the development of a population of a particular species. A similar phenomenon likely applies to the excystment of tintinnids as well.

Kamiyama & Aizawa (1992) examined effects of temperature on excystment of tintinnid ciliates from marine sediments incubated at different temperatures. Although higher temperatures were associated with overall higher excystment rates, there were species-specific temperatures for excystment (Fig. 7.5). Interestingly, the temperature range suitable for excystment for a given species corresponded to the range of near bottom temperature when each plankton species appears in the water column (Fig. 7.6). The results suggested that water-column populations may be "seeded" by excystment of each species.

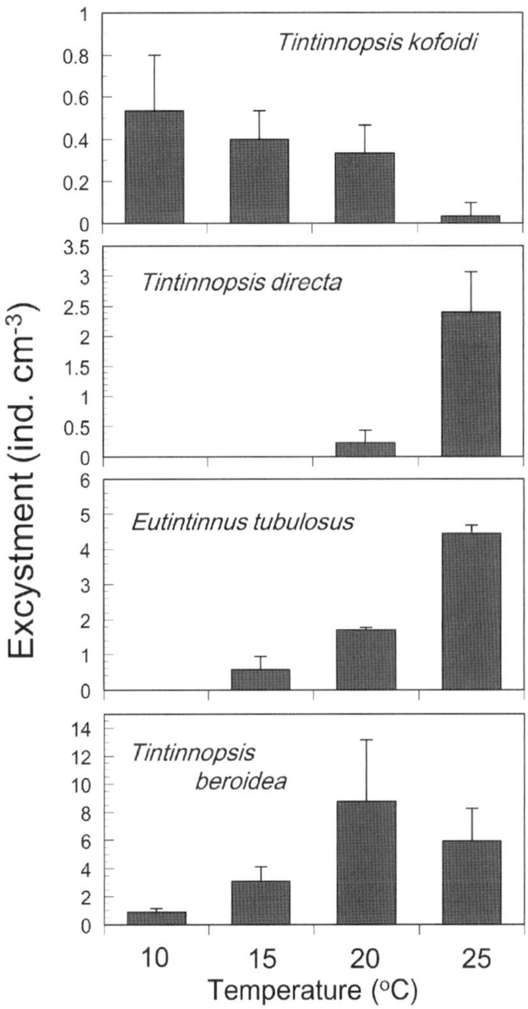

Fig. 7.5 Effects of temperature on excystment of each tintinnid species in marine sediments. Excystment indicates the cumulative number of tintinnids excysted from sediments over 5 days. Vertical bars indicate the standard deviation of the mean ($n = 3$). (Data source: Kamiyama & Aizawa 1992). Note that *Tintinnopsis kofoidi* may be a synonym of *T. cylindrica* (Agatha & Reidel-Lorjé 2006).

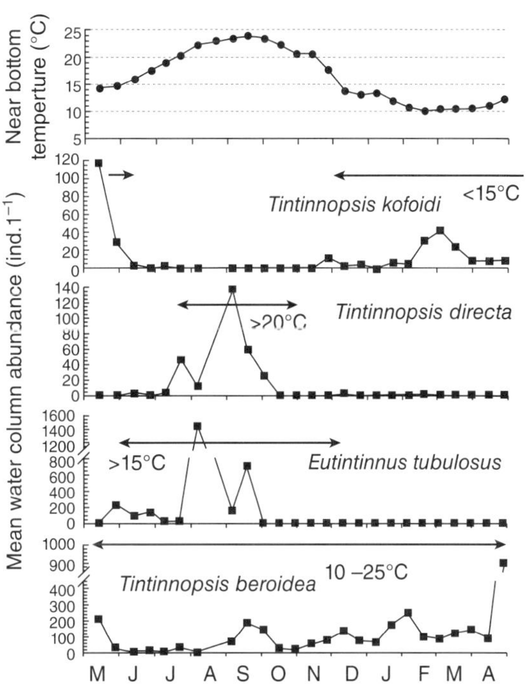

Fig. 7.6 Seasonal variation of near bottom temperature (1 m above the bottom) and abundance of four tintinnid species in Hiroshima Bay, the Seto Inland Sea of Japan. Arrows denotes the period during which near-bottom temperature was suitable for excystment of each tintinnid species based on data in Fig. 7.5 (data from Kamiyama & Aizawa 1992). Note that *Tintinnopsis kofoidi* may be a synonym of *T. cylindrica* (Agatha & Reidel-Lorjé 2006).

In marine oligotrichs, Kim & Taniguchi (1995, 1997) examined effects of environmental conditions on excystment of cysts collected from sediments. Their results also indicated that the most important factor triggering excystment is temperature. For example, *Strombidium conicum* excysted in the range from 15 to 20 °C, whereas excystment of *Cyrtostrombidium boreale* was limited to between 10 and 15 °C (Kim et al. 2002b). Furthermore, the temperature to which cysts were exposed influenced the ability and patterns of excystment of *S. conicum*, suggesting a triggered endogenous factor for excystment.

Light as a factor has also been examined. Dark conditions decreased excystment of tintinnids and the oligotrich *Strombidium conicum* from sediments (Kamiyama & Aizawa 1992; Kim & Taniguchi 1995). Kamiyama et al. (1995) reported that the threshold level of irradiance causing inhibition of excystment was between 2.5 and 6.9 μmol photon $m^{-2} s^{-1}$ (Fig. 7.7). These studies showed, however, that some excystment of ciliates occurred even under dark conditions, implying that the effects of light on excystment are a supplementary or secondary factor (Kim & Taniguchi

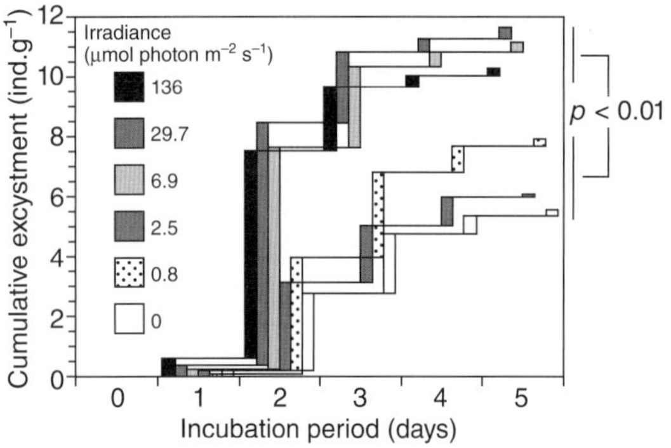

Fig. 7.7 Effects of irradiance on tintinnid excystment from marine sediments during 5 days of incubation. The cumulative number of excystment for 5 days is significantly higher at irradiance at or above 6.9 μmol photon $m^{-2} s^{-1}$ than at or below 2.5 μmol photon $m^{-2} s^{-1}$ (data from Kamiyama et al. 1995).

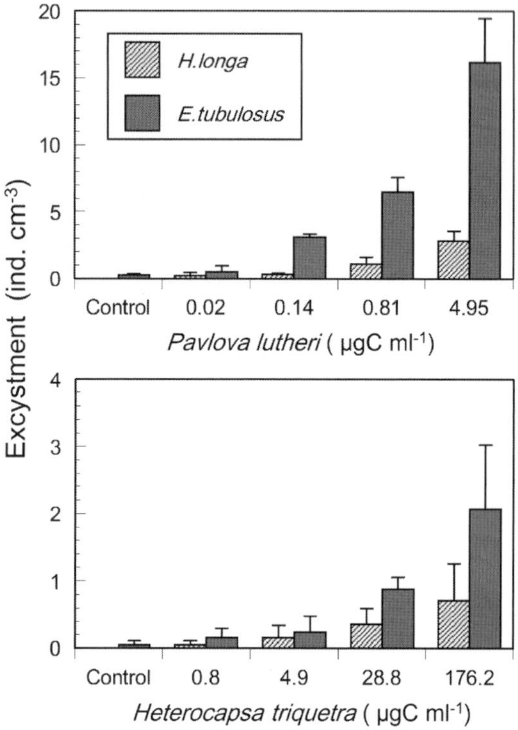

Fig. 7.8 Effects of phytoplankton concentration on excystment of two tintinnid species, *Helicostomella longa* and *Eutintinnus tubulosus*, from marine sediments. Original excystment data are from Kamiyama (1997a). Phytoplankton (*Pavlova lutheri* and *Heterocapsa triquetra*) concentration was standardized as carbon concentration based on cell concentration and cellular carbon content of each species, and the cellular carbon contents were estimated from cell dimension (Kamiyama & Arima 2001) and volume-carbon conversion factor (Verity et al. 1992). Controls denote no addition of phytoplankton. Vertical bars, standard deviation ($n = 3$).

1995). Alternatively, the exposure to weak light for a few minutes during experimental manipulation used in some studies possibly diminished any darkness-related inhibition of excystment. Among phytoplankton taxa, light can trigger excystment of, for example, dinoflagellate cysts (Endo & Nagata 1984; Anderson et al. 1987; Kremp & Anderson 2000). Light exposure for only a few seconds can stimulate germination of phytoplankton cysts (Binder & Anderson 1986).

The effect of oxygen concentration on excystment has not been investigated in tintinnids, but has in the oligotrich *Strombidium conicum*. Excystment rates were slightly lower at low dissolved oxygen concentration (about 4 ml l^{-1}) compared with a "normal" dissolved oxygen concentration (about 10 ml l^{-1}). Low dissolved oxygen concentration can inhibit germination of dinoflagellates cysts (Montani et al. 1995; Kremp & Anderson 2000; Ishikawa & Taniguchi 1994). Low light intensity and low dissolved oxygen concentration are more likely to co-occur in the benthic zones, implying that cysts deep in the sediment (dark and anaerobic conditions) cannot excyst. Hence, suspension of bottom sediments owing to turbulence or bioturbation is possibly an indirect trigger for excystment of dinoflagellates and ciliates in coastal regions.

Low nutrient concentrations delay germination of dinoflagellate cysts (Binder & Anderson 1986), indicating that solute concentrations can influence excystment. There is evidence for some tintinnid species showing that dissolved substances originating from phytoplankton are an important factor promoting excystment under suitable temperature conditions. Kamiyama (1994a, 1997a) found that excystment of two species of hyaline tintinnids, *Eutintinnus tubulosus* and *Helicostomella longa*, from bottom sediments is significantly increased by co-incubation with several phytoplankton species (Fig. 7.8), and found that this effect is owing to the presence of dissolved extracellular products of the phytoplankton.

For the oligotrich *Strombidium conicum*, excystment was inhibited when an antibiotic substance was added to surrounding seawater of the cysts and promoted when the extract of bacteria (*Escherichia coli*) was added, suggesting that bacterial action or a bacterial product can influence excystment of *S. conicum* (Kim & Taniguchi 1995). Excystment of terrestrial ciliates may be also stimulated by trace elements chemical matters (e.g., Matsusaka 2006). At present, it is unknown how the dissolved substances stimulate excystment of ciliates. Although a ciliate cyst is protected by a wall that can resist strong acid and prevent the penetration of some chemicals (Reid 1987), the cyst wall nonetheless must be permeable to some dissolved substances. Alternatively, essential environmental factors such as temperature may primarily change the permeability of the cyst wall, and then dissolved substances and/or bacterial action possibly act as additional factors.

Endogenous factors

Excystment of protists can be controlled fully or partly by an endogenous rhythm. For example, among terrestrial ciliates, a fixed period of the cyst stage seems to characterize *Burusaria truncatellau* (Beers 1948) and *Histriculus cavicola* (Nakamura & Matsusaka 1991). In the freshwater oligotrich *Pelagostrombidium* sp., cyst age, as well as temperature, is an important factor for excystment, so that excystment occurred without temperature stimulation in cysts older than 3 months (Müller 2002). Germination of some species of dinoflagellate cysts requires a certain period of maturation after encystment. They do not excyst during this period, even if the environmental conditions are suitable. This period can be fixed or determined by endogenous factors. Depending on the species, the duration ranges from several weeks to several months and is influenced by environmental conditions, especially temperature in some cases. On the other hand, some species have fixed periods for maturation. The duration of the fixed period in *Scrippsiella trochoidea* is about 25 days and is not influenced by temperature (Binder & Anderson 1986).

For planktonic marine oligotrichs and tintinnids, a cyst maturation period, if it exists, can be short. As described above, encystment and excystment of *Helicostomella subulata* can be observed in the same culture series, and *Strombidium conicum* excysted only a few days after encystment, suggesting no, or a very short, resting period before excystment.

So far, there are few reports clearly showing evidence of a biological clock with regard to excystment of marine protists other than certain dinoflagellates. For example, Anderson & Keafer (1987) demonstrated that a biological clock probably controls excystment of the dinoflagellate *Alexandrium tamarense*, because germination of the cysts shows a seasonal cycle in an area where bottom environmental conditions (temperature and day length) were seasonally invariant. Similarly, seasonal rhythms in excystment were observed in the oligotrich *Pelagostrombidium* sp. in lake water (Müller 2002). As mentioned above, in the marine oligotrich *Strombidium conicum*, excystment is largely controlled by the temperature to which the cyst has been exposed. However, a small portion of the cysts were unaffected by temperature, implying that some endogenous factor may also exist (C. Nakaie, unpublished data).

It is also worth recalling that Faure-Fremiet (1948), Jonsson (1994), and Montagnes et al. (2002a) reported that excystment and encystment of the oligotrich *Strombidium oculatum* living in tide pools corresponds to the tidal rhythm. They considered encystment and excystment to be controlled by endogenous factors because the tidal rhythm of encystment and excystment occurs *in vitro* without any possible direct or indirect tidal stimuli.

Although temperature is an important factor triggering excystment of ciliates, it can also influence the endogenous rhythm of ciliate excystment, which is determined by temperature that the cysts have experienced. Kim & Taniguchi (1995, 1997) found that the excystment pattern (increase of cumulative success rates of excystment during the incubation period) of *Strombidium conicum* varied seasonally in Onagawa Bay, Japan. They found three patterns (Fig. 7.9). A "rapid pattern" was observed in late summer to early winter (September to January), in which excystment takes place abruptly within a few days and cumulative success rate reaches more than 50%. A "delayed pattern" appeared from late winter to early spring (March), in which excystment is inactive during the entire period of incubation (100 days). Finally, a "transitional pattern" occurred twice a year: once in mid-winter (February) and again from spring to mid-summer (April–August), in which excystment proceeded steadily after about a 1-month delay. This seasonality was observed cyclically during two years (Kim & Taniguchi 1997). The factor governing the seasonality in excystment patterns was probably the temperature the cysts experienced in the sediment (Kim & Taniguchi 1997). Such a synchronized annual excystment pattern has been reported for cysts of the dinoflagellates *Scrippsiella*

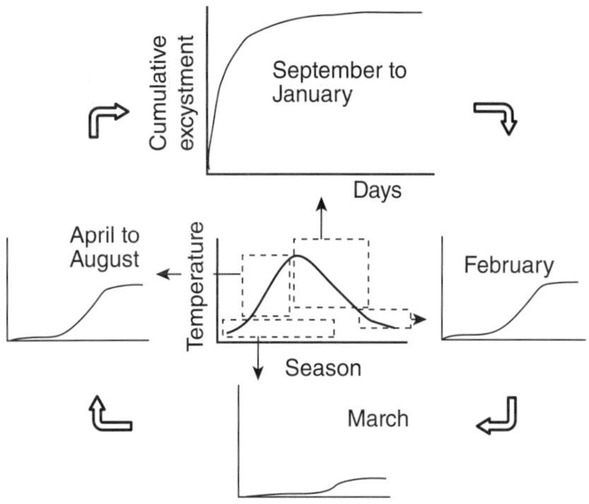

Fig. 7.9 Illustration of cyclic change in the excystment pattern of *Strombidium conicum* cysts under natural conditions in Onagawa Bay, Japan. Three cumulative curves denote the rapid pattern (September–January), transitional pattern (February, and April–August), and delayed pattern (March); data from Kim & Taniguchi (1997).

trochoidea (Kim & Han 2000) and *Alexandrium tamarense* (Kim et al. 2002a). Hence, seasonality of excystment pattern responding to temperature is possibly common to the other ciliates and dinoflagellates in coastal waters.

7.4 ECOLOGICAL FUNCTION OF CYSTS

Distribution of tintinnid cysts in coastal sediments

Cysts with higher specific gravity than water rapidly sink to the bottom sediments. The distribution of cysts is generally associated with that of vegetative cells serving as their source. Hence, the distribution pattern and abundance of cysts (quantitative cyst maps) can correspond with the density of the source populations as well as its spatial distribution. For harmful algal species, such patterns have been established (e.g., Anderson et al. 1982). However, there are very few reports on the relationships between the distributions of cysts and vegetative cells of ciliates, although ciliate cysts have been often found in seawater and sediment traps. This is probably due to difficulty of cyst identification of each ciliate species and, even if it were possible, sorting of cysts from sediment samples is very time-consuming and hard work. Consequently, although there are some data on cyst abundance in sediment or sediment trap samples based on cyst morphology of oligotrichs (Kim 1995; Müller & Wünsch 1999; Müller et al. 2002), information on horizontal distribution of ciliate cysts, including tintinnids, is largely non-existent.

The most probable number (MPN) method was developed and used for enumeration of viable bacteria. This method, also referred to as the "extinction dilution method", has been used for the enumeration of phytoplankton cysts in marine sediments (Imai et al. 1984; Imai & Itakura 1991). Kamiyama (1996) applied this method for the enumeration of viable tintinnid cysts (here, defined as cysts that can be excysted in laboratory conditions) exploiting knowledge of environmental conditions known to promote the excystment of tintinnid cysts. This method is far easier and less labor-intensive than quantitative counting procedures. It allows estimation, however, only of viable cysts, not all cysts in sediment samples. Furthermore, the incubation conditions used may not be appropriate for all species.

The MPN method was used to examine the distributional patterns of tintinnid cysts in Hiroshima Bay, Japan. The abundance of viable cysts was high in the sediments in the inner parts of the bay where vegetative cells occur in the highest concentrations (Fig. 7.10). Also, this area is the region where a close coupling between cyst abundance and vegetative cells of the harmful alga *Heterosigma akashiwo* is also observed, suggesting that limited water exchange and high concentrations of vegetative forms are linked to high abundances of cysts of these organisms in the sediment. In the northern parts of the bay, depletion of dissolved oxygen occurs near the bottom in summer and high sulfide concentrations are observed in sediment in this area (Itaoka & Tamai 1993). However, tintinnid cysts may survive such adverse conditions and remain able to excyst when the environmental conditions become favorable for the growth of tintinnids.

Densities of total viable tintinnid cysts as MPN observed in Hiroshima Bay ranged from 0.3 to 4.5 cysts per gram of sediment. In comparison, Kim (1995) found that the maximum cyst densities of two tintinnid species *Favella taraikaensis* and *Helicostomella subulata* were 7 and 13 cysts per gram of sediment, respectively, determined by direct count of the cysts in sediment

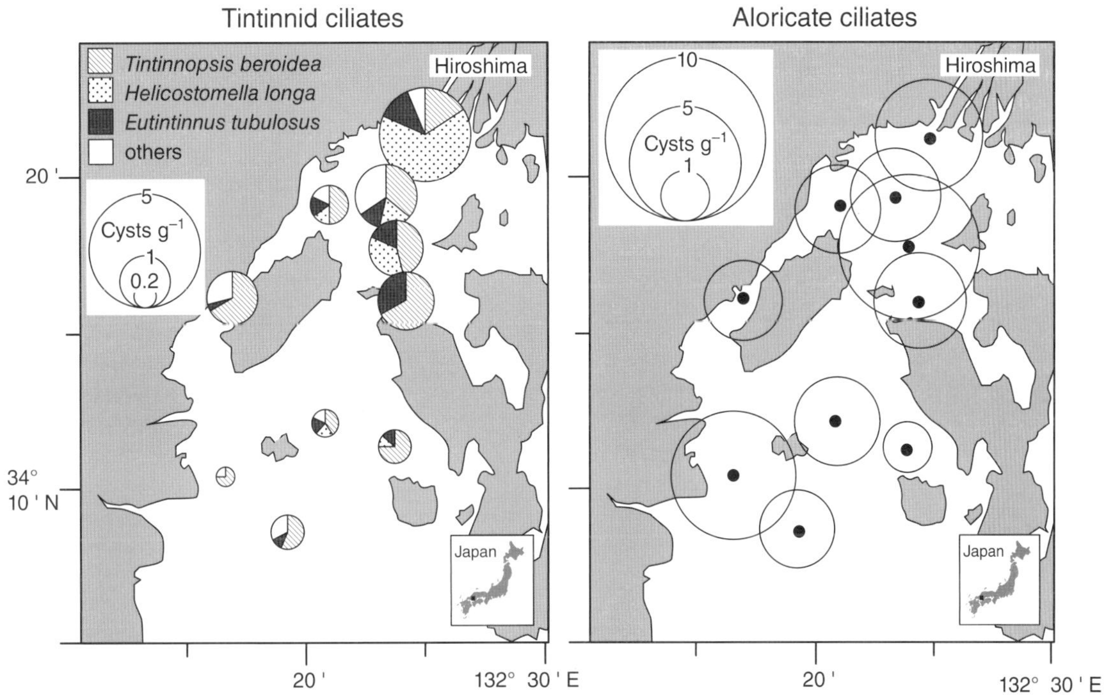

Fig. 7.10 Abundance of viable cysts of tintinnid ciliate and aloricate ciliate in the sediment estimated by the MPN method and the distributions in Hiroshima Bay, the Seto Inland Sea of Japan, in February1993 (data from Kamiyama 1996).

collected monthly from Onagawa Bay over 1.5 years. The maximum density of either species in Onagawa Bay was higher than total cysts in Hiroshima Bay, which is a eutrophic embayment where planktonic tintinnids are abundant. This is probably due to the difference between the MPN method, which estimates only viable cysts, and the direct count method of all cysts. Similar differences between cyst densities of Raphidophycea and *Chattonella* spp., in sediments using both methods have been reported (Itakura et al. 1991; Imai et al. 1998). To clarify the ecology of ciliates in marine coastal waters, data on the horizontal distribution and seasonal variation of species-specific cyst abundance would be of great value. Ichinomiya et al. (2004) proposed a simple detection method of ciliate cysts in sediments by exploiting the autofluorescence of glutaraldehyde-fixed samples visible using epifluorescence microscopy. This may be a practical method to estimate abundances of ciliate cysts in coastal regions, if ciliate cysts can be reliably distinguished and identified.

Although cysts are resistant, allowing survival in the benthos, there a few reports from laboratory studies suggesting that benthic fauna may consume cysts. The cysts of *Alexandrium* spp. are ingested both by a polychaete and a bivalve, with the bivalve reducing cyst concentrations (Tsujino & Uchida 2004). Interestingly, cysts found in fecal matter of both the polychaete and the bivalve excysted at rates similar to cysts that had not been ingested. The eastern oyster ingests and apparently destroys a large variety of dinoflagellate cysts and undescribed "ciliate cysts" (Persson & Smith 2009).

Influence of excystment and encystment on population dynamics

The germination of cysts is thought to play a role as a trigger for population development of some dinoflagellate species (see, for example, Steidinger 1975). Such a role for bloom initiation may apply to the excystment of ciliates as well. However, excystment may not be necessarily a major influence on population dynamics, as excystment peaks of dinoflagellates are not consistently related to peak water column concentrations or population growth rates (Ishikawa & Taniguchi 1996). Ultimately, population development in the water column depends on the growth of the vegetative cells.

Based on the few existing reports on encystment of ciliates, the relationships with population dynamics also appear inconsistent. Encystment of *Helicostomella subulata* occurred during the development of the population bloom (Paranjape 1980; Kim 1995), whereas formation of *Favella taraikaensis* cysts peaked at the end of the bloom (Kim 1995). In laboratory experiments, encystment of *Favella taraikaensis* occurred when the population growth rates decreased (Stoecker et al. 1983). These reports show that encystment of tintinnids can correspond with both peak population abundance and declining populations.

Quantitatively, a substantial part of the tintinnid population can be implicated in encystment. In Bedford Basin (Nova Scotia, Canada) the cyst density of *Helicostomella subulata* in the water accounted for up to 40% of the vegetative cells. The "efficiency of encystment" (the ratio of cyst flux estimated from data of sediment trap samples to concentration of vegetative cells) in Onagawa Bay, Japan, reached 40% for *H. subulata*. In comparison, for the oligotrich *Strombidium capitatum*, the peak value was 27% (Kim 1995). These estimates suggest that the encystment can be an important ultimate fate of a bloom population of tintinnids or oligotrichs. However, once again, diverse patterns exist. In the oligotrich *Strombidium conicum*, a dominant species in Onawaga Bay, cyst formation occurs at a very low rate, an efficiency of only 0.07% (Kim 1995). Furthermore, in the *Pelagostrombidium* population of Lake Mondsee, some individuals remain active throughout the winter months, whereas others encyst in late fall/autumn (Müller et al. 2002).

All ciliates, including tintinnids, will fully exploit their growth potential under suitable environmental conditions (e.g., of temperature and prey availability) and increase population density. Probably, the ultimate goal is to exchange genes with other individuals through conjugation. Development of high population densities increases opportunities for conjugation. However, if the ciliate population overruns its prey, it will create unfavorable conditions of low prey concentration. If encystment follows conjugation, occurrence of encystment during population development, before peak density is reached, is a reasonable strategy to avoid near-future unfavorable conditions. In ciliates of temperate areas, many species bloom in summer to fall/autumn and over-winter as cysts. However, some species encyst during the period when conditions appear suitable for population growth. *Strombidium crassulum* encysts in early summer when temperature is increasing and prey appear to be abundant. Reid (1987) suggested that this strategy may serve to avoid competition with other microzooplankton grazers and crustacean predators (copepods). After the spring bloom, the abundance of copepods increases, but with a time-lag of about 30 days. During the intervening period, *S. crassulum* populations can increase and then shift from active growth mode to an encystment stage just before the copepod population increases. A similar strategy for escaping from copepod predation may explain the encystment of the freshwater oligotrich *Limnostrombidium viride* after the spring bloom (Müller & Wünsch 1999). Although it is tempting to generalize, it is worth recalling that there is little ecological information on relationships between population dynamics of vegetative cells and cysts. Furthermore, diverse responses of excystment and encystment to environmental factors appear to be characteristic of ciliates and these differences may reflect very different strategies.

7.5 KEY POINTS

1. Historical overview of tintinnid ciliate cysts. The many kinds of tintinnid cyst in seawater and sediments which have been reported suggest that a cyst is probably a common life-history stage for coastal species, although early reports may have confused parasite development with cyst formation in some tintinnid species.

2. Morphology of tintinnid cysts. Although a variety of presumptive cyst types have been reported, excystment has been observed only in the flask-shaped cysts of *Favella* and cylindrical or spherical cysts formed by *Helicostomella*, *Eutintinnus*, and *Tintinnopsis*.

3. Encystment. Based on observations of *Favella*, encystment begins with a rapid circular motion of the cell inside the lorica and complete cyst formation takes about 6 hours. Precise knowledge of factors inducing encystment in tintinnids is lacking.

4. Excystment. Among environmental factors, temperature appears to be the most important.
5. Distribution of tintinnid cysts. Two methods are used to estimate cyst abundance in sediment samples: direct counts and a "most probable number (MPN)" method. Direct counts enumerate both living and dead cysts, whereas the MPN method, which relies on triggering excystment, may underestimate concentrations of living cysts.
6. Encystment of field populations. Encystment of tintinnids can correspond with peak population abundance in the water column or declining populations. In either case, a substantial part of the tintinnid population (up to about 40%) can be implicated in encystment.

ACKNOWLEDGMENTS

Thanks are extended to Dr Toshikazu Suzuki (Nagasaki University) for his identification of *Strombidium chlorophilum* by observation of protargol-stained samples, and to Dr Young-Ok Kim (Korean Ocean Research and Development Institute) for her suggestion to explain flask-like cysts of ciliates in sediment samples.

Chapter 8

FOSSIL TINTINNIDS

Jere H. Lipps[1], *Thorsten Stoeck*[2], *and Micah Dunthorn*[2]

[1]John D. Cooper Archaeology and Paleontology Center, Santa Ana and University of California Museum of Paleontology, Berkeley, CA, USA
[2]University of Kaiserslautern, Kaiserslautern, Germany

The Biology and Ecology of Tintinnid Ciliates: Models for Marine Plankton, First Edition. Edited by John R. Dolan, David J.S. Montagnes, Sabine Agatha, D. Wayne Coats, and Diane K. Stoecker.

8.1 INTRODUCTION: A PAUCITY OF DATA

Ciliates are difficult to observe and interpret, even when seen through the lenses of powerful microscopes. Although they have a relatively high number of morphological characters compared with other microbial eukaryotes, this number is nevertheless extremely limited and their simplified structures may easily be converged upon by distantly related lineages (Dunthorn & Katz 2008). The processes of fossilization and the slow decay of time commonly affect those few morphological characters that allow us to place fossils into extant taxa with any reliable resolution, or even to accurately identify them as ciliates. The very characters that we use to circumscribe extant species and larger clades – the patterns of somatic and oral cilia, the arrangements of kinetostome-associated microtubules and fibers, and morphogenesis – are obscured or lost in ciliate fossils. Soft cell parts may be preserved in ciliates, but only under exceptional conditions such as in amber (Ascaso et al. 2005; Martín-González et al. 2008; Schmidt et al. 2001; Schönborn et al. 1999; Waggoner 1994).

Tintinnids, with their more easily fossilized loricae, therefore offer the best hope of tracing ciliate evolution throughout the long history that is preserved in the rock record. However, these loricate fossils only contain two pieces of evidence to both determine they are tintinnid ciliates and to place them somewhere in the tintinnid taxonomy: their shape and their composition. With these two pieces of evidence, numerous paleontologists and taxonomists have analyzed putative tintinnid fossils since their discovery in the late 19th century by Rüst (1885). These "tintinnids" have been reported from Proterozoic through Recent marine deposits. The only other group of heterotrophic plankton that has nearly such an early beginning is the Radiolaria, which appeared in the early Cambrian period (Braun et al. 2007; Nazarov 1973) and diversified throughout it (Dong et al. 1997; Won & Below 1999) to the present. Proterozoic microplankton chiefly includes a variety of acritarchs, which are cysts of phytoplankton (Lipps 2006). Benthic heterotrophic microbial eukaryotes may have been present in the Proterozoic (Huldtgren et al. 2011) and certainly began to radiate in the Cambrian–Ordovician (Lipps 2006), but the other major groups of microplankton, including diatoms, foraminifera, coccolithophorids, and silicoflagellates, appeared in the Mesozoic era and are extant today. Against this general background, the presence of tintinnids in Proterozoic and Paleozoic rocks cannot be predicted.

Known, extant tintinnids now number about 1200 described species in 75 genera both in marine and brackish- to freshwater environments (Agatha 2009); only about 10 species live in freshwaters (Tappan 1993). They occur in today's oceans in five separate biotas: cosmopolitan occurrences of hyaline loricate forms, neritic occurrences of organic and agglutinated loricae in nearshore waters, warm-water containing the most numerous species of tintinnids, and boreal and austral forms in high northern and southern areas (Chapter 10). Tintinnids may be especially abundant in nearshore eutrophic waters (Xu et al. 2011) and were even found in caves on land that had been penetrated by marine waters (van Hengstum & Scott 2011).

Fossil tintinnids are almost exclusively marine and number far fewer described taxa. The fossils are difficult to observe and interpret, and mistakes have commonly been made. Many of these so-called tintinnids from the rock record have been described or reported from cross-sections observed in petrographic thin-sections of the rock. There these fossils may resemble tintinnid cup-like loricae, but so do cross-sections many other tiny fossils or parts of fossils, especially in limestones of the Paleozoic and Mesozoic, making positive identification trying, if not impossible. Others of these tintinnid fossils are demonstrably the remains of other kinds of eukaryotes that have undergone diagenetic changes. Still others have thickly calcified skeletons, unknown in modern forms that have been assigned to other groups, although many workers accept them as tintinnids. For these reasons, fossil tintinnids have been controversial and hard to relate to modern forms and to the evolutionary history of the group and to ciliates in general. Figure 8.1 shows all claimed tintinnid occurrences in the geologic record. However, all such reports require special scrutiny. Extensive historical summaries of these analyses are found elsewhere (Tappan & Loeblich 1968; Tappan 1993), and we summarize more recent work here.

8.2 PROTEROZOIC TINTINNIDS: NO EVIDENCE OF A BEGINNING

Molecular clock analyses place the origin of ciliates sometime in the Proterozoic (Berney & Pawlowski 2006; Douzery et al. 2004; Parfrey et al. 2011; Wright & Lynn 1997). If these estimates are correct, then

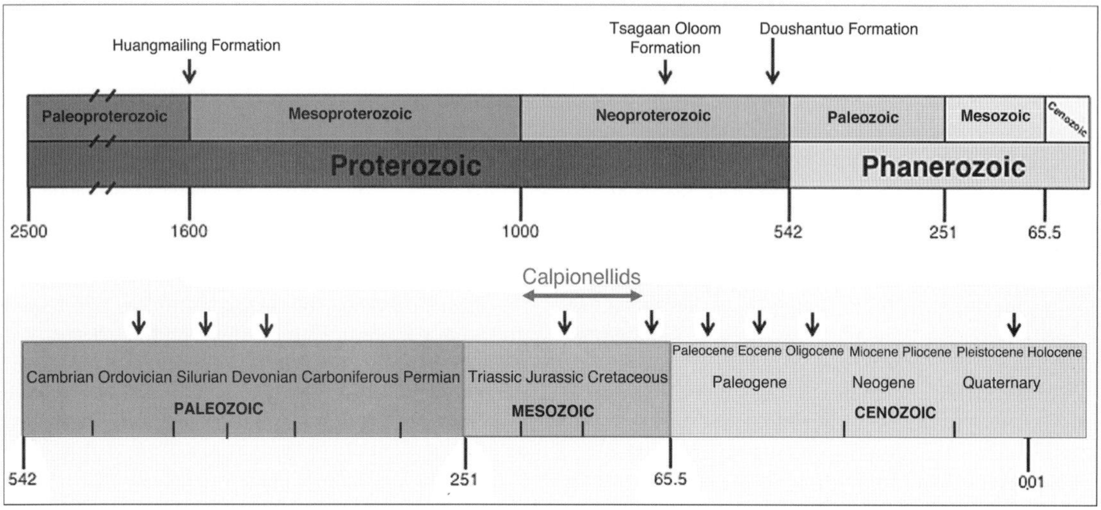

Fig. 8.1 Timeline showing geological eons, eras, periods, and epochs relative to reports of fossil tintinnids (vertical arrows). The lower panel shows the subdivisions of the Phanerozoic eon. The numbers indicate millions of years before present; note the nonlinear time scale.

ciliate fossils dating from those ancient times are possible. Because tintinnids appear to relatively evolved ciliates, we might not expect them to be found among the earliest plankton. However, three sets of fossils have been described and interpreted as Proterozoic tintinnids so far.

The oldest supposed fossil tintinnids were reported from the very old (Figs. 8.1 and 8.2) Mesoproterozoic Huangmailing Formation, China, dated at 1600 million years ago (MYA) (Li & Zhan 2006; Li et al. 2009). Eight species described in six genera have the general shape of unaltered and undistorted modern tintinnid loricae (Fig. 8.2). If these are truly tintinnid fossils, then the date for the origins of ciliates would be pushed back much further in time than any molecular clock estimate, and would imply an even older origin for the ciliates in general. However, because they come from high-grade metamorphic rocks, termed "granulite" and "leptite" by Li & Zhan (2006) and Li et al. (2009), these specimens should have been affected to some degree by metamorphism. Given the absence of any of those kinds of metamorphic effects, these so-called fossils are more likely to be contaminates or mineral shards from the metamorphic rock itself. They could also have come from parts of the rock that were not affected by any metamorphic processes; however, Li & Zhan (2006) and Li et al. (2009) are not clear about whether these unaltered areas in the rocks exist.

The second set of fossils (Figs. 8.1 and 8.3) comes from limestone in the Tsagaan Oloom Formation in Mongolia, dating to 635–715 MYA (Bosak et al. 2011). Many of the over 100 organic fossils appear flask-shaped with an apical blunt-end collar, tapering to a bowl-shaped base. However, in some of the figures (Bosak et al. 2011), the fossils are not flask-shaped; rather, the basal end looks like a slightly larger version of the apical blunt end (Fig. 8.3b). On the surface of some of the fossils are alveolar spheroids (not to be confused with the flattened alveolar sacs found in ciliate cell membranes) that are 2–5 µm in diameter; these spheres are potentially similar to those found in some tintinnids during lorica formation (Gold & Morales 1976b).

An additional character shape described from these fossils is the central invaginations in the apical collars (Fig. 8.3c) that are "consistent with a cemented or collapsed aperture" (Bosak et al. 2011). Presumably, they are composed of the same organic material as the rest of the fossils. However, in extant tintinnids, no such

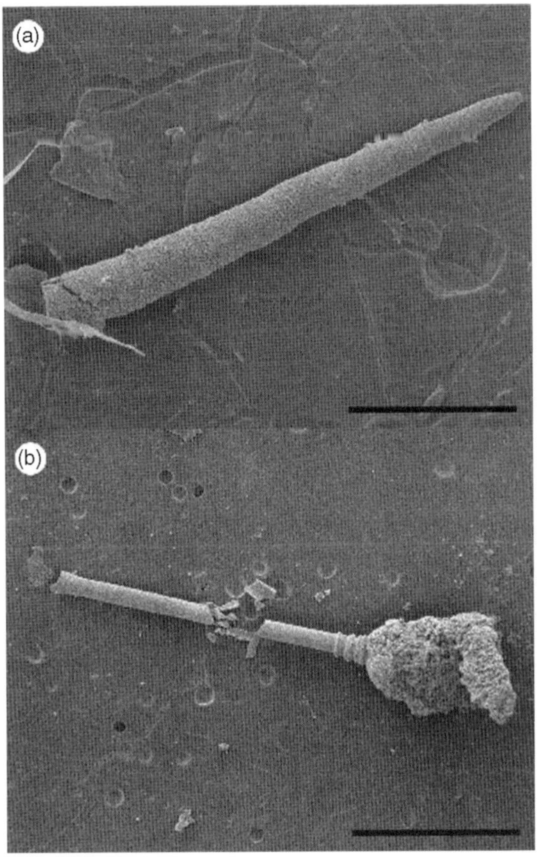

Fig. 8.2 Putative tintinnid fossils from the 1600 MYA Mesoproterozoic Huangmailing Formation of China from Li & Zhan (2006) and Li et al. (2009). (a) *Alpingella acuminata*; scale bar, 300 µm. (b) *Eutintinnus frakmoi*; scale bar, 1 mm.

central-invagination-shaped structures occur in or on the loricae, and the collars are broadly opened to allow the cells to feed. Although closing apparatuses occur in some tintinnids such as *Codonella*, these structures are found at the base of the collar, and their membranelles, which are 0.1–0.2 µm thick, are arranged like an iris diaphragm (Agatha 2010a). The membranes may be secreted by the cell, but their composition is different from the loricae. Such thin membranes are unlikely to be preserved in any fossil. The overall organic compositions of the Mongolian fossils (Bosak et al. 2011) do support a tintinnid relationship. In addition, silicon was also found in these fossils, an element not found in extant ciliates without agglutinated loricae; whether this silicon was included during the fossilization process is not known. Given that extant tintinnids do not have central invaginations of the collar of their loricae and silicon is lacking in non-agglutinated species, these fossils are unlikely to be tintinnids. In addition, given the lack of such central invaginations in the tests of testate amoebae, reliable taxonomic placement within the eukaryotes is currently impossible.

The youngest set comes from the 580 MYA Wengan Phosphate Member of Doushantuo Formation of China (Figs. 8.1 and 8.4). These rocks, formerly thought to be marine but now considered a lake deposit (Bristow et al. 2009), preserve an array of microfossils said to include sponge and other smaller animal embryos and larval bilaterians (Chen et al. 2009); embryonic stages larger than 16 cells were attributed to taphonomic factors that eliminated them, perhaps because of their fragility (Dornbos et al. 2005). These so-called animal embryos are neither animals nor embryos, but more likely represent other microbial eukaryotes (Huldtgren et al. 2011). The bilaterian fossils have been dismissed as well because the putative cell layers are diagenetic overgrowths on interior and exterior surfaces (Bengtson & Budd 2004) likely of and on acritarch cysts (Dunthorn et al. 2010). From these same rocks, two new tintinnid taxa, *Eotintinnopsis pinniforma* and *Wujiangella beidoushanese* (Fig. 8.4), were described based on single specimens (Li et al. 2007). These fossils were interpreted as putatatively cup-shaped organisms with organic loricae, characters that are consistent with extant tintininids. However, their similarity to tintinind shape ends there. In the specimen of *Eotintinnopsis* (Fig. 8.4a–c), multi-tiered apical and a subapical feather-like structures were described. These apical features are unknown in any extant ciliate, and, like the misidentified bilaterians from the same rocks, this fossil likely represents the remains of two acritarchs that were altered, distorted, and compacted together during and after fossilization (Dunthorn et al. 2010). Of the two putative Doushantuo fossil tintinnids, *Wujiangella* (Fig. 8.4d–f) is the most similar to extant ciliates (Li et al. 2007). However, its so-called somatic cilia are twice as wide as they should be, and they lack any kinety arrangement that would normally be found in extant tintinnids (Dunthorn et al. 2010). Rather than being cilia, the projections on *Wujiangellai* are most likely spines on diagenetically and taphonomically

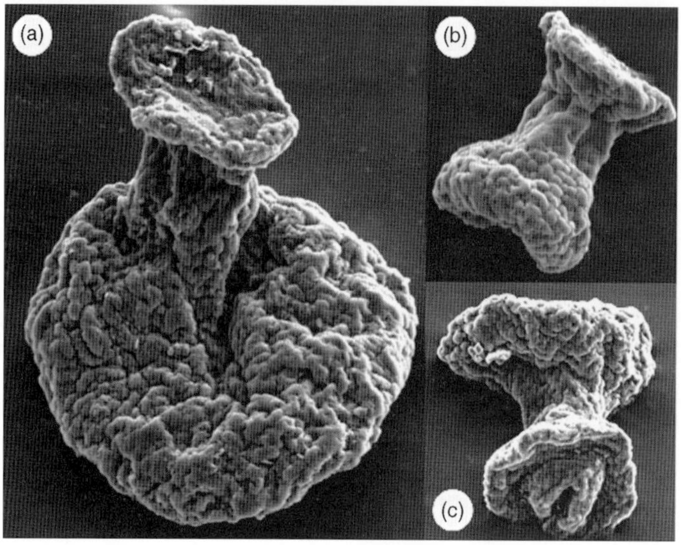

Fig. 8.3 Putative tintinnid fossils from the 635–715 MYA Tsagaan Oloom Formation of Mongolia from Bosak et al. (2011). (a, b) Side view with deflated round bowls, constricted necks, and centrally invaginated collar. a is ~130 μm wide; b is 70 μm long. (c) View of apical central invagination. c is 75 μm long. © Geological Society of America.

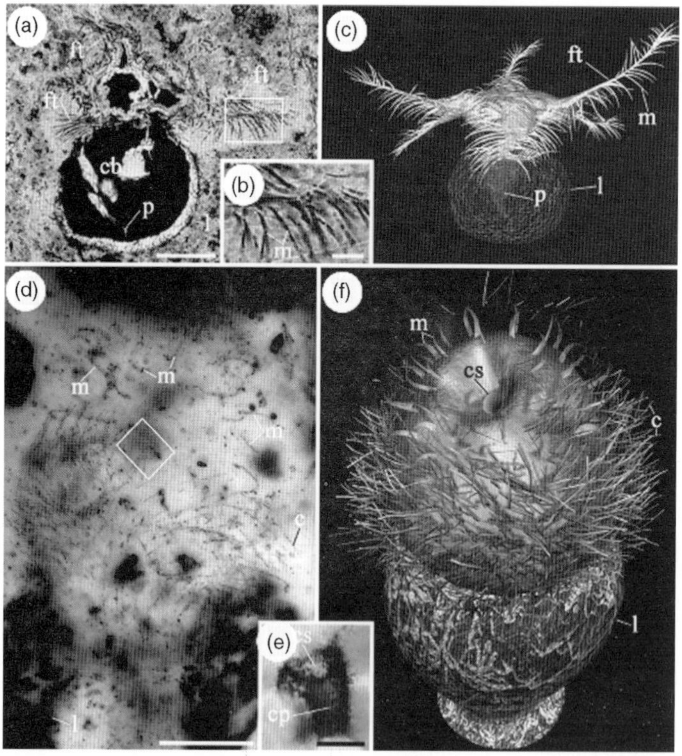

Fig. 8.4 Putative tintinnid fossils from the 580 MYA Wengan Phosphate Member of Doushantuo Formation of China from Li et al. (2007). (a) *Eotintinnopsis pinniforma*, longitudinal section; scale bar, 50 μm. (b) Close up of feather-like membranelles on *E. pinniforma*; scale bar, 10 μm. (c) Reconstruction of *E. pinniforma*.(d) *Wujiangella beidoushanese*, longitudinal section; scale bar, 50 μm. (e) Close up of cytostome and cytopharynx of *W. beidoushanese*; scale bar, 10 μm. (f) Reconstruction of *W. beidoushanese*; c, somatic cilia; cb, cell body; cp, cytopharynx; cs, cytostome; ft, feather-like membranelles; l, lorica; m, membranelles; p, peduncle. © The Geological Society, London.

degraded acritarchs (Dunthorn et al. 2010). A third ciliate, reportedly a suctorian, was also described from one specimen as *Yonyangella ovalis*. It too is a diagenetically and taphonomically altered acritarch. This evidence and the freshwater depositional environment of the Doushantou Formation indicate that these fossils are not tintinnids: rather they are acritarch cysts of algal microplankton.

Thus, given the alternative interpretations by Dunthorn et al. (2010) and herein of the fossils of Li et al. (2007), Bosak et al. (2011), Li & Zhan (2006), and Li et al. (2009), no solid evidence of Proterozoic tintinnids or other ciliates comes from the Precambrian rock record. Darwin's dilemma of the lack of fossils for this ancient age (Schopf 2001; Knoll 2004) therefore still holds for at least the ciliates. If there are tintinnid fossils from this ancient time, they have yet to be discovered.

8.3 PALEOZOIC TINTINNID REPORTS: TOO THIN TO KNOW

In the Phanerozoic eon, tintinnid fossils have been reported rarely in Paleozoic rocks, but are much more commonly reported in the Mesozoic era where they can be quite abundant. Whether or not any of these fossils are tintinnids is much debated.

Tintinnids have been reported from thin sections in Ordovician carbonate rocks (Fig. 8.5) of the Sahara of Africa (Chennaux 1968), the Silurian of Spain (Hermes 1966), the Lower Devonian of North Africa (Cuvillier & Sacal 1963) and of Canada (Murray & Taylor 1965), and the Mississippian of Teverga in northern Spain (Cuvillier & Barreyre 1964). From only one of these occurrences, the Devonian, was a new genus and species described. Thus Tappan & Loeblich (1968), although noting the reports of tintinnids in Ordovician, Silurian, and Devonian, did not include them in their review because new taxa were not described. The Mississippian occurrence (Cuvillier & Barreyre 1964) are not tintinnids and are more likely calcispheres (Tappan & Loeblich 1968), which are spherical calcite objects found commonly in carbonate rocks and believed to be algal in origin. The other records include two-dimensional cross-sections of spheres and cups, some with an opening and flange. Although the cross-sections do resemble those of Mesozoic calpionellids, they have not been confirmed as tintinnids by the study of free specimens. So far, no Paleozoic tintinnids are known with certainty.

8.4 MESOZOIC TINTINNIDS AND CALPIONELLIDS: SAME SHAPE, DIFFERENT COMPOSITION

Tintinnids and the calpionellids have been described from the upper Triassic to lower Cretaceous sedimentary rocks from both onshore and offshore settings

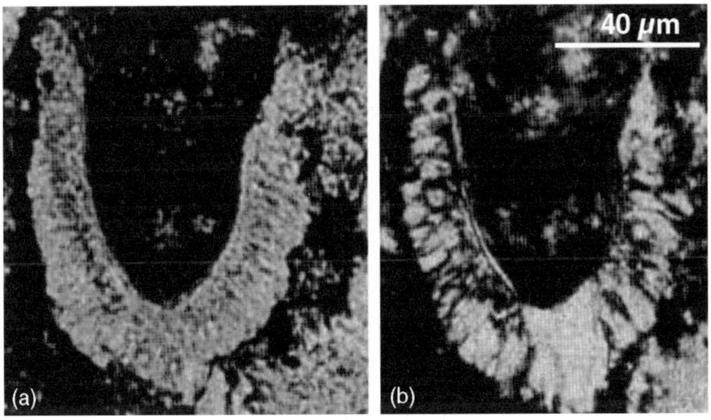

Fig. 8.5 The presence of tintinnids in the Ordovician is based primarily on a report by Chennaux (Tappan & Loeblich 1968). He documented what appears to be a fossilized tintinnid lorica from an Ordovician formation in Algeria. The right panel (a) is brightfield image of a thin section, the left panel (b) a polarized light view. Only one specimen was found and there were there no co-occurring fossils of any type. Images from Chennaux (1968) with micrometer scale bar added.

(Tappan 1993). These tintinnids and calpionellids are used or have potential for biostratigraphic correlation with their parent rocks throughout the Tethyan Sea (Tappan 1993), being widespread from Australia (Brunnschweiler 1951) through Russia and Asia (Bugrova 2003), Iran and Morocco (Benzaggagh et al. 2010) in the east and across the Atlantic Ocean to Mexico in the west (Colom 1955, 1965), and in South Dakota, Wyoming, and Colorado (Eicher 1965) in central USA. Some of these fossils resemble modern species in their morphology and construction materials (Fig. 8.6). However, central in these debates about what fossils are tintinnids are the calcareous calpionellids. Calpionellids range from the late Tithonian, in the upper Jurassic, to the early Valanginian, in the lower Cretaceous (Tappan 1993; Tappan & Loeblich 1968; Remane 1985). Because of their cup-shaped loricae, calpionellids have commonly been assumed to be tintinnids. With this view, some taxonomists have interspersed calpionellids into modern families based on lorica shape (Kofoid & Campbell 1939; Campbell 1954; Tappan & Loeblich 1968; Corliss 1979; Lynn 2008). Others have separated the calpionellids into their own tintinnid subgroup (see, e.g., Bonet 1956).

The central problem with identifying and describing calpionellids as tintinnids – either interspersing them into extant families or in their own taxon – is the composition of their loricae. Although all extant tintinnids have organic or agglutinated loricae (Chapter 3), calpionellid fossils are composed of calcite (Tappan 1993). Calpionellid fossils would then include the so-called calcareous tintinnids in the Paleozoic, Mesozoic, and Cenozoic. No modern classification separates the calpionellids into their own group.

Colom (1934, 1948) suggested that calcite may have replaced the original organic material during the fossilization process, but this seems highly unlikely as it would have had to occur in many fossils from many different rocks from many parts of the world (Tappan 1993). This difference in lorica composition between extant ciliates and fossil calpionellids led Remane (1971, 1985) to declare that calpionellids are not tintinnids and not even ciliates.

With the current evidence, Remane's (1971, 1985) view of a non-ciliate origin of calpionellids seems to be the most likely. That does not mean that sometime in the Jurassic a lineage of tintinnids evolved from making organic loricae to calcareous ones. This monophyletic group would then have gone extinct in the Cretaceous. However, no evidence exists for such an evolutionary scenario linking calpionellids with tintinnids.

Whatever calpionellids are – tintinnids or not – they do make excellent markers for use in biostratigraphy of upper Jurassic to lower Cretaceous rocks (Grün & Blau 1997; Remane 1985). Over the years, several schemes of zonal and subzonal divisions using calpionellids have been proposed (e.g., Altiner & Özkan 1991; Cantu-Chapa 1996; Grün & Blau 1996; Le Hégarat & Remane 1968; Pop 1976; Remane 1963, 1964, 1971; Remane et al. 1986). Grün & Blau's (1997) latest calpionellid biochronology distinguishes six zones and 19 subzones (these zones also correlate with zonation developed for Mediterranean ammonites).

Genuine tintinnids, defined by having agglutinated loricae, occur in the Jurassic and Cretaceous (Tappan 1993). They have only one occurrence in the Jurassic (Rüst 1885) but are abundant in the shallow, neritic waters of the Cretaceous mid-continental seaway that flooded a central corridor through the center of North America from the Gulf of Mexico to the Arctic Ocean. In this environment, agglutinated tintinnids (Fig. 8.6) were abundant (Eicher 1965).

8.5 CENOZOIC TINTINNIDS: SPARSE

Tintinnids have been described from Paleocene, Eocene, and Oligocene marine deposits, and from Pleistocene lake sediments of the Cenozoic over the past 50 or so years by several authors (summarized by Tappan 1993; Bignot & Poignant 2010). Over 60 species (Figs 8.7–8.9) occur in the Paleogene rocks from the Gulf Coast of the USA, across Europe from Spain to Poland and Romania, and in Ukraine, Uzbekistan, and Turkmenistan (Tappan & Loeblich 1968; Tappan 1993; Bugrova 1983, 2003; Bignot & Poignant 2010). These occurrences are all in the ancient Tethys Seaway, an equatorial ocean that extended from the modern Gulf of Mexico region, across southern Europe and Asia, to Australia and southeast Asia. This seaway gradually closed as plate tectonic motions moved Africa and India into Europe and Asia, causing mountains to form, and oceanography to change in the rest of the world. The Tethys Seaway was a unique warm-water environment containing an abundance of fossil planktic and benthic organisms.

The tintinnids described from the ancient Tethyan marine rocks are mostly calcareous cup-shaped cones

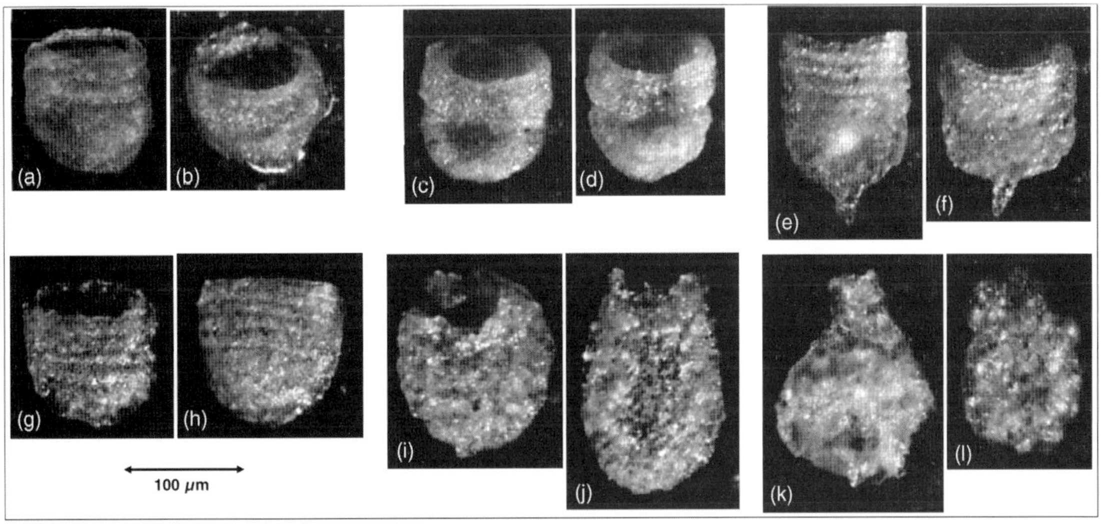

Fig. 8.6 Lower and Upper Cretaceous agglutinated tintinnid loricae from Colorado and Wyoming, USA, referred to modern genera (from Plate 1 of Eicher 1965). (a, b) *Dicloepella borealis*. (c, d) *Codonella bojiga*. (e, f) *Coxiella coloradoensis*. (g, h) *Coxiella atricollium*. (i, j) *Tintinnopsis parovalis*. (k, l) *Tintinnopsis ampullula*. The assignment to modern genera is considered tentative here. Courtesy of Micropaleontology Project.

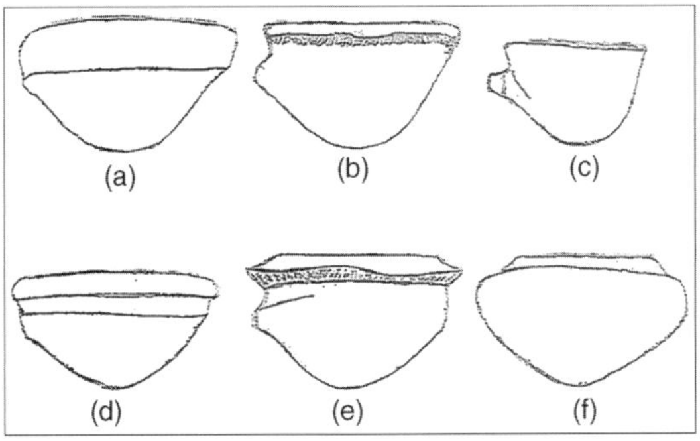

Fig. 8.7 Pseudoarcellinid (Paleogene in age) cup morphology showing flat cups with various protruberances. Reprinted from Bignot, G. & Poignant, A. (2010) Les Pseudarcellinae, nouvelle sous-famille des Codonellidae (Tintinnida) – le genre *Pseudarcella* et l'espèce *Pseudarcella rhumbleri* Spandel, 1909 – description de nouvelles espèces. Revue de Micropaléontologie, 53, 107–119. Figure 4, with permission from Elsevier.

with various shapes, invaginations, and protrubrances (Figs 8.8 and 8.9) that have been placed in the new subfamily Pseudarcellinae of the tintinnid family Codonellidae (Bignot & Poignant 2010). Other similar forms (Fig. 8.10) also occur in Paleogene rocks in these regions (Tappan & Loeblich 1968) that are not considered pseudarcellinids. Because no modern species of tintinnids are known to secrete $CaCO_3$, these also are unlikely to be tintinnids. The cups are mostly quite shallow and not deep; most are wider than they are long. Many of these might be larval-shelled invertebrates or other microbial eukaryotes.

8.6 ORGANIC AND AGGLUTINATED PHANEROZOIC FOSSILS: THE MOST LIKELY CANDIDATES

Tintinnids with organic loricae or loricae agglutinated with foreign particles gathered from the water column (Chapter 3), such as with coccoliths, silt, and clay attached to it, are abundantly preserved (Fig. 8.11) in some neritic marine sediments (Harman 1972; Echols & Fowler 1973), but they are uncommon in the fossil record. This dichotomy indicates that loricae cannot readily survive the processes associated with lithification of sediments into rocks. The oldest known tintinnid with an agglutinated organic lorica occurs in the Jurassic (Rust 1885). Others, placed in modern genera although this is likely incorrect, are abundant in the Cretaceous (Eicher 1965).

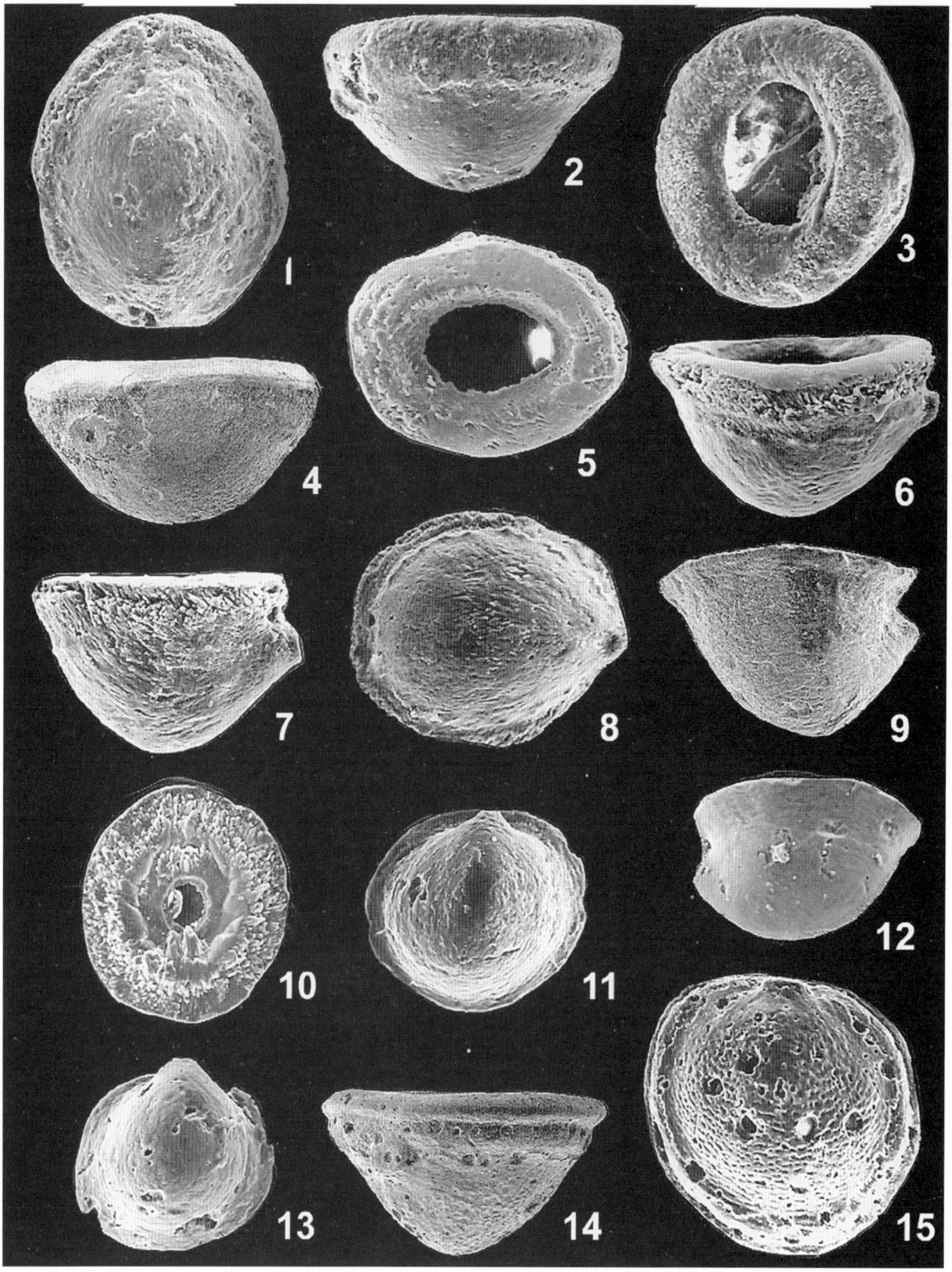

Fig. 8.8 Pseudoarcellinids from France and Belgium, reprinted from Bignot, G. & Poignant, A. (2010) Les Pseudarcellinae, nouvelle sous-famille des Codonellidae (Tintinnida) – le genre *Pseudarcella* et l'espèce *Pseudarcella rhumbleri* Spandel, 1909 – description de nouvelles espèces. Revue de Micropaléontologie, 53, 107–119. Plate 1, with permission from Elsevier. 1–4, *Pseudoarcella rhumbleri*, diameters = 380 μm; height = 250 μm. 5–8, *P. szczechurae*, large diameter = 330 μm; small diameter = 270 μm; height = 250 μm. 9–13, *P. belgica*, diameter = 260 μm; height = 190 μm. 14, 15, *P. konenkovae*, diameter = 450 μm; height = 260 μm.

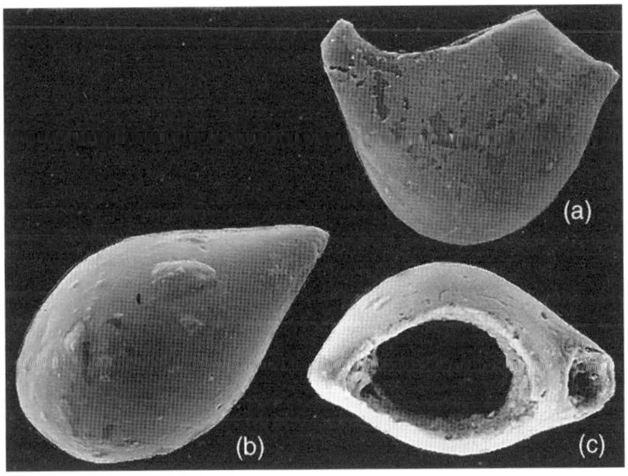

Fig. 8.9 Three views of *Pseudoarcella sp.* (length = 170 μm; height = 230 μm) from the Chattian of the Aquitain Basin (reprinted from Bignot, G. & Poignant, A. (2010) Les Pseudarcellinae, nouvelle sous-famille des Codonellidae (Tintinnida) – le genre *Pseudarcella* et l'espèce *Pseudarcella rhumbleri* Spandel, 1909 – description de nouvelles espèces. Revue de Micropaléontologie, 53, 107–119. Plate 2, figs 12, 14, 15, with permission from Elsevier).

A gap in the occurrence of agglutinated loricae exists after the Cretaceous until they reappear in sediment cores that penetrate the Upper Pleistocene (Echols & Fowler 1973). Agglutinated and organic loricae are common in modern biotas in both open-ocean and neritic environments.

8.7 CONCLUSIONS AND PERSPECTIVES

The calcareous calpionellids and other forms may not be tintinnids at all. This view was common in the past when calpionellids were compared with a variety of microbial eukaryotes including thecameobians, foraminifera, and others. Some, especially species of *Pseudarcella*, could even be small or larval invertebrate shells. Thus the reliable fossil record of tintinnids that can be used in calibrating molecular phylogenetic analyses is sparse and very incomplete. The first organic tintinnid from the Jurassic appears to be a fully developed form, thus indicating a previous undiscovered history.

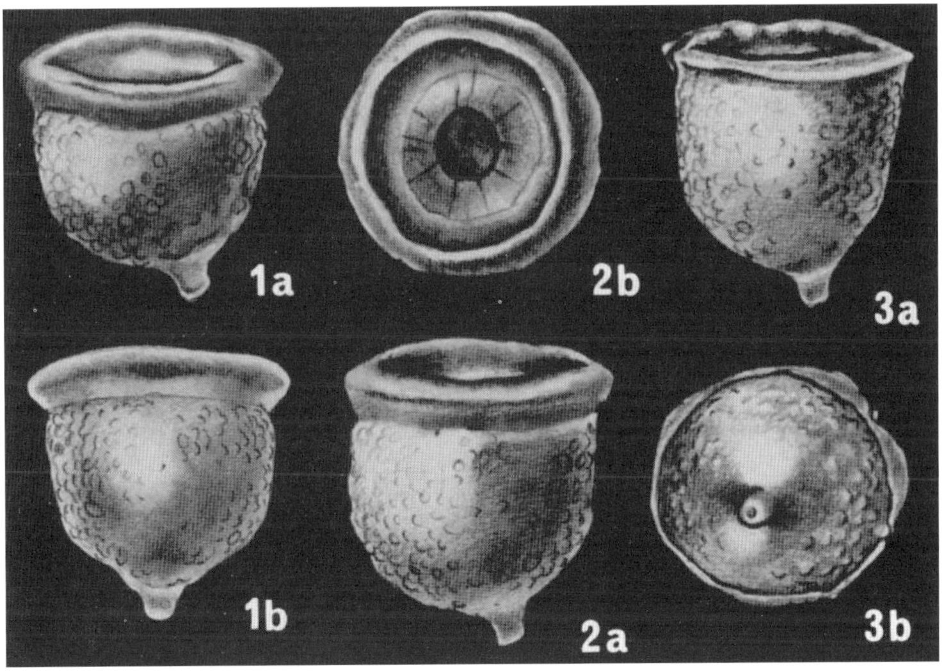

Fig. 8.10 *Tytthocorys coronula* from the Upper Eocene at Jackson, Mississippi, USA (from plate 167, figs 1–3 in Tappan & Loeblich 1968). These specimens are of secreted $CaCO_3$ and show a characteristic tintinnid shape of the bowls (width = ~200 μm) with aboral horns. © Society for Sedimentary Geology.

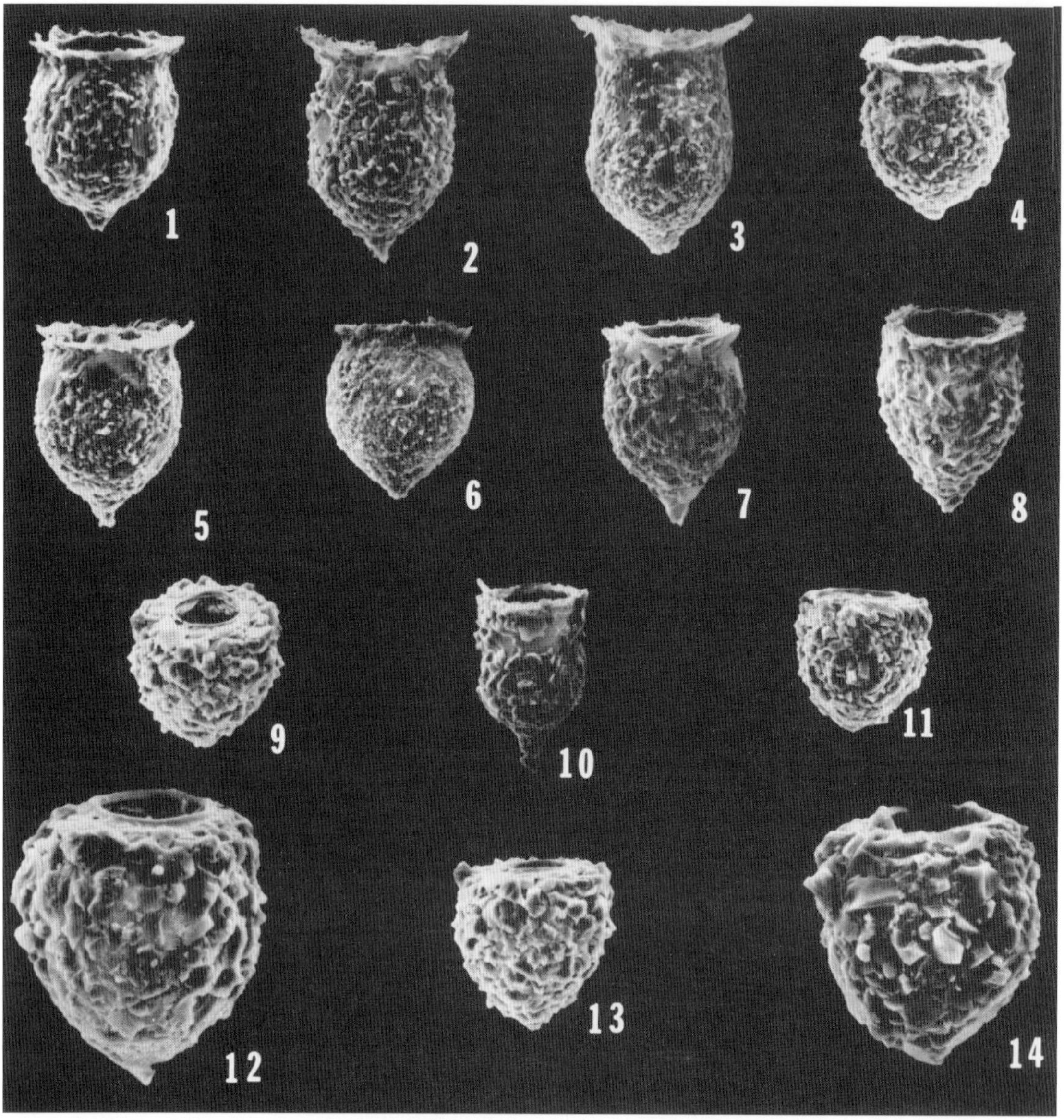

Fig. 8.11 Agglutinated tintinnid loricae from modern sediments from the continental shelf off northern Siberia, eastern Alaska, the Kara Sea, and the shelf off Washington and Oregon (Plate 1 from Echols & Fowler 1973). Similar forms (*Tintinnopsis fimbriata*) are found in Pleistocene sediments cored from the Chukchi Sea, Arctic Ocean. (1–8, ~85 μm wide. 9–13, 61–68 μm wide. 10, 50 μm wide.) Courtesy of Micropaleontology Project.

Molecular data suggest that tintinnid classifications suffer from morphologic convergence of forms with dissimilar sequences, and some forms that are morphologically quite similar have distinct molecular sequences (Chapter 3). The fossil record of tintinnids does not provide a way to determine which of these convergences and divergences may be deciphered. Indeed, because of these molecular sequence data, no classification of the fossil tintinnids is attempted in this summary. Any such classification would be quite artificial.

8.8 KEY POINTS

1. Fossils normally leave few clues to their ciliate origin, except their shape and composition.
2. All putative tintinnids from the Precambrian and Proterozoic differ in shape, composition, or both from all known extant tintinnids, and are not tintinnids.
3. Calpionellids are likely not tintinnids because their composition differs from all known extant tintinnids.
4. The history of reliable tintinnids in the rock record is very incomplete, starting even with the first occurrence in the Jurassic, which indicates an even longer history.
5. Tintinnid classifications of both modern and fossil species are artificial and do not show evolutionary relationships.
6. Not only is more work necessary on the molecular phylogenetics of tintinnids, the fossil record must also be clarified and better documented with positive determinations that fill in the history of the group.

ACKNOWLEDGMENTS

We thank the editors of this book, especially John Dolan and Sabine Agatha. We also thank Alberto E. Patiño Douce, University of Georgia, for comments on the Mesoproterozoic Huangmailing Formation fossils. Funding came from the Alexander von Humboldt Foundation to M.D., and the Deutsche Forschungsgemeinschaft (grant STO414/3-1) to T.S. This is University of California Museum of Paleontology publication number 2033.

Chapter 9

TINTINNIDS IN MICROZOOPLANKTON COMMUNITIES

George B. McManus and Luciana F. Santoferrara

University of Connecticut, Groton, CT, USA

The Biology and Ecology of Tintinnid Ciliates: Models for Marine Plankton, First Edition. Edited by John R. Dolan, David J.S. Montagnes, Sabine Agatha, D. Wayne Coats, and Diane K. Stoecker.

9.1 ECOLOGY OF MICROZOOPLANKTON

Microzooplankton as a functional trophic category

Depending on one's perspective, tintinnids can be studied for their taxonomic affiliation (they are spirotrich ciliates), their type of metabolism (they are heterotrophs), their mode of living (they are members of the plankton), or any one of several other categories to which they may be assigned. In this chapter we will consider them as members of a group called the microzooplankton. This category was created by oceanographers to include all "animal-like" organisms (i.e., they have to eat to survive) that drift with ocean currents and are in the size range of 20–200 µm. Some tintinnids are larger than this, and many have long slender shapes that make assignment of a size somewhat arbitrary. But given that the original intent of creating the microzooplankton as a group for study was to include all the zooplankton that would pass through a standard 200 µm mesh plankton net, it is safe to say that virtually all tintinnids are members of this group.

Realizing that taxonomic affiliations among planktonic organisms were often less useful indicators of ecological function than their relative sizes, Dussart (1965) proposed a formal size-based classification of plankton (Table 9.1). Sieburth et al. (1978) refined the system by adding a category for "femtoplankton" (principally viruses), and pointed out that the smallest size groups (pico-, nano-, and microplankton) contained both heterotrophs and autotrophs. Although they introduced other terms to account for this (e.g., "protozooplankton", "bacterioplankton", etc.), many contemporary planktologists simply use an additional prefix to indicate trophic status, as in "microphytoplankton" for autotrophs and "microzooplankton" for heterotrophs. The original Dussart categories and the consensus contemporary system are shown in Table 9.1. The great majority of tintinnids are in the microplankton size category and, as strict heterotrophs, are hence designated as microzooplankton.

Long before Dussart formalized the system, early planktologists introduced the categorization of plankton by size (Lohmann 1911), but systematic attempts to quantify abundances and biomasses of microzooplankton in relation to phytoplankton and larger zooplankton began in earnest during the 1960s and 1970s (Beers & Stewart 1969; Sorokin & Kogelschatz 1979; Burkill 1982; Sorokin et al. 1985; Sorokin, et al. 1998). The rationale for using size categories in plankton studies is both practical and theoretical. On the one hand, plankton nets and membrane filters

Table 9.1 Size categorization of plankton groups.

Size range (µm)	Dussart (1965) category	Current consensus usage	Organisms included
0.02–0.2	Not included	Femtoplankton	Viruses
0.2–2	Nannoplankton	Picoplankton	Heterotrophic and autotrophic prokaryotes; the smallest protists
2–20	Ultramicroplankton	Nanoplankton	Flagellates and other small protists
20–200	Microplankton	Microplankton	Autotrophic and heterotrophic protists; some metazoan larvae
200–2000	Mesoplankton	Mesoplankton	Copepods and other small metazoans
>2000	Megaplankton	Macroplankton	Larger crustaceans; some jellyplankton

with precise pore sizes allow us to easily separate size classes of organisms into fractions that can subsequently be weighed, counted under the microscope, or used for experiments; on the other hand, there is a large body of theory going back to Elton's "pyramid of numbers" (Elton 1927) that describes the dependence of abundance and metabolism, including feeding, respiration, and growth, on organism size (Platt 1985; Fenchel 1987). The latter arguments and supporting observations have been used to suggest that zooplankton of 20–200 μm size (microzooplankton) are, in aggregate, quantitatively more important as consumers and remineralizers of primary production than the larger and better-studied mesozooplankton (see, for example, Johannes 1965; Sieburth et al. 1978) (See Box 9.1). The allometric arguments remain important to our understanding of the structure of planktonic food webs because methods for the direct measurement of microzooplankton grazing, growth, and respiration in natural assemblages are limited, and they led to a great increase in the number of studies focused on microzooplankton from the 1970s through the present. For this chapter, we will review this material, emphasizing studies of natural communities in which the role of tintinnids within the microzooplankton is specifically quantified. Other chapters in this volume focus on the many laboratory studies of feeding, growth, and mortality in tintinnids (see Chapters 3, 5, and 6). Seasonal changes in tintinnid communities are treated in Chapter 10.

Box 9.1 An allometric argument for the importance of the microzooplankton

Relationship between body size and metabolism
I = Ingestion rate = grams eaten/grazer/day
B = body mass = grams/grazer
a = a constant

$$I = aB^{0.75}$$

note: $\log(I) = \log(a) + 0.75 \log(B)$

$$B_{cop} = 1000\,B_{cil}$$

$$I_{cop} = aB_{cop}^{0.75}$$

$$= a(1000B_{cil})^{0.75}$$

$$= 1000^{0.75}aB_{cil}^{0.75}$$

$$= 178\,aB_{cil}^{0.75} = 178\,I_{cil}$$

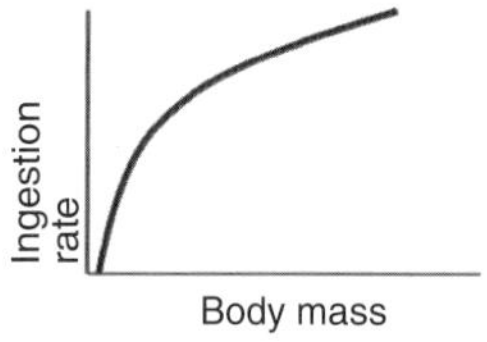

It is well known that physiological rates increase with body mass in a decelerating fashion, as in the graph above. A larger animal such as an elephant eats more than a bird in absolute amount of food, but less as a fraction of its body mass. This probably has to do with the fact that many physiological processes (e.g., gas or heat exchange) are surface-related, whereas body mass is volume-related; so metabolism increases as a square of linear dimension, while mass increases as its cube. The consequent argument that physiological rates should increase as the ⅔ power of mass has been referred to as "Rubner's Law" (Fenchel 1987; Fenchel & Finlay 1983). Observations confirm the decelerating rate, but show that the exponent is closer to the ¾ power.

In the present case, if we are trying to evaluate the relative importance of microzooplankton and mesozooplankton in ingesting phytoplankton, we have to consider both relative metabolic rates and abundances. For example, if microzooplankton are about one-tenth the size of mesozooplankton in linear dimension (20–200 μm compared with 200–2000 μm; here represented as a ciliate and a copepod, respectively), then they would individually be about one-thousandth the biomass (B_{cop} = 1000 times B_{cil}). In the example above, that means that I_{cop}, the ingestion rate of an individual copepod mesozooplankter (the elephant), is 178 times that of an individual ciliate microzooplankter (I_{cil}, the bird in this case). However, because microzooplankton abundance in coastal waters is about 1000-fold greater than that of mesozooplankton (10^3–10^4 individuals per liter compared with 1–10 individuals per liter), the implication is that the two size classes have about equal biomass in aggregate (the 1000 times smaller ciliate is 1000 times more abundant). Thus, although the individual copepod eats 178 times as much, in aggregate the 1000 times more abundant ciliates eat over five times as much total phytoplankton. Note that the same argument could be made for other metabolic processes, including respiration, excretion, etc., all of which are just as critical for food web ecology as ingestion rate.

Composition of microzooplankton assemblages

The microzooplankton can be divided into four broad groups that roughly correspond to phylum-level affiliations: (1) ciliates, including the tintinnids; (2) dinoflagellates and other large flagellates; (3) rotifers; and (4) the larvae of microcrustaceans, principally copepod nauplii. The size fraction approach to studying ecology of plankton has some advantages, but it also obscures biological differences among groups of organisms that may employ different strategies for success in a pelagic realm where food is dilute and predation pressure is high. Thus, many investigators have focused on individual groups within the microzooplankton and this has led, to a certain degree, to some divergent perspectives on the overall role of microzooplankton in pelagic food webs. Although the subject of this chapter is tintinnids in the microzooplankton, we will try to balance the approach by giving an overview including other members of the group.

Ciliates

Ciliates have often been considered as the main component of microzooplankton assemblages, largely owing to the conspicuousness of tintinnids and their ability to survive concentration in plankton nets, as well as the lack of appreciation for other microzooplankton groups, in particular heterotrophic dinoflagellates. Planktonic ciliate abundance and biomass is usually dominated by the Spirotrichea subclasses Oligotrichia and Choreotrichia (Montagnes & Lynn 1991; Pierce & Turner 1992; Fenchel 1988). The latter group includes the tintinnids and some of their aloricate relatives, principally in the families Strobilidiidae and Strombidinopsidae. A variety of other ciliates may be represented in the plankton at times, especially in nearshore, shallow waters. These include *Tiarina fusus* and *Balanion comatum* (Prostomatea), which are not difficult to cultivate and hence have been the subject of feeding and ecophysiological studies (Jakobsen & Hansen 1997; Jakobsen & Montagnes 1999; Stoecker et al. 1986; Jeong et al. 2002); scuticociliates (Oligohymenophorea), which are opportunistic bacterivores that often grow up in seawater enrichments; and the haptorids (Litostomatea) *Askenasia* spp. and *Mesodinium* spp. The last group is an assemblage of species that includes both heterotrophic and mixotrophic forms, with the well-studied *Mesodinium rubrum* essentially being an autotroph, though it does require cryptophyte food to replace its enslaved chloroplasts (Gustafson et al. 2000; Johnson et al. 2006). It now appears that there are multiple mixotrophic, as well as heterotrophic, species within this group, and the ecological roles of the different species may diverge (Moestrup et al. 2012). Peritrichs and other particle-associated ciliates are found in many marine environments attached to large diaphanous aggregates (marine snow) and as epibionts (Song & Wilbert 2002; Caron, et al. 1982). They probably have been undersampled because of the rarity of large particles in the pelagic environment and the destruction of marine snow in net collections.

Dinoflagellates and other large flagellates

Dinoflagellates are the most problematic group in the microzooplankton because the trophic roles of individual species are generally not known (Lessard & Swift 1986; Lessard 1991). Although some groups are wholly heterotrophic (e.g., *Protoperidinium* spp.) it is likely that most, if not all, dinoflagellates are capable of ingestion (Stoecker 1999). Furthermore, it is clear that individual species vary greatly over time in the relative importance of auto- and heterotrophic metabolism, depending on environmental conditions (Bockstahler & Coats 1993; Adolf, et al. 2006; Smalley & Coats 2002; Smalley, et al. 2003).

There have been several studies that compared abundances of heterotrophic dinoflagellates to those of other microzooplankton groups (e.g. Verity et al. 1993, 1996). In general, the most abundant strict heterotrophs are forms smaller than 20 µm, probably bacterivores and grazers on the smallest phytoplankton (Verity et al. 1996). Large heterotrophs, including pallium feeders capable of ingesting prey much larger than themselves, appear to be most abundant during blooms of chain-forming diatoms (Sherr and Sherr 2008b). A full understanding of the significance of dinoflagellates as microzooplankton awaits development of new methods that will allow measurement of *in situ* grazing activities by forms traditionally considered to be autotrophic.

Rotifers

Rotifers are a significant component of microzooplankton assemblages in many systems, especially lakes, rivers, and estuaries (see, for example, Dolan & Gallegos 1991, 1992; Holst et al. 1998; Heinbokel et al. 1988; Neumann-Leitao, et al. 1992; Barria de Cao et al. 2003; Rougier et al. 2005). In brackish-water systems, they can reach densities above 10^3 individuals per liter and be the dominant zooplankton grazer (Sellner et al. 1993; Park & Marshall 2000), but they are apparently not well adapted to oceanic or open coastal environments (Heinbokel et al. 1988). Some aloricate forms do not preserve well, especially in aldehydes, so they are often undercounted in plankton surveys (Holst et al. 1998).

Copepod nauplii and other crustacean larvae

The principal orders of pelagic copepods in the sea are the Calanoida and the Cyclopoida. As adults, they range from about 0.5–5 mm in length, and their eggs are typically in the 50–250 µm size range. They hatch as first instar nauplii that are about the same size as the eggs and proceed through twelve molt stages to adulthood, passing from the micro- to the mesozooplankton size class in the process. Although there have been some studies of feeding and other metabolic processes in early-stage (i.e., microzooplanktonic) nauplii, a comprehensive understanding of their trophic impact is lacking. In some cases, early naupliar stages may not feed at all (Mauchline 1998), relying on energy reserves retained from the egg. Some tracer studies suggest a diet similar to that of later stages, including adults (Finlay & Roff 2004). At times, they can dominate the biomass of microzooplankton assemblages (McManus et al. 2007), but it is likely that their mass-specific metabolic rates are quite a bit lower than those of their protist counterparts, so their overall impact is likely less. Because they are not sampled quantitatively with coarse-mesh nets and are not numerous enough to be counted accurately in the small volumes concentrated for counts of protists, they are often excluded from microzooplankton surveys.

One negative consequence of lumping plankton into sized-based trophic groups is that it ignores many of the interesting biological differences among them. It is clear that dinoflagellates and ciliates, for example, despite their phylogenetic relationship as members of the Alveolata, have very different behaviors and physiological adaptations. Ciliates swim rapidly, have short generation times, feed by filtration, and practice autotrophy only when they can use chloroplasts from ingested autotrophs. Dinoflagellates have more complex modes of ingestion (peduncle, pallium, or direct engulfment), feed by direct interception of prey, swim and grow more slowly and, when mixotrophic, occupy a wider portion of the spectrum between strict autotrophy and strict heterotrophy and can either use ingested chloroplasts or contain their own permanent chloroplasts. Many dinoflagellates are capable of synthesizing toxins, including those effective against ciliates and other microzooplankton (see Chapters 3 and 4). Generalizations about metazoans feeding on dinoflagellates have thus been difficult to make because of species-specific toxic interactions that cause some dinoflagellates to be essentially immune from grazing by some metazoans (see, for example, Turriff, et al. 1995; Huntley et al. 1986; Sykes & Huntley 1987; Colin & Dam 2003). Different groups within the microzooplankton show different allometric relationships for physiological processes as well (Hansen et al. 1997). It is thus important to keep in mind that categorization of plankton by size is only a rather crude tool, necessary for making generalizations about carbon flow and comparisons among ecosystems, but not sufficient for a true understanding of microbial food webs.

Grazing impact of microzooplankton

Early efforts to determine the ecological role of microzooplankton focused strongly on grazing, especially of phytoplankton. A variety of methods have been used to estimate the grazing impact of microzooplankton *in situ*. Capriulo & Carpenter (1983) preconcentrated the microzooplankton size fraction using nylon sieves and incubated the concentrate with the natural phytoplankton assemblage, quantifying community grazing by the subsequent decrease in chlorophyll-*a*. Their use of sieves to concentrate the microzooplankton may have led to underestimation of the importance of aloricate ciliates, which can pass through even very fine meshes. Later workers used radioisotopes or fluorescent dyes to label phytoplankton assemblages and then quantified incorporation of label in the grazers to estimate grazing impact (Lessard & Swift 1985; Roman & Rublee 1981; Rublee & Gallegos 1989). Tracer methods have the advantage of requiring short incubation times, hence minimizing containment artifacts, and have the ability to provide grazer-specific rates if different grazers (dinoflagellates, ciliates, nauplii) are subsequently isolated or examined separately under the microscope or if labeled nucleic acids in predators can be recovered after exposure to labeled prey (Frias-Lopez et al. 2009). Tracer methods also require knowledge of the time scale of digestion so that rates can be estimated before egestion of tracer particles or isotopes occurs (McManus & Okubo 1991). This can require laborious time series sampling.

A few studies have attempted to apply the gut fluorescence method to estimate microzooplankton grazing. In this method, ingestion is estimated from the product of measured ingested chlorophyll inside the grazers and an estimate of the gut (or vacuole) turnover rate (Tsuda, et al. 1989). For microzooplankton, this method is especially difficult because of the overlap in sizes of microzooplankton and their food organisms and problems associated with trying to separate these fractions mechanically in order to measure the chlorophyll in each. In addition, it cannot easily be used to measure ingestion of non-pigmented food items such as bacteria or heterotrophic flagellates. More recently, Johnson et al. (2003) measured disappearance of fluorescently labeled prey to study microzooplankton grazing on individual phytoplankton species. This method is especially useful where monospecific blooms are occurring or the phytoplankton assemblage is dominated by only a few species.

By far the most commonly applied technique for estimating microzooplankton grazing *in situ* is the dilution method originally developed by Landry & Hassett (1982). Net growth of phytoplankton (as chlorophyll-*a*, or in some cases cell counts via flow cytometer; Landry, et al. 1995) is compared in a series of samples that differ only in the degree to which they have been diluted by filtered seawater. If grazing rates are food-concentration dependent, then net phytoplankton growth in more dilute treatments will be greater because of lower mortality due to dilution of grazers. This method has the advantage of requiring only minimal manipulation of microzooplanton, many of which are delicate. On the other hand, a relatively long incubation period of about 24 hours is usually needed to measure significant changes in phytoplankton abundance and because of the diel cycle in phytoplankton growth. Some variations of the basic design have been proposed to expand the method and to deal with situations where grazing rates are food-saturated (Redden, et al. 2002). There have also been several studies describing limitations of this method and suggestions for improvement; these are beyond the scope of this chapter (Dolan & McKeon 2004; Landry & Calbet 2004; Dolan et al. 2000; Gallegos 1989; Landry, et al. 1995).

Calbet & Landry (2004) summarized 20 years' worth of dilution experiments by various investigators. They compiled the results from 788 individual measurements covering most areas of the World Ocean. Overall, microzooplankton grazing accounted for the fate of about two-thirds of phytoplankton production, with modest variation in relation to onshore/offshore or latitudinal gradients (Table 9.2). Given the presumed strong differences in the structure of the planktonic food web in different habitats, it is remarkable that the microzooplankton size fraction is so consistent in its role as the primary herbivorous pathway in the sea. One suggested explanation for this is that microzooplankton have growth rates similar to those of their phytoplankton prey and hence are closely coupled in time with changes in their abundance.

Although the dilution method can be cumbersome to use, and does not allow for particularly good spatial or temporal resolution of grazing rates, it is clear now from the aggregate number of studies that the earlier allometric arguments about the importance of microzooplankton were correct. However, one unfortunate drawback of the method is that although impact of the microzooplankton assemblage can be assessed for different phytoplankton groups (for example through the use of taxon-specific pigments, flow cytometric signatures or microscopy; Burkill et al. 1987; Strom & Welschmeyer 1991; McManus & Ederington-Cantrell

Table 9.2 Summary data on the percentage of phytoplankton production grazed by microzooplankton in various habitats (± standard error). The full data set consists of 788 individual measurements of intrinsic phytoplankton growth and grazing mortality. The reduced dataset contains only those experiments that accounted for possible nutrient depletion artifacts in the incubation bottles (392 measurements). Table from Calbet, A. & Landry, M.R. (2004) Phytoplankton growth, microzooplankton grazing, and carbon cycling in marine systems. Limnology and Oceanography, 49, 51–57. Copyright the Association for the Sciences of Limnology and Oceanography, Inc.

Habitat	Full data	Reduced data
Oceanic	69.6 ± 1.5	78.0 ± 1.8
Coastal	59.9 ± 3.3	56.6 ± 2.9
Estuarine	59.7 ± 2.7	38.6 ± 2.5
Tropical/ subtropical	74.5 ± 2.0	71.3 ± 2.3
Temperate/ subpolar	60.8 ± 1.8	68.6 ± 2.3
Polar	59.2 ± 3.3	65.2 ± 3.7

1992; Landry, et al. 1995) it is not possible to partition grazing impact among different groups of microzooplankton. We thus have much less information about the relative importance of tintinnids, other ciliates, dinoflagellates, rotifers, and crustacean larvae in the fate of planktonic production than we do about the aggregate impact of the whole microzooplankton assemblage.

Nutrient regeneration by microzooplankton

As the primary herbivores in the sea, microzooplankton are likely the primary regenerators of inorganic nutrients as well. At an average gross growth efficiency of about 30% (Fenchel 1987; Straile 1997), unicellular microzooplankton such as ciliates and dinoflagellates can return about 70% of ingested food to the dissolved or microparticulate pool for subsequent use by phytoplankton or bacterioplankton. Johannes (1965) recognized this idea, in part from allometric considerations, and it was also supported empirically using stable isotope tracers of nitrogen (Caperon et al. 1979; Glibert 1982), though it seems that more than half of the total grazer regeneration comes from small flagellates and other protists smaller than 10 µm ("nanozooplankton"; Probyn 1987), rather than from larger microzooplankton, *sensu stricto*, such as tintinnids. The difficulty of making these measurements and the paucity of observations suggests caution in advancing any definitive conclusions about nutrient regeneration by tintinnids or other large microzooplankton, but the evidence of their importance based on feeding and respiration rates alone is strong (Dolan 1997; Calbet & Landry 2004). A recent review by Calbet (2008) emphasizes the need for new methods and many more observations in this area.

Microzooplankton as food for higher trophic levels

In tandem with observations that uncovered the importance of microzooplankton as grazers and mineralizers of inorganic nutrients, there have been many studies and reviews documenting their importance as food for larger organisms (Tiselius 1989; Sanders and Wickham 1993; Stoecker & Capuzzo 1990; Gifford & Dagg 1991; Gifford & Dagg 1988; Figueiredo et al. 2005; Nakagawa et al. 2004). A review by Calbet & Saiz (2005) suggests that ciliates alone account for some 30%, on average, of the daily ration of copepods. They are thus universally acknowledged as constituting a critical link in planktonic food webs. Some of the separate lines of evidence that support this include the following: (1) direct experimental measurements of mesozooplankton (see, for example, Bouley & Kimmerer 2006; Fileman, et al. 2010; Fessenden & Cowles 1994), larval fish (see, for example, Stoecker & Govoni 1984), or jelly plankton (Olesen, et al. 1996) feeding on microzooplankton; (2) models of trophic flows that require an important microzooplankton pathway in order to be balanced (Pace, et al. 1984); and (3) experimental manipulations of natural assemblages to quantify trophic pathways (Calbet & Landry 1999).

One commonly used method to evaluate trophic links through microzooplankton is to manipulate abundances of copepods or other mesozooplankton so that a gradient of predation pressure is created. This approach was pioneered by Lehman & Sandgren (1985) to study mesozooplankton grazing of phytoplankton, but it has been applied subsequently to study indirect effects of mesozooplankton on phytoplankton via their grazing on microzooplankton. Using this method, Sipura et al. (2003) found strong direct effects of copepods grazing on microzooplankton, including tintinnids, in 241 estuarine mesocosms. Many such "cascade" experiments show evidence of multiple trophic transfers within the microzooplankton (Sarnelle 1997). Stibor et al. (2004) and Sundt-Hansen et al. (2006) suggested that microzooplankton play a significant role in transferring the effects of higher trophic level grazing down through the food chain, especially when small algae, inedible to mesozooplankton, are dominant in the phytoplankton assemblage. Froneman (2002) observed strong cascading effects of mesozooplankton on cells smaller than 2 µm (picophytoplankton) that were mediated by microzooplankton.

The trophic connection between microzooplankton and mesozooplantkon can be quite important, but also variable. For example, Liu et al. (2005) showed that microzooplankton constituted 30% of the diet of the copepod *Neocalanus cristatus* during phytoplankton blooms, but 70% in non-bloom conditions. Gifford & Dagg (1988) showed similar variability by season in a coastal environment. The sparse data in this area has been reviewed recently (Calbet 2008), but much more work needs to be done, in particular to document the differential impacts of grazers on ciliates, dinoflagellates, and other microzooplankters, and to evaluate intraguild predation (microzooplankton eating other microzooplankton). Regarding the latter, there are many observations of dinoflagellates feeding on ciliates and vice versa. These kinds of interactions may even be species-specific (Jacobson & Anderson 1996; Sjoqvist & Lindholm 2011), suggesting that the pathways of energy flow through microzooplankton maybe quite tortuous.

9.2 QUANTITATIVE IMPORTANCE OF TINTINNIDS IN MICROZOOPLANKTON ASSEMBLAGES

Abundance and biomass

Although estimates of the abundance and biomass of tintinnids are numerous in the literature, making comparisons among different studies is difficult because methods are not uniform. Some studies, for example, include empty loricae in tintinnid counts, although most do not. In converting lorica volumes measured under the microscope to cell volumes, different conversion factors may be used for the fraction of the lorica that is occupied by the cell. Also, in evaluating the relative importance of tintinnids compared with other microzooplankton groups, some studies include only a portion of the 20-200 µm heterotrophs in the total. Most, for example, still do not include mixotrophic dinoflagellates, and many either overlook or undersample copepod nauplii. Some include *Mesodinium rubrum* among "total ciliate microzooplankton" even though it is effectively a fulltime autotroph. The conspicuousness of tintinnids due to their loricae may have led in some cases to overestimates of their relative abundance, but on the other hand naïve observers may confuse loricae with detritus, especially for highly agglomerated forms. Finally, there are many more studies that report abundances of tintinnids or the percentage of total ciliate abundance that is composed of tintinnids, but relatively few studies report on tintinnid biomass separately from that of other ciliates or microzooplankton as a whole.

Table 9.3 and Fig. 9.1 provide a summary of 39 studies that report on the abundance of tintinnids in proportion to the whole ciliate assemblage. Of the 638 total data points, 42% were from coastal environments, including estuaries, 32% were from open shelf environments, and 26% were from oceanic environments. The modal percentage of ciliates composed of tintinnids is 0–5% across all environments, but there are some differences evident among them. For example, tintinnids are rarely more than half of all ciliates in the shelf and oceanic samples (13 out of 371 samples), whereas they were often that abundant near the coast (90 out of 267 samples). It should be kept in mind that as a rare taxon, tintinnids may be undercounted in shelf and oceanic samples for which small volumes are examined, leading to an increase in the number of zeroes (tintinnid abundances below the detection limit). For example, Santoferrara & Alder (2009a) compared 30 small volume (0.5 l) and large volume (200 l of water concentrated over a 20 µm mesh) samples from the same stations. They found that when the tintinnid *Helicostomella subulata* was present in the large volume samples it was not detected in two-thirds of the corresponding small volume samples, although its abundance was significantly underestimated in the mesh concentrates. For this reason, when presence/absence is being evaluated, as in biogeographic studies of tintinnids, most investigators still use nets to collect them (Dolan et al. 2007).

Overall, the frequency distribution is flatter in the nearshore data, with tintinnids being 95–100% of the total ciliates in fully 9% of those samples. This is likely due to the episodic nature of phytoplankton blooms in nearshore waters and to patchiness of resources, which cause overall ciliate abundance to be highly variable in space and time (Rassoulzadegan 1977).

For all data taken together, the abundance of tinins is correlated significantly with that of other ciliates, but the correlation is not particularly strong (Fig. 9.2). In fact, there are many examples in the data where tintinnid abundance was zero, even when other ciliates were abundant, and vice versa. The median tintinnid abundance observed in estuaries and coastal environments in the studies we reviewed was 500 individuals per liter; for the shelf and oceanic environments it was 21 individuals per liter.

There are many fewer studies where the biomass of tintinnids has been quantified relative to that of other microzooplankton groups. Most often, when biomass is estimated, tintinnids are lumped with other ciliates. When tintinnid abundance is estimated separately, it is usually in studies where only ciliate biomass has been measured and other microzooplankton were not counted. We did find 15 studies where tintinnid biomass was separately estimated and where total microzooplankton biomass was reported. A summary

Table 9.3 Summary of 39 studies that report the percentage of the total ciliate assemblage comprised by tintinnids. Values are grouped in bins of 5 percentage points, except that the first bin is just the data for studies that reported a value of zero for tintinnid abundance. The original data can be found in Andersen & Sorensen 1986, Barría De Cao et al. 2003, Bojanić 2001, Dennett et al. 1999, Dolan et al. 2002, Fileman & Leakey 2005, Fileman et al. 2002, Froneman & Mcquaid 1997, Froneman & Perissinotto 1996a, Froneman & Perissinotto 1996b, Gómez & Gorsky 2003, Graham & Strom 2010, Hall et al. 2008, Huang et al. 2011, Ichinomiya et al. 2009, Johansson et al. 2004, Johnson et al. 2003, Jyothibabu et al. 2006, Klaas 1997, Leakey et al. 1996, Lugomela et al. 2001, Martin & Montagnes 1993, McManus et al. 2007, Modigh 2001, Nakamura & Hirata 2006, Olson & Strom 2002, Pitta & Giannakourou 2000, Quinlan et al. 2009, Sakka Hlaili et al. 2007, Santoferrara & Alder 2009, Santoferrara & Alder 2012, Santoferrara et al. 2011, Seuthe et al. 2011, Stoecker et al. 1989, Stoecker et al. 1996, Tamigneaux et al. 1997, Vargas & Martínez 2009, Yang et al. 2008, Zhang & Wang 2000.

	All	Coast	Shelf	Ocean
References	39	21	7	11
Total data	638	267	205	166
***T* (%)**	**Frequency**	**Frequency**	**Frequency**	**Frequency**
0	64	39	14	11
1–5	211	32	123	56
6–10	78	19	21	38
11–15	42	14	9	19
16–20	40	15	10	15
21–25	28	14	5	9
26–30	28	11	8	9
31–35	24	17	5	2
36–40	10	7	3	0
41–45	10	9	1	0
46–50	24	20	2	2
51–55	8	8	0	0
56–60	11	9	1	1
61–65	7	5	1	1
66–70	9	8	1	0
71–75	5	3	1	1
76–80	3	3	0	0
81–85	1	0	0	1
86–90	7	6	0	1
91–95	5	5	0	0
96–100	23	23	0	0
	638	267	205	166

is shown in Fig. 9.3. In all cases, at least one other group of microzooplankton besides ciliates was measured, usually either heterotrophic dinoflagellates or copepod nauplii, or both. Because of the difficulty in distinguishing heterotrophic from mixo- or autotrophic dinoflagellates, some studies report only total dinoflagellates. Where that was the case, we did not include dinoflagellates in the biomass estimates. The 15 studies ranged across many habitats and latitudes, from the tropics to the arctic, and comprised oceanic to estuarine systems.

There is clearly a great deal of variability within studies, especially when tintinnid biomass becomes more than a few percent of the total. Values for the tintinnid portion of total microzooplankton biomass as low as zero and as high as 81% were reported. In general, oceanic regions had a lower contribution by tintinnids to the total microzooplankton biomass. This is in line with early data indicating that the tintinnid portion of microzooplankton is higher near the coast (Beers et al. 1980) and the general perception that tintinnids are mainly estuarine and coastal organisms.

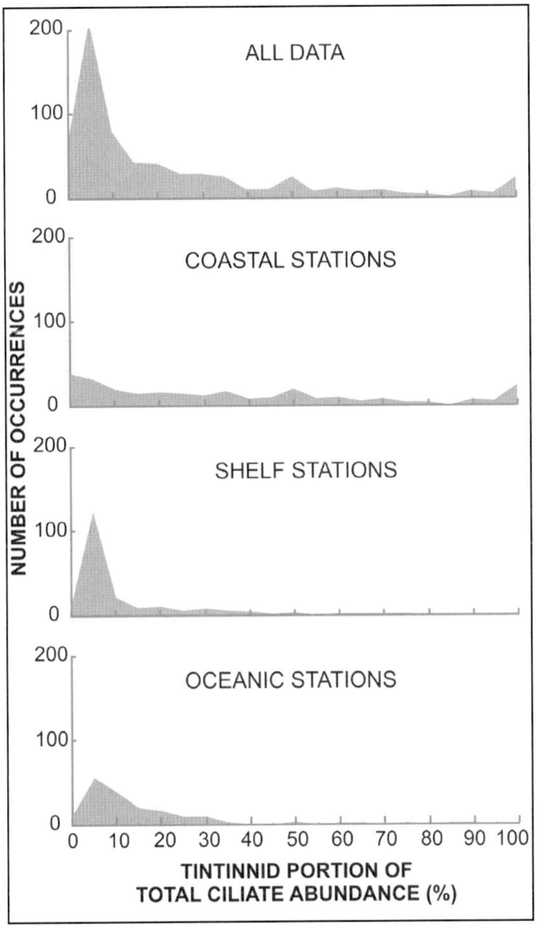

Fig. 9.1 Frequency distributions for the percentage of total ciliate abundance composed of tintinnids. Results were compiled from 39 studies (638 total data points) in which separate estimates of tintinnid abundance were made. Data have been aggregated into bins five percentage points wide, and the values for zero (no tintinnids) are in a separate bin (as in Table 9.3).

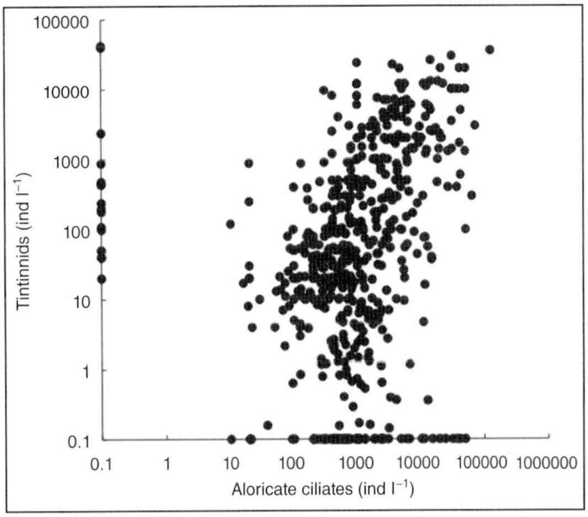

Fig. 9.2 Relationship between tintinnid abundance (individuals per liter) and that of all other ciliates in the same sample. Least-squares linear regression: tintinnids = 0.17 × aloricate ciliates + 688 (R^2 = 0.16). For the plot (not the regression), 0.1 was added to all zero values to avoid log (0), so all values of zero abundance appear on the graph as 0.1.

Some oceanic and open coastal environments occasionally showed high tintinnid biomass, however, especially at mid to high latitudes (see, for example, Fileman & Leakey 2005). Despite occasional high tintinnid contributions to microzooplankton biomass, however, none of the studies had average contributions as high as one-quarter of the total. The real contribution of tintinnids may be smaller than that, however, because most studies ignored at least one major group of microzooplankton, and the estimate of dinoflagellate contributions is probably too small in most studies, even when they are included, because many putative wholly autotrophic forms likely engage in phagotrophy as well (Stoecker 1999). Thus, something on the order of 20% or less of the total microzooplankton biomass is in the form of tintinnids, and their overall trophic impact is likely commensurate with this.

One underappreciated aspect of tintinnid biomass distributions is the potential for very small species to have a significant ecological impact. Aloricate oligotrich ciliates as small as 7 μm have been observed to be abundant in coastal and estuarine waters, and their role as bacterivores and grazers of picophytoplankton has been suggested (Sherr et al. 1986; Burkill 1982; Beers, et al. 1980). Tintinnids, on the other hand, have generally been considered to be among the larger microzooplankton, and many earlier studies of their abundance used fine-mesh nets to sample them quantitatively, eliminating the possibility of observing small forms. There are numerous small tintinnids, however.

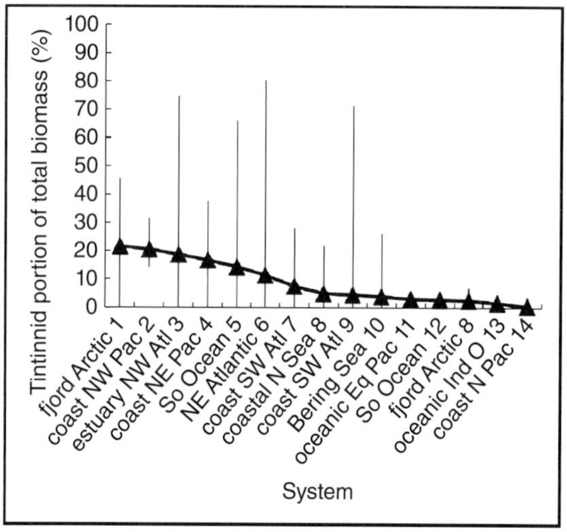

Fig. 9.3 Tintinnids as a percentage of the total microzooplankton biomass. Data were compiled from 15 studies in which tintinnid biomass was estimated separately from that of all other ciliates and at least one of the two other major microzooplankton groups (dinoflagellates and copepod nauplii) was included. Means are ordered from greatest to least tintinnid biomass, and error bars are the ranges within each study. Data are from [1]Seuthe et al. 2011; [2]Yang et al. 2008; [3]Johnson et al. 2003; [4]Beers et al. 1980; [5]Safi et al. 2007; [6]Fileman & Leakey 2005; [7]McManus et al. 2007; [8]Levinsen et al. 2002; [9]Santoferrara et al. 2011; [10]Olsen et al. 2002; [11]Stoecker et al. 1996; [12]Burkill et al. 1987; [13]Dennett et al. 1999; [14]Strom et al. 2007.

Tintinnopsis nana, for example, has a lorica that is cylindrical and roughly 15 μm × 20 μm in size (Lohmann 1908). This gives it an equivalent spherical diameter of just under 20 μm. If we consider that the cell occupies only about half the lorica volume, its equivalent spherical diameter would be less than 15 μm. *Tintinnopsis minuta* is only slightly larger (Gold & Morales 1975). In the annual study of Capriulo & Carpenter (1983), these two species together comprised 29–99% of all tintinnids in Long Island Sound in summer and fall. *T. minuta* was also the most abundant species in Narragansett Bay, ranging as high as 7×10^4 individuals per liter (Verity 1986), and it was the fastest growing tintinnid, with rates of 0.2–2.3 per day (converted from the reported doublings per day for consistency within this chapter). Other small *Tintinnopsis* species, such as *T. beroidea*, *T. parva*, and *T. rapa*, can also be highly abundant at times. In our own 9-year study of microzooplankton abundance and biomass in Long Island Sound (unpublished observations), we have found that tintinnids with lorica equivalent spherical diameter less than 20 μm can be up to 100% of total tintinnids by number. Over the whole dataset ($n = 374$ samples), these "nano-tintinnids" averaged only 4% of total tintinnid biovolume, but considering that their growth and mass-specific metabolic rates should be higher than those of larger forms, their ecological impact may be significant. They probably graze on the smallest phytoplankton and possibly on heterotrophic bacteria and hence constitute a link in the microbial loop. More studies of these smallest tintinnids need to be done to elucidate their role in coastal and estuarine planktonic food webs.

Grazing

The great majority of observations on microzooplankton grazing impact do not separately measure rates of different taxonomic groups or they do not separately estimate tintinnid impact from that of other ciliates. However, there are some studies where this has been done. Table 9.4 summarizes twenty such studies. We included only those studies where tintinnids from natural assemblages were used. Each study used one of three methods to evaluate tintinnid impact: (1) incubations of natural assemblages in which tintinnids dominated the microzooplankton assemblage or were isolated from the assemblage by micropipette; (2) tracer studies in which uptake of surrogates (starch grains, yeast cells) or labeled foods (fluorescent stains or radioisotopes) was quantified; or (3) food vacuole contents and turnover rates were measured.

Tintinnid clearance rates in these studies ranged from less than 1 to about 200 μl per tintinnid per hour, and ingestion from less than 1 to 15 phytoplankton cells per tintinnid per hour. The overall impact on phytoplankton of consumption by tintinnids was correspondingly variable. Just the tintinnid component of microzooplankton assemblages could at times graze up to 70% of primary productivity (see, for example, Verity 1986; Pitta et al. 2001; Karayanni et al. 2005). In many studies, however, tintinnid grazing was minimal. This variability is likely due to the temporal variabilty of tintinnid populations and to species-specific interactions. For example, Kamiyama & Matsuyama (2005) and Kamiyama et al. (2001) found that overall grazing mortality of the bloom-forming

Table 9.4 Grazing impact of tintinnids on natural particle assemblages. GR, grazing rate; IR, ingestion rate; CR, clearance rate; PSS, phytoplankton standing stock; PP, primary production. Only studies that specifically measured tintinnid feeding are included. References: 1, Capriulo & Carpenter 1980; 2, Capriulo 1982; 3, Rassoulzadegan & Etiennne 1981; 4, Verity 1986; 5, Spittler 1973; 6, Heinbokel 1978; 7, Cleven 1996; 8, Kamiyama 2000; 9, Kamiyama et al. 2001; 10, Karayanni et al. 2005; 11, Stoecker et al. 2000; 12, Watras et al. 1985; 13, Neuer & Cowles 1995; 14, Kopylov & Tumantseva 1987; 15, Pitta et al. 2001.

Area (reference)	Type of observation	Feeding rate	Grazing impact
Long Island Sound (1)	Incubation, natural food	CR = 0.45–1.05 μl $cell^{-1} h^{-1}$	15-41% PSS d^{-1}
Long Island Sound (2)	Incubation, natural food	CR = 2–65 μl $cell^{-1} h^{-1}$	—
Mediterranean coast (3)	Incubation, natural food	IR = $4.6 \times 10^4 - 4 \times 10^5$ μm^3 $cell^{-1} d^{-1}$	5.1% PSS d^{-1}
Narragansett Bay (4)	Incubation, natural food	GR = 0–2.2 pg chl-*a* $L^{-1} d^{-1}$ (<10 μm)	62% <10 μm PP
Baltic Sea (5)	Tracers: starch grains, yeast cells	CR = 0.5–8.5 μl $cell^{-1} h^{-1}$	—
California Bight (6)	Tracers: starch grains	CR = 10 μl $cell^{-1} h^{-1}$	—
Lake Constance (7)	Fluorescently labeled prey	CR = 1.3–6.3 μl $cell^{-1} h^{-1}$	0.6-9.7% of heterotrophic nanoflagellate production
Hiroshima Bay (8)	Fluorescently labeled prey	IR = 1.08 ± 0.16 cells $cell^{-1} h^{-1}$	—
Hiroshima Bay (9)	Fluorescently labeled prey	IR = 0.16–15 cells $cell^{-1} h^{-1}$	Species-specific
Northeast Atlantic (10)	Fluorescently labeled prey	CR = 0.5–11.7 μl $cell^{-1} h^{-1}$	up to 69% PP
Chesapeake Bay (11)	Fluorescently labeled prey	CR = 0.2–1 μl $cell^{-1} h^{-1}$	—
Cape Cod (12)	Radio-labeled prey	CR = 3.7–45 μl $cell^{-1} h^{-1}$	—
Oregon Upwelling (13)	Radio-labeled prey	CR = 0–7.2 μl $cell^{-1} h^{-1}$	
Peru upwelling (14)	Vacuole contents		<4% PP
Mediterranean Sea (15)	Vacuole contents	IR = 0.61 algae $cell^{-1} h^{-1}$ 0.41 *Synechococcus* $cell^{-1} h^{-1}$	Tintinnids IR = 5 × aloricate ciliates

dinoflagellate *Heterocapsa circularisquama* was low when the tintinnid *Eutintinnus tubulosus* was abundant but was maximal when *Favella* spp. tintinnids were abundant. At present, there is not enough information to evaluate whether the grazing impact of tintinnids is proportional to their abundance or biomass, for which a great deal more data are available.

Tintinnids as prey for other zooplankton

This topic is covered more thoroughly in the chapter on predation (Chapter 5), so we will only touch on it here in the context of overall microzooplankton mortality. Most studies of natural populations in the field do not separately quantify the grazing mortality of tintinnids, instead lumping them in with other ciliates or all microzooplankton. For example, of 19 studies reviewed by Calbet & Saiz (2005) in which copepod grazing on microzooplankton was quantified, none reported separate clearance rates on tintinnids, and only one separately quantified them at all (Bollens & Penry 2003). Thus the amount of data on mortality in natural populations of tintinnids is very small compared with the many studies on laboratory cultures.

Qualitatively, early studies of gut contents and fecal pellets showed that tintinnids could be found among phytoplankton and other ingested foods in copepods and larval fish (Capriulo & Ninivaggi 1982; Govoni, et al. 1983; Turner 1984). This led to efforts to document feeding rates and selectivity, using experimental methods. Robertson (1983), for example, found clearance rates of 95–289 ml per copepod per day when the common estuarine copepod *Acartia tonsa* fed on the tintinnids *Favella panamensis* or *Tintinnopsis tubulosa*. Stoecker & Egloff (1987) found a similar result, with the copepod clearing up to 200 ml per copepod per day for large tintinnids; they further established that clearance was higher on tintinnids than it was on phytoplankton prey. Selective ingestion of tintinnids by copepods when alternative prey are offered has also been shown (Stoecker & Sanders 1985). Larval fish also preferentially ingested tintinnids when presented with different microzooplankton prey (Stoecker & Govoni 1984). Gifford & Dagg (1988) studied the feeding of natural populations of the copepod *Acartia tonsa* on natural microplankton assemblages. They found clearance rates comparable to those from laboratory studies (approximately 3–7 ml per copepod per hour), and found strong preferences for microzooplankton, including tintinnids, compared with phytoplankton. Nakagawa et al. (2004) found that the euphausiid *Euphausia pacifica* ingested tintinnids with higher clearance rates than on phytoplankton and suggested that poor escape abilities of tintinnids led them to be caught more easily than aloricate ciliates.

Olson et al. (2006) performed incubation experiments to evaluate grazing pressure on a variety of microplankton, including especially the toxin-producing diatoms *Pseudonitzschia* spp. They found strong preference for feeding on microzooplankton, especially aloricate ciliates and heterotrophic dinoflagellates. Although tintinnids were not abundant in most of their experiments ($\leq 10^3 l^{-1}$), when they were present clearance of tintinnids by large and medium-sized copepods ranged from undetectable to 3.3 ml per copepod per hour. Interestingly, electivity indices usually showed selection *against* tintinnids (four out of five cases) in their experiments, even when other microzooplankton were preferentially ingested over phytoplankton. Bouley & Kimmerer (2006) found a similar pattern in the cyclopoid copepod *Limnoithona tetraspina*, which readily ingested aloricate ciliates, but not tintinnids. Bollens & Penry (2003) found instances of both positive and negative selection for tintinnids in experiments with natural populations of copepods in San Francisco Bay, CA, USA.

Other microzooplankton can be predators of tintinnids as well. Smalley et al. (1999) found that tintinnids constituted up to 10% of the food vacuole contents of the mixotrophic dinoflagellate *Ceratium furca* and that the dinoflagellate appeared to ingest tintinnids without their loricae. Robertson (1983) found that the large tintinnid *Favella panamensis* could clear the smaller *Tintinnopsis tubulosa* at rates up to 130 µl per ciliate per hour.

In situ growth

There have been many laboratory studies of growth in tintinnid species that have been brought into culture (reviewed in Chapter 4). Field studies with natural populations under *in situ* conditions of food, temperature, etc. are not as plentiful. Some studies have reported a few measurements in conjunction with other experiments (e.g., Dolan et al. (2000) measured growth of tintinnids during two dilution experiments). Others measured growth of all ciliates, but either found tintinnid concentrations to be too low, or failed to report tintinnid growth rates separately (Carrick et al. 1992; Leakey et al. 1994a; Nielsen & Kiørboe 1994). We found several studies that measured tintinnid growth and met the following criteria: they worked on natural tintinnid assemblages under natural conditions; they eliminated or accounted for predators; and they reported growth rates of individual species. Their results are summarized in Fig. 9.4.

Stoecker et al. (1983) pioneered the study of tintinnid growth *in situ*, using plexiglass "cages" that isolated natural populations of tintinnids but allowed for the exchange of water and dissolved nutrients during incubations for 24 h. Of 16 total measurements across three different tintinnid species (*Favella* sp., *Eutintinnus pectinis*, and *Tintinnopsis kofoidi*), she found negative growth rates four times. This may have been due in

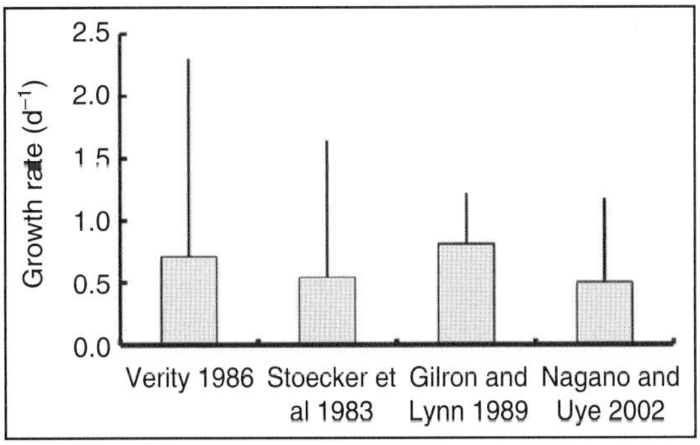

Fig. 9.4 Four studies where specific growth rates of individual tintinnid species were measured on natural popuations under *in situ* conditions. Means (bars) and maxima (vertical lines) are shown. Rates were converted from doublings reported in Verity (1986), and negative growth rates observed by Stoecker et al. (1983) were not included in the mean for consistency with Nagano & Uye (2002), who reported only significant positive growth rates.

part to parasitism by the dinoflagellate *Duboscquella* sp., which could not be excluded from the cages. In a comparison with food-replete laboratory cultures, she found *in situ* growth to be much lower. Verity (1986), Gilron & Lynn (1989b), and Nagano & Uye (2002) reported somewhat larger data sets. Verity measured growth rates for 22 different species, including 240 total measurements over two years in Narragansett Bay, an estuary in the Northwest Atlantic. His overall average growth rate (an average of his reported annual averages by species, converted from his calculation of doubling rates) was 0.71 per day, with a range of 0–2.3. Nagano & Uye (2002) found somewhat lower rates (mean 0.50 per day, for 107 measurements), and Gilron & Lynn (1989b), working at a tropical North Atlantic site, found an average growth rate of 0.81 per day for 41 measurements. All four studies suggested limitation of tintinnid growth either by food quantity or quality, as field growth rates were often less than maximum rates obtained for the same species in culture. In addition, Verity (1986) suggested that the episodic presence of high concentrations of a toxic raphidophyte were associated with low growth in tintinnids, indicating a kind of species-specific interaction that is independent of food concentration.

The magnitude of tintinnid growth rates in nature, and the fact that they are generally less than those under unlimited food in the laboratory, indicate that they have a large growth *potential*, which may be important in the termination of phytoplankton blooms. In a review of laboratory studies, Banse (1982) suggested that ciliates in general have the ability to outgrow both diatoms and dinoflagellates, their potential prey organisms. Ciliates probably rarely reach their maximum growth rates (about one to three divisions per day, depending on size), he argued, because they are limited by food availability. Thus, their rapid growth rates represent a potential to respond quickly to increases in food concentration. Such situations occur when phytoplankton bloom, especially in coastal waters, and some studies have found that tintinnids appear to be most abundant during or immediately after diatom blooms (see, for example, Gomez & Gorsky 2003). Thus, the high variability observed in time series of tintinnid abundance may be a reflection of their ability to respond rapidly to changes in abundances of their phytoplankton food. This, in turn, would give them some ability to control those blooms, including those of harmful algae. So tintinnids may be viewed as part of a "rapid response team", capable of much faster growth than other microzooplankton and hence potentially capable of preventing or terminating microalgal blooms.

9.3 CHARACTERISTICS THAT SET TINTINNIDS APART FROM OTHER MICROZOOPLANKTON

The lorica

One unique tintinnid property that seems to affect its susceptibility to predation relative to other microzooplankton groups is the possession of a lorica. The construction, composition, and function of this structure are covered in greater detail in Chapter 2. We only mention it here in passing to contrast tintinnids with other microzooplankton groups, including the so-called "naked" ciliates. The lorica has been proposed to function as a defense against predation. Capriulo et al. (1982) suggested that its purpose was to protect the cell from predators while also facilitating escape by rapid sinking. It has also been suggested to function as a buoyancy aid (Corliss 1979) or a "sea anchor", facilitating filter feeding (Fenchel 1986; Jonsson et al. 2004). Broglio et al. (2001) established experimentally that the tintinnid *Metacylis* sp. lacked the ability to escape predation by the rapid "jumping" behavior that is found in the closely related Strobilidiidae and other planktonic ciliates (e.g., the haptorid *Mesodinium rubrum*). The lack of a jumping behavior may relate to the possession of a lorica, which thus protects the tintinnid from being eaten while at the same time limiting its ability to escape. Margalef (1982) suggested that the need to build a lorica diverts energy from growth and speculated that tintinnids thus represented "*K*-adapted" ciliates (slower growth but lower mortality) in contrast to the "*r*-adapted" aloricate forms (rapid growth when resources are abundant). As discussed above, the laboratory data indicate that tintinnids show rapid growth when not food-limited. As of now, there are not enough experimental data to test Margalef's idea that, in general, tintinnids grow more slowly than aloricate forms.

Obligate planktonic habitat

Although they are close phylogenetic cousins to the oligotrichs, the choreotrichs (including the tintinnids) clearly have some distinctive ecological properties. For example, although they may occasionally be found in sediment samples, tintinnids are very rare in interstitial benthic habitats and probably only occur there as accidentals (Carey 1992). In his extensive study, Fenchel (1969) reported that members of the oligotrich genus *Strombidium*, which has many planktonic representatives, were almost always found as important herbivores in the surface layers of sublittoral sands, but he did not report significant presence of tintinnids in any of the benthic habitats he surveyed. Possibly, the possession of a lorica restricts the movement of tintinnids through interstices of fine sands and thus tintinnids are limited to the pelagic habitat only. Other groups commonly found in the microzooplankton, including dinoflagellates, rotifers, and copepod nauplii, all have non-planktonic representatives.

Strict heterotrophy?

Since the mid-1980s, many cases of chloroplast symbiosis ("kleptoplasty") have been documented among oligotrich ciliates, a sister clade to the choreotrichs, the subclass that includes tintinnids. Several oligotrichs have been shown to be obligate mixotrophs, ingesting microalgae and digesting all but the chloroplasts, which can be used to provide a carbon subsidy by photosynthesis (Stoecker et al. 2009). Tintinnids, however, have been assumed to be strictly heterotrophic, and this has been affirmed by the observations of many tintinnidologists. There have been some published observations of epibiotic phototrophs living inside tintinnid loricae, but not within the cytoplasm. Foster et al. (2006) and Carpenter & Foster (2003), for example, observed epibionts, in this case unicellular cyanobacteria, within the loricae of oceanic tintinnids. Presumably, there is some mutualistic benefit to this association. Mary Silver and Robin Lothrop Pinto (personal communication) have observed fluorescent intracytoplasmic inclusions in wild populations of oceanic tintinnids (Fig. and Plate 9.5), but there have been no experimental studies to establish whether these are chloroplasts, endocytic algae, or parasites.

The significance of possible symbiotic associations between oceanic tintinnids and microalgae is not clear at this time. It is an established fact from many observations that mixotrophy is widespread in non-tintinnid planktonic ciliates and absent from most tintinnids. This suggests that tintinnids may be even more important in grazing and remineralization than their abundance would suggest because they do not receive any growth subsidy by photosynthesis.

9.4 KEY POINTS

1. Tintinnids are but one part of a taxonomicaly diverse functional category, the microzooplankton. This category of planktonic heterotrophs is responsible for grazing most of the algal production in both open water and coastal systems as well as most of the nutrient regeneration and comprises an important food resource for higher trophic levels.

2. Tintinnids are usually a relatively minor component of the microzooplankton community, averaging about 7% of total ciliate concentrations, though at times this can be much higher. The abundance of tintinnids is positively but weakly related to total ciliate abundance.

3. There are few estimates of grazing impact of discrete components of the microzooplankton. However, tintinnids have been reported on occasion to be responsible for grazing most of the primary production.

4. Growth rates estimated for tintinnids in natural populations, although variable, suggest average values

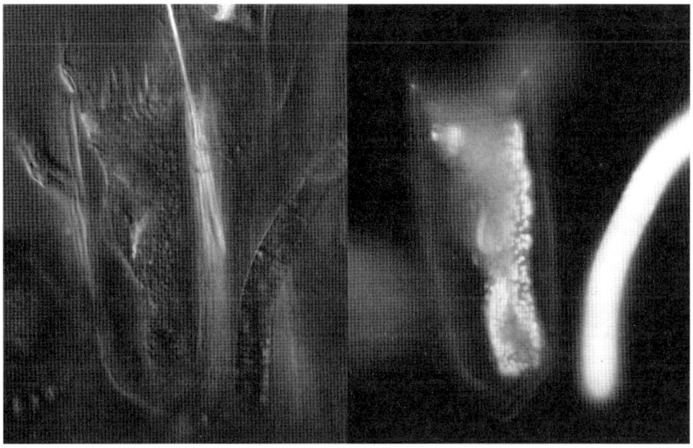

Fig. 9.5 Differential interference contrast (left) and fluorescence (right) images of a large tintinnid collected 40 km north of Hawaii in 1983 by Mary Silver of the Unversity of California at Santa Cruz. Bright red autofluorescent bodies (white in the right panel), indicative of chlorophyll, were seen inside the tintinnid cell, and their distribution within the cytoplasm indicates that they are not food vacuole contents. The shape of the lorica suggests that this is a member of the genus *Favella* or *Metacylis*. For scale, the filament to the right of the tintinnid is a trichome of the cyanobacterium *Trichodesmium* sp., about 10 μm in width. Image courtesy of Dr Mary Silver. For color version, see Plate 9.5.

of generation times of 1–2 days in different coastal systems.

ACKNOWLEDGMENTS

We thank Mary Silver and Robin Lothrop Pinto for the images and information about tintinnids with possible chloroplast symbioses, and we thank fellow chapter authors for helpful comments. Students, post-docs, and research assistants too numerous to list contributed to the Long Island Sound microzooplankton counts. G.B.M. acknowledges the late Dr Gerry Capriulo for transmission of a small portion of his boundless enthusiasm for tintinnids to a curious new graduate student.

Chapter 10

DIVERSITY AND DISTRIBUTIONS OF TINTINNIDS

John R. Dolan[1] *and Richard W. Pierce*[2]

[1]Laboratoire d'Océanographie de Villefranche-sur-Mer, CNRS Université Paris VI, Villefranche-sur-Mer, France
[2]P.O. Box 132, North Attleboro, MA, USA

The Biology and Ecology of Tintinnid Ciliates: Models for Marine Plankton, First Edition. Edited by John R. Dolan, David J.S. Montagnes, Sabine Agatha, D. Wayne Coats, and Diane K. Stoecker.

10.1 INTRODUCTION

The diversity and distributions of planktonic organisms are a subject that has intrigued oceanographers for well over a century. Early observations emphasized the differences in species assemblages that characterize different areas, perhaps corresponding to different water masses. For example, by the early 1900s the distinct character of the polar compared with temperate assemblages in both hemispheres was recognized as well as the similarity of the polar assemblages. These findings led to the suggestion that perhaps undercurrents joined the polar seas (e.g. Cleve 1900). Seasonal changes in the species found in coastal systems were also well known. Seasonality, first attributed to changes in changes in currents and water masses, was shown to likely result from shifts in local conditions of light, temperature, and nutrients (e.g. Brandt 1905). Today, the distributional patterns of planktonic organisms remain an area of very active research because, even though some patterns can be documented from space, such as those of phytoplankton, identifying the factors controlling their dynamics remains elusive (see, for example, Kudela 2010).

This chapter explores the biogeography and patterns of species diversity of tintinnids. The basic geographic assemblages and patterns are defined. Seasonal changes of tintinnid assemblages of coastal systems are documented and the relationships between abundance and distribution both temporal and spatial are examined. When possible, the patterns of tintinnids are compared with those of other groups of planktonic organisms.

10.2 BIOGEOGRAPHY

Global biogeography

The first attempt to quantify large-scale ciliate distributions was by Entz (1884), who tabulated presence/absence of each species of the known catalogue of marine planktonic species, calculating a percentage of total species for each sea. Not surprisingly, he found that Northern European waters, from which most species had been described, appeared the most species-rich. The distributions presented here are also based on simple species records but of a considerably larger database than that available to Entz. The tintinnid distributions are an updated version that of Pierce & Turner (1993), increased to include 302 references, 1800 sampling locations, and references up to the year 2011. This has yielded a fairly extensive geographic coverage (Fig. 10.1).

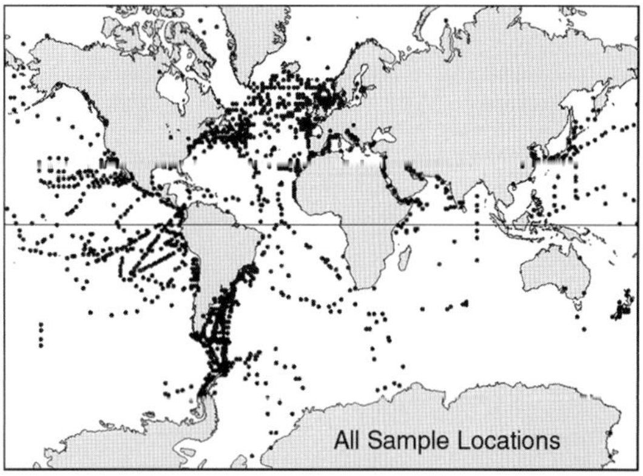

Fig. 10.1 The 1800 locations of tintinnid species records used to map distributions.

The distributions are based on literature reports of species occurrences. Data recorded were species names along with latitude and longitude of sample collection as reported, or determined from sample site names, or maps in the report. The references are indicated in the reference list with an asterisk. The database was used to plot occurrences on world maps for each genus that met a minimum occurrence criterion of containing a species reported in at least four publications by two different authors. Here, the term "species" in relation to distributional patterns refers to morpho-species.

The relationship between morpho-species and the generally accepted biological definition of "species" in ciliates is very far from clear as morphology, mating types, and genetics can be discordant (Hall & Katz 2011). However, although data are very limited at present, it does suggest that at least in terms of genetics, morpho-species of tintinnids, rather than grouping cryptic species, may more often be different phenotypes of the same genetic species. First, it appears that geographically distant populations of planktonic ciliates can be genetically identical (Agatha et al. 2004a; Katz et al. 2005). Thus there is *a priori* no reason to associate geographic separation in species of planktonic ciliates with genetic isolation. Secondly, although genetic studies of tintinnid are very sparse, identical sequences have been found for different "species" of *Tintinnopsis* (Li et al. 2009) *Petalotricha* and *Cyttarocylis* (Bachy et al. 2012) as well as *Favella* (Kim et al. 2010). Recently some

Table 10.1 Biogeographic distribution patters of common tintinnid genera; genera considered were those that included species reported in at least four publications by two different authors.

Cosmopolitan	Neritic	Warm water	Boreal	Austral
Acanthostomella	*Favella*	*Amplectella*	*Parafavella*	*Cymatocylis*
Amphorellopsis	*Helicostomella*	*Ascampbelliella*	*Ptychocylis*	*Laackmanniella*
Amphorides	*Leprotintinnus*	*Brandtiella*		
Codonella	*Metacylis*	*Canthariella*		
Codonellopsis	*Stenosemella*	*Climacocylis*		
Dadayiella	*Stylicauda*	*Codonaria*		
Dictyocysta	*Tintinnidium*	*Cyttarocylis*		
Eutintinnus	*Tintinnopsis*	*Daturella*		
Parundella		*Epicancella*		
Protorhabdonella		*Epiplocylis*		
Salpingacantha		*Epiplocyloides*		
Salpingella		*Petalotricha*		
Steenstrupiella		*Poroecus*		
		Proplectella		
		Rhabdonella		
		Rhabdonellopsis		
		Undella		
		Undellopsis		
		Xystonella		
		Xystonellopsis		

ambiguities in establishing species boundaries have been found (Santoferrara et al. 2012a; Xu et al. 2012), but no cryptic species have been discovered so far.

We largely follow the tintinnid genus distributions established by Pierce & Turner (1993): cosmopolitan, neritic, warm–temperate, boreal, and austral. These distributional patterns, or categories, have long been known to characterize many if not most groups of planktonic organisms (see, for example, Backus 1986), including well-studied protist groups such as foraminifera (see, for example, Bé & Tolderlund 1971). As Taylor (1987) stated for marine dinoflagellates, "The essence . . . is a modified latitudinal cosmopolitanism: the occurrence of the same (morpho-) species around the world within broad latitudinal limits, the boundaries of which approximate to particular upper-layer temperatures." Within the broad latitudinal bands, neritic or coastal forms can be distinguished from oceanic species among dinoflagellates (Taylor 1987) and the same large-scale patterns characterize diatoms as well (Guillard & Kilham 1977).

Cosmopolitan

We designate as "cosmopolitan" those genera containing species recorded from the Arctic, through the Tropics and into the Southern Ocean, and not restricted to nearshore areas (Figs 10.2 and Fig. 10.3). Most, but not all, of the genera are forms that have purely hyaline loricae. The exceptions are the genera *Acanthostomella*, *Dictyocysta*, *Codonella*, and *Codonellopsis*, all of which include "agglutinating species". However, the particles used by species of these genera are mainly remains of diatoms and coccolithophorids, which are also quite widespread in the world ocean.

Neritic

We define neritic genera as those containing species that are very largely restricted to nearshore waters (Fig. 10.4). These are the most familiar forms as fieldwork has usually focused on nearshore environments and most laboratory studies have used species from coastal waters. Neritic genera are a morphologically heterogeneous set including both genera formed of species with agglutinated loricae (*Tintinnopsis*, *Tintintinnidium*, *Stenosemella*, *Leprotintinnus*, *Stylicaudata*) as well as genera with hyaline loricae (*Metacylis*, *Helicostomella*, *Favella*). The size range of species is also very large including some of the largest tintinnids known (*Favella* spp.) and the smallest (*Tintinnopsis* spp.). The genera all occupy a nearly pole-to-pole latitudinal

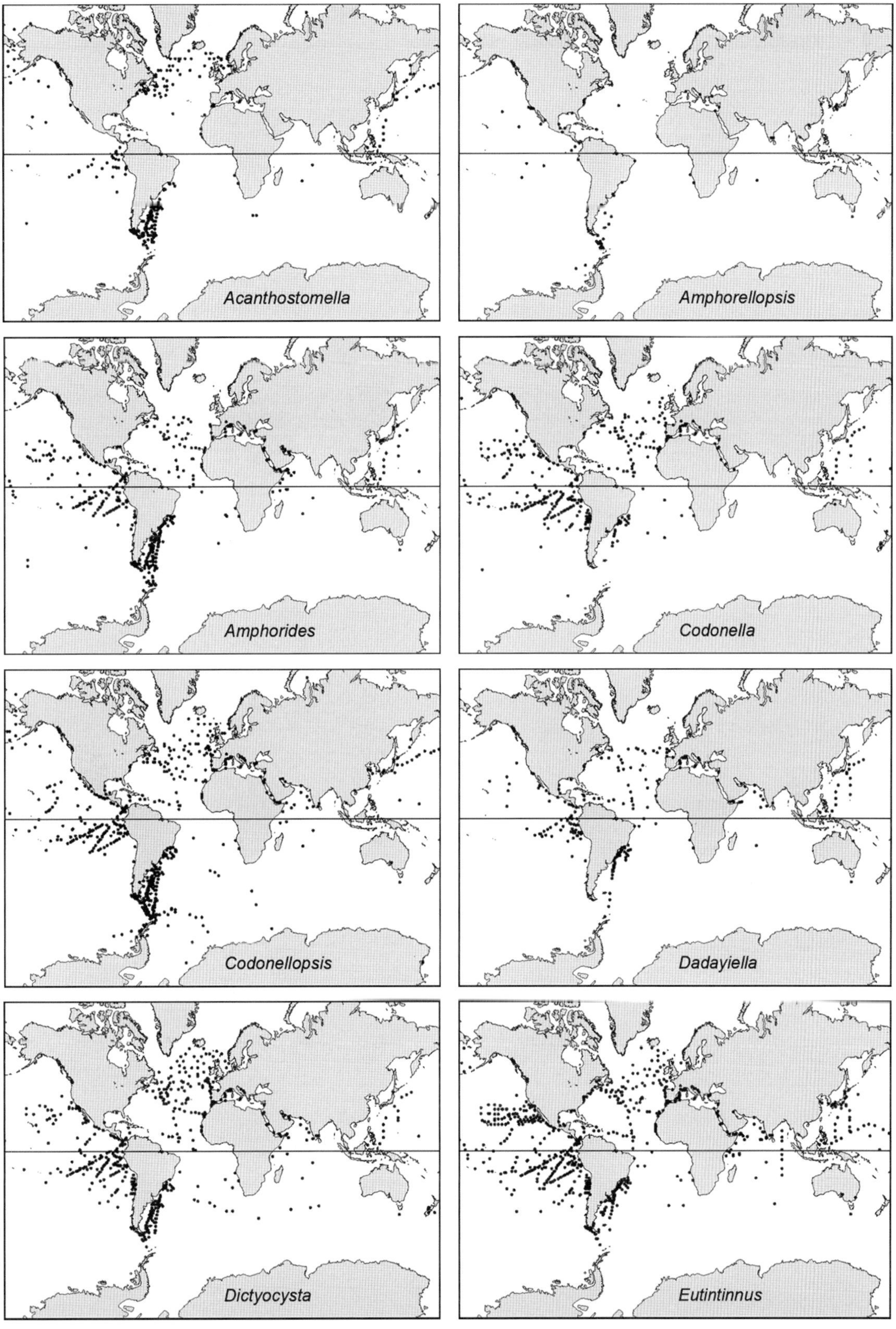

Fig. 10.2 The cosmopolitan distributions of the genera *Acanthostomella*, *Amphorellopsis*, *Amphorides*, *Codonella*, *Codonellopsis*, *Dadayiella*, *Dictyocysta*, and *Eutintinnus*. Species of these genera have been reported from neritic and oceanic areas from Arctic to Antarctic waters.

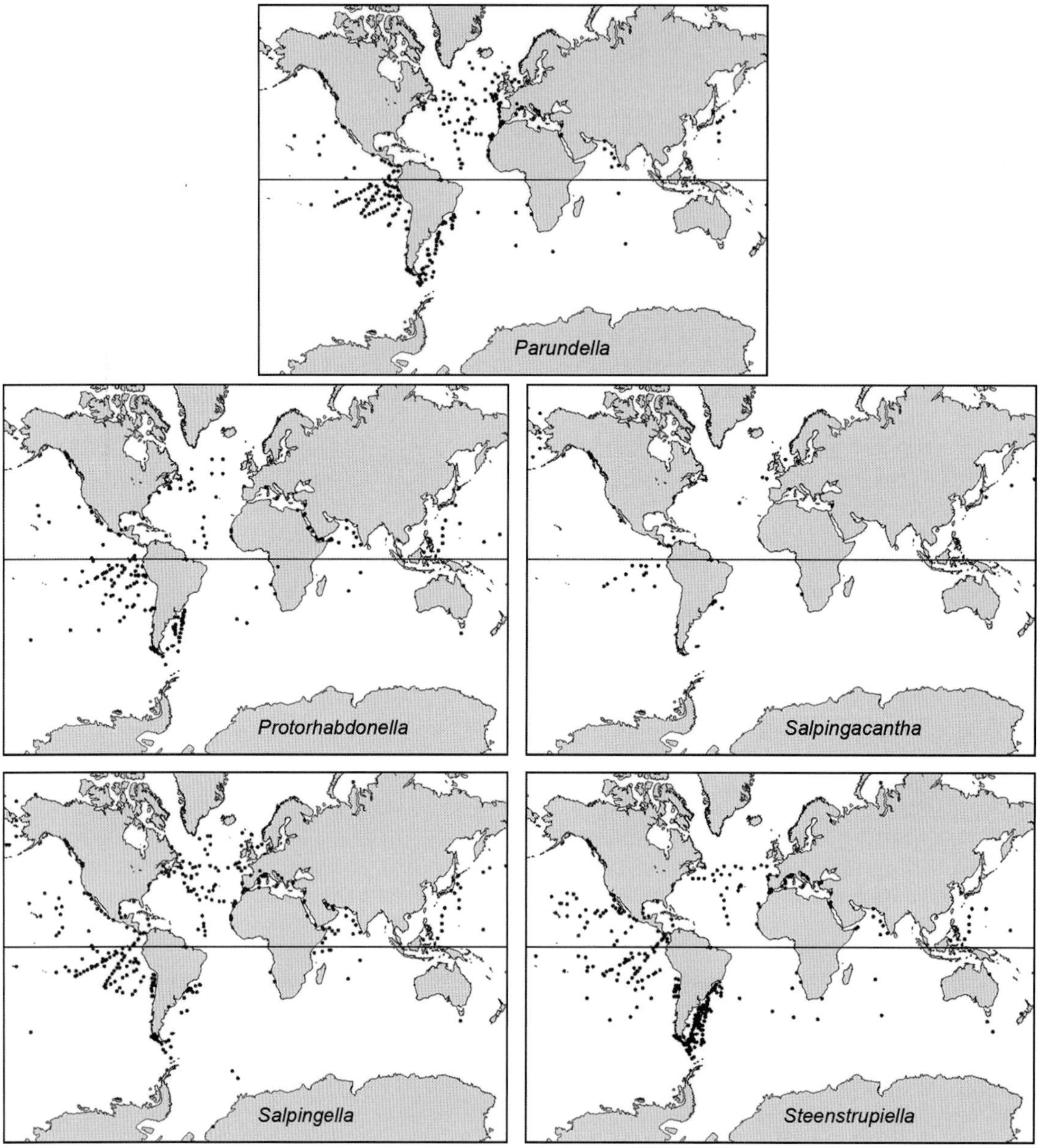

Fig. 10.3 The cosmopolitan distributions of the genera *Parundella*, *Protorhabdonella*, *Salpingacantha*, *Salpingella*, and *Steenstrupiella*. Species of these genera have been reported from neritic and oceanic areas from Arctic to Antarctic waters.

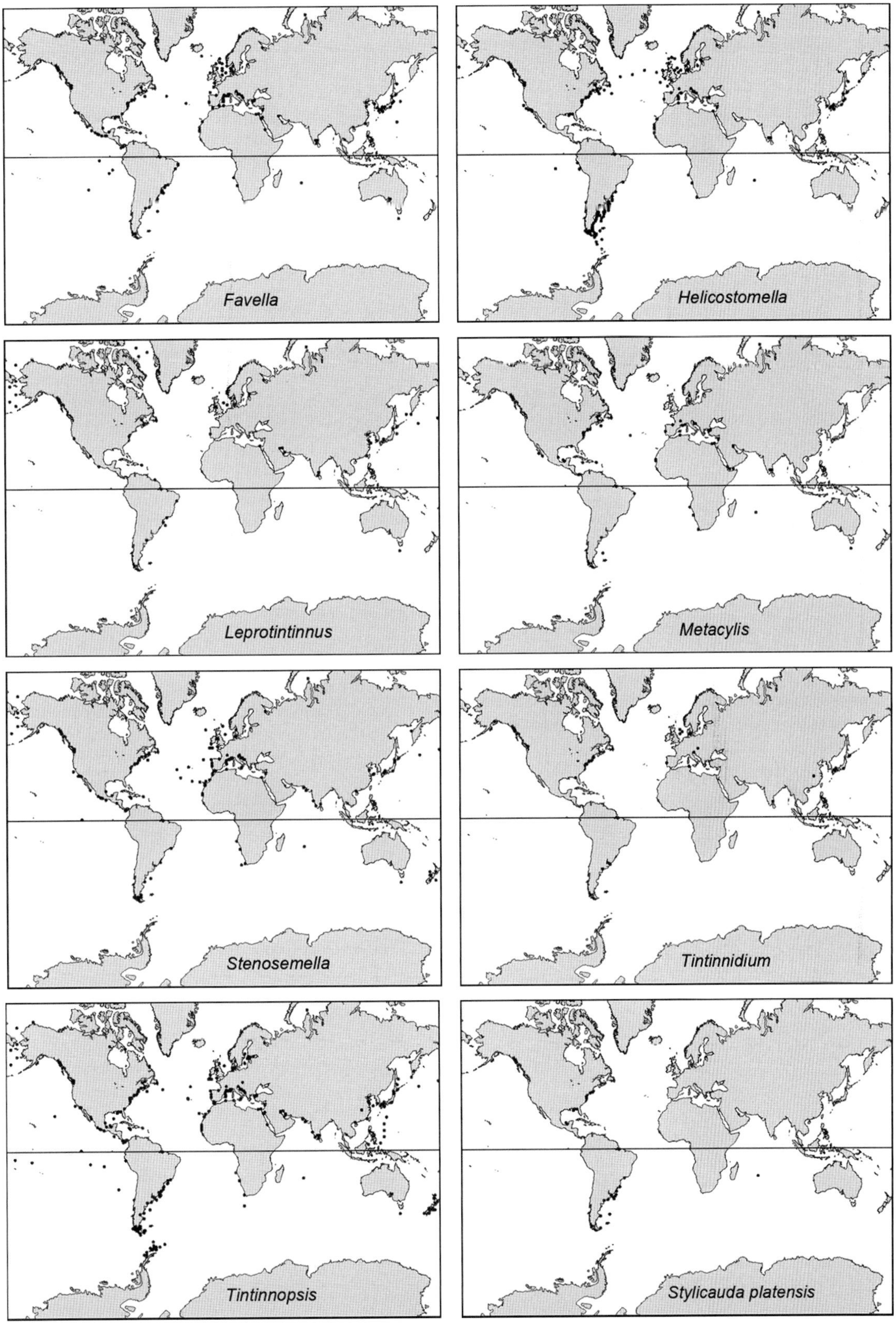

Fig. 10.4 Neritic distributions characterize species of the genera *Favella*, *Helicostomella*, *Leprotintinnus*, *Metacylis*, *Stenosemella*, *Tintinnidium*, *Tintinnopsis*, and *Stylicaudata*. Although some records are from open waters, most species records are from coastal waters.

distribution; thus the limit is to coastal waters, rather than a temperature zone.

The restriction of the genera *Tintinnopsis*, *Stenosemella*, and *Tintinnidium* to nearshore areas probably reflects a requirement for small mineral particles used to form loricae in all three genera. Waters with sufficient concentrations of small mineral particles are likely limited to areas that are relatively shallow and turbulent. The restriction of the forms with hyaline loricae (*Favella*, *Helicostomella*, and *Metacylis*) to coastal waters is difficult to explain. Cyst formation is known in species of *Favella*, *Helicostomella*, and *Leprotintinnus* (Reid & John 1978); it may be that a cyst stage is an important component in their life cycles, restricting distribution to shallow waters.

Warm–temperate

Warm–temperate genera are those observed in both coastal systems and open waters throughout the world ocean (Figs 10.5–10.7). This group differs from cosmopolitan only in that species of these genera are absent from sub-polar and polar waters. It is the largest biogeographic group in terms of the number of genera and species. Although their distributions are nearly cosmopolitan, many of the genera such as *Brandtiella*, *Xystonellopsis*, *Codonaria*, and *Climacocylis* are characteristic of tropical and sub-tropical waters and are rarely reported from temperate waters. None of the genera contains species with agglutinated loricae.

Boreal

The genera *Parafavella* and *Ptychocylis* (Fig. 10.8) have long been known to be restricted to Arctic and Subarctic waters. Notably, species of both *Parafavella* and *Ptychocylis* appear to produce loricae that are variable in overall length and shape but consistent in oral diameter (Burkovsky 1973; Davis 1978, 1981). Both genera comprise species with relatively large oral diameters of about 50 μm, suggesting exploitation of similar-sized prey.

Austral

The genera *Cymatocylis* and *Laackmanniella* (Fig. 10.9) are restricted to Antarctic and Subantarctic waters. In common with boreal genera, species of both *Cymatocylis* and *Laackmanniella* have variable lorica characteristics, especially length. Unlike the boreal genera, the two austral genera group species of distinctly different oral diameters. *Cymatocylis* species average about 100 μm whereas *Laackmanniella* species have considerably smaller oral diameters of 30–35 μm, indicating that likely different-sized prey are exploited by *Cymatocylis* and *Laackmanniella* spp.

Distributions of commonly reported species

Although the cosmopolitan genera contain the most common species, none of the individual species within the genera is cosmopolitan. For example, the most widespread species of *Acanthostomella*, *A. norvegica*, appears to be excluded from warm waters (Fig. 10.10). The distribution of *A. norvegica* is reminiscent of the "bi-polar" distribution known among a wide range of planktonic taxa, for example radiolaria and medusozoa (Stepanjants et al. 2006) as well in foraminifera (Bé & Tolderlund 1971; Darling et al. 2004) and dinoflagellates (Montresor et al. 2003). In the specific cases of the foraminfera *Neogloboquadrina pachyderma* and the dinoflagellate *Polarella glacialis*, there is evidence of genetic differentiation between the northern and southern populations (Darling et al. 2004; Montresor et al. 2003). This may be the case for *A. norvegica* but it has yet to be examined.

Globally, the four most widely distributed and most-often reported species are *Amphorides quadrilineata*, *Dadayiella ganymedes*, *Eutintinnus apertus*, and *Steenstrupiella steenstrupii*. These species have been reported both from neritic and open waters yet appear to be absent from far northern and southern seas (Fig. 10.10). Thus, there appears to be no species that is truly cosmopolitan. Furthermore, within cosmopolitan genera there are species that, though frequently reported, have restricted distributions and appear to be endemics. For example, among *Codonellopsis* species, *C. lagunela* is found only in the North Atlantic, *C. gaussi* only in the Southern Ocean, whereas *C. ecuadata* has been recorded only from the Indian and tropical Pacific Oceans (Fig. 10.11). These species of *Codonellopsis* are rather large and distinctive, so it is unlikely that they were frequently overlooked or mistaken for other species.

General considerations in global biogeography of tintinnids

It should be noted that there are, in principle, opposing views on the biogeography of protists in general. One posits no protist biogeography: "everything is

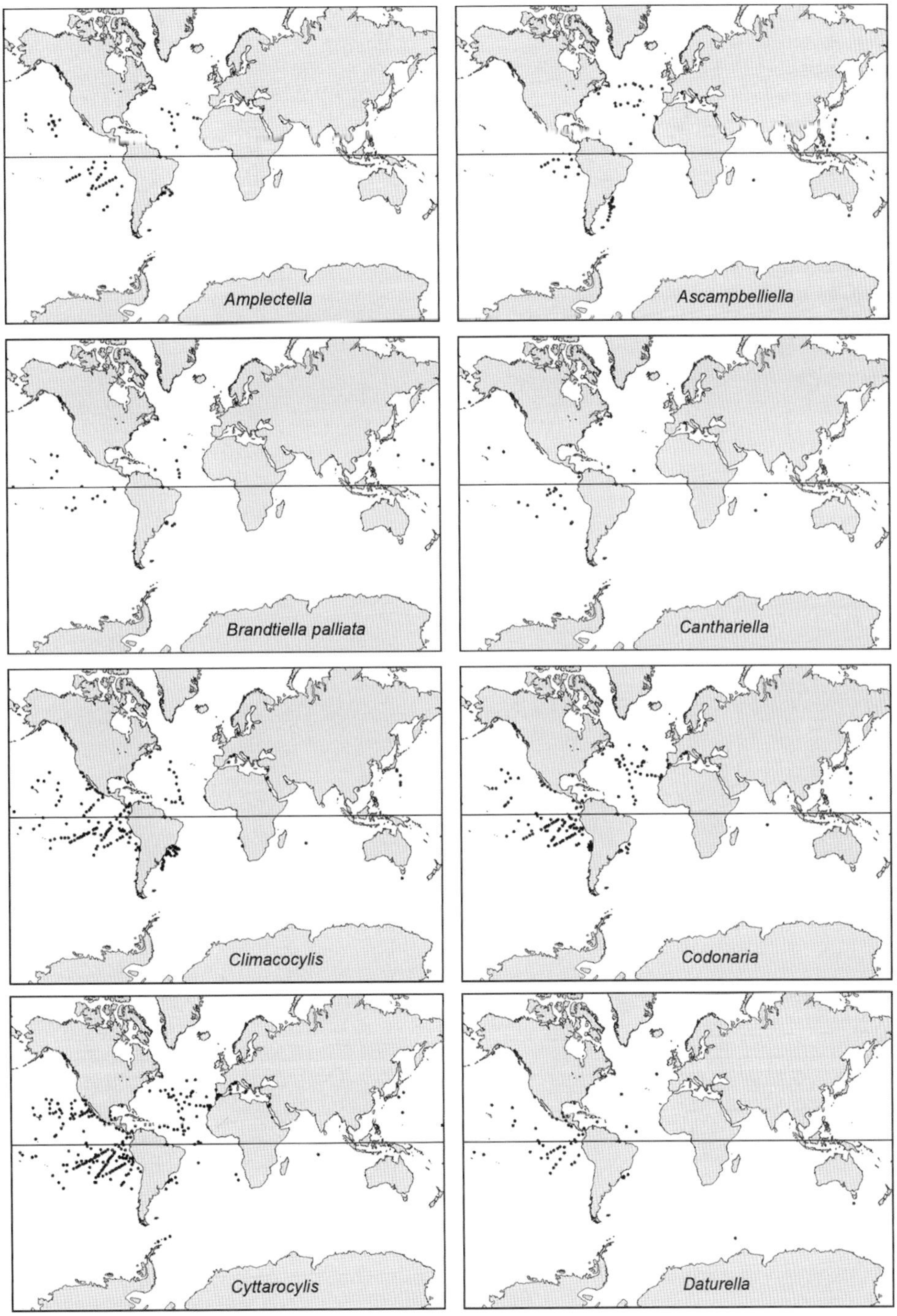

Fig. 10.5 Warm–temperate distributions are shown by *Amplectellopsis*, *Ascampbelliella*, *Brandtiella*, *Canthariella*, *Climacocylis*, *Codonaria*, *Cyttarocylis*, and *Daturella*.

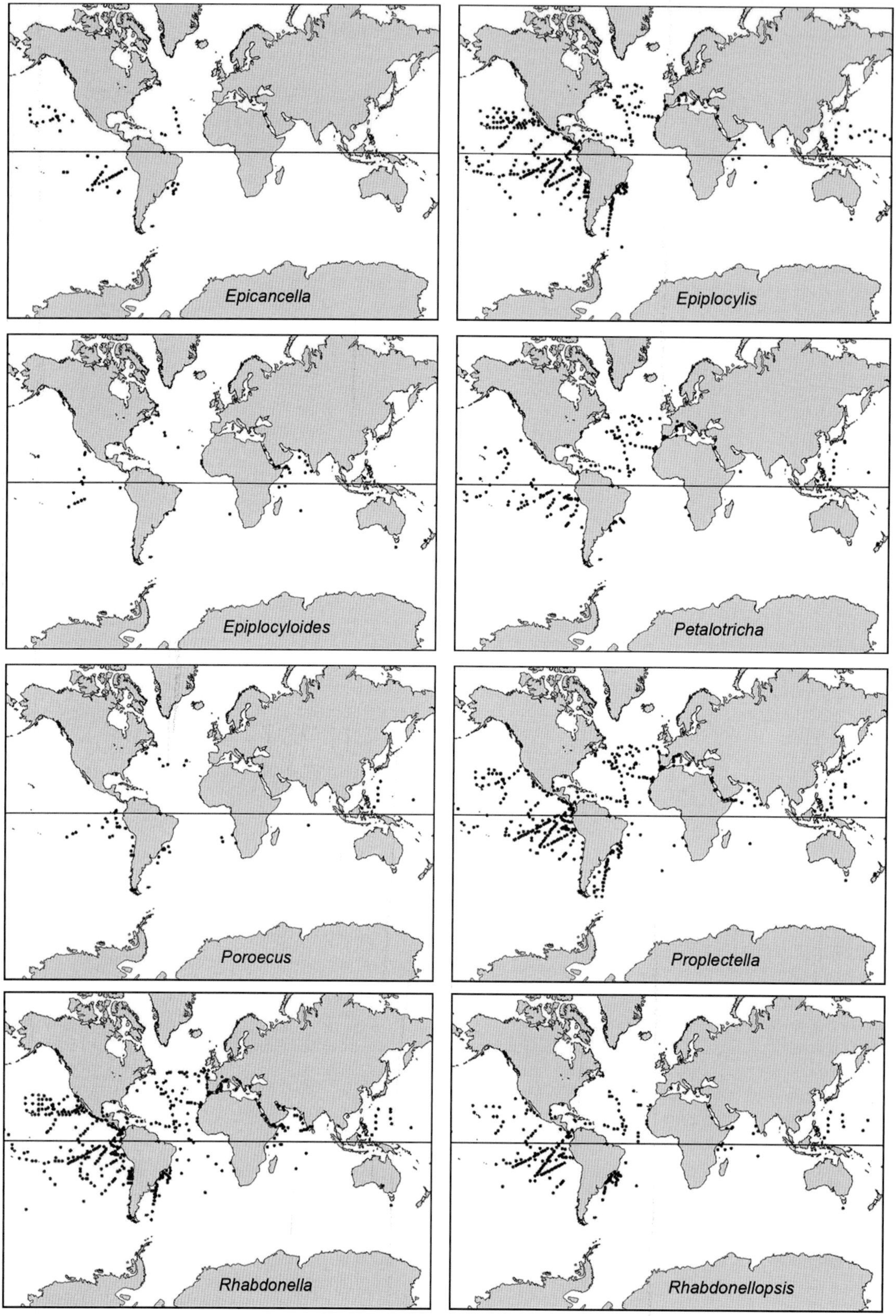

Fig. 10.6 Warm–temperate distributions are shown by *Epicancella, Epiplocylis, Epiplocyloides, Petalotricha, Poroecus, Proplectella, Rhabdonella,* and *Rhabdonellopsis*.

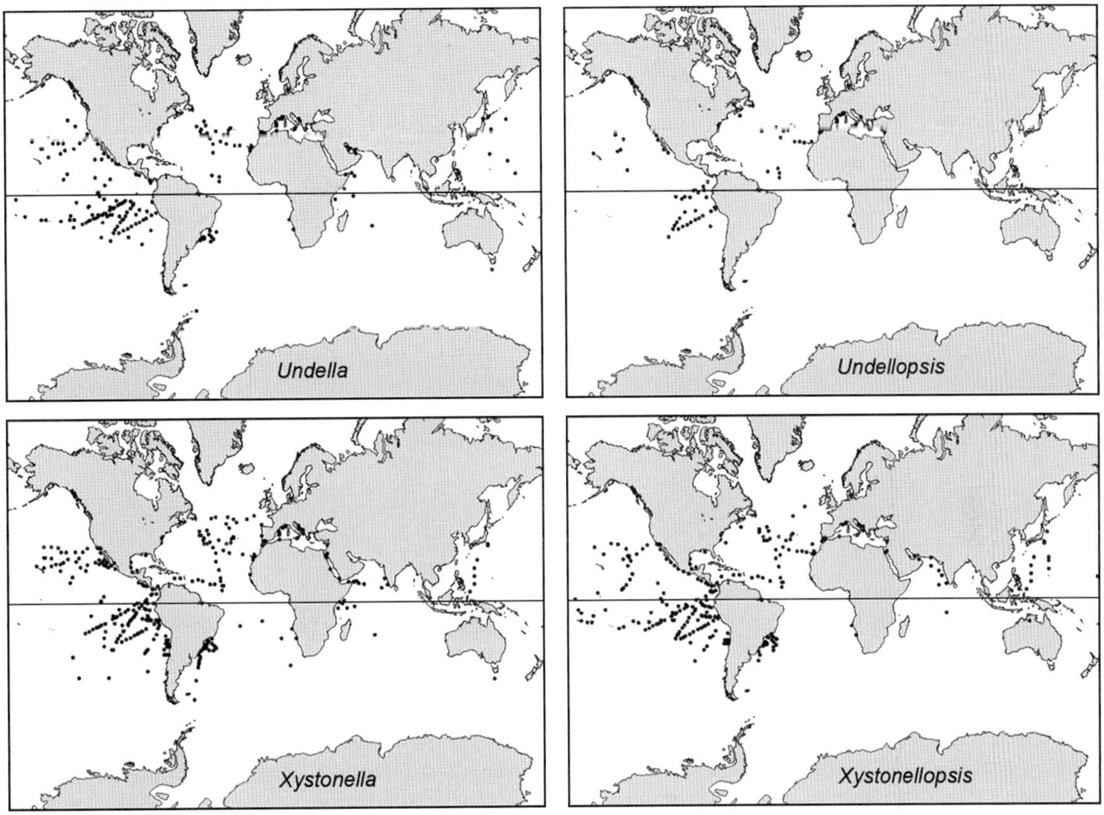

Fig. 10.7 Warm–temperate distributions are shown by *Undella*, *Undellopsis*, *Xystonella*, and *Xystonellopsis*.

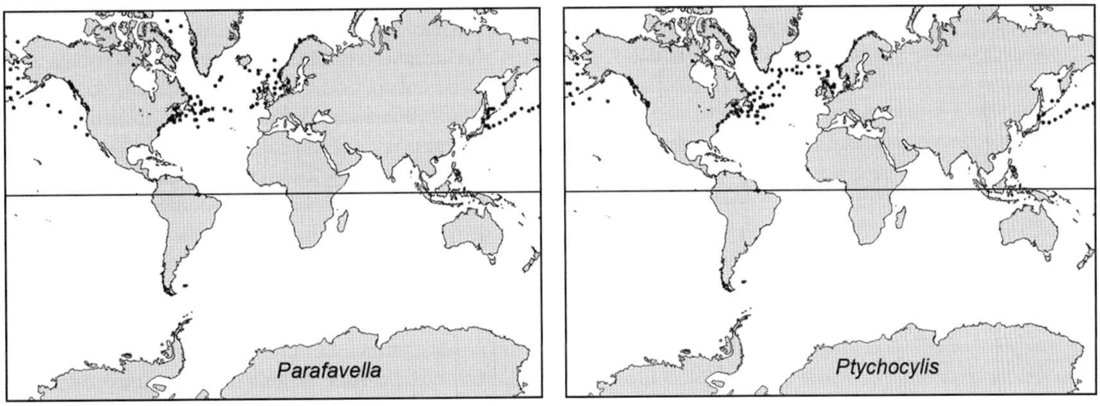

Fig. 10.8 Boreal. Boreal distributions, restricted to the far north but reported from both neritic and open waters, are shown by the genera *Parafavella* and *Ptychocylis*.

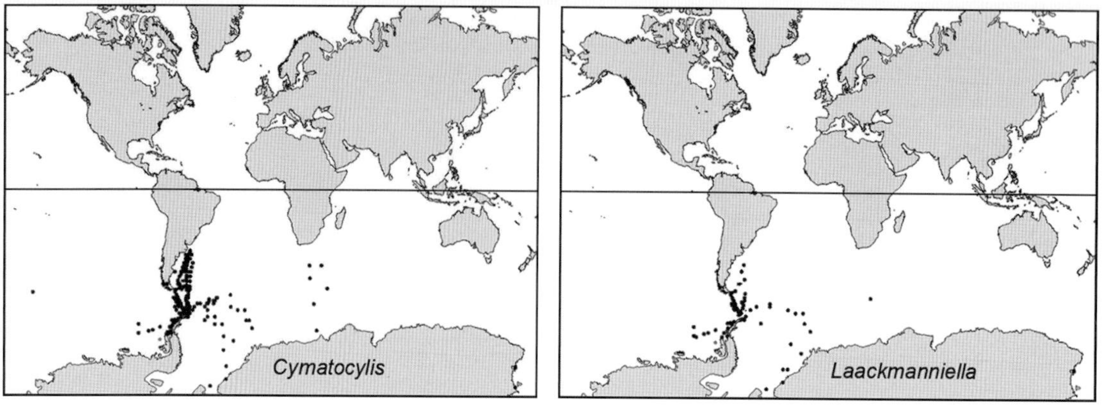

Fig. 10.9 Austral. Austral distributions, restricted to the far south but reported from both neritic and open waters, are shown by the genera *Cymatocylis* and *Laackmanniella*.

everywhere" because protists are effectively not dispersion-limited. The combination of large global population sizes, short generation times, and lack of geographic barriers is thought to prevent endemism among protists (see, for example, Fenchel & Findlay 2004). Indeed many protists form cysts, including tintinnids (see Chapter 7), which can be transported actively by other organisms or passively by water or wind, giving some forms a potentially worldwide distribution. The opposing view is that varying degrees of endemism exist among protists and that few, if any, "species" of protists are truly cosmopolitan (see, for example, Foissner et al. 2009). This view has in recent years become established as the dominant school of thought (see, for example, Fontaneto & Brodie 2011).

In examining the biogeographic patterns of tintinnids, there are some tintinnid "specificities" worth recalling. Many investigations have relied on the same taxonomic references (i.e., Kofoid & Campbell 1929, 1939) regardless of the area under study. This is understandable as region-specific works are often nonexistent. However, reliance on the same references, combined with a natural tendency to assign taxa established names, likely increases the chance that species found will be a subset of those illustrated in Kofoid & Campbell (1929). Thus, the distributions of some species may be artificially expanded.

The fact that lorica morphology is the basis of tintinnid identification, and lorica morphology can be variable, also complicates biogeographic studies. For example, tintinnids of diverse genera (*Favella*, *Epiplocylis*, *Xystonella*) can produce an alternative, "coxlielliform" lorica, which was only recognized as an abberant lorica form in the 1970s (Laval-Peuto 1977, 1981). Some lorica characteristics, typically length, can be quite variable. One morphometric study of the genus *Cymatocylis* (Williams et al. 1994) found that 45 lorica morphologies described as separate species could only be reliably sorted into five distinct morphotypes. Often, species identifications are made based on one or a few loricae, which may not be typical forms. Lorica-based species identifications are not always unambiguous. Partly for this reason, the distributions presented here are largely for genera.

A third consideration is the possibility of relatively recent changes in distribution due to phenomena such as climate change, ecosystem changes, construction of canals, or transport of tintinnids through ballast water (Pierce et al. 1997). This is very difficult to assess as many areas have not been well sampled and the lack of a report of a species cannot be taken as proof that the species was not already present. Range expansion of some species may have occurred, for example into the Black Sea, with major ecosystem shifts (Gavrilova & Dolan 2007).

The "invasive species" among tintinnids have received no attention as tintinnids, unlike dinoflagellates and diatoms, do not form harmful blooms nor are they likely to shift trophic food-web structure as known from certain ctenophores. However, at least one

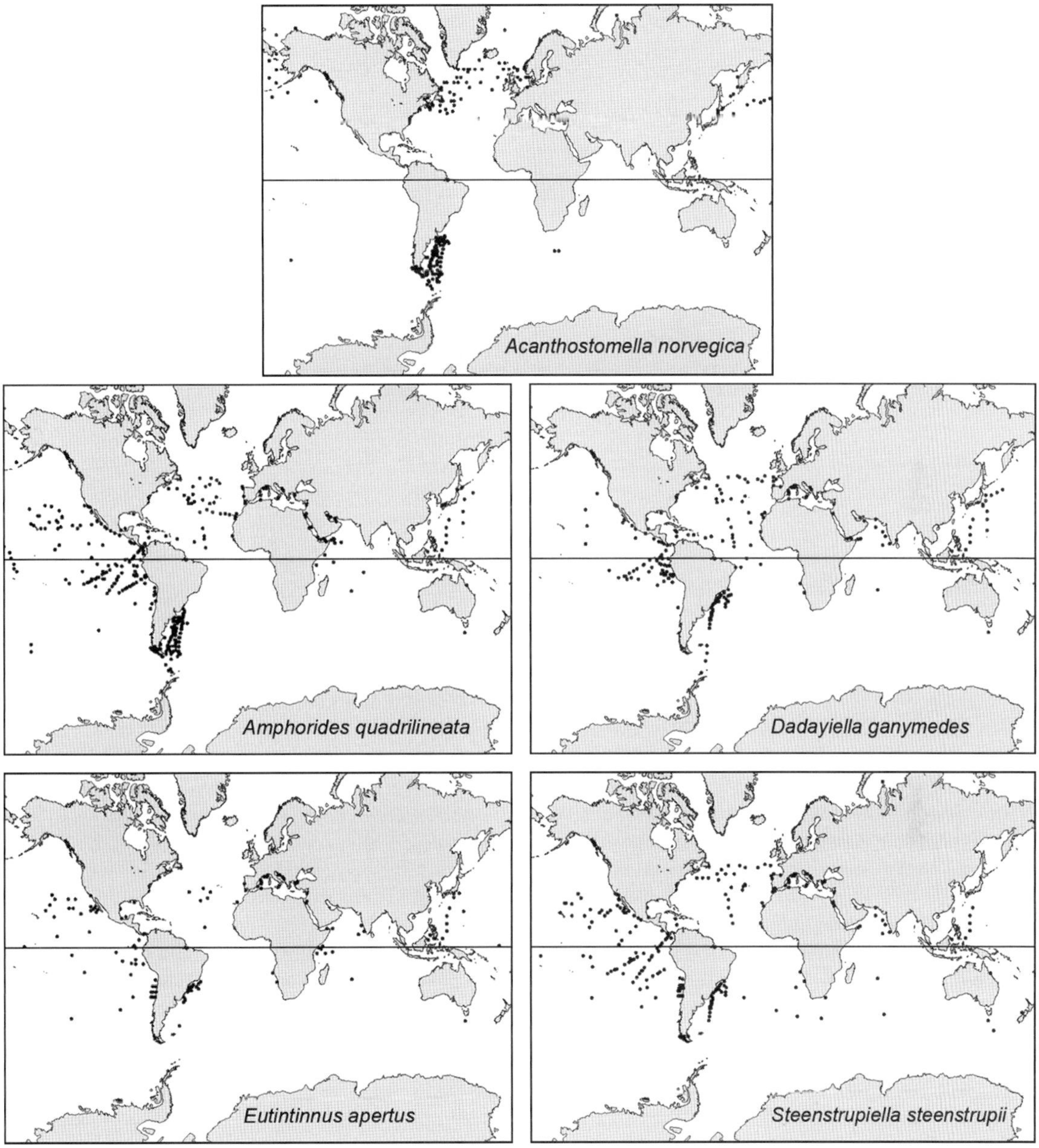

Fig. 10.10 The most common species of the cosmopolitan genera show restricted distributions. *Acanthostomella norvegica* is absent from warm waters. The four most commonly reported and widely distributed tintinnid species (*Amphorides quadrilineata, Dadayiella ganymedes, Eutintinnus apertus, Steenstrupiella steenstrupii*) are absent from far northern and southern seas.

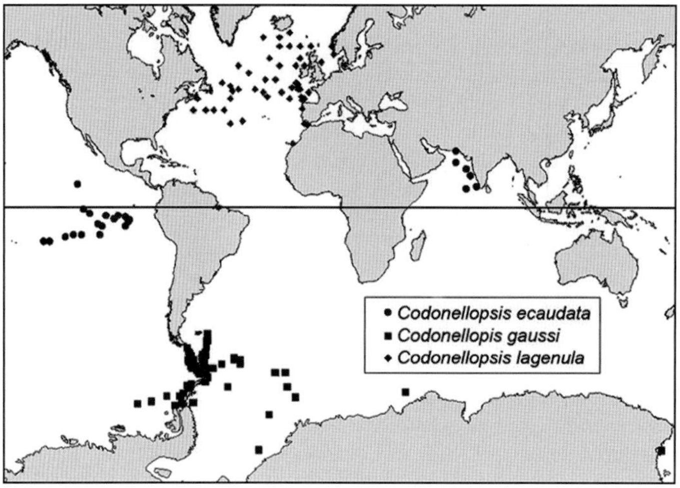

Fig. 10.11 Some frequently reported species show endemic distributions or highly restricted distributions. *Codonellopsis ecaudata* has been reported only from the Indian and central Pacific Oceans. *Codonellopsis gaussi* is found only in the Southern Ocean and *C. lagenula* in the North Atlantic Ocean.

tintinnid species may have expanded its range in recent years, perhaps through transport of ballast water. It is a species most commonly known as *Tintinnopsis corniger*, a large tintinnid with a distinctive, branching, hyaline aboral extension, unmistakable when seen. Such an unusual *Tintinnopsis*, with a large hyaline branching aboral horn, was first mentioned by Busch in 1925 and described as a form of *T. karajensis* from the Strait of Makassar in Indonesia. A very similar tintinnid, from Peter the Great Bay in the northern, subpolar sector of the Sea of Japan, was described as *Rhizodomus tagatzi* by Strelkow and Wirketis in 1950 without reference to Busch (1925). The Strelkow and Wirketis description, though, was in a Russian-language publication, so that when Hada (1964) found the species in Hiroshima Bay in the Inland Sea of Japan, he described it as a new species, *Tintinnopsis corniger*, again with no reference to Busch 1925. It is very unlikely that Hada would have missed the species in his earlier studies of tintinnids in Japanese waters (Hada 1932a, b, c, 1937). It is now routinely reported from Japanese waters (Nakashima & Kimoto 1987; Nomura et al. 1992; Kamiyama 1994a; Kamiyama & Tsujino 1996; Akizawa et al. 1998; Nakamachi & Iwasaki 1998; Godhantaraman & Uye 2003; Nakane et al. 2008) and Korean waters (Yoo et al. 1988; Yoo and Kim 1990).

Outside of Asia, *Tintinnopsis corniger* was first reported in the Gulf of Mexico and Caribbean Sea (Balech 1968; Lubel 1974) then in the Arabian Gulf (Sharaf 1995). Later the species was found in coastal waters of the western Mediterranean Sea where it was once again described as a new species, *Tintinnopsis nudicauda* (Paulmier 1997) without reference to Busch (1925), Strelkow & Wirketis (1950), or Hada (1964). It was found in the North Sea in Dutch coastal waters in 2004 (Verweij et al. 2005). In recent years, *T. corniger* has been reported from Mission Bay, California (Elliott & Kaufmann 2007). In the Mediterranean it has been found in a saline lake near the Straits of Messina in the Central Mediterranean (Saccà et al. 2008), in the Eastern Mediterranean in Lebanese coastal waters (Abbou-Abi Saab 2008), Damietta Harbor, Egypt (Dorgham et al. 2009), and most recently in the Sea of Marma (Durmus et al. 2011) and the Black Sea (Gavrilova 2010). Tintinnids have been found in the ballast water of cargo ships (Pierce et al. 1997, Chandrasekera & Fernando 2009), including *T. corniger* (David et al. 2007). The far-flung and disjunct distribution (Fig. 10.12) is very suggestive of trans-

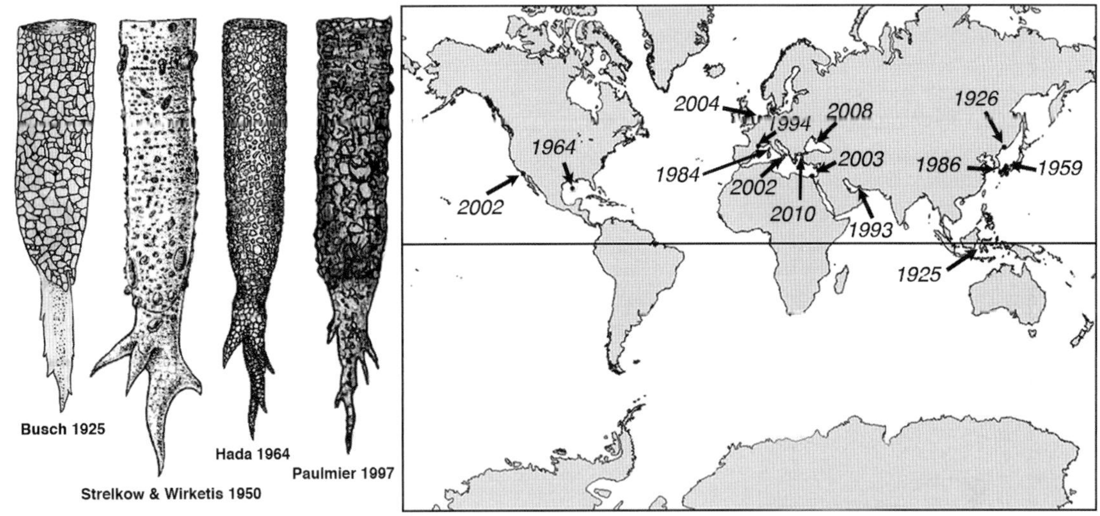

Fig. 10.12 *Tintinnopsis corniger* is a large, distinctive species which apparently has been described four times and reported from widely scattered locations suggestive of ballast-water transport or other artificial means of distribution. The left panel shows illustrations from published descriptions. The right panel shows locations from which the species has been reported; the years refer to when samples were collected: 1925 = Busch (1925), 1926 = Strelkow and Wirketis (1950), 1959 = Hada (1964), 1964 = Balech (1968), 1984 = Paulmier, G. (1997) Tintinnides (Ciliophora, Oligotrichida, Tintinnina) de l'Atlantique Boréal, de l'Océan Indien et quelques mers adjactentes: Méditerranée, Mer Caraibe, Mer Rouge. Inventaire et distribution. Observations basée sur les loricas. Rapport IFREMER DRV/RH/97-17, Brest, France. © IFREMER, 1994 = Lam Hoai et al. (1997), 2002 Mediterranean = Saccà et al. 2008), 2002 N. America = Elliot et al. (2007), 2003 = Dorgham et al. (2009), 2004 = Verweij et al. (2005), 2008 = Gavrilova (2010), 2010 = Durmus et al. (2011). We use *T. corniger* as it is the common name in the literature. Saccà et al. (2012) have re-described the species as *Rhizodomus tagatzi*.

port by ballast water (all the locations are near major shipping channels) or other human activities.

Distribution–abundance relationships

The global biogeography of tintinnid taxa clearly indicates a large heterogeneity in distributions. There are a few very widespread taxa, whereas most appear to have a "patchy" distribution, whether they be neritic or open-water forms. Do these differences reflect any quantitative differences? A correlation of average abundance and spatial distribution is a very common phenomenon when comparing ecologically and phylogenetically related taxa (see, for example, Brown 1984). The positive relationship between abundance and distribution is a fundamental ecological pattern, the precise nature of which appears to be variable, perhaps related to the scales considered (see, for example, Holt et al. 2002; Blackburn et al. 2006) and many mechanisms have been proposed to account for the pattern (Borregaard & Rahbek 2010). The interested reader is directed to the reviews by Blackburn et al. (2006), Borregaard & Rahbek (2010), and Gaston & He (2011) for detailed considerations of the relationship. Here we will be concerned only with the deceptively simple question: are widespread tintinnids found in higher concentrations compared with forms with apparently restricted or patchy distributions?

The interest in examining this question is twofold. First, it is to establish if indeed tintinnids conform to the general pattern of abundant species being more widespread than rare or less-abundant species. The relationship has not been examined for organisms of the marine plankton. Among terrestrial organisms, exceptions to the general distribution–abundance rule have been found among taxa with great dispersal capacity, specifically some groups of birds (Symonds &

Johnson 2006) and butterflies (Paivinen et al. 2005). Second, if distribution does not reflect abundance, then restricted or patchy distributions could be artifacts of inadequate sampling effort.

Answering the simple question of the relationship between distribution and abundance requires detailed data from studies on large geographic scales, which are rarely reported. However, the few data available clearly show a positive relationship between spatial distribution and average abundance among tintinnid species (Fig. 10.13). The data are from different systems sampled on distinct geographic scales: the tropical Southeast Pacific Ocean, 22 stations along a cruise track of 8000 km (Dolan et al. 2007), and the Sub-Polar Straits of Magellan, 18 stations along a 500 km cruise track (Fonda Umani et al. 2011). In both data sets, "distribution", as the number of stations from which a species was recorded, was positively related to the overall average abundance of the species. Similarly, the distribution of a tintinnid species relative to other species, its "rank distribution", was correlated with the overall abundance rank of the species. Thus, spatial distribution does appear to be positively related to abundance in tintinnids.

Another component of distribution–abundance relationships is temporal rather than spatial occurrence. The question then becomes, "are temporally persistent tintinnids found in higher concentrations

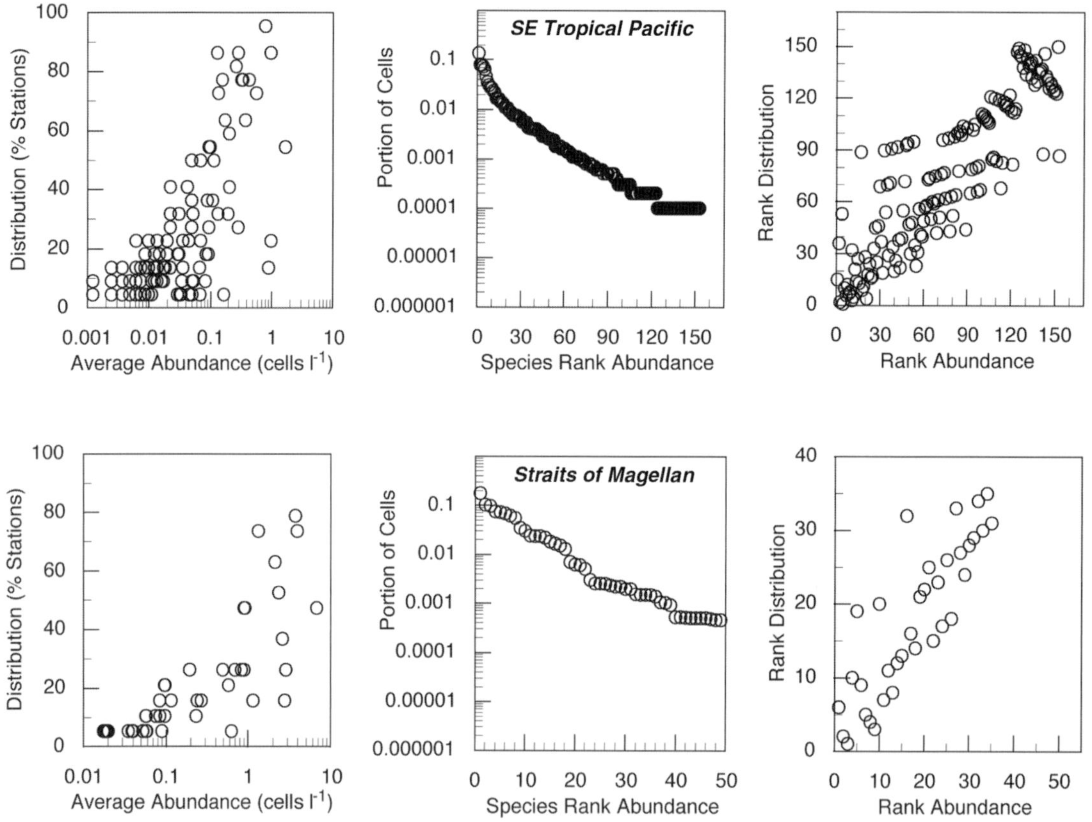

Fig. 10.13 The common pattern of a positive relationship between abundance and the extent of spatial distribution is shown by tintinnid species. Data from the southeast tropical Pacific Ocean are from 22 stations sampled in 2004 along a cruise track of about 8000 km between the Marquise Islands and the coast of Chile (Dolan et al. 2007). The Straits of Magellan data are from 18 stations sampled in 1991 along a 500 km cruise track (Fonda Umani et al. 2011). For both data sets, average abundance represents abundance averaged over all the stations sampled and species rank abundance for a pooled data set.

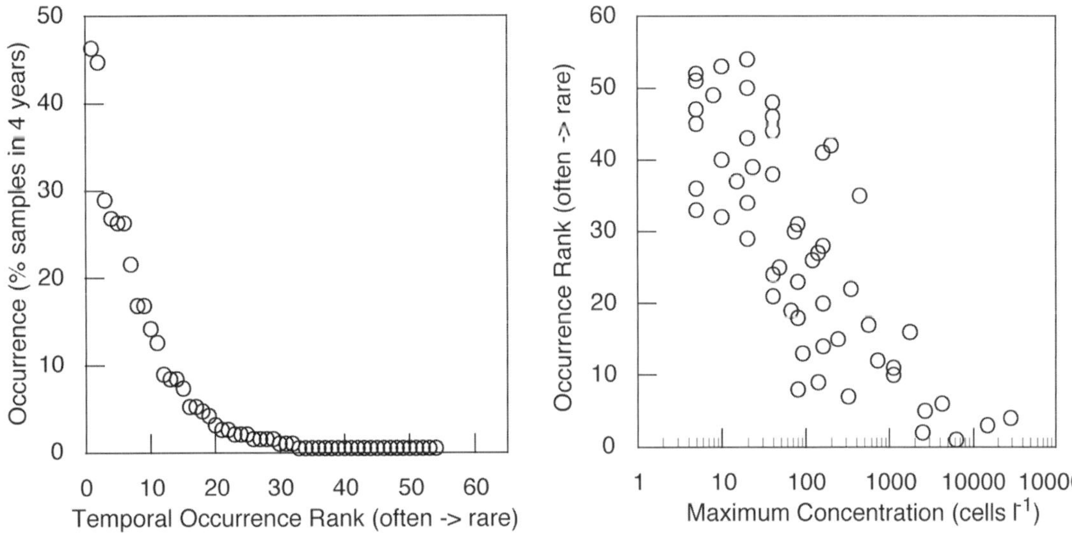

Fig. 10.14 Tintinnid species that are the most often present in a time-series are also those species that reach the highest abundances. Data from weekly sampling of the Gulf of Naples (Italy) over 4 years (Modigh & Castalado 2002).

compared with forms with apparently ephemeral distributions?" The answer requires time-series data. Perhaps the most complete of these is for a coastal system, the Gulf of Naples, which was sampled weekly for 4 years (Modigh & Castalado 2002). The most perennial species was present in less than half the weekly samples, indicative of seasonal changes in the species composition of the tintinnid assemblage (Fig. 10.14). Nonetheless, maximum abundance of a given species was positively related to its frequency of occurrence.

Overall, tintinnids that are widely distributed spatially are also the most abundant forms. However, it should be noted that global patterns are not the same as local patterns. The most widespread species globally are not necessarily the most common within a given system. For example, the most widespread and abundant species in the Straits of Magellan in the 1991 cruise was *Acanthostomella norvegica*, which is restricted to polar and sub-polar waters. Likewise, the species found most often in the Naples time-series was *Tintinnopsis beroidea*, a species found in coastal waters.

Latitudinal diversity gradient

Common to many groups of organisms, both terrestrial and marine, is the distinct increase in species diversity from high to low latitudes. This pattern, the latitudinal diversity gradient, is perhaps the most familiar and studied of all phenomena in biodiversity. The gradient has been described as "remarkable for its pervasiveness, its lack of a generally agreed explanation, and for the plethora of explanations put forward" (Williamson 1997). One author has gone so far as to title a chapter in a book on diversity "The holy grail of ecology: latitudinal gradients" (Adams 2009). Observations of the marked diversity of pelagic tropical fauna and flora compared with the temperate zone go back to the 19th century (see, for example, Haeckel 1893). For tintinnids, Kofoid remarked that a sample from tropical waters will contain few individuals but lots of species compared with a sample from polar waters with lots of individuals of very few species (Kofoid 1930). Nonetheless, not long ago the existence of latitudinal diversity gradients in the sea was questioned (Clarke 1992). Now it is recognized that many marine organisms, if not most groups, do exhibit the gradient. A latitudinal diversity gradient characterizes marine organisms ranging from fish and macroplankton (Angel 1997) to benthic invertebrates (Macpherson 2002), bacteria (Fuhrman et al. 2008), and perhaps even viruses (Angly et al. 2006). The details of the marine gradients have been reviewed by Hillebrand (2004) and the relative importance of the

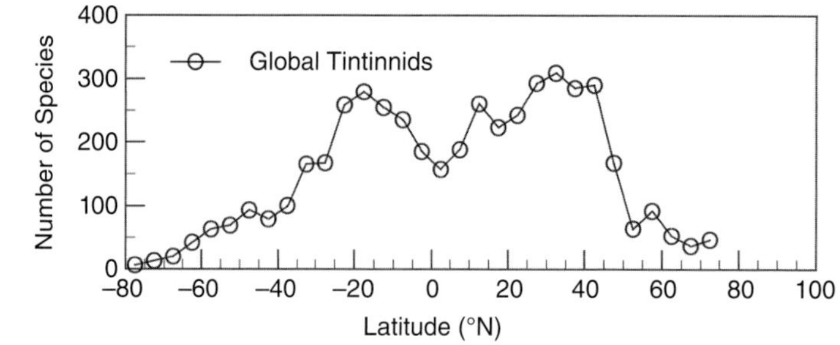

Fig. 10.15 Tintinnids show a typical latitudinal diversity gradient. Total species richness in bands of 5° of latitude from the database of the species records in Pierce & Turner (1993) updated for this chapter.

different mechanisms responsible remains an active area of debate (see, for example, Buzas et al. 2002; Allen & Gillooly 2006; Corliss et al. 2009; Schemske et al. 2009). Here the discussion will be limited to the general pattern shown by tintinnids as it compares with those known from other planktonic groups, along with the mechanisms proposed to explain latitudinal diversity gradients in the plankton.

The tintinnid pattern described here is based on records of morphologically defined taxa, as are virtually all the known latitudinal gradients. However, it should be mentioned that distributions of morphologically defined taxa may not completely correspond with patterns of genetically defined taxa. Cryptic species are known from a large range of planktonic groups: foraminifera (de Vargas et al. 1999), nanoflagellates (Slapeta et al. 2006), diatoms (Kooistra et al. 2008), dinoflagellates (Darling et al. 2004), and copepods (Goetz & Ohman 2010). As previously mentioned among tintinnids, cryptic species, genetically distinct groups of morphologically identical organisms, are not known. Plastic species, those that display very different morphologies and have been described as distinct taxa, are known from studies of lorica development in *Favella* (Laval-Peuto 1977), sequence data in *Petalotricha*/*Cyttarocylis* (Bachy et al. 2012) and are suspected from morphometric studies of *Cymatocylis* (Williams et al. 1994). How diversity patterns might differ between morphologically defined and genetically defined species is, unfortunately, unclear at present.

The stereotypical latitudinal diversity gradient is one of low species richness near the poles, increasing toward the equator with a peak at about 15–20° both north and south, and a slight inflection or decrease around the equator. This "textbook" pattern has been described for tintinnids (Dolan & Gallegos 2001; Dolan et al. 2006) and is shown in Fig. 10.15, based on the global species records in Pierce & Turner (1993) and updated for this chapter.

Among other planktonic groups, the typical pattern of tintinnids also characterizes copepod diversity (Fig. 10.16). For copepods and another group of zooplankton, the foraminifera, the latitudinal trend in species richness has been statistically linked to annual average sea surface temperature by a positive curvilinear relationship (Rutherford et al. 1999; Woodd-Walker et al. 2002; Rombouts et al. 2009), but somewhat distinct explanations have been proposed for the nature of the mechanism. For Rutherford et al. (1999), average sea surface temperature is a proxy measure of the average depth of the surface mixed layer, thus representing habitat size. According to this explanation, deep surface layers contain more species because they contain more spatial niches. Alternatively, areas of distinct sea surface temperature characterize areas that differ in temporal patterns and variability of primary production and thus represent areas of different habitat stability or seasonality (Woodd-Walker et al. 2002; Rombouts et al. 2009). The areas of marked variability in water column structure and primary production are lower in species richness, presumably because such systems are poorer in the number of distinct, stable, niches owing to temporal variability of the environment. It should be noted that the explanations of habitat size versus stability are not mutually exclusive. Although the commonality of patterns among tintinnids, copepods, and foraminifera suggests a single "zooplankton" gradient, the same cannot be said for phytoplankton.

Based on the few reports that exist, phytoplankton show distinct group-specific and ocean-basin-specific patterns. Different patterns for diatoms and coccolitho-

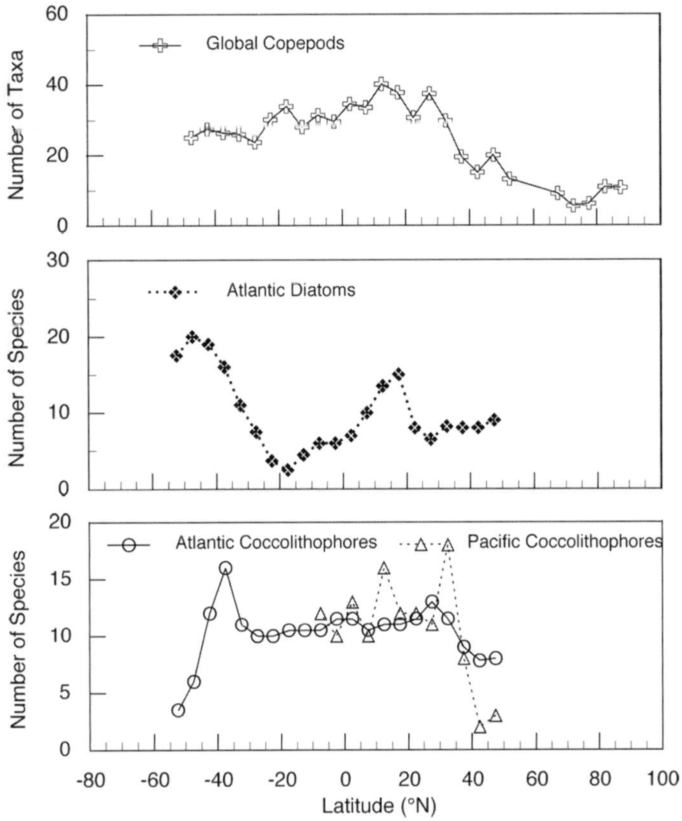

Fig. 10.16 Latitudinal diversity gradients of different groups of planktonic organisms can be similar to tintinnids, such as copepods, whereas those of diatoms and coccolithophores appear distinct. The copepod data were taken from figure 2a in Rombouts et al. (2009) and averaged in bands of 5° of latitude. The *y*-axis is labeled "taxa" as some species were grouped by genus. The Atlantic diatom and coccolithophore data were taken from figure 1c in Cermano et al. (2008). The Pacific coccolithophore data were taken from figure 6 in Honjo & Okada (1974), surface layer, and averaged in bands of 5° of latitude.

phores have been described for the Atlantic Ocean (Cermano et al. 2008). Diatom diversity appears to be highest in high latitudes and areas of upwelling, and is lower in tropical latitudes, suggesting a positive relationship with mixed water columns and high nutrient conditions (thus opposite from that of tintinnids and other zooplankton). Coccolithophore diversity, while sharing the same peaks with diatoms in nutrient-rich zones, remains relatively high throughout the Central Atlantic. The distinct patterns were explained as reflecting diatom adaptation to high-energy, high-nutrient conditions, and coccolithophores as adapted to low-energy, low nutrient conditions. Coccolithophore diversity in the Central and Northern Pacific Ocean was also described by Honjo & Okada (1974). The same basic pattern as that described for the Atlantic is evident with a relatively high diversity across the Central Pacific Ocean although the North Pacific appears to be species-poor relative to the North Atlantic Ocean (Fig. 10.16).

In reality, it is somewhat unclear whether or not tintinnids are actually similar to, or differ from, other groups in large-scale patterns of diversity because descriptions are invariably based on analysis of distinct data sets, derived from sampling on different time or space scales, potentially confounding patterns.

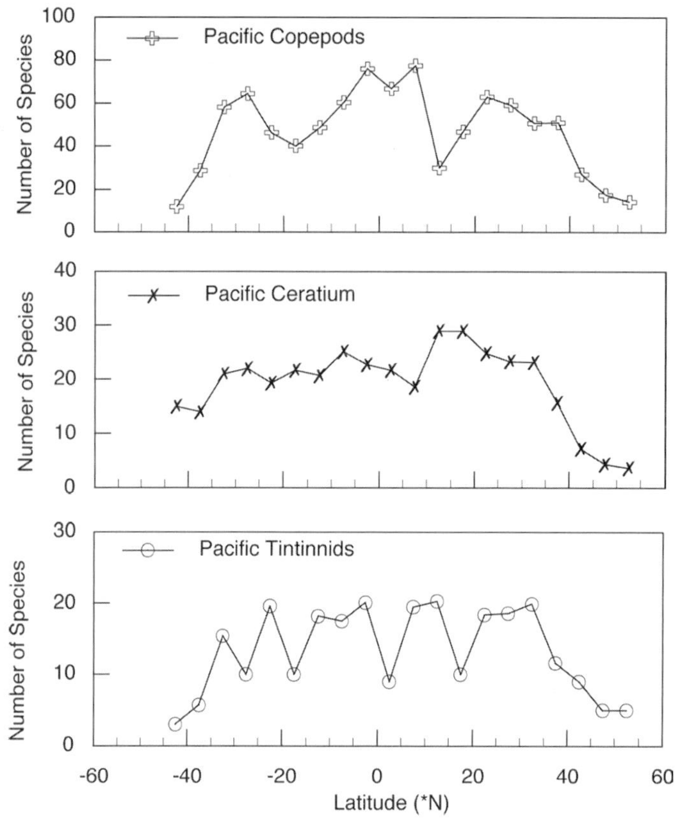

Fig. 10.17 Latitudinal diversity gradient in the Pacific Ocean of copepods, *Ceratium* species, and tintintinids are similar. Based on data from Cruise VII of the Carnegie (Campbell 1942; Wilson 1942; Graham & Bronikovsky 1944). Species richness was averaged in bands of 5° of latitude.

However, data from the Cruise VII of the Carnegie in 1928–1929 represent an exception as different groups of the zooplankton and phytoplankton (tintinnids, copepods, and species of the dinoflagellate genus *Ceratium*), were catalogued from the same surface layer samples, allowing direct comparison of species abundance patterns. The complete data on the three groups suggest similar global latitudinal diversity gradients (Dolan 2011). Here the patterns of the Pacific Ocean based on the Carnegie reports (Campbell 1942; Graham & Bronikovsky 1944; Wilson 1942) will be considered because data on the Pacific populations are often sparse (see, for example, Rutherford et al. 1999; Rombouts et al. 2009) and the Carnegie data are dominated by Pacific Ocean stations (126 of the 160 stations).

A very similar latitudinal gradient of the Pacific Ocean populations of tintinnids, copepods, and *Ceratium* is evident but is relatively irregular for tintinnids compared with copepods and *Ceratium* (Fig. 10.17). The patterns for Pacific Ocean tintinnids, copepods, and *Ceratium* all resemble closely that of the "global" pattern for zooplankton in which diversity is thought to be inversely related to variability in primary production and/or positively related to average surface layer depth. Thus, resource stability and/or habitat size are associated with high diversity. Notable is the fact that *Ceratium* in the phytoplankton shares the pattern as tintinnids and copepods. *Ceratium* may not be typical of the phytoplankton. Some species of *Ceratium* are known to be mixotrophic, although mixotrophy appears to be a relatively minor nutritional mode

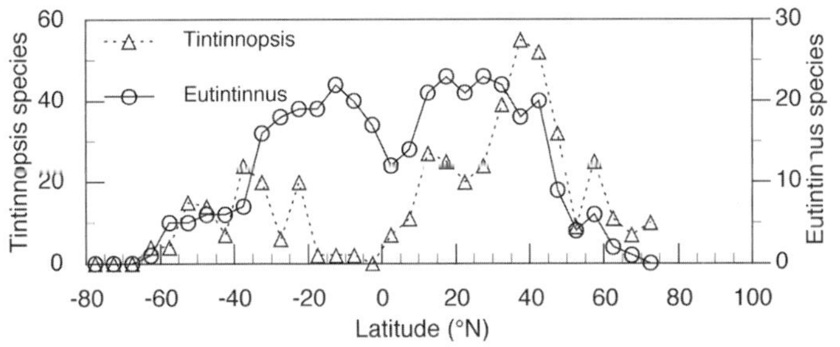

Fig. 10.18 Latitudinal diversity gradients of the genera *Tintinnopsis* and *Eutintinnus* are distinct, reflecting adaptation to different types of system. Total species richness in bands of 5° of latitude are from the database of the species records in Pierce & Turner (1993) updated for this chapter.

(Bockstahler & Coats 1993). There may be no general pattern characteristic for different phytoplankton taxa. It appears in comparing diversity trends of Atlantic Ocean diatoms and coccolithophores that individual groups of phytoplankters differ considerably in their large-scale diversity patterns.

Just as different major phytoplankton taxa show distinct large-scale patterns of diversity, patterns of specific groups within a given taxon can be distinct as a lower level taxon can be adapted to, or characteristic of, a particular type of environment. This is easily seen in tintinnids by comparing two species-rich genera: *Tintinnopsis* and *Eutintinnus*. *Tintinnopsis* species, with their loricae agglutinated with small mineral particles, are essentially restricted to neritic shallow waters and usually dominate tintinnid assemblages in bays and estuaries, especially those in temperate climates. *Eutintinnus* species, although found among neritic assemblages, are very commonly found in open waters from temperate to tropical zones and are often the dominant form in tropical assemblages. The two genera display very different biogeographies (Fig. 10.4) and show very different latitudinal diversity gradients (Fig. 10.18). *Tintinnopsis* species richness is very low in the tropics and peaks in the temperate zones of both the southern and the northern hemispheres at about 40°. In contrast, *Eutintinnus* species richness resembles the general global tintinnid pattern, being highest in the tropics and sub-tropical areas.

Among other zooplankton, the same phenomenon of distinct differences in biogeographic patterns among particular taxa is also known. For example, different patterns characterize calanoid and cyclopoid copepods in the estuaries of eastern North America. Species richness is highest in temperate systems for calanoid copepods whereas cyclopoid copepods are more species-rich in sub-tropical estuaries. The differences were explained as likely reflecting calanoid copepod adaptation to the large seasonal changes typical of temperate systems (Turner 1981).

Tintinnids, as a whole, show a latitudinal diversity gradient similar to that of copepods and *Ceratium*. It is reasonable to expect that the same mechanism explains the pattern in all three groups and that it can be linked directly or indirectly with sea surface temperature, as shown for copepods and foraminfera (Rutherford et al. 1999; Woodd-Walker et al. 2002; Rombouts et al. 2009). However, the precise nature of the mechanism (e.g., exactly how environmental stability promotes diversity or conversely how environmental variability inhibits diversity), remains to be identified.

10.3 ASSEMBLAGES OF COASTAL SYSTEMS

Seasonal changes in species richness

Very early in the 20th century, under the influence of a need to understand fisheries dynamics, studies of plankton shifted focus from taxonomy to community and population biology, with particular attention devoted to annual plankton cycles (Mills 1989). Temporal variability of planktonic organisms in coastal systems was already well known; Karl Brandt (1901) remarked, "To know the plankton of a body of water, it is necessary to repeat sampling at the shortest interval of time possible over at least a year". It was soon recognized that seasonal changes in species compositions were not due to shifts in the locations of waters masses containing different assemblages but rather the waxing and waning of different species within the system. Some of the earliest studies on seasonal changes in planktonic populations concerned or included tintintinnids (see, for example, Laackmann 1908; Lohmann 1908; Lebour 1917). Given the long history of European studies, it is perhaps not surprising that although tintinnid assemblages have been studied in a very wide range of coastal systems, Mediterranean and North Atlantic systems have received the greatest attention (Table 10.2).

Reports from nearly every system describe very distinct seasonal changes in the occurrences of individual species and species richness. Perhaps the most intensive time-series available, the Gulf of Naples sampled weekly for 4 years, showed that none of the 10 most common species were found even half the time and most less than a third of the time (Fig. 10.14). Differences in the assemblages and patterns in different systems have led to the suggestion that these constitute distinct "tintinnid community fingerprints" for each system (Modigh & Castalado 2002). Although no two systems appear identical in detail, quite similar qualitative patterns have been reported from disparate systems.

An apparently common pattern is that species with hyaline loricae constitute the bulk of the community in the summer months whereas species with agglutinated loricae dominate in winter. Thus, the summer communities of the tropical São Sebastião Channel in Brazil and a Mangrove system in southeast India are dominated by species with hyaline loricae (*Eutintinnis, Dadayiella, Amphopellopsis*) and the winter assemblage by species with agglutinated loricae (*Tintinnopsis, Stenosemella, Codonellopsis*). Such a pattern, first reported for the Kiel Bight (Laackmann 1908), has been reported for a surprising variety of coastal systems: the Bay of Fundy (Middlebrook et al. 1987), Narragansett Bay (Verity 1987), Mission Bay in Southern California (Elliott & Kaufman 2007), the Gulf of Naples (Modigh & Castalado 2002), and Hiroshima Bay (Kamiyama & Tsujino 1996). The pattern appears to extend from the Arctic waters of the White Sea, where *Helicostomella* dominates in mid-summer (Burkovsky 1976), to Antarctic sub-polar waters, where the springtime community in the Straits of Magellan was dominated by *Tintinnopsis* spp. and the autumn assemblage by *Acanthostomella* (Fonda Umani et al. 2011).

Exceptions to the general pattern of a seasonal shift from species with hyaline loricae to those with agglutinated loricae have been reported. A nearly invariant, *Tintinnopsis*-dominated assemblage characterized the Bahia Blanca Estuary in Argentina. In waters near the Isle of Man in the northern Irish Sea, the tintinnid assemblage is dominated year-round by species of *Stenosemella* or *Tintinnopsis* (Graziano 1989). Species with hyaline loricae, which appear in other coastal systems of the North Atlantic in the summer months, were virtually absent. Species with agglutinated loricae may be better adapted to turbulent conditions, which generally correspond with the winter season. The small mineral particles required to build their loricae may not be present in sufficient concentrations in the water column during the summer months in most coastal systems. The year-round dominance of species with agglutinated loricae may then be because some systems are also high-energy, turbulent systems year-round.

Another characteristic common to many different coastal systems is that the transitional periods between winter and summer correspond to the period when species richness is highest (Fig. 10.19). These periods are characterized by assemblages containing species both with hyaline and agglutinate loricae. In contrast to the commonality of such transitional periods representing peak species richness, the period of minimum species richness varies in different regions. In the temperate systems of the Atlantic, the winter months correspond with periods of low species richness, whereas in Mediterranean systems the summer months are the most species-poor.

A variety of temporal patterns in species richness has been reported from tropical and sub-tropical systems, some related to occurrence of the rainy or dry periods. In the mangrove system of southeast India, peak abundances during dry months correspond with peak species richness and abundance (Godhantaraman 2002). In New Caledonia, species richness was strongly correlated with tintinnid abundance (Dolan et al. 2006). These relationships suggest that species richness may be related to overall tintinnid abundance. Comparing numbers of species found and concentrations of tintinnids in some of the seasonal studies (those in which the data could be easily extracted) shows a positive, albeit weak, relationship across different coastal systems (Fig. 10.20).

A general positive relationship between tintinnid total abundance and species richness suggests conditions that allow the development of a dense population

Table 10.2 Studies of seasonal changes in tintinnid assemblages of coastal systems.

System	Location	Reference
Mediterranean Region		
Black Sea	Romanian coast	Petran 1958
Black Sea	Crimean coast	Dogopolskaya 1940
Damietta Harbor	Egypt	Dorgham et al. 2009
Jounieh Bay	Lebanon	Abboud-Abi Saab 1989
Palm Island	Lebanon	Abboud-Abi Saab 2002
Bozcaada Island	Aegean Sea	Balkis & Wasik 2005
Bay of Buyucekmece Bay	Marmara Sea	Balkis 2004
Mali Ston Bay	Eastern Adriatic	Krsinic 1980
Kastela Bay	Eastern Adriatic	Bojanic 2001
Ionian Sea	Mediterranean Sea	Sitran et al. 2007
Naples	Southern Italy	Modigh & Castaldo 2002
Bay of Villefranche	Southern France	Balech 1959; Rassoulzadegan 1979; Dolan et al. 2006
San Remo	Northeast Italy	Rampi 1948
Gulf of Marseille	Southern France	Balech 1959, Travers & Travers 1971
Thau Lagoon	Southern France	Lam-Hoai et al. 1997
Blanes	Southeast Spain	Margalef 1957
Gulf of Valencia	Southeast Spain	Duran 1951, 1953
Bay of Algers	Algeria	Vitello 1964
North Atlantic		
Bay of Fundy	New Brunswick, Canada	Middlebrook et al. 1987
Bedford Basin	Nova Scotia, Canada	Paranjape 1987a
Long Island Sound	New York, USA	Capriulo & Carpenter 1983
New York Bight	New York, USA	Gold & Morales 1975
Narrangansett Bay	Rhode Island, USA	Hargraves 1981; Verity 1987
Buzzards Bay	Massachusetts, USA	Pierce & Turner 1994
Damariscotta Estuary	Michcigan, USA	Sanders 1987
Gullmar Fjord	Western Sweden	Hedin 1974
Kiel Bight	Baltic, Germany	Laackmann 1908
Isle of Man	North Sea	Graziano 1989
Southampton Water	UK	Leakey et al. 1993
Bay of Cascais	Portugal	Silva 1950
Obidos Lagoon	Portugal	Silva 1953
Nervion Estuary	Northeast Spain	Urrutxurtu 2004
South Atlantic		
São Sebastião	São Paulo, Brazil	Eskinazi-Sant'anna & Bjornberg 2006
Bahia Blance Estuary	Argentina	Barria de Cao 1992
Mauritania and Senegal	West Africa	Duran 1965
Others		
White Sea	Arctic	Burkovsky 1976
Signy Island	Antarctica	Leakey et al. 1994
Hiroshima Bay	Japan	Kamiyama & Tsujino 1996
Akkeshi Bay	Japan	Taguchi 1976
Tokyo Bay	Japan	Nomura et al. 1992
Funka Bay	Japan	Dohi 1982
Gulf of Elat	Red Sea	Kimor & Golandsky 1981
Kuwait	Arabian Gulf	Skryabin & Al-Yamani 2006
Parangipettai	Southeast India	Godhantaraman 2002
Mission Bay	California, USA	Elliott & Kaufman 2007
New Caledonia	South Pacific	Dolan et al. 2006

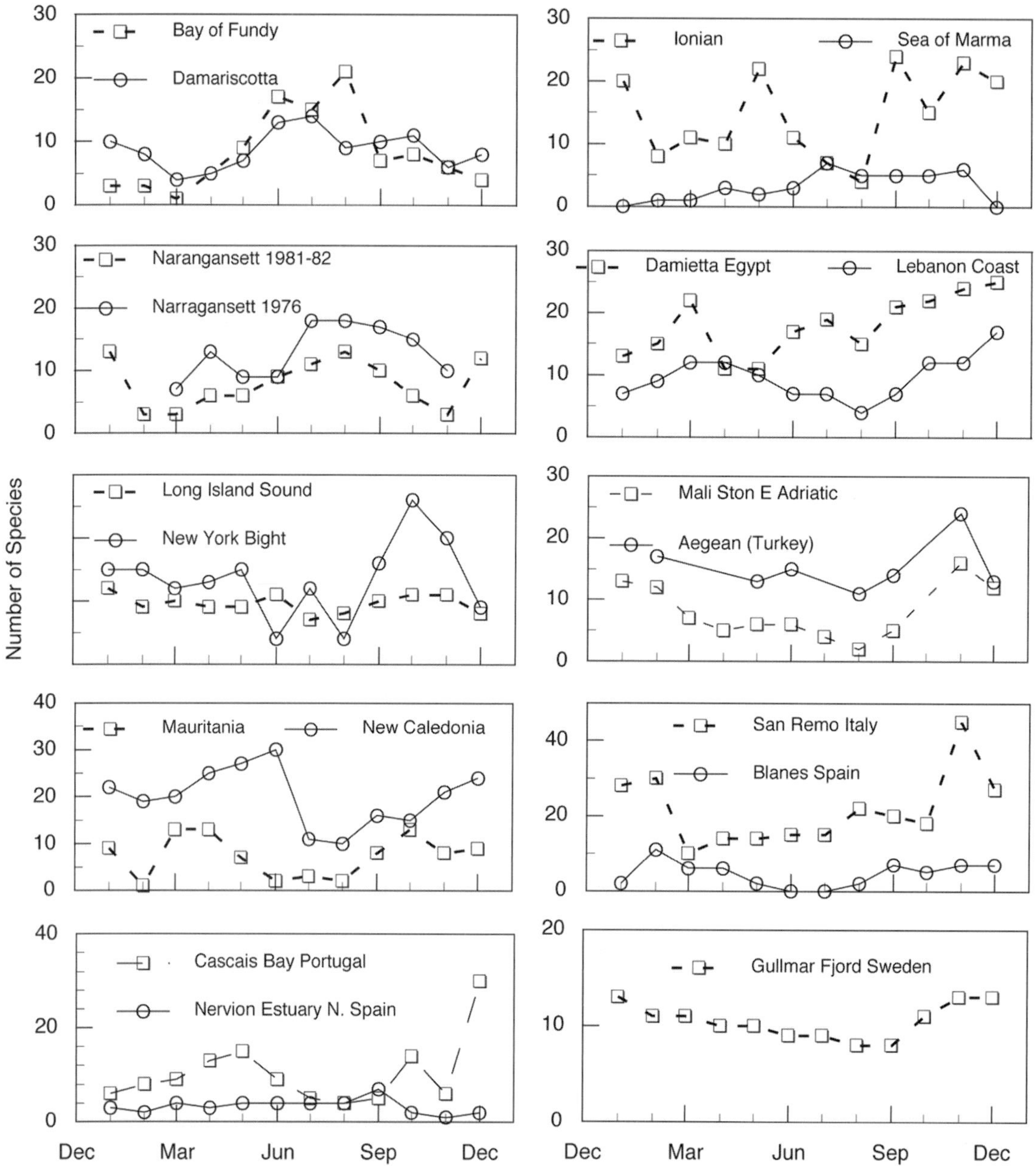

Fig. 10.19 Coastal systems usually show distinct seasonal changes in the species richness of tintinnid assemblages. Data are from sources listed in Table 10.1.

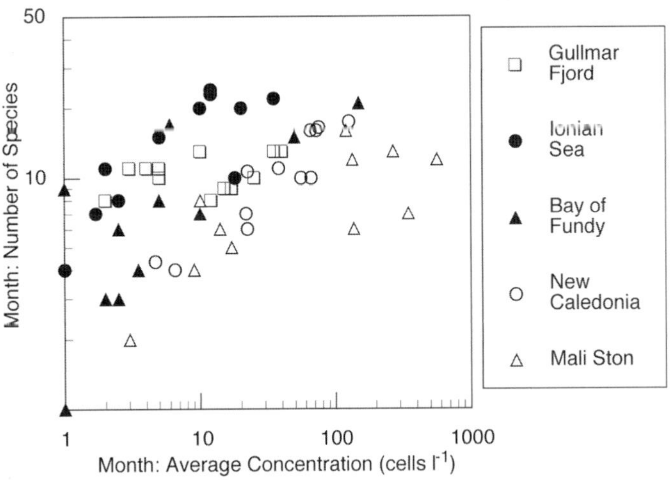

Fig. 10.20 Species richness of tintinnid assemblages in coastal systems is often highest when overall tintinnid concentrations are high. Data are shown by plotting species richness by month against average tintinnid concentration for a wide range of systems. Data sources are references for the systems given in Table 10.2.

of any particular tintinnid species will allow the development of detectable populations of other species. Conversely, conditions that prevent any particular species from becoming abundant are most often poor conditions for all other species as well. The seasons, then, of minimum tintinnid species richness reflect low overall population size corresponding with seasonal minima in primary production, which in temperate systems is in winter and in Mediterranean systems in summer.

Whether the seasonal patterns in coastal systems shown by tintintinids are peculiar to them or are shared with other planktonic groups is unclear. For other planktonic ciliates, owing to the difficulty in precisely identifying ciliates other than tintinnids, there are very few reliable data on seasonal changes in species richness. A perhaps singular exception is the study of Leakey et al. (1993) on the ciliates of Southampton Water, a eutrophic temperate estuary on the southern coast of the United Kingdom in which quantitative protargol staining was used (Montagnes & Lynn 1987) and taxonomic expertise provided by Denis Lynn. Seasonal changes in species richness of ciliates (pooling all taxa) showed the typical tintinnid pattern of North Atlantic coastal systems: that is, a distinct summer peak and winter minimum in species richness. Thus, it would appear reasonable to assume that temporal trends in species richness are similar for tintinnids and other planktonic ciliates.

There are also some data on seasonal trends of species richness in tintinnids compared with other groups of the "microplankton": diatoms, and dinoflagellates. Many of the earlier reports on seasonal changes in coastal plankton communities focused not on tintinnids but microplankton with, in reality, a primary interest in patterns of diatoms and dinoflagellates. Fortuitously, some of these studies were conducted by investigators with a wide range of taxonomic expertise (i.e., Rampi, Margalef, and Duran), allowing comparison with some confidence of temporal trends in the species richness of tintinnids compared with the other groups of the microplankton.

Seasonal changes in species richness of different groups of the microplankton in three coastal sites of the northwest Mediterranean (San Remo (Italy), Blanes (Spain), and the Gulf of Valencia (Spain)) are shown in Fig. 10.21. The three sites all showed the "Mediterranean" pattern of tintinnid species richness: that is, peaks in spring and autumn with minimum species numbers in winter. Furthermore, all three sites show parallel trends for species richness of diatoms and *Ceratium*. Species richness of all three groups co-varied closely in the San Remo and Gulf of Valencia populations. However, diatom and *Ceratium* species

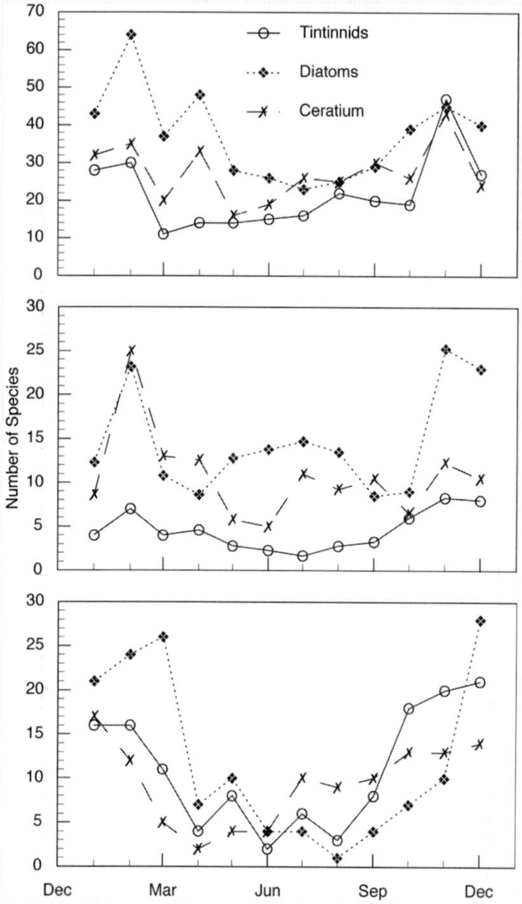

Fig. 10.21 Temporal changes in the species diversity of tintinnids can parallel, or be distinct from, species richness trends in other groups such as diatoms or *Ceratium*. Monthly species richness of tintinnids, diatoms, and *Ceratium* in three coastal systems of the northwest Mediterranean Sea. Top panel, data of Rampi (1948) for San Remo, Italy; middle panel, data (monthly averages) from Margalef (1957) for Blanes, Spain; bottom panel, data from Duran (1953) for the Gulf of Valencia.

richness differed from the tintinnids in the Blanes assemblages. Interestingly, Margalef noted that the seasonal trends of the Blanes populations, with a summer phytoplankton bloom, were unusual for the northwest Mediterranean (Margalef 1957). Nonetheless, it is clear that seasonal changes in the species richness of tintinnids can be distinct from temporal trends in diatoms or *Ceratium* species richness, thus indicating that different mechanisms govern the diversity of zooplankton and phytoplankton at least in northwest Mediterranean systems.

Seasonal changes in morphotypes

As tintinnid species are distinguished by characteristics of their loricae, changes in species composition of an assemblage corresponds with changes in the loricae of the assemblages. Among the morphological characters of tintinnids, lorica oral diameters have received special attention. For a given species, the oral diameter is a relatively invariant taxonomic characteristic (Gold & Morales 1975, 1976; Laval-Peuto & Brownlee 1986), one that is closely related to both maximum and optimum prey size, among other ecological characteristics (see Chapter 4 and Dolan 2010).

The morphological character "lorica oral diameter" has been used to assess seasonal changes in community characteristics of tintinnid assemblages. It has been examined in different manners, somewhat complicating comparisons. Community average lorica diameter was used to describe assemblages of the Bay of Fundy (Middlebrook et al. 1987), Naragansett Bay (Verity 1987), and Kingston, Jamaica (Gilron et al. 1991). The modal and largest diameters were reported for the Long Island Sound assemblages (Capriulo and Carpenter 1983), median size for assemblages of New Caledonia (Dolan et al. 2006) and the Ionian Sea (Sitran et al. 2007), and the size–frequency distribution of the most abundant species for Funka Bay (Dohi 1982). Nonetheless, all the studies documented changes in the morphotypes constituting the tintinnid community specifically in terms of lorica oral diameters.

An example of a seasonal change in "community average lorica oral diameter" is shown in Fig. 10.22 from Verity's study of Narragansett Bay tintinnids over 3 years (Verity 1987). Verity documented an annual cycle showing a seasonal variability of about 50%, which appears to be typical based on the results of other studies. In Verity's study, community oral diameter was significantly, albeit weakly, related to the portion of chlorophyll-*a* in the size fraction between 10 and 150 µm. Metrics of community oral diameter, whether average, modal, or median, have not been easily and unambiguously related to characteristics of the phytoplankton crop. This is likely because not only phytoplankton resources, but also competition from

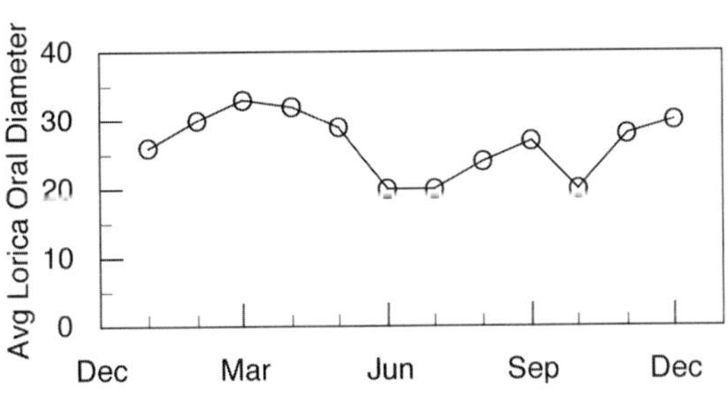

Fig. 10.22 Seasonal change in tintinnid assemblages is reflected in the average lorica oral diameter of the assemblage. The graph shows average monthly values of Narragansett Bay tintinnids from Verity (1987). Changes in community values of lorica oral diameter (average, mode, or median) have been related to changes in phytoplankton size-composition and concentration.

other microplankton grazers as well as species-specific tintinnid mortality from metazoan grazers or parasites, can influence the size of the dominant oral diameter.

Tintinnid assemblages of coastal systems are usually dominated by one species, which more or less defines the community oral diameter. Most other species present will be those with oral diameters distinctly different from that of the dominant species. An example illustrating the distribution of lorica oral diameters in the spring and autumn assemblages of contrasting system, the Gullmar Fjord in Sweden and the Great Lagoon of New Caledonia, is shown in Fig. 10.23. In these assemblages, abundant species are either alone in a size-class of lorica oral diameter or share the size-class with but one other species. In the New Caledonia assemblages, the three most abundant species are in distinct size-classes of lorica oral diameter. It appears to be common that in a given assemblage, the most abundant species are often of distinct oral diameters (compared with the other species present), presumably exploiting prey of distinct sizes (see, for example, Boltovskoy & Alder 1992; Dolan et al. 2009), which is suggestive of resource partitioning among different tintinnid species.

10.4 ASSEMBLAGES OF OPEN WATERS

In contrast to the communities of bays and estuaries, little is known of seasonal changes in tintinnid assemblages in open-water systems. Very few open-water systems have been studied in different seasons and even fewer by the same investigators, assuring a certain comparability among samplings and sample analysis. Detailed quantitative data are completely lacking, with most reports giving little more than brief lists of abundant forms in winter and summer. However, such "occurrence data" do suggest that open-water systems may not exhibit the large seasonal changes in species composition that characterize coastal systems.

For example, the typical seasonal change from a winter assemblage dominated by species with agglutinated loricae to a summer assemblage dominated by hyaline species can be absent. In both winter and summer, the assemblage of the Inland Sea of Japan appears to be dominated by *Tintinnopsis beroidea* and *Stenosemella nivalis* (Godhantaraman & Uye 2001). Likewise, assemblages of "large" tintinnids of the Southern Yellow Sea are dominated by *Stenomella steini*, *Tintinnopsis karajacensis*, and *T. radix* in winter and by *Codonellopsis mobilis* and *T. karajacensis* in the summer. These were both studies of temperate, open, but relatively shallow, waters of the western Pacific Ocean and may be thought exceptional. However, relatively detailed data from a contrasting system, the subtropical Gulf of Mexico, also suggest that seasonal changes in species composition in deep open waters, in which agglutinated species are a minor component, are also slight compared with coastal systems.

The microplankton of the Gulf of Mexico was studied by Balech (1967a, b), who collected samples from a series of 10 cruises in 1964 and 1965 in the Northeast Gulf of Mexico. Of the 10 cruises, four were summer cruises (June–September) and four were winter cruises (December to early March). The pooled summer cruise data represent species lists from 40 stations, and the winter cruise data represent 46 stations, allowing a rough comparison of summer and winter tintinnid assemblages. Furthermore, the occurrences of *Ceratium* species, among other dinoflagellates, were also recorded, allowing a comparison with the species-rich group of phytoplankters known to share temporal patterns of species abundances in the northwest Mediterranean Sea (see, for example, Rampi, 1948; Raybaud et al. 2009).

Balech did not report quantitative data but rather species lists for each of the stations; consequently, occurrence or "occupancy" – describing how widespread or rare was each species – could be estimated simply from the percentage of stations from which a

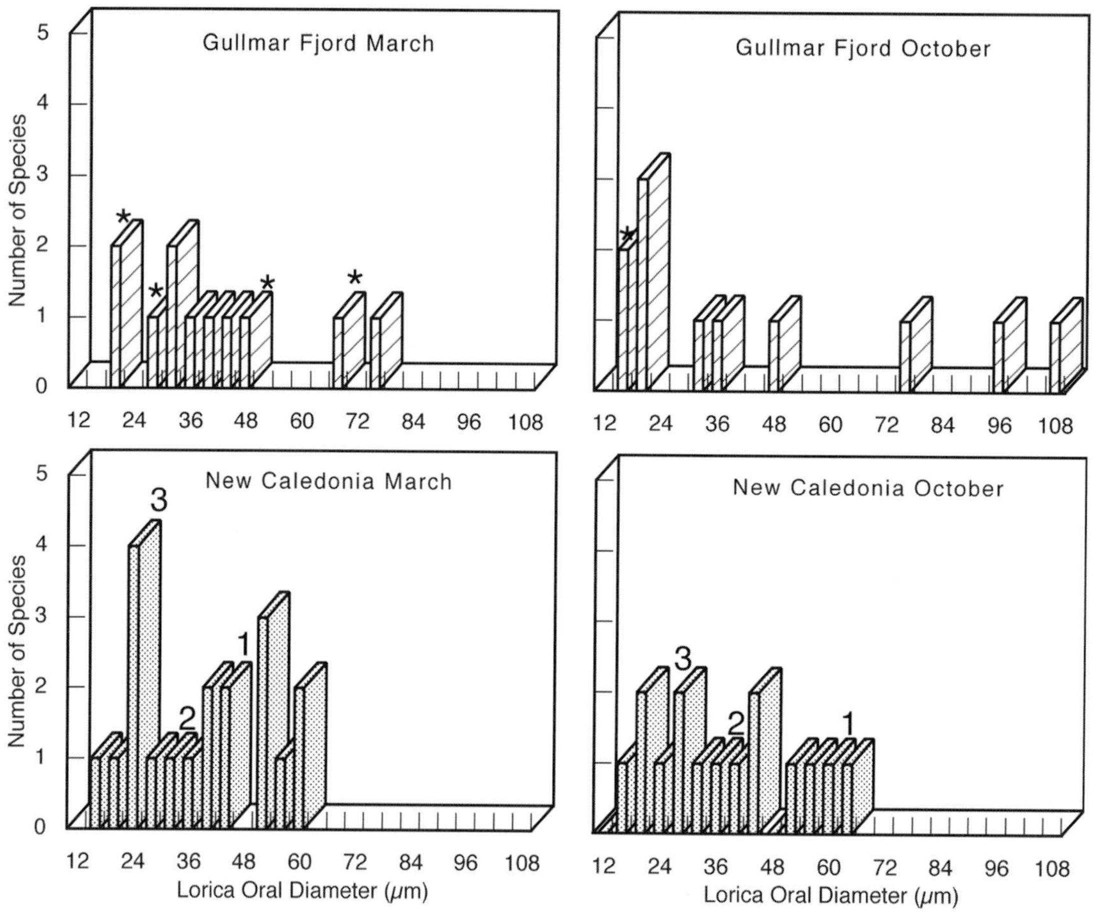

Fig. 10.23 Co-occurring abundant species usually differ in lorica oral diameter. The graphs show the distribution of tintinnid species in size-classes of lorica oral diameter in March and October in contrasting systems: the Gullmar Fjord in western Sweden (Hedin 1974) and the Great Lagoon of New Caledonia (Dolan et al. 2006). In the Gullmar Fjord graphs, asterisks denote size-classes containing species described as abundant. In the New Caledonia graphs, numbers 1, 2, and 3 denote the size-classes of the first, second, and third most abundant species.

species was recorded. In general, within groups of ecologically similar organisms, occurrence is closely correlated with abundance, meaning the more often a species occurs in a set of samples, the higher its abundance relative to other species (Brown 1984; Gaston & He 2011). Overall, the winter and summer assemblages of tintinnids and *Ceratium* were surprisingly similar in species composition as well in the occurrence patterns of the species.

The tintinnid species catalogue for summer and winter combined numbered 112. Out of the 112, 64 were common to the summer and winter lists. Species found in the winter, but not in the summer, numbered but 14 of the 78 found. Of the 98 summer species, only 32 did not occur in the winter samples. Besides the large overlap in the lists of summer and winter species, species occurrences were also very similar. The most widespread species in the summer were, by and large, also widespread in winter (Table 10.3). The rank-occupancy patterns for the entire summer and winter assemblages (ranking each species in order of the portion of stations from which it was recorded) were

Table 10.3 Tintinnids of the Gulf of Mexico in winter and summer differed little in terms of the identities of the most abundant species: the 15 most widespread species in the summer and winter and their corresponding ranks in the two seasons (data extracted from Balech (1967a)); the most widespread summer species are denoted in bold.

Winter rank	**Top 15 species** Summer **and winter**	Summer **rank**
15	***Amphorides amphora***	1
10	***Amphorides quadrilineata***	11
6	***Ascampbelliella urceolata***	9
31	***Climacocylis scalaroides***	14
11	*Codonaria cistelula*	17
22	***Codonellopsis orthoceras***	15
13	*Dadayiella ganymedes*	16
2	***Dictyocysta lepida***	8
4	***Epiplocyloides reticulata***	4
36	***Epiplocylis undella***	6
27	***Eutintinnus fraknoii***	12
14	***Eutintinnus medius***	6
1	***Eutintinnus tenuis***	10
8	***Proplectella claparedei***	2
3	***Protorhabdonella simplex***	7
7	*Rhabdonella corncopia*	33
12	***Rhabdonellopsis apophysata***	3
9	*Steenstrupiella steenstrupi*	24
5	***Xystonella treforti***	13

similar log-normal or log-series patterns of a very few widespread species, and most species were found in fewer than 20% of the stations sampled (Fig. 10.24). The similarity of the summer and winter assemblages was also reflected in the correlation between a given species occurrence rank in the summer list and its occurrence rank in the winter list (Fig. 10.23).

The *Ceratium* species, which totaled 55 for summer and winter combined, showed the same overall patterns as the tintinnids when comparing the summer and winter assemblages. Most of the species were present in summer and winter samples. The summer assemblage of 49 species included only four not found in the winter samples. Of the 50 species in the winter assemblage, only six did not occur in the summer samples. The species occupancy patterns of the summer and winter assemblages were also similar, characterized by a very few widespread species and many species found in a minority of the stations sampled. Like tintinnid species, the occupancy ranks of a *Ceratium* species among the summer stations was correlated with its occupancy rank in the winter assemblage but with a stronger correlation than that found with the tintinnids.

Overall, the assemblages of tintinnids and *Ceratium* in the Gulf of Mexico were nearly the same in the summer and in the winter. The apparent lack of a large seasonal change in assemblages contrasts with the studies of nearshore systems, whether temperate, subtropical, or tropical, all of which show marked differences. The apparent similarity of summer and winter species assemblages of both tintinnids and *Ceratium* could be the result of pooling data from several cruises sampling different areas, and thus an effect of sampling scale. Alternatively, stability of species assemblages may be an attribute of large open-water systems. Seasonality remains to be investigated in other open-water systems.

10.5 KEY POINTS

1. Global biogeography. Tintinnid geographic distributions at the generic level fall into one of five patterns: cosmopolitan, neritic, warm–temperate, boreal, and austral. These distributional patterns, or categories, roughly correspond to a modified latitudinal cosmopolitanism with a dichotomy between coastal and open-water forms. Although there are tintinnid genera with cosmopolitan distributions, no single species, or more precisely "morpho-species", is known to have a cosmopolitan distribution. In contrast, some species do show the very restricted distributions of endemics. The records of at least one species suggest recent artificial dispersal.

2. Distribution–abundance relationships. In common with many groups of organisms, there is a positive relationship between abundance and distribution among tintinnid species. Within a given system or biome, the population size of a species is positively related to its spatial and temporal occurrence. Conversely, rare species of tintinnids are both low in abundance as well as infrequently found.

3. Latitudinal diversity gradient. Tintinnids show a latitudinal gradient of species richness, one that is similar to that described for foraminifera and copepods. Thus, there may be a common "zooplankton latitudinal diversity gradient"; the pattern has been linked to average sea surface temperature, thought to reflect water column characteristics. The tintinnid pattern is also shared by dinoflagellates of the genus

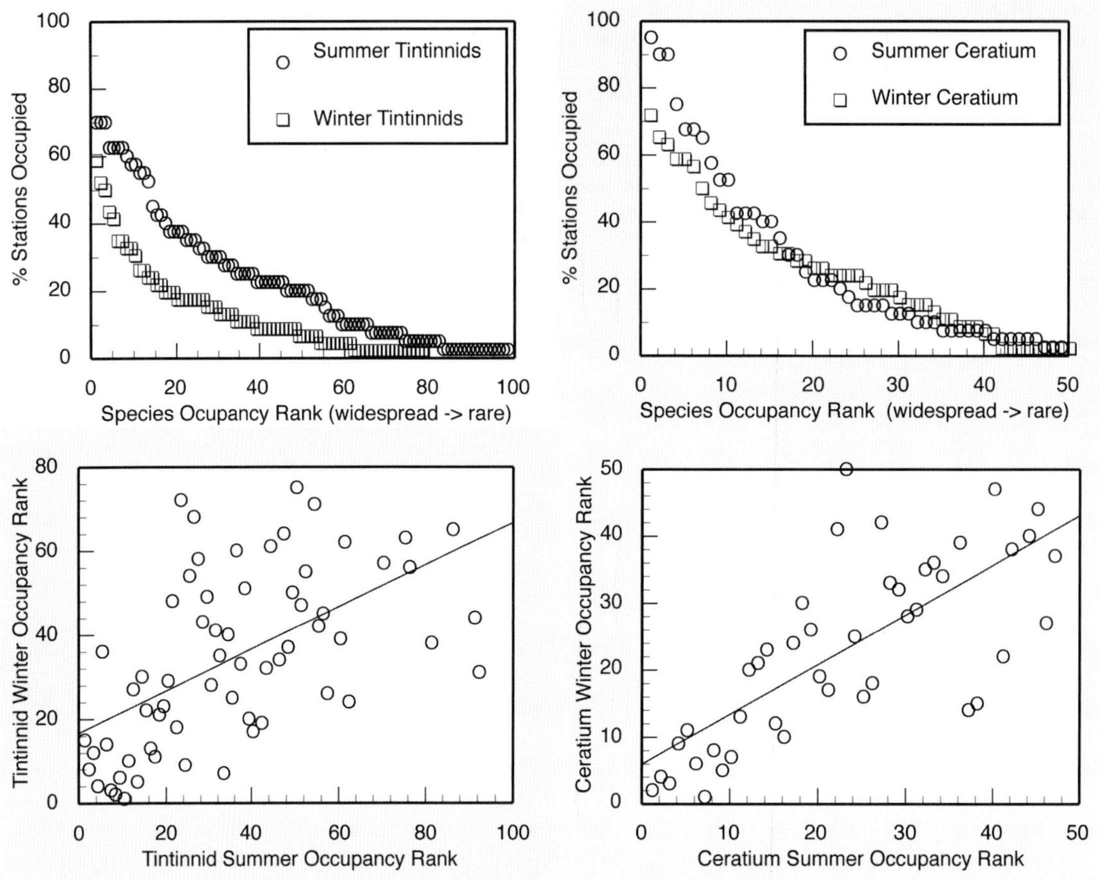

Fig. 10.24 Occupancy patterns of summer and winter assemblages of tintinnids and *Ceratium* in the Gulf of Mexico are both show little seasonal differences (based on data from Balech (1967a)). Top panels, geographic ranks, from widespread to rare, of each species in the summer and winter assemblages; bottom panels, the relationship between the geographic rank of species found both in the summer and winter. In summer, the most widespread tintinnid species, *Amphorides amphora*, was found in 70% of stations sampled; in winter, the most widespread species, *Eutintinnus tenuis*, was found in about 60% of the stations sampled. All the species assemblages showed a pattern of species occupancy or occurrence of most species in just a few stations. For the species that occurred in both summer and winter, summer and winter ranks were significantly related (for tintinnids: $n = 64$, $r = 0.57$; for *Ceratium*: $n = 44$, $r = 0.74$).

Ceratium. However, different latitudinal gradients characterize other phytoplankton groups, diatoms, and coccolithophorids. As there appears to be no single latitudinal diversity gradient for marine plankton, different mechanisms likely act on different groups of the plankton.

4. Seasonal changes in species richness of coastal systems. Distinct seasonal patterns in the occurrences of individual species of tintinnids and species richness characterize most coastal ecosystems. A common pattern is species with hyaline loricae constitute the bulk of the community in the summer months whereas species with agglutinated loricae dominate during the remainder of the year. The period of highest species richness generally corresponds to periods of high overall tintinnid abundance. Within a coastal system, patterns of seasonal changes in the species richness of tintinnids can parallel those of other planktonic groups such as aloricate ciliates, diatoms, or *Ceratium*, but can also be distinct.

5. Seasonal changes in morphotypes in coastal systems. In tintinnids the lorica oral diameter, which is relatively invariant for a given species, is related to the size of prey upon which the species feeds. In an assemblage, the most abundant species are usually of distinct oral diameters (compared with the other species present), presumably exploiting prey of distinct sizes. Various metrics of a "tintinnid community oral diameter" have been used to assess seasonal changes in community characteristics of tintinnid assemblages in terms of morphology, and seasonal changes in these metrics have been loosely related to changes in the characteristics of the phytoplankton crop.

6. Assemblages of open waters. Little is known of seasonal changes in tintinnid assemblages in open waters. However, it appears that open-water systems may lack the large seasonal change in assemblages known from coastal systems.

ACKNOWLEDGMENTS

The comments and contributions of Sabine Agatha, Urania Christaki, Wayne Coats, Takashi Kamiyama, George McManus, Dave Montagnes, Alessandro Saccà, and Diane Stoecker are gratefully acknowledged. Funding for this work was provided by the French Agence National de la Recherche and Pole Mer PACA through the ANR-Biodiversité AQUAPARADOX project.

REFERENCES

*Abboud-Abi Saab, M. (1989) Distribution and ecology of tintinnids in the plankton of Lebanese coastal waters (eastern Mediterranean). *Journal of Plankton Research*, **11**, 203–222.

*Abboud-Abi Saab, M. (2002) Annual cycle of the microzooplankton communities in the waters surrounding the Palm Island Nature Reserve (north Lebanon), with special attention to tintinnids. *Mediterranean Marine Science* **3/2**, 55–76.

Abboud-Abi Saab, M. (2008) *Tintinnids of the Lebanese Coastal Waters (Eastern Mediterranean)*. CNRS-Lebanon/UNEP/MAP/RAC/SPA, Lebanon, 192pp.

Adams, J. (2009) The Holy Grail of ecology: latitudinal gradients. In *Species Richness: Patterns in the Diversity of Life*, pp. 47–95, Springer, Berlin.

Adl, S.M., Simpson, A.G., Farmer, M.A., Andersen, R.A., Anderson, O.R., Barta, J.R., Bowser, S.S., Brugerolle, G., Fensome, R.A., Fredericq, S., James, T.Y., Karpov, S., Kugrens, P., Krug, J., Lane, C.E., Lewis, L.A., Lodge, J., Lynn, D.H., Mann, D.G., McCourt, R.M., Mendoza, L., Moestrup, O., Mozley-Standridge, S.E., Nerad, T.A., Shearer, C.A., Smirnov, A.V., Spiegel, F.W. & Taylor, F.J.R. (2005) The new higher level classification of eukaryotes with emphasis on the taxonomy of protists. *Journal of Eukaryotic Microbiology*, **52**, 399–451.

Admiraal, W. & Venekamp, L.A.H. (1986) Significance of tintinnid grazing during blooms of *Phaeocystis pouchetii* (Haptophyceae) in Dutch coastal waters. *Netherlands Journal of Sea Research*, **20**, 61–66.

Adolf, J.E., Stoecker, D.K. & Harding, L.W. (2006) The balance of autotrophy and heterotrophy during mixotrophic growth of *Karlodinium micrum* (Dinophyceae). *Journal of Plankton Research*, **28**, 737–751.

Aelion, C.M. & Chisholm, S.W. (1985) Effect of temperature on growth and ingestion rates of *Favella* sp. *Journal of Plankton Research*, **7**, 821–830.

Aescht, E. (2001) Catalogue of the generic names of ciliates (Protozoa, Ciliophora). *Denisia*, **1**, 1–350.

Agassiz, A. (1906) Reports on the scientific results of the Expedition to the Eastern Tropical Pacific, in charge of Alexander Agassiz, by the U.S. Fish Commision steamer "Albatross" from October 1904, to March, 1905, Lieut Commander L.M. Garrett, U.S.N., commanding. V. General report of the expedition. *Memoires of the Mueum of Comparative Zoology Harvard*, **33**, 1–75.

Agatha, S. (2003) Redescription of *Strombidinopsis minima* (Gruber, 1884) Lynn et al., 1991 (Protozoa, Ciliophora), with notes on its ontogenesis and distribution. *European Journal of Protistology*, **39**, 233–244.

Agatha, S. (2004a) Evolution of ciliary patterns in the Oligotrichida (Ciliophora, Spirotricha) and its taxonomic implications. *Zoology*, **107**, 153–168.

Agatha, S. (2004b) A cladistic approach for the classification of oligotrichid ciliates (Ciliophora: Spirotricha). *Acta Protozoologica*, **43**, 201–217.

*Agatha, S. (2008) Redescription of the tintinnid ciliate *Tintinnopsis fimbriata* Meunier, 1919 (Spirotricha, Choreotrichida) from coastal waters of Northern Germany. *Denisia*, **23**, 261–272.

Agatha, S. (2009) Oligotrich planktonic ciliates. In *A Course in Protozoology* (ed. R. Röttger, R. Knight & W. Foissner), pp. 185–193, Protozoological Monographs **4**, 2nd revised edn., Shaker Verlag, Aachen.

*Agatha, S. (2010a) A light and scanning electron microscopic study of the closing apparatus in tintinnid ciliates

Asterisks before a reference denote references used in Chapter 10 but not necessarily cited in the chapter text. The reference provided data used in mapping the distributions of tintinnids.

The Biology and Ecology of Tintinnid Ciliates: Models for Marine Plankton, First Edition. Edited by John R. Dolan, David J.S. Montagnes, Sabine Agatha, D. Wayne Coats, and Diane K. Stoecker.

(Ciliophora, Spirotricha, Tintinnina): a forgotten synapomorphy. *Journal of Eukaryotic Microbiology*, **57**, 297–307.

*Agatha, S. (2010b) Redescription of *Tintinnopsis parvula* Jörgensen, 1912 (Ciliophora: Spirotrichea: Tintinnina), including a novel lorica matrix. *Acta Protozoologica*, **49**, 213–234.

Agatha, S. (2011) Updated hypothesis on the evolution of oligotrichid ciliates (Ciliophora, Spirotricha, Oligotrichida) based on somatic ciliary patterns and ontogenetic data. *European Journal of Protistology*, **47**, 51–56.

Agatha, S. & Foissner, W. (2009) Conjugation in the spirotrich ciliate *Halteria grandinella* (Müller, 1773) Dujardin, 1841 (Protozoa, Ciliophora) and its phylogenetic implications. *European Journal of Protistology*, **45**, 51–63.

*Agatha, S. & Riedel-Lorjé, J.C. (2006) Redescription of *Tintinnopsis cylindrica* Daday, 1887 (Ciliophora: Spirotricha) and unification of tintinnid terminology. *Acta Protozoologica*, **45**, 137–151.

Agatha, S. & Simon, P. (2012) On the nature of tintinnid loricae (Ciliophora: Spirotricha: Tintinnina): a histochemical, enzymatic, EDX, and high-resolution TEM study. *Acta Protozoologica*, **51**, 1–19.

Agatha, S. & Strüder-Kypke, M.C. (2007) Phylogeny of the order Choreotrichida (Ciliophora, Spirotricha, Oligotrichea) as inferred from morphology, ultrastructure, ontogenesis, and SSrRNA gene sequences. *European Journal of Protistology*, **43**, 37–63.

Agatha, S. & Strüder-Kypke, M.C. (2012) Reconciling cladistic and genetic analyses in choreotrichid ciliates (Protists, Spirotricha, Oligotrichea). *Journal of Eukaryotic Microbiology*, DOI: 10.1111/j.1550-7408.2012.00623.x.

*Agatha, S. & Tsai, S. (2008) Redescription of the tintinnid *Stenosemella pacifica* Kofoid and Campbell, 1929 (Ciliophora, Spirotricha) based on live observation, protargol impregnation, and scanning electron microscopy. *Journal of Eukaryotic Microbiology*, **55**, 75–85.

Agatha, S., Strüder-Kype, M.C. & Beran, A. (2004a) Morphologic and genetic variability in the marine planktonic ciliate *Laboea strobila* Lohmann 1908 (Ciliophora, Oligotrichia), with notes on its ontogenesis. *Journal of Eukaryotic Microbiology*, **51**, 267–281.

Agatha, S., Strüder-Kypke, M.C., Beran, A. & Lynn, D.H. (2005) *Pelagostrobilidium neptuni* (Montagnes and Taylor, 1994) and *Strombidium biarmatum* nov. spec. (Ciliophora, Oligotrichea): phylogenetic position inferred from morphology, ontogenesis, and gene sequence data. *European Journal of Protistology*, **41**, 65–83.

*Akizawa, K., Iwasaki, N. & Ueda, H. (1998) List of microplankton in Uranouchi Inlet, Kochi, Japan during the summer of 1997. *Bulletin of Marine Sciences and Fisheries Kochi University*, **18**, 57–63.

Akselman, R. & Santinelli, N. (1989) Observaciones sobre dinoflagelados parasitos en el litoral Atlantico sudoccidental. *Physis, B. Aires, Sec. A* **47**, 43–44.

Alder, V.A. (1999) Tintinnoinea. In *South Atlantic Zooplankton* (Boltovskoy, D. ed.), pp. 321–384, Backhuys Publishers, Leiden, Holland.

Alldredge, A.L. & Madin, L.P. (1982) Pelagic tunicates: Unique herbivores in the marine plankton. *Bioscience*, **32**, 655–663.

Allen, A.P. & Gillooly, J.F. (2006) Assessing latitudinal gradients in speciation rates and biodiversity at the global scale. *Ecology Letters*, **9**, 947–954.

Altiner, D. & Özkan, S. (1991) Calpionellid zonation in northwestern Anatolia (Turkey) and claibration of the stratigraphic ranges of some benthic formanifera at the Jurassic-Cretaceous boundary. *Geologica Romana*, **27**, 215–235.

Alvarez-Cadena, J.N. (1993) Feeding of the chaetognath *Sagitta elegans* Verrill. *Estuarine, Coastal and Shelf Science*, **36**, 195–206.

*Andersen, P. & Sørensen, H.M. (1986) Population dynamics and trophic coupling in pelagic microorganisms in eutrophic coastal waters. *Marine Ecology Progress Series*, **33**, 99–109.

Anderson, D.M. (1998) Physiology and bloom dynamics of toxic *Alexandrium* species, with emphasis on life cycle transitions. In *Physiological Ecology of Harmful Algal Blooms* (Anderson D.M., Cembella A.D., Hallegraeff G.M. eds) pp. 29–48, Springer, Berlin.

Anderson, D.M. & Keafer, B.A. (1987) An endogenous annual clock in the toxic marine dinoflagellate *Gonyaulax tamarensis*. *Nature*, **325**, 616–617.

Anderson, D.M. & Pfiester, L.A. (1987) Dinoflagellate reproduction. In *The Biology of Dinoflagellates*, (F.J.R. Taylor ed.) pp. 611–648, Blackwell, Oxford.

Anderson, D.M., Aubery, D.G., Tyler, M.A. & Coats, D.W. (1982) Vertical and horizontal distributions of dinoflagellate cysts in sediments. *Limnology and Oceanography*, **27**, 757–765.

Anderson, D.M., Cembella, A.D. & Hallegraeff, G.M. (1998) *Physiological Ecology of Harmful Algal Blooms*. Springer-Verlag, Berlin.

Anderson, D.M., Taylor, C.D. & Armbrust, E.V. (1987) The effects of darkness and anaerobiosis on dinoflagellate cyst germination. *Limnology and Oceanography*, **32**, 340–351.

Anderson, O.R. (1983) *Radiolaria*. Springer-Verlag, New York. 355 pp.

Anderson, O.R. (1993) The trophic role of planktonic Foraminifera and Radiolaria. *Marine Microbial Food Webs*, **7**, 31–51.

Angel, M.V. (1997) *Pelagic Biodiversity. Marine Biodiversity: patterns and processes* (Ormond R.F.G., Gage J.D., Angel M.V. eds), pp. 35–68, Cambridge University Press, Cambridge.

Angly, F.E., Felts, B., Breitbart, M., Salamon, P., Edwards, R.A., Carlson, C., Chan, A.M., Haynes, M., Kelley, S., Liu, H., Mahaffy, J.M., Mueller, J.E., Nulton, J., Olson, R., Parsons, R., Rayhawk, S., Suttle, C.A. & Rohwer, F. (2006) The marine viromes of four oceanic regions. *PLoS Biology*, **4**, 2121–2131.

Apple, J.K., Strom, S.L., Palenik, B. & Brahamsha, B. (2011) Variability in protist grazing and growth on different marine *Synechococcus* isolates. *Applied and Environmental Microbiology*, **77**, 3074–3084.

Apstein, C. (1893) Ein Fall von Conjugation bei Tintinnen. *Schriften des Naturwissenschaftlichen Vereins für Schleswig-Holstein*, **10**, 95–98.

Arashkevich, Y.G. (1969) The food and feeding of copepods in the northwestern Pacific. *Oceanology*, **9**, 695–705.

*Araujo, H.M.P., Nascimento-Viera, D.A., Neumann-Leitão, S., Schwamborn, R., Lucas, A.P.O. & Alves, J.P.H. (2008) Zooplankton community dynamics in relation to the seasonal cycle and nutrient inputs in an urban tropical estuary in Brazil. *Brazilian Journal of Biology*, **68**, 751–762.

Armstrong, H. & Brasier, M.D. 2005. *Microfossils*, 2nd ed., Blackwell, Oxford.

Arthur, D.K. (1976) Food and feeding of larvae of three fishes occurring in the California current, *Sardinops sagax*, *Engraulis mordax*, and *Trachurus symmetricus*. *Fishery Bulletin*, **74**, 517–530.

Ascaso, C., Wierzchos, J., Speranza, M., Gutiérrez, J.C., González, A.M., de los Ríos, A. & Alonso, J. (2005) Fossil protists and fungi in amber and rock substrates. *Micropaleontology*, **51**, 59–72.

Atkinson, D., Ciotti, B.J. & Montagnes, D.J.S. (2003) Protists decrease in size linear with temperature: ca. 2.5 °C^{-1} *Proceedings of the Royal Society (London) B*, **270**, 2605–2611.

*Ayukai, T. (1987) Predation by *Acartia clausi* (Copepoda: Calanoida) on two species of tintinnids. *Marine Microbial Food Webs*, **2**, 45–52.

*Ayukai, T. (1991) Standing stock of microzooplankton on coral reefs: a preliminary study. *Journal of Plankton Research*, **13**, 895–899.

Azam, F., Fenchel, T. Field, J.G., Meyer-Reil, R.A. & Thingstad, T.F. (1983) The ecological role of water column microbes in the sea. *Marine Ecology Progress Series*, **10**, 257–263.

*Aziz, N.E.A. (2006) Vertical migration of zooplankton groups in an eutrophic bay, Dekhaila Harbor, Alexandria, Egypt. *Egyptian Journal of Aquatic Research*, **32**, 135–157.

Bachvaroff, T.R. & Place, A.R. (2008) From stop to start: tandem gene arrangement, copy number and trans-splicing sites in the dinoflagellate *Amphidinium carterae*. *PLoS ONE*, **3**, e2929.doi:10.1371/journal.pone.0002929.

Bachvaroff, T.R., Handy, S.M., Place, A.R. & Delwiche, C.F. (2011) Alveolate phylogeny using ribosomal proteins. *Journal of Eukaryotic Microbiology*, **58**, 223–233.

Bachvaroff, T.R., Kim, S., Guillou, L., Delwiche, C.F. & Coats, D.W. (2012) Diversity of the syndinian genus *Euduboscquella* based on single cell PCR. *Applied and Environmental Microbiology*, (in press).

Bachy, C., Gómez, F., López-García, P., Dolan, J.R. & Moreira, D. (2012) Molecular phylogeny of tintinnid ciliates (Tintinnida, Ciliophora). *Protist*, **163**, in press.

Backus, R.H. (1986) Biogeographic boundaries in the open ocean. *UNESCO Technical Papers in Marine Science*, **49**, 9–13.

Bainbridge, V. (1958) Some observations on *Evadne nordmanni* Loven. *Journal of the Marine Biological Association of the United Kingdom*, **37**, 349–370.

Baker, J.R. (1958) Note on the use of bromophenol blue for the histochemical recognition of protein. *Quarterly Journal of Microscopical Science*, **99**, 459–460.

Bakker, C. & Phaff, W.J. (1976) Tintinnida from coastal waters of the S.W.-Netherlands I. The genus *Tintinnopsis* Stein. *Hydrobiologia*, **50**, 101–111.

Baldwin, B.S. & Newell, R.I.E. (1991) Omnivorous feeding by planktotrophic larvae of the eastern oyster *Crassostrea virginica*. *Marine Ecology Progress Series*, **78**, 285–301.

*Balech, E. (1942) Tintinnoineos del estrecho Le Marie. *Physis, Buenos Aires*, **19**, 242–254.

*Balech, E. (1944) Contribucion al conocimiento del plancton de Lennox y Cabo de Hornos. *Physis, Buenos Aires*, **19**, 423–446.

*Balech, E. (1945) Tintinnoinea de Quequen. *Physis, Buenos Aires*, **20**, 1–15.

*Balech, E. (1947) Contribucion al conocimiento del plancton antartico. Plancton del Mar de Bellingshausen. *Physis, Buenos Aires*, **20**, 75–91.

*Balech, E. (1948) Tintinnoinea de Atlantida (R.O. del Uruguay). *Comunicaciones del Museo Argentino de Ciencias Naturales "Bernardino Rivadavia", Serie Ciencias Zoologicas*, **7**, 1–23.

*Balech, E. (1957) Dinoflagellés et Tintinnides de la Terre Adélie (Secteur français antarctique). *Vie et Milieu*, **8**, 382–408.

*Balech, E. (1958) Plancton de la Campaña Antartica Argentina. *Physis, Buenos Aires*, **21**, 75–108.

*Balech, E. (1959a) The changes in the phytoplankton population off the California Coast. *CALCOFI Report* **1**, 127–132.

*Balech, E. (1959b) Tintinnoinea del Mediterraneo. *Trabajos del Instituto Español de Oceanografia*, **28**, 1–88.

*Balech, E. (1962) Tintinnoinea y Dinoflagellata del Pacifico segun material de las expediciones Norpac y Downwind del Instituto Scripps de Oceanographia. *Revista del Museo Argentino de Ciencias Naturales "Bernardino Rivadavia" Instituto Nacional de Investigacion de las Ciencias Naturales*, **7**, 3–249.

Balech, E. (1967a) *Microplankton of the Gulf of Mexico and the Caribbean Sea.* Texas A&M University Research Foundation, A&M Project 382, Texas A & M University, Department of Oceanography, College Station, Texas.

*Balech, E. (1967b) Dinoflagellates and tintinnids in the northeastern Gulf of Mexico. *Bulletin of Marine Science*, **17**, 280–298.

Balech, E. (1968) Algunas especies nuevas o interesantes de tintinnidos del Golfo de Mexico y Caribe. *Revista del Museo Argentino de Ciencias Naturales "Bernardino Rivadavia" e*

Instituto Nacional de Investigacion de las Ciencias Naturales, Hidrobiología, **2**, 165–197.

*Balech, E. (1971) Microplancton de la Campana Oceanografica Productividad III. *Revista del Museo Argentino de Ciencias Naturales "Bernardino Rivadavia" Instituto Nacional de Investigacion de las Ciencias Naturales, Hidrobiologia*, **3**, 1–202.

*Balech, E. (1973) Segunda contribucion al conocimiento del plancton del Mar de Bellingshausen. *Contribucion del Instituto Antartico Argentino*, **107**, 3–63.

*Balech, E. & El-Sayed, S.Z. (1965) Microplankton of the Weddell Sea. In *Biology of the Antarctic Seas*, pp. 107–124. American Geophysical Union, Washington D.C.

*Balech, E. & Souto, S. (1980) Los tintinnidos de la campana oceanografica "Productividad IV". Parte I. *Physis, Buenos Aires*, **39**, 1–8.

*Balkis, N. (2004) Tintinnids (Protozoa: Ciliophora) of the Büyükçekmece Bay in the Sea of Marmara. *Scientia Marina*, **68**, 33–44.

*Balkis, N. & Toklu-Aliçh, B. (2009) Tintinnid (Protozoa: Ciliophora) species in the Edremit Bay. *IUFS Journal of Biology*, **68**, 47–53.

Balkis, N. & Wasik, A. (2005) Species composition of the tintinnids found in the neritic water of Bozcaada Island, Aegean Sea, Turkey. *Fresenius Environmental Bulletin*, **14**, 327–333.

Ball, G.H. (1969) Organisms living on and in protozoa. In *Research in Protozoology*, Vol. **3**. (T.-T. Chen ed.), pp. 565–718. Pergamon Press, Ltd, Oxford.

Båmstedt, U., Gifford, D.J., Irigoien, X., et al. (2000) Feeding. In *ICES Zooplankton Methodology Manual* (R.P. Harris, P.H. Wiebe, J. Lenz, H.R. Skjodal and M. Huntley eds), pp. 297–399, Academic Press, New York.

Banse, K. (1962) Net zooplankton and total zooplankton. *Rapports et Process-Verbaux Réunion de la Conseil International pour Exploration de la Mer*, **153**, 211–215.

Banse, K. (1982) Cell volumes, maximal growth rates of unicellular algae, and the role of ciliates in the marine pelagial. *Limnology and Oceanography*, **27**, 1059–1071.

Barbaree, J.M., Fields, B.S., Feeley, J.C., Gorman, G.W. & Martin, W.T. (1986) Isolation of protozoa from water associated with a legionellosis outbreak and demonstration of intracellular multiplication of *Legionella pneumophila*. *Applied and Environmental Microbiology*, **51**, 422–424.

*Barría de Cao, M.S. (1986) Contribucion al conocimiento de Tintinnina (Protozoa, Ciliophora) de la zona de Bahia Blanca, II (Argentina). *Boletin de Institucion Español Oceanografia*, **3**, 143–150.

Barria de Cao, M.S. (1992) Abundance and species composition of tintinnina (Ciliophora) in Bahia Blanca Estuary, Argentina. *Estuarine, Coastal and Shelf Science*, **34**, 295–303.

*Barria de Cao, M.S., Pettigrosso, R., Parodi, E. & Ruben, F. (2003) Abundance and Species Composition of Planktonic Ciliophora from the Wastewater Discharge Zone in the Bahia Blanca Estuary, Argentina. *Iheringia, Seria Zoologica, Porto Alegre*, **93**, 229–236.

*Barría de Cao, M.S., Beigt, D. & Piccolo, C. (2005) Temporal variability of diversity and biomass of tintinnids (Ciliophora) in a southwestern Atlantic temperate estuary. *Journal of Plankton Research*, **27**, 1103–1111.

Barth, D., Krenek, S., Fokin, S.I. & Berendonk, T.U. (2006) Intraspecific genetic variation in *Paramecium* revealed by mitochondrial cytochrome c oxidase I sequences. *Journal of Eukaryotic Microbiology*, **53**, 20–25.

Basu, S., Mohan, M.L., Luo, X., Kundu, B., Kong, Q. & Singh, N. (2007) Modulation of proteinase K-resistant prion protein in cells and infectious brain homogenate by redox iron: implications for prion replication and disease pathogenesis. *Molecular Biology of the Cell*, **18**, 3302–3312.

Bathmann, U.V., Peinert, R., Non, T.T. & Bodungen, B.V. (1990) Pelagic origin and fate of sedimenting particles in the Norwegian Sea. *Progress in Oceanography*, **24**, 117–125.

Bé, A.W.H. & Tolderlund, D.S. (1971) Distribution and ecology of living planktonic foraminifera in surface waters of the Atlantic and Indian Oceans. In *The Micropalaeontology of Oceans* (B.M. Funnell & W.R. Riedel ed.), pp. 105–150, Cambridge University Press, Cambridge.

Beers, C.D. (1948) Excystment in the ciliate *Bursaria truncatella*. *Biological Bulletin*, **94**, 86–98.

*Beers, J.R. & Stewart, G.L. (1967) Micro-zooplankton in the euphotic zone at five locations across the California Current. *Journal of the Fisheries Research Board of Canada*, **24**, 2053–2068.

*Beers, J.R. & Stewart, G.L. (1969) The vertical distribution of microzooplankton and some ecological observations. *Journal du Conseil International pour l'Exploration de la Mer*, **33**, 30–44.

*Beers, J.R. & Stewart, G.L. (1970) Numerical abundance and estimated biomass of microzooplankton. In 'The ecology of the plankton off La Jolla, California, in the period April through September, 1967'. *Bulletin of the Scripps Institution of Oceanography*, **17**, 67–87.

*Beers, J.R. & Stewart, G.L. (1971) Micro-zooplankters in the plankton communities of the upper waters of the eastern tropical Pacific. *Deep-Sea Research*, **18**, 861–883.

Beers, J.R. (1982) An introduction and historical overview. *Annales de l'Instutut Océanographique Paris*, **58** (S), 5–14.

*Beers, J.R., Reid, F.M.H. & Stewart, G.L. (1975) Microplankton of the North Pacific Central Gyre. Population structure and abundance, June 1973. *Internationale Revue gesamten Hydrobiologie*, **60**, 607–638.

Beers, J.R., Reid, F.M.H. & Stewart, G.L. (1980). Microplankton population structure in southern California nearshore waters in late spring. *Marine Biology*, **60**, 209–226.

Bell, G. (1988) Sex and death in protozoa: the history of an obsession, Cambridge University Press, Cambridge.

Belmonte, G., Miglietta, A., Rubino, F. & Boero, F. (1997) Morphological convergence of resting stages of planktonic organisms: a review. *Hydrobiologia*, **355**, 159–165.

Bengtson, S. & Budd, B. (2004) Comment on "Small bilaterian fossils from 40 to 55 million years before the Cambrian". *Science*, **306**, 1291a.

Benzaggagh, M., Cecca, F. & Rouget, I. (2010) Biostratigraphic distribution of ammonites and calpionellids in the Tithonian of the internal Prerif (Msila area, Morocco). *Paläontologische Zeitschrift*, **84**, 301–315.

Berger, H. (1999) *Monograph of the Oxytrichidae (Ciliophora, Hypotrichia)*. Monographiae Biologicae 78, Springer, Dordrecht.

Berger, H. (2006) *Monograph of the Urostyloidea (Ciliophora, Hypotricha)*. Monographiae Biologicae 85, Springer, Dordrecht.

Berger, H. (2008) *Monograph of the Amphisiellidae and Trachelostylidae (Ciliophora, Hypotricha)*. Monographiae Biologicae 88, Springer, Dordrecht.

Berges, J. (1997) Ratios, regression statistics, and "spurious" correlations. *Limnology and Oceanography*, **42**, 1006–1007.

Berk, S.G., Faulkner, G., Garduño, E., Joy, M.C., Ortiz-Jimenez, M.A. & Garduño, R.A. (2008) Packaging of live *Legionella pneumophila* into pellets expelled by *Tetrahymena* spp. does not require bacterial replication and depends on a Dot/Icm-mediated survival mechanism. *Applied and Environmental Microbiology*, **74**, 2187–2199.

Berkes, F. (1976) Ecology of euphausiids in Gulf of St Lawrence. *Journal of the Fisheries Research Board of Canada*, **33**, 1894–1905.

*Bernard, C. & Rassoulzadegan, F. (1993) The role of picoplankton (cyanobacteria and plastidic picoflagellates) in the diet of tintinnids. *Journal of Plankton Research*, **15**, 361–373.

Bernatzky, G., Foissner, W. & Schubert, G. (1981) Rasterelektronenmikroskopische und biometrische Untersuchungen über die Variabilität der Form, Struktur und Größe des Gehäuses einiger limnischer Tintinnina (Protozoa, Ciliophora). *Zoologica Scripta*, **10**, 81–90.

Berney, C., Farhrni, J. & Pawlowski, J. (2004) How many novel eukaryotic 'kingdoms'? Pitfalls and limitations of environmental DNA surveys. *BioMed Central Biology*, **2**, 13.

Berney, C. & Pawlowski, J. (2006) A molecular time-scale for eukaryote evolution recalibrated with the contimuous microfossil record. *Proceedings of the Royal Society (London) B*, **273**, 1867–1872.

*Biancalana, F., Barria de Cao, M.S. & Hoffmeyer, M. (2007) Micro and mesozooplankton composition during winter in Ushuaia and Golondrina Bays (Beagle Channel, Argentina). *Brazilian Journal of Oceanography*, **55**, 83–95.

Biernacka, I. (1952) Studia nad rozrodem nicktórych gatunków rodzaju *Tintinnopsis* Stein. (Studies on the reproduction of some species of the genus *Tintinnopsis* Stein). *Annales Universitatis Mariae Curie-Sklowdowska, Sec. C*, **6**, 211–247.

Biernacka, I. (1965) Ausscheidung gehäusebildender Substanzen durch reife Formen gewisser Arten der Gattung *Tintinnopsis* Stein. *Acta Protozoologica*, **3**, 265–268.

Biernacka, I. (1968) Influence de la salinité et de la thermique sur les protistes de la mer Baltique et de la Lagune de la Vistule. *Ekologia Polska Seria A*, **16**, 261–278.

*Bigelow, H.B. (1928) Plankton of the offshore waters of the Gulf of Maine. *Bulletin of the Unites States Bureau of Fisheries*, **40**, 1–501.

Bignot, G. & Poignant, A. (2010) Les Pseudarcellinae, nouvelle sous-famille des Codonellidae (Tintinnida) – le genre *Pseudarcella* et l'espèce *Pseudarcella rhumbleri* Spandel, 1909 – description de nouvelles espèces. *Revue de Micropaléontologie*, **53**, 107–119.

Binder, B.J. & Anderson, D.M. (1986) Green light-mediated photomorphogenesis in a dinoflagellate resting cyst. *Nature*, **322**, 659–661.

*Biyu, S. (2000a) A comparative study on planktonic ciliates in two shallow mesotrophic lakes (China): species composition, distribution and quantitative importance. *Hydrobiologia*, **427**, 143–153.

*Biyu, S. (2000b) Planktonic protozooplankton (ciliates, heliozoans and testaceans) in two shallow mesotrophic lakes in China – a comparative study between a macrophyte-dominated lake (Biandantang) and an algal lake (Houhu). *Hydrobiologia*, **434**, 151–163.

*Blackbourn, D.J. (1974) The Feeding Biology of Tintinnid Protozoa and Some Other Inshore Microzooplankton. Ph.D. Thesis. University of British Columbia, Vancouver, BC, Canada.

Blackburn, N. & Fenchel, T. (1999) Modelling of microscale patch encounter by chemotactic protozoa. *Protist*, **150**, 337–343.

Blackburn, T.M., Cassey, P. & Gaston, K.J. (2006) Variations on a theme: sources of heterogeneity in the form of the interspecific relationship between abundance and distribution. *Journal of Animal Ecology*, **75**, 1426–1439.

Blackford, J.C., Allen, J.I. & Gilbert, F.J. (2004) Ecosystem dynamics at six contrasting sites: a generic modelling study. *Journal of Marine Systems*, **52**, 191–215.

Blatterer, H. & Foissner, W. (1990) Beiträge zur Ciliatenfauna (Protozoa: Ciliophora) der Amper (Bayern, Bundesrepublik Deutschland). *Archiv für Protistenkunde*, **138**, 93–115.

Bockstahler, K.R. & Coats, D.W. (1993) Spatial and temporal aspects of mixotrophy in Chesapeake Bay dinoflagellates. *Journal of Eukaryotic Microbiology*, **40**, 49–60.

*Boikova, E. (1984) Ecological character of protozoans (Ciliata, Flagellata) in the Baltic Sea. *Ophelia, Supplement* **3**, 23–32.

*Bojanic, N. (2001) Seasonal distribution of ciliated protozoa in Katela Bay. *Journal of the Marine Biological Association of the United Kingdom*, **81**, 383–390.

*Bojanic, N. (2007) Seasonal variability in abundance and biomass of tintinnids in Katela Bay (Middle Adriatic Sea). *Rapports de la Commission pour l'Exploration Scientifique de la mer Méditerranée*, **38**, 438.

Boltovskoy, D. & Alder, V.A. (1992) Microzooplankton and tintinnid species-specific assemblage structures: patterns of distribution and year-to-year variations in the Weddell Sea (Antarctica). *Journal of Plankton Research*, **14**, 1405–1423.

*Boltovskoy, D., Alder, V.A. & Spinelli, F. (1989) Summer Weddell Sea microplankton: Assemblage structure, distribution and abundance, with special emphasis on the Tintinnina. *Polar Biology*, **9**, 447–456.

*Boltovskoy, D., Dinofrio, E.O. & Alder, V.A. (1990) Intraspecific variability in Antarctic tintinnids: the *Cymatocylis affinis/convallaria* species group. *Journal of Plankton Research*, **12**, 403–413.

Boltovskoy, D., Alder, V.A. & Abelmann, A. (1993) Annual flux of Radiolaria and other shelled plankters in the eastern equatorial Atlantic at 853 m: seasonal variations and polycystine species-specific responses. *Deep-Sea Research I*, **40**, 1863–1895.

Boltovskoy, D., Uliana, E. & Wefer, G. (1996) Seasonal variation in the flux of microplankton and radiolarian assemblage compositions in the northeastern tropical Atlantic at 2.195 m. *Limnology and Oceanography*, **41**, 615–635.

Bonet, F. (1956) Zonificación microfaunística de las calizas cretácica del este de México. XX International Geolological Congress, 3–102.

Borregaard, M.K. & Rahbek, C. (2010) Causality of the relationship between geographic distribution and species abundance. *Quarterly Review of Biology*, **85**, 3–25.

*Borror, A.C. (1962) Ciliate protozoa of the Gulf of Mexico. *Bulletin of Marine Science*, **12**, 333–349.

Bosak, T., Macdonald, F., Lahr, D. & Matys, E. (2011) Putative Cryogenian ciliates from Mongolia. *Geology*, **39**, 1123–1126.

Bouley, P. & Kimmerer, W.J. (2006) Ecology of a highly abundant, introduced cyclopoid copepod in a temperate estuary. *Marine Ecology Progress Series*, **324**, 219–228.

Bowser, S.S. & Bernhard, J.M. (1993) Structure, bioadhesive distribution and elastic properties of the agglutinated test of *Astrammina rara* (Protozoa: Foraminiferida). *Journal of Eukaryotic Microbiology*, **40**, 121–131.

Box, G.E.P. & Draper, R.D.N. (1987) *Empirical Model-Building and Response Surfaces*, John Wiley & Sons, New York.

Bradbury, P.C. (1987) Protozoan adaptations for survival. In *Survival and Dormancy of Microorganisms*. Henis, Y. (ed), pp. 267–299. Wiley & Sons, New York.

*Brandini, F.P. (1993) Phytoplankton biomass on an Antarctic coastal environment during stable water conditions – implications for the iron limitation theory. *Marine Ecology Progress Series*, **93**, 267–275.

Brandt, K. (1901) Life in the ocean. Annual Report of the Board of Regents of the Smithsonian Institution 1901, pp. 493–506.

Brandt, K. (1905) On the production and conditions of production in the sea. *Rapports et Procès-Verbaux des Réunions, Conseil International pour l'Exploration de la Mer*, **3**, appendix D 12 pp.

Brandt, K. (1906) Die Tintinnodeen der Plankton-Expedition. Tafelerklärungen nebst kurzer Diagnose der neuen Arten. In *Ergebnisse der Plankton-Expedition der Humboldt-Stiftung* 3 La (ed. V. Hensen), pp. 1–33 + Plates 1–70. Lipsius and Tischer, Kiel, Leipzig.

Brandt, K. (1907) Die Tintinnodeen der Plankton-Expedition. Systematischer Teil. In *Ergebnisse der Plankton-Expedition der Humboldt-Stiftung* 3 La (ed. V. Hensen), pp. 1–488, Lipsius and Tischer, Kiel, Leipzig.

Braun, A., Chen, J., Waloszek, D. & Maas, A. (2007) First Early Cambrian Radiolaria. Geological Society, London, *Special Publications*, **286**, 143–149.

Brewer, G.D. & Kleppel, G.S. (1986) Diel vertical distribution of fish larvae and their prey in nearshore waters of southern California. *Marine Ecology Progress Series*, **27**, 217 226.

Bristow, T.F., Kennedy, M.F., Derkowski, A., Droser, M.L., Jiang, G. & Creaser, R.A. (2009) Mineralogical constraints on the paleoenvironments of the Ediacaran Doushantuo Formation. *Proceedings of the National Academy of Sciences of the United States of America*, **106**, 13190–13195.

Broglio, E., Johansson, M. & Jonsson, P.R. (2001) Trophic interaction between copepods and ciliates: effects of prey swimming behavior on predation risk. *Marine Ecology Progress Series*, **220**, 179–186.

Broglio, E., Jónasdóttir, H., Calbet, A., Jakobsen, H.H. & Saiz, E. (2003) Effect of heterotrophic verus autotrophic food on feeding and reproduction of the calanoid copepod *Acartia tonsa*: relationship with prey fatty acid composition. *Aquatic Microbial Ecology*, **31**, 267–278.

Broglio, E., Saiz, E., Calbet, A., Trepat, I. & Alcaraz, M. (2004) Trophic impact and prey selection by crustacean zooplankton on the microbial communities of an oligotrophic coastal area (NW Mediterranean Sea). *Aquatic Microbial Ecology*, **35**, 65–78.

Brown, J.H. (1984) On the relationship between abundance and distribution of species. *American Naturalist*, **124**, 255–279.

*Brownlee, D.C. (1977) The Significance of Cytological Characteristics as Revealed by Protargol Silver Staining in Evaluating the Systematics of the Ciliate Suborder Tintinnina. M.S. Thesis, University of Maryland, College Park MD, USA.

Brownlee, D.C. (1982) Measuring the Secondary Production of Marine Planktonic Tintinnine Ciliates. Ph.D. Thesis. University of Maryland, College Park MD, USA.

*Brownlee, D.C. & Jacobs, F. (1987) Mesozooplankton and microzooplankton in the Chesapeake Bay. In *Contaminant Problems and Management of Living Chesapeake Bay Resources*, (S.K. Majumdar L.W. Hall and & H.M. Austin eds), pp. 217–269, Pennsylvannia Academy of Science, Philadelphia, PA, USA.

Brunnschweiler, R.O. (1951) Discovery of the Late Jurassic genus Calpionella Lorenz (Oligotrich Infusoria) in Australia. *Australian Journal of Science*, **14**, 94.

Bugrova, E.M. (1983) New records of tintinnids (Infusoria) from the Eocene of the southern part of the USSR. *Paleontological Journal*, **8**, 78–81.

Bugrova, E.M. (2003) Paleogene ciliates and problematic microfossils from the northeastern margins of the Tethys Basin. *Paleontological Journal*, **37**, 1–7.

Buitenhuis, E.T., Rivkin, R.B., Sailley, S. & Le Quéré, C. (2010) Biogeochemical fluxes through microzooplankton. *Global Biogeochemical Cycles*, **24**, DOI: 10.1029/2009GB003601.

Burkill, P.H. (1982) Ciliates and other microplankton components of a near-shore food-web: Standing stocks and production processes. *Annales Institut Océanographique Paris*, **58** (S), 335–350.

Burkill, P.H., Mantoura, R.F.C., Llewellyn, C.A. & Owens, N.P.J. (1987) Microzooplankton grazing and selectivity of phytoplankton in coastal waters. *Marine Biology*, **93**, 581–590.

Burkovsky, I.V. (1973) Variability of *Parafavella denticulata* in the White Sea. *Zoologicheskii Zhurnal*, **52**, 1277–1285.

*Burkovsky, I.V. (1976) New data on tintinnids (Ciliata) of the Arctic and revision of fauna. *Zoologicheskii Zhurnal*, **55**, 325–336.

*Burkovsky, I.V., Zamyshlyak, E.A. & Poskryakova, N.P. (1974) Revision of the Tintinnida (Ciliata) fauna of the White Sea. *Zoologicheskii Zhurnal*, **53**, 1757–1766.

*Burns, D.A. (1983) The distribution and morphology of tintinnids (ciliate protozoans) from coastal waters around New Zealand. *New Zealand Journal of Marine and Freshwater Research*, **17**, 387–406.

*Bursa, A.S. (1961) The annual oceanographic cycle at Igloolik in the Canadian Arctic. II. Phytoplankton. *Journal of the Fisheries Research Board of Canada*, **18**, 563–615.

Busch, W. (1920) Dauercysten bei *Cyttarocylis edentata* var. *parumdentata* Brandt. *Archiv für Hydrobiologie*, **12**, 756–758.

Busch, W. (1925) Beitrag zur Kenntnis der Gehäusebildung bei den Tintinnidae und zur Kenntnis mariner Ciliaten. *Archiv für Protistenkunde*, **53**, 183–190.

Busch, W. (1948) Mitteilung von Beobachtungen an marinen Tintinnoinea. *Mitteilungen aus dem Museum für Naturkunde und Vorgeschichte und dem naturwissenschaftlichen Arbeitskreis zu Magdeburg*, **1**, 121–135.

Buskey, E.J. & Stoecker, D.K. (1988) Locomotory patterns of the planktonic ciliate *Favella* sp.: Adaptations for remaining within food patches. *Bulletin of Marine Science*, **43**, 783–796.

Buskey, E.J. & Stoecker, D.K. (1989) Behavioral responses of the marine tintinnid *Favella* sp. to phytoplankton: influence of chemical, mechanical and photic stimuli. *Journal of Experimental Marine Biology and Ecology*, **132**, 1–16.

Buskey, E.J., Coulter, C. & Strom, S. (1993) Locomotory patterns of microzooplankton – potential effects on food selectivity of larval fish. *Bulletin of Marine Science*, **53**, 29–43.

Bussers, J.C. (1984) Les appareils protecteurs des infusoires ciliés. In *Traité de Zoologie, Ciliés, Tome II, Fasicule I* (P.P. Grassé ed.), pp. 439–455 Masson, Paris.

Bussers, J.C. & Jeuniaux, C. (1974) Recherche de la chitine dans les productions métaplasmatiques de quelques ciliés. *Protistologica*, **10**, 43–46.

Butler, G.H., Kotani, H., Kong, L. et al. (1991) Identification and characterization of proteinase K-resistant proteins in members of the class Mollicutes. *Infection and Immunity*, **59**, 1037–1042.

Bütschli, O. (1887–1889) Erster Band. Protozoa. III. Abtheilung. Infusoria und System der Radiolaria. In *Klassen und Ordnungen des Thier-Reichs, wissenschaftlich dargestellt in Wort und Bild* (H.G. Bronn ed.), pp. 1089–2035 (1887: 1089–1280, 1888: 1281–1584, 1889: 1585–2035). C.F. Winter'sche Verlagshandlung, Leipzig.

Buzas, M.A., Collins, L.S. & Culvert, S.J. (2002) Latitudinal difference in biodiversity caused by higher tropical rate of increase. *Proceedings of the National Academy of Sciences of the United States of America*, **99**, 7841–7843.

Cachon, J. (1964) Contribution a l'étude des péridiniens parasites. Cytologie, cycles évolutifs. *Annales des Sciences Naturelles – Zoologie et Biologie Animale*, **6**, 1–158.

Cachon, J. & Cachon, M. (1987) Parasitic dinoflagellates. In *The Biology of Dinoflagellates* (F.J.R. Taylor ed.), pp. 571–610, Blackwell, Oxford.

Cai, S., Song, W., Xu, D. & Chiang, K. (2006) Morphological studies on the infraciliature of a planktonic ciliate, *Tintinnopsis brasiliensis* (Ciliophora: Tintinina). *Journal of Ocean University of China*, **5**, 55–57.

Calado, A.J., Craveiro, S.C., Daugbjerg, N. & Moestrup, Ø. (2009) Description of *Tyrannodinium* gen. nov., a freshwater dinoflagellate closely related to the marine *Pfiesteria*-like species. *Journal of Phycology*, **45**, 1195–1205.

Calbet, A. (2008) The trophic roles of microzooplankton in marine systems. *ICES Journal of Marine Science*, **65**, 325.

Calbet, A. & Landry, M.R. (1999) Mesozooplankton influences on the microbial food web: Direct and indirect trophic interactions in the oligotrophic open ocean. *Limnology and Oceanography*, **44**, 1370–1380.

Calbet, A. & Landry, M.R. (2004) Phytoplankton growth, microzooplankton grazing, and carbon cycling in marine systems. *Limnology and Oceanography*, **49**, 51–57.

Calbet, A. & Saiz, E. (2005) The ciliate–copepod link in marine ecosystems. *Aquatic Microbial Ecology*, **38**, 157–167.

*Calderon-Aragon, L.D.C. & Lopez-Ochoterena, E. (1973) Protozarios ciliados de Mexico. XX Sistematica y morfologia de algunas especies del orden Tintinnida Kofoid y Campbell, del Noroeste del Golfo de Mexico. *Revista de la Sociedad Mexicana de Historia Natural*, **34**, 71–86.

Calvo, P., Fernandez-Aliseda, M.C., Garrido, J. & Torres, A. (2003) Ultrastructure, encystment and cyst wall composition of the resting cyst of the peritrich ciliate *Opisthonecta henneguyi*. *Journal of Eukaryotic Microbiology*, **50**, 49–56.

Campbell, A. (1954) *Treatise on Invertebrate Paleontology. Tintinnina. Protista 3* (A. Campbell, A. and R.C. Moore eds), pp. 166–180, Kansas University Press, Lawrence.

*Campbell, A.S. (1926) On *Tintinnus neriticus*, sp. nov., from San Francisco Bay. *University of California Publications in Zoology*, **29**, 237–239.

*Campbell, A.S. (1926) The cytology of *Tintinnopsis nucula* (Fol) Laackmann with an account of its neuromotor apparatus, division, and a new intranuclear parasite. *University of California Publications in Zoology*, **29**, 179–236.

*Campbell, A.S. (1927) Studies on the marine ciliate *Favella* (Jörgensen), with special regard to the neuromotor apparatus and its rôle in the formation of the lorica. *University of California Publications in Zoology*, **29**, 429–452 + Plates 21, 22.

Campbell, A.S. (1942) *The Oceanic Tintinnoina of the Plankton Gathered during the Last Cruise of the Carnegie*. Scientific Results of Cruise VII of the Carnegie during 1928–1929 under Command of Captain J. P. Ault, Biology II. Publication **537**, Carnegie Institution of Washington, Washington, D.C.

Canter, H.M. & Dick, M.W. (1994) *Eurychasmopsis multisecunda* gen. et sp. nov., a parasite of the suctorian ciliate Podophrya. *Mycological Research*, **98**, 105–117.

Canter, H.M., Heaney, S.I. & Lund, J.W.G. (1990) The ecological significance of grazing on planktonic populations of cyanobacteria by the ciliate Nassula. *New Phytologist*, **114**, 247–263.

Cantu-Chapa, A. (1996) Qué hacer con las calpionelas del límite Jurásico-Cretácio? *Revista Espangol de Micropaleontogica*, **28**, 57–65.

Caperon, J., Schell, D., Hirota, J. & Laws, E. (1979) Ammonium excretion rates in Kaneohe Bay, Hawaii, measured by a 15-N isotope dilution technique. *Marine Biology*, **54**, 33–40.

Capriulo, G.M. (1982) The Feeding Ecology of Temperate Zone Coastal Water Tintinnids. Ph.D. Thesis, State University of New York at Stony Brook, Stony Brook, NY.

*Capriulo, G.M. (1982) Feeding of field collected tintinnid microzooplankton on natural food. *Marine Biology*, **71**, 73–86.

Capriulo, G.M. (1990) *Ecology of Marine Protozoa*, Oxford University Press, Oxford.

Capriulo, G.M. & Carpenter, E.J. (1980) Grazing by 35 to 202 um micro-zooplankton in Long Island Sound. *Marine Biology*, **56**, 319–326.

*Capriulo, G.M. & Carpenter, E.J. (1983) Abundance, species composition and feeding impact of tintinnid micro-zooplankton in Central Long Island Sound. *Marine Ecology Progress Series*, **10**, 277–288.

*Capriulo, G.M. & Ninivaggi, D.V. (1982) A comparison of the feeding activities of field collected tintinnids and copepods fed identical natural particle assemblages. *Annales de L'institute Oceanographique Paris*, **58** (S), 325–334.

Capriulo, G.M., Gold, K. & Okubo, A. (1982) Evolution of the lorica in tintinnids: a possible selective advantage. *Annales Institut Océanographique Paris*, **58** (S), 319–324.

*Capriulo, G.M., Taveras, J. & Gold, K. (1986) Ciliate feeding: effect of food presence or absence on occurrence of striae in tintinnids. *Marine Ecology Progress Series*, **30**, 145–158.

*Cardinal, A., Lafleur, P.-E. & Bonneau, E. (1977) Les tintinnides (Ciliata: Tintinnida) des eaux marines et saumatres du Quebec. I. Formes hyalines. *Acta Protozoologica* **16**, 15–22.

Carey, P.G. (1992) *Marine Interstitial Ciliates*. Chapman and Hall, London.

*Cariou, J.-B., Dolan, J.R. & Dallot, S. (1999) A preliminary study of tintinnid diversity in the NW Mediterranean Sea. *Journal of Plankton Research*, **21**, 1065–1075.

Carlsson, P., Graneli, E. & Olsson, P. (1990) Grazer elimination through poisoning: one of the mechanisms behind *Chrysochromulina polylepis* blooms. In *Toxic marine phytoplankton* (E. Granéli, B. Sundström, L. Edler and D.M. Anderson eds), pp. 116–122, Elsevier, New York.

Caron, D.A. & Bé, A.W.H. (1984) Predicted and observed feeding rates of the spinose planktonic foraminifer *Globigerinoides sacculifer*. *Bulletin of Marine Science*, **35**, 1–10.

Caron, D.A. & Swanberg, N.R. (1990) The ecology of planktonic sarcodines. *Reviews in Aquatic Sciences*, **3**, 147–180.

Caron, D.A., Davis, P.G. & Madin, L.P. (1982) Heterotrophic bacteria and bacterivorous protozoa in oceanic macroaggragates. *Science*, **218**, 795–797.

Carpenter, E.J. & Foster, R.A. (2003) Marine cyanobacterial symbioses. In *Cyanobacteria in Symbiosis* (A.N. Rai, B. Bergmann and U. Rasmussen eds), pp. 11–17, Kluwer, Amsterdam.

*Carrick, H.J. (2005) An under-appreciated component of biodiversity in plankton communities: the role of protozoa in Lake Michigan (a case study). *Hydrobiologia*, **551**, 17–32.

Carrick, H.J., Fahnenstiel, G.L. & Taylor, W.D. 1992. Growth and production of planktonic protozoa in Lake Michigan: in situ versus in vitro comparisons and importance to food web dynamics. *Limnology and Oceanography*, **37**, 1221–1235.

*Casagranda, C. & Boudouresque, C.F. (2010) A first quantification of the overall biomass, from phytoplankton to birds, of a Mediterranean brackish lagoon: revisiting the ecosystem of Lake Ichkeul (Tunisia). *Hydrobiologia* **637**, 73–85.

Cermano, P., Dutkiewicz, S., Harris, R.P., Follows, M., Schofield, O. & Falkowski, P.G. (2008) The role of nutricline depth in regulating the ocean carbon cycle. *Proceedings of the National Academy of Sciences of the United States of America*, **105**, 20344–20349.

Chambouvet, A., Alves-de-Souza, C. Cueff, V., Marie, D. Karpov, S. & Guillou, L. (2011) Interplay between the parasite *Amoebophrya* sp. (Alveolata) and the cyst formation of the red tide dinoflagellate *Scrippsiella trochoidea*. *Protist*, **162**, 637–649.

Chandrasekera, W.U. & Fernando, M.A.S.T. (2009) Accidental introduction of alien plankton into the Sr Lankan coastal zone through ballast water of cargo ships. *Sri Lanka Journal of Aquatic Sciences*, **14**, 87–103.

*Chang, F.H. (1983) Winter phytoplankton and microzooplankton populations off the coast of Westland, New Zealand, 1979. *New Zealand Journal of Marine and Freshwater Research*, **17**, 279–304.

*Chang, F.H. (1990) Quantitative distribution of microzooplankton off Westland, New Zealand. *New Zealand Journal of Marine and Freshwater Research*, **24**, 187–195.

Chatton, E. (1920) Les peridiniens parasites. Morphologie, reproduction, ethologie. *Archives Zoologie Experimentale Generale*, **59**, 1–475.

Chatton, E. (1952) Classe des Dinoflagellés ou Péridiniens. In *Traité de Zoologie*, Vol. 1 (ed. P.P. Grassé, P.P.), pp. 309–390, Masson et Cie, Paris.

Chatton, E. & Biecheler, B. 1936. Documents nouveaux relatives aux Coccidinides (Dinoflagellés parasites). La sexualité du *Coccidinium mesnili* n. sp. *Comptes Rendues Academie Sciences, Paris*, **203**, 573–576.

Chatton, E., Lwoff, A., Lwoff, M. & Monod, J.L. (1931) La formation de l'ébauche buccale postérieur chez les ciliés en division et ses relation de continuité topographique et génétique avec la bouche antérieure. *Comptes Rendus de la Société de Biologie, Paris*, **107**, 540–548.

Chen, J.-Y., Bottjer, D.J., Li, G., Hadfield, M.G., Gao, F., Cameron, A.R., Zhang, C.-Y., Xian, D.-C., Tafforeau, P., Liao, X. & Yin, Z.-J. (2009) Complex embryos displaying bilaterian characters from Precambrian Doushantuo phosphate deposits, Weng'an, Guizhou, China. *Proceedings of the National Academy of Sciences of the United States of America*, **106**, 19056–19060.

*Chen, Y. & Yang, Y. (2009) Characteristics of the microzooplankton community in Jiaozhou Bay, Qingdao, China. *Chinese Journal of Oceanology and Limnology*, **27**, 435–442.

Chennaux, G. (1968) Présence de tintinnoidiens dans l'Ordovcien du Sahara. *Comptes Rendues de l'Academie de Science (France), serie D*, **266**, 86–87.

*Chester, A.J. (1978) Microzooplankton in the surface waters of the Strait of Juan de Fuca. NOAA Technical Report ERL 403-PMEL 30.

*Choi, J.K., Coats, D.W., Brownlee, D.C. & Small, E.B. (1992) Morphology and infraciliature of three species of *Eutintinnus* (Ciliophora; Tintinnina) with guidelines for interpreting protargol-stained tintinnine ciliates. *Journal of Protozoology*, **39**, 80–92.

Claparède, É. & Lachmann, J. (1858) Études sur les infusoires et les rhizopodes. *Mémoires de l'Institut National Genevois*, **5** (year 1857), 1–260 + Plates 1–13.

Clarke, A. (1992) Is there a latitudinal diversity cline in the sea? *Trends in Ecology and Evolution*, **7**, 286–287.

*Cleve, P.T. (1899) Plankton collected by the Swedish expedition to Spitzbergen in 1898. *Kongl Svenska Vetenskaps-Akademiens Handlingar* **32**, 51 p.

*Cleve, P.T. (1900a) Plankton from the southern Atlantic and the sourthern Indian Ocean. Ofversigt af Kongl. *Vetenskaps-Akademiens Förhandlingar*, **57**, 919–938.

*Cleve, P.T. (1900b) Some Atlantic Tintinnodea. Ofversigt af Kongl. *Vetenskaps-Akademiens Förhandlingar*, **56**, 969–975.

Cleven, E-J. (1996) Indirectly fluorescently labelled flagellates (IFLF): a tool to estimate the predation on free-living heterotrophic flagellates. *Journal of Plankton Research*, **18**, 429–442.

Clough, J. & Strom, S. (2005) Effects of *Heterosigma akashiwo* (Raphidophyceae) on protist grazers: laboratory experiments with ciliates and heterotrophic dinoflagellates. *Aquatic Microbial Ecology*, **39**, 121–134.

*Coats, D.W. (1988) *Duboscquella cachoni* N. Sp., a parasitic dinoflagellate lethal to its tintinnine host *Eutintinnus pectinis*. *Journal of Protozoology*, **35**, 607–617.

Coats, D.W. (1999) Parasitic life styles of marine dinoflagellates. *Journal of Eukaryotic Microbiology*, **46**, 402–409.

*Coats, D.W. & Heinbokel, J.F. (1982) A study of reproduction and other life cycle phenomena in planktonic protists using an acridine orange fluorescence technique. *Marine Biology*, **67**, 71–79.

*Coats, D.W. & Heisler, J.J. (1989) Spatial and temporal occurrence of the parasitic dinoflagellate *Dubosquella cachoni* and its tintinnine host *Eutintinnus pectinis* in Chesapeake Bay. *Marine Biology*, **101**, 401–409.

Coats, D.W., Bockstahler, K.R., Berg, G.M. & Sniezek, J.H. (1994) Dinoflagellate infections of *Favella panamensis* from two North American estuaries. *Marine Biology*, **119**, 105–113.

*Coats, D.W., Kim, S., Bachvaroff, T.R., Handy, S.M. & Delwiche, C.F. (2010) *Tintinnophagus acutus* n. g., n. sp. (Phylum Dinoflagellata), an ectoparasite of the ciliate *Tintinnopsis cylindrica* Daday 1887, and its relationship to *Duboscquodinium collini* Grassé 1952. *Journal of Eukaryotic Microbiology*, **57**, 468–482.

Coats, D.W., Bachvaroff, T.R. & Delwiche, C.F. (2012) Revision of the family Duboscquellidae with description of *Euduboscquella crenulata* n. gen., n. sp. (Dinoflagellata, Syndinea), an intracellular parasite of the ciliate *Favella panamensis* Kofoid and Campbell, 1929. *Journal of Eukaryotic Microbiology*, **85**, in press.

Colin, S.P. & Dam, H.G. (2003) Effects of the toxic dinoflagellate *Alexandrium fundyense* on the copepod *Acartia hudsonica*: a test of the mechanisms that reduce ingestion rates. *Marine Ecology Progress Series*, **248**, 55–65.

Colin, S.P., Costello, J.H., Graham, W.M. & Higgins III, J. (2005) Omnivory by the small cosmopolitan hydromedusa *Aglaura hemistoma*. *Limnology and Oceanography*, **50**, 1264–1268.

Colom, G. (1934) Estudios sobre las Calpionela. *Boletín de la Sociedad Española de Historia Natural*, **34**, 379–388.

Colom, G. (1948) Fossil tintinnids: loricated infusoria of the order of the oligotricha. *Journal of Paleontology*, **22**, 233–263.

Colom, G. (1955) Jurassic-Cretaceous pelagic sediments of the western Mediterranean zone and the Atlantic Ocean. *Micropaleontology*, **1**, 109–124.

Colom, G. (1965) Essais sur la biologie, la distribution geographique et stratigraphique des tintinnoidiens fossiles. *Ecolgae Geologica Helvetica*. **58**, 319–334.

Conover, R.J. (1982) Interactions between microzooplankton and other plankton organisms. *Annales de l'Institut Océanographique, Paris*, **58** (Supplement), 31–46.

Conway, D.V.P. (1980) The food of larval blue whiting, *Micromesistius poutassou* (Risso), in the Rockall area. *Journal Fish Biology*, **16**, 709–723.

*Cordeiro, T.A., Brandini, F.P. & Martens, P. (1997) Spatial distribution of the Tintinnina (Ciliophora, Protista) in the North Sea, spring of 1986. *Journal of Plankton Research*, **19**, 1371–1383.

Corliss, B.H., Brown, C.W., Sun, X. & Showers, W.J. (2009) Deep-sea benthic diversity linked to seasonality of pelagic productivity. *Deep-Sea Research I* **56**, 835–841.

Corliss, J.O. (1973) Protozoan ecology: a note on its current status. *American Zoologist*, **13**, 145–148.

Corliss, J.O. (1979) *The Ciliated Protozoa. Characterization, Classification and Guide to the Literature*, 2nd edition. Pergamon Press, Oxford.

Corliss, J.O. & Esser, S.C. (1974) Comments on the role of the cyst in the life cycle and survival of freeliving protozoa. *Transactions of the American Microscopical Society*, **95**, 578–593.

*Cosper, T.C. (1972) The identification of tintinnids (Protozoa: Ciliata: Tintinnida) of the St. Andrew Bay system, Florida. *Bulletin of Marine Science*, **22**, 391–418.

Costello, J.H. & Coverdale, R. (1998) Planktonic feeding and evolutionary significance of the lobate "body" plan within the Ctenophora. *Biological Bulletin*, **195**, 247–248.

Courtois, R. & Dodson, J.J. (1986) Régime alimentaire et principaux facteurs influençant l'alimentation des larves de capelan (*Mallotus villosus*), d'éperlan (*Osmerus mordax*) et de hareng (*Clupea harengus harengus*) dans un estuaire partiellement mélangé. *Canadian Journal of Fisheries and Aquatic Science*, **43**, 968–979.

Cromer, L., Gibson, J.A.E., Swadling, K.M. & Ritz, D.A. (2005) Faunal microfossils: indicators of Holocene ecological change in a saline Antarctic lake. *Paleogeography, Palaeoclimatology, Palaeoecology*, **221**, 83–97.

Cunha, A.M. da & Fonseca, O. da (1917) O microplancton do Atlantico nas imediações de Mar del Plata. *Memorias do Instituto Oswaldo Cruz*, **9**, 140–142.

Curds, C.R. & Wu, I.C.H. (1983) A review of the Euplotidae (Hypotrichida, Ciliophora). *Bulletin of the British Museum of Natural History (Zoology)*, **44**, 191–247.

Cuvillier, J. & Barreyre, M. (1964) Presence de tintinnoidiens dans le Viséen des Asturias. *Revue de Micropaleontologie*, **7**, 80–81.

Cuvillier, J. & Sacal, V. (1963) Presence de tintinnoidiens dans le Devoinian superieur du Sahara septronial. *Revue de Micropaleontologie*, **6**, 73–75.

Daday, E. von (1886) Kenntnis der Infusorien-Fauna des Golfes von Neapel. *Mittheilungen aus der Zoologischen Station zu Neapel*, **6**, 33–498.

Daday, E. von (1887) Monographie der Familie der Tintinnodeen. *Mittheilungen aus der Zoologischen Station zu Neapel*, **7**, 473–591.

*Dakin, W.J. & Colefax, A.N. (1940) The plankton of the Australian coastal waters off New South Wales. Part 1. Publications of the University of Sydney. Department of zoology. Monograph no. l, 215 pp.

*Dale, T. (1987) Diel vertical distribution of planktonic ciliates in Lindaspollene, western Norway. *Marine Microbial Food Webs*, **2**, 15–28.

*Dale, T. (1988) Oil pollution and plankton dynamics. VI. Controlled ecosystem experiments in Lindaspollene, Norway, June 1981: Effects on planktonic ciliates following nutrient addition to natural and oil-polluted enclosed water columns. *Sarsia*, **73**, 179–191.

*Dale, T. (1989) Sedimentation of loricas of tintinnids (Ciliata, Protozoa) and shells of bivalve larvae in the landlocked fjord, Lindaspollene, Norway. *Archiv für Protistenkunde*, **137**, 45–55.

*Dale, T. & Dahl, E. (1987) Mass occurrence of planktonic oligotrichous ciliates in a bay in southern Norway. *Journal of Plankton Research*, **9**, 871–879.

Damodara, N.W. (1981) Studies on Tintinnids (Protozoa: Ciliata) of Proto Novo Region, S. India. Ph.D. Thesis, Annamalai University, Tamil Nadu, India.

Darling, K.F., Kucera, M., Pudsey, C.J. & Wade, C.M. (2004) Molecular evidence links cryptic diversification in polar planktonic protists to Quaternary climate dynamics. *Proceedings National Academy of Sciences the United States*, **101**, 7657–7662.

Darwin, C. (1859) *On the Origin of Species by Means of Natural Selection, or the Preservation of Favoured Races in the Struggle for Life*. John Murray, London.

Davenport, J., Smith, R.J.J.W. & Packer, M. (2000) Mussels *Mytilus edulis*: significant consumers and destroyers of mesozooplankton. *Marine Ecology Progress Series*, **198**, 131–137.

David, M., Gollasch, S., Cabrini, M., Perkovic, M., Bosnjak, D. & Virgilio, D. (2007) Results from the first ballast water sampling study in the Mediterranean Sea – the Port of Koper study. *Marine Pollution Bulletin*, **54**, 53–65.

*Davis, C.C. (1978) Variations of the lorica in the genus *Parafavella* (Protozoa: Tintinnida) in northern Norway waters. *Canadian Journal of Zoology*, **56**, 1822–1827.

Davis, C.C. (1981) Variations of lorica shape in the genus *Ptychocylis* (Protozoa: Tintinnina) in relation to species identification. *Journal of Plankton Research*, **3**, 433–443.

*Davis, C.C. (1982) A preliminary quantitative study of the zooplankton from Conception Bay, insular Newfoundland, Canada. *Internationale Revue gesamten Hydrobiologie*, **67**, 713–747.

Davis, C.C. (1985) *Acanthostomella norvegica* (Daday) in Insular Newfoundland waters, Canada (Protozoa: Tintinnina). *Internationale Revue der Gesamten Hydrobiologie*, **70**, 21–26.

*Davis, C.C. (1986) A comparison of the zooplankton in two Newfoundland bays with differing influences from major currents. *Internationale Revue gesamten Hydrobiologie*, **71**, 11–47.

*de Faria, J.G. & Cunha, A.M. (1917) Estudios sobre o microplancton da ba¡a Rio de Janero e suas imediatoes. *Memorias do Instituto Oswaldo Cruz*, **9**, 68–93.

De Vargas, C., Norris, R., Zaninetti, L., Gibb, S.W. & Pawlowski, J. (1999) Molecular evidence of cryptic speciation in planktonic foraminifera and their relation to oceanic provinces. *Proceedings of the National Academy of Sciences of the United States of America*, **96**, 2864–2868.

Dennett, M.R., Caron, D.A., Murzov, S.A., Polikarpov, I.G., Gavrilova, N.A., Georgieva, L.V. & Kuzmenko, L.V. 1999. Abundance and biomass of nano- and microplankton during the 1995 Northeast Monsoon and Spring Intermonsoon in the Arabian Sea. *Deep-Sea Research II*, **46**, 1691–1717.

Deroux, G. (1974) Quelques précisions sur *Strobilidium gyrans* Schewiakoff. *Cahiers de Biologie Marine*, **15**, 571–588.

Detwyler, R. & Houde, E.D. (1970) Food selection by laboratory-reared larvae of the scaled sardine *Harengula pensacolae* (Pisces, Clupeidae) and the bay anchovy *Anchoa mitchilli* (Pisces, Engraulidae). *Marine Biology*, **7**, 214–222.

*Diodato, S.L. & Hoffmeyer, M.S. (2008) Contribution of planktonic and detritic fractions to the natural diet of mesozooplankton in Bahia Blanca Estuary. *Hydrobiologia*, **614**, 83–90.

Dobson, A., Lafferty, K.D., Kuris, A.M., Hechinger, R.F. & Jetz, W. (2008) Homage to Linnaeus: How many parasites? How many hosts? *Proceedings of the National Academy of Sciences of the United States of America* **105** (Supplement 1), 11482–11489.

Dobzhansky, T. (1964) Biology, molecular and organismic. *American Zoologist*, **4**, 443–452.

Dodge, J.D. & Lee, J.J. (2000). Phylum Dinoflagellata (Bütschli, 1885). In *An Illustrated Guide to the Protozoa* (J.J. Lee, G.F. Leedale, G.F. and P. Bradbury eds), pp. 565–689, Allen Press Inc., Lawrence, Kansas.

Dogopolskaya, M.A. (1940) Zooplankton of the Black Sea near Karadag. *Transactions of the Karadag Biological Station*, **1940**, 57–110.

Doherty, M., Tamura, M., Vriezen, J.A.C., McManus, G.B. & Katz, L.A. (2010) Diveristy of Oligotrichia and Choreotrichia ciliates in coastal marine sediments and in overlying plankton. *Applied and Environmental Microbiology*, **76**, 3924–3935.

*Dohi, K. (1982) Seasonal changes of tintinnid community in Funka Bay. *Bulletin of the Plankton Society of Japan*, **29**, 77–87.

Dolan, J.R. (1991) Microphagous ciliates in mesohaline Chesapeake Bay waters – estimates of growth rates and consumption by copepods. *Marine Biology*, **111**, 303–309.

Dolan, J.R. (1997) Phosphorus and ammonia excretion by planktonic protists. *Marine Geology*, **139**, 109–122.

*Dolan, J.R. (2005) Introduction to the biogeography of aquatic microbes. *Aquatic Microbial Ecology*, **41**, 39–48.

Dolan, J.R. (2010) Morphology and ecology in tintinnid ciliates of the marine plankton: correlates of lorica dimensions. *Acta Protozoologica*, **49**, 235–344.

Dolan, J.R. (2011) The legacy of the last cruise of the Carnegie: a lesson in the value of dusty old taxonomic monographs. *Journal of Plankton Research* **33**, 1317–1324.

*Dolan, J.R. & Coats, D.W. (1991) Changes in fine-scale vertical distributions of ciliate microzooplankton related to anoxia in Chesapeake Bay waters. *Marine Microbial Food Webs*, **5**, 81–93.

Dolan, J.R. & Gallegos, C.L. (1991) Trophic coupling of rotifers, microflagellates, and bacteria during fall months in the Rhode River Estuary. *Marine Ecology Progress Sereries*, **77**, 147–156.

Dolan, J.R. & Gallegos, C.L. (1992) Trophic role of planktonic rotifers in the Rhode River Estuary, spring–summer 1991. *Marine Ecology Progress Series*, **85**, 187–199.

*Dolan, J.R. & Gallegos, C.L. (2001) Estuarine diversity of tintinnids (planktonic ciliates). *Journal of Plankton Research*, **23**, 1009–1027.

Dolan, J.R. & McKeon, K. (2004) The reliability of grazing rate estimates from dilution experiments: Have we overestimated rates of organic carbon consumption? *Ocean Science*, **1**, 1–7.

Dolan, J.R. & Stoeck, T. (2011) Repeated sampling reveals differential variability in measures of species richness and community composition in planktonic protists. *Environmental Microbiology Reports*, **3**, 661–666.

*Dolan, J.R., Gallegos, C.L. & Moigis, A. (2000) Dilution effects on microzooplankton in dilution grazing experiments. *Marine Ecology Progress Series*, **200**, 127–139.

Dolan, J.R., Clauste, H., Carlotti, F., Plounevez, S., Moutin, T. (2002) Microzooplankton diveristy: relationships of tintinnid ciliates with reources, competitors and predators from the Atlantic Coast of Morocco to the Eastern Mediterranean. *Deep-Sea Research I*, **49**, 1217–1232.

Dolan, J.R., Sall, N., Metcalfe, A. & Gasser, B. (2003) Effects of turbulence on the feeding and growth of a marine oligotrich ciliate. *Aquatic Microbial Ecology*, **31**, 183–192.

*Dolan, J.R., Jacquet, S. & Torreton, J.-P. (2006) Comparing taxonomic and morphological biodiversity of tintinnids (planktonic ciliates) of New Caledonia. *Limnology and Oceanography*, **51**, 950–958.

*Dolan, J.R., Lemée, R., Gasparini, S., Mousseau, L. & Heyndrickx, C. (2006) Probing diversity in the plankton: using patterns in Tintinnids (planktonic marine ciliates) to identify mechanisms. *Hydrobiologia*, **555**, 143–157.

*Dolan, J.R., Ritchie, M.R. & Ras, J. (2007) The neutral community structure of planktonic herbivores, tintinnid ciliates of the microzooplankton, across the SE Pacific Ocean. *Biogeosciences*, **4**, 397–410.

*Dolan, J.R., Ritchie, M.E., Tunin-Ley, A. & Pizay, M.-D. (2009) Dynamics of core and occasional species in the marine plankton: tintinnid ciliates in the north-west Mediterranean Sea. *Journal of Biogeography*, **36**, 887–895.

Dolven, J.K., Lindqvist, C., Albert, V.A., Bjørklund, K.R., Yuasa, T., Takahashi, O. & Mayama, S. (2007) Molecular diversity of alveolates associated with neritic North Atlantic radiolarians. *Protist*, **158**, 65–76.

Dong, X., Knoll, A.H. & Lipps, J.H. (1997) Late Cambrian Radiolaria from Hunan, China. *Journal of Paleontology*, **71**, 753–758.

*Dorgham, M.M., El-Ghobashy, E.A. & El-Tohamy, S.W. (2009) Preliminary study on protozoan community in Damietta Harbor, Egypt. *Global Veterinaria*, **3**, 495–502.

Dornbos, S.Q., Bottjer, D.J., Chen, J.-Y., Oliveri, P., Gao, F. & Li, C.-W. (2005) Precambrian animal life: Taphonomy of phosphatized metazoan embryos from southwest China. *Lethaia*, **38**, 101–109.

Douzery, E.J.P., Snell, E.A., Bapteste, E., Delsuc, F. & Philippe, H. (2004) The timing of eukaryotic evolution: does a relaxed molecular clock reconcile proteins and fossils? *Proceedings of the National Academy of Sciences of the United States of America*, **101**, 15386–15391.

Duboscq, O. & Collin, B. (1910) Sur la reproduction sexuée d'un protiste parasite des tintinnides. *Comptes Rendus de l'Académie des Sciences de Paris*, **51**, 340–341.

*Duff, R.J., Ball, H. & Lavrentyev, P.J. (2008) Application of combined morphological–molecular approaches to the identification of planktonic protists from environmental samples. *Journal of Eukaryotic Microbiology*, **55**, 306–312.

Dujardin, F. (1841) *Histoire Naturelle des Zoophytes. Infusoires, comprenant la Physiolgie et la Classification de ces Animaux et La Maniere des les Etudier à l'Aide du Microscope*. Librarie Encyclopédique de Roret, Paris.

Duka, L.A. & Gordina, A.D. (1973) Abundance of ichthyoplankton and feeding of fish larvae in the western Mediterranean and adjacent areas of the Atlantic Ocean. *Hydrobiological Journal*, **9**, 54–59.

Dunthorn, M. & Katz, L.A. (2008) Richness of morphological hypotheses in ciliate systematics allows for detailed assessment of homology and comparisons with gene trees. *Denisia*, **23**, 389–394.

Dunthorn, M., Lipps, J.H. & Stoeck, T. (2010) Reassessment of putative ciliate fossils *Eotintinnopsis, Wujiangella*, and *Yonyangella* from the Neoproterozoic Doushantuo Formation in China. *Acta Protozoologica*, **49**, 139–144.

*Dupuy, C., Vaquer, A., Lam-Hoai, T., Rougier, C., Mazouni, N., Lautier, J., Collos, Y. & Le Gall, S. (2000) Feeding rate of the oyster *Crassostrea gigas* in a natural planktonic community of the Mediterranean Thau Lagoon. *Marine Ecology Progress Series*, **205**, 171–184.

Durán, M. (1951) Contribución al estudio de los tintínidos del plancton de las costas de Castellón (Mediterráneo occidental). *Publicaciones del Instituto de Biología Aplicada (Barcelona)*, **5**, 101–122.

Durán, M. (1953) Contribución al estudio de los tintínidos del plancton de las costas de Castellón (Mediterráneo occidental). Nota II. *Publicaciones del Instituto de Biología Aplicada (Barcelona)*, **12**, 79–95.

*Durán, M. (1957) Nota sobre algunos tintinnoineos del plancton de Puerto Rico. *Investigacion Pesquera, Barcelona*, **8**, 97–120.

Durán, M. (1965) Tintinnoineos de las costas de Mauritania y Senegal. *Trabajos del Instituto Español de Oceanografía*, **32**, 1–32 + Plates 1–11.

Durmus, T., Balci, M. & Balkis, N. (2011) Species of genus *Tintinnopsis* Stein 1867 in Turkish coastal wates and new record of *Tintinnopsis cornigera* Hada, 1964. *Pakistan Journal of Zoology*, **44**, 383–388.

Dussart, B.H. (1965). Les différentes catégories de plancton. *Hydrobiologia* **26**, 72–74.

Echols, R.J. & Fowler, G.A. (1973) Agglutinated tintinnid loricae from some Recent and Late Pleistocene shelf sediments. *Micropaleontology*, **19**, 431–443.

Edgcomb, V., Orsi, W., Bunge, J., Jeon, S., Christen, R., Leslin, C., Holder, M., Taylor, G.T., Suarez, P. Varela, R. & Epstein, S. (2011) Protistan microbial observatory in the Cariaco Basin, Caribbean. 1. Pyrosequencing vs Sanger insights into species richness. *ISME*, **5**, 1344–1356.

Ehrenberg, C.G. (1832) Uber die Entwicklung und Lebensdauer der Infusionsthiere, nebst ferneren Beitragen zueiner Vergleichung ihrer organischen Systeme. *Königliche Akademie der Wissenschaften zu Berlin Physikalische Abhandlungen*, **1831**, 1–154.

Ehrenberg, C.G. (1841) Über die Entwickelung und Lebensdauer der Infusionthiere; nebst ferneren Beitrâgen zu einer Vergleichung ibrer organischen Systeme. Physikalische Abhandlungen der Königlichen Akademie der Wissenschaften zu Berlin. *Aus dem Jahre* **1831**, 1–154.

Eicher, D.L. (1965) Cretaceous tintinnids from the western interior of the United States: *Micropaleontology* **11**, 449–456.

*Elliott, D.T. & Kauffmann, R.S. (2007) Spatial and temporal variability of mesozooplankton and tintinnid Ciliates in a seasonally hypersaline estuary. *Estuaries and Coasts*, **30**, 418–430.

Elton, C.S. (1927) *Animal Ecology*. University of Chicago Press, Chicago.

Emlet, R.B. & Strathmann, R.R. (1985) Gravity, drag and feeding currents of small zooplankton. *Science*, **228**, 1016–1017.

Endo, T. & Nagata, H. (1984) Resting and germination of cysts of Peridinium sp. (Dinophyceae). *Bulletin of the Plankton Society of Japan*, **31**, 23–33.

Endo, Y. (2005) Method of Chloroplast Sequestration and Induction of Encystment in the Planktonic Ciliate *Strombidium conicum*. Ph.D. Thesis, Tohoku University, Sendai, 56pp.

Endo, Y. & Taniguchi, A. (2006) Conjugation and encystment of the planktonic choreotrich ciliate *Strombidium conicum*. *Japanese Journal of Protozoology*, **39**, 98–99.

Entz, G., Sr (1884) Über Infusorien des Golfes von Neapel. *Mittheilungen aus der Zoologischen Station zu Neapel*, **5**, 289–444 + Plates 20–25.

Entz, G., Sr (1885) Zur näheren Kenntnis der Tintinnoden. *Mittheilungen aus der Zoologischen Station zu Neapel*, **6**, 185–216 + Plates 13, 14.

Entz, G., Jr (1909a) Die Süsswasser-Tintinniden. *Mathematische und naturwissenschaftliche Berichte aus Ungarn*, **25**, 197–225.

Entz, G., Jr (1909b) Studien über Organisation und Biologie der Tintinniden. *Archiv für Protistenkunde*, **15**, 93–226 + Plates 8–21.

Entz, G., Jr (1937) Fibrillen in *"Favella Ehrenbergii"* Jörgensen ("Ciliata, Oligotricha"). International Congress of Zoology 1937, Lisboa, 1446–1448.

Erickson, J.S., Hashemi, N., Sullivan, J.M., Weidemann, A.D. & Ligler, F.S. (2011) In situ phytoplankton analysis: there's plenty of room at the bottom. *Analytical Chemistry*, DOI: 10.1021/ac201623k.

*Eskinazi-Sant'anna, E.M. & Björnberg, T.K.S. (2006) Seasonal dynamics of microzooplankton in the São Sebastião Channel (SP, Brazil). *Brazilian Journal of Biology*, **66**, 221–231.

Fauré-Fremiet, E. (1908) Le *Tintinnidium inquilinum*. *Archiv für Protistenkunde*, **11**, 225–251 + Plate 12.

Fauré-Fremiet, E. (1912) Sur le mode de division du *Strombidium sulcatum* Cl. et Lach. *Bulletin de la Société Zoologique de France*, **36** (year 1911), 207–208.

Fauré-Fremiet, E. (1924) Contribution a la connaissance des infusoires planktoniques. *Bulletin Biologique de France et de Belgique, Suppl.* **6**, 1–171.

Fauré-Fremiet, E. (1948) Le rythme de marée du *Strombidium oculatum* Gruber. *Bulletin Biologique de France et de Belgique*, **82**, 3–23.

Fauré-Fremiet, E. (1953) La bipartition énantiotrope chez les ciliés oligotriches. *Archives d'Anatomie microscopique et de Morphologie Expérimentale*, **42**, 209–225.

Fauré-Fremiet, E. (1970) Remarques sur la systématique des ciliés Oligotrichida. *Protistologica*, **5** (year 1969), 345–352.

Fauré-Fremiet, E. & Ganier, M.-Cl. (1970) Structure fine du *Strombidium sulcatum* Cl. et L. (Ciliata Oligotrichida). *Protistologica* **6**, 207–223.

Felsenstein, J. (2004) *Inferring Phylogenies*. Sinauer Associates, Sunderland, Massachusetts.

Fenchel, T. (1969) The ecology of marine microbenthos: 4. Structure and function of the benthic ecosystem, its chemical and physical factors and the microfauna communities with special reference to the ciliated Protoza. *Ophelia*, **6**, 1–182.

Fenchel, T. (1980a) Suspension feeding ciliated protozoa: structure and function of feeding organelles. *Archive für Protistenkunde*, **123**, 239–260.

Fenchel, T. (1980b) Suspension feeding ciliated protozoa: functional response and particle selection. *Microbial Ecology*, **6**, 1–11.

Fenchel, T. (1980c) Suspension feeding ciliated protozoa: feeding rates and thier ecological significance. *Microbial Ecology* **6**, 13–25.

Fenchel, T. (1980d) Suspension feeding ciliated protozoa: relation between particle size and clearance in suspension feeding ciliates. *Limnology and Oceanography*, **25**, 733–738.

Fenchel, T. (1982a) Ecology of heterotrophic microflagellates 1. Some important forms and their functional ecology. *Marine Ecology Progress Series*, **8**, 211–223.

Fenchel, T. (1982b) Ecology of heterotrophic microflagellates II. *Bioenergetics and growth Marine Ecology Progress Series*, **8**, 225–231.

Fenchel, T. (1982c) Ecology of heterotrophic microflagellates III: Adaptations to heterogenous environments. *Marine Ecology Progress Series*, **9**, 25–33.

Fenchel, T. (1982d) Ecology of heterotrophic microflagellates IV. Quantitative occurrence and importance as bacterial consumers. *Marine Ecology Progress Series*, **9**, 35–43.

Fenchel, T. (1986) Protozoan filter feeding. *Progress in Protistology*, **1**, 65–113.

Fenchel, T. (1987) *Ecology of Protozoa: the Biology of Free-living Phagotrophic Protists*. (ed. T.D. Brock). Brock/Springer Series in Contemporary Bioscience, Berlin.

Fenchel, T. (1988) Marine plankton food chains. *Annual Review of Ecology and Systematics*, **19**, 19–38.

Fenchel, T. & Blackburn, N. (1999) Motile chemosensory behaviour of phagotrophic protists: mechanisms for and efficiency in congregating at food patches. *Protist*, **150**, 325–336.

Fenchel, T. & Findlay, B.J. (2004) The ubiquity of small species: patterns of local and global diversity. *Bioscience*, **54**, 777–784.

Fenchel, T. & Findlay, B.J. (1983) Respiration rates in heterotrophic, free-living Protozoa. *Microbial Ecololgy*, **9**, 99–122.

Fenchel, T. & Jonsson, P.R. (1988) The functional biology of *Strombidium sulcatum*, a marine oligotrich ciliate (Cilophora, Oligotrichina). *Marine Ecology Progress Series*, **48**, 1–15.

Fensome, R.A., Taylor, F.J.R., Norris, G., Sargeant, W.A.S., Wharton, D.I. & William, G.L. (1993) A Classification of Living and Fossil Dinoflagellates. *Micropaleontology*, Special Publication no. 7. Sheridan Press, Hanover, Pennsylvania.

Fenton, A., Spencer, M. & Montagnes, D.J.S. (2010) Parameterising variable assimilation efficiency in predator prey models. *Oikos*, **119**, 1000–1010.

*Fernandes, L.F. (1999) Tintinnids (Ciliophora – Suborder Tintinna) from subantarctic and antarctic waters between Argentine and Antarctic Peninsula (35°8–62°8) (November/1992). *Brazilian Journal of Oceanography*, **47**, 155–171.

*Fernandes, L.F. (2004a) Tintininos (Ciliophora, Tintinnina) de águas subtropicais na região Sueste-Sul do Brasil. I. Famílias Codonellidae, Codonellopsidae, Coxliellidae, Cyttarocylidae, Epiplocylidae, Petalotrichidae, Ptychocylidae, Tintinnididae e Undellidae. *Revista Brasiliera de Zoologia*, **21**, 551–576.

*Fernandes, L.F. (2004b) Tintinnids (Ciliophora, Tintinnina) from subtropical waters of the Southern Brazil. II. Families Dictyocystidae, Rhabdonellidae, Tintinnidae and Xystonellidae. *Revista Brasiliera de Zoologia*, **21**, 605–628.

Fernandez-Leborans, G. (1983) Nouvelles observations sur les deux ciliés Oligotriches: *Halteria grandinella* (la morphologie et la morphogenèse) et *Strombilidium gyrans* (la morphologie). *Acta Protozoologica*, **22**, 251–256 + Plates 1–3.

Fessenden, L. & Cowles, T.J. (1994) Copepod predation on phagotrophic ciliates in Oregon coastal waters. *Marine Ecology Progress Series*, **107**, 103–111.

Fields, B.S., Shotts, E.B. Jr, Feeley, J.C., Gorman, G.W. & Martin, T.M. (1984) Proliferation of *Legionella pneumophila* as an intracellular parasite of the ciliated protozoa *Tetrahymena pyriformes*. *Applied and Environmental Microbiology*, **47**, 467–471.

Figueiredo, G.M., Montagnes, D.J.S. & Nash, R.D.M. (2005) The role of the generally unrecognised microprey source as food for larval fish in the Irish Sea. *Marine Biology*, **148**, 395–404.

Figueiredo, G.M., Nash, R.D.M. & Montagnes, D.J.S. (2007) Do protozoa contribute significantly to the diet of larval fish in the Irish Sea? *Journal of the Marine Biological Association of the United Kingdom*, **87**, 843–850.

Figueiredo, G.M., Montagnes, D.J.S. & Nash, R.D.M. (2009) The importance of protozooplankton as prey for copepods in the coastal areas of the central Irish Sea. *Hydrobiologia*, **628**, 227–239.

Fileman, E.S. & Leakey, R.J.G. (2005) Microzooplankton dynamics during the development of the spring bloom in the north-east Atlantic. *Journal of the Marine Biological Association of the United Kingdom*, **85**, 741–753.

Fileman, E.S., Cummings, D.G. & Llewellyn, C.A. (2002) Microplankton community structure and the impact of microoplankton grazing during *Emiliania huxleyi* bloom, off the Devon coast. *Journal of the Marine Biological Association of the United Kingdom*, **82**, 359–368.

Fileman, E.S., Petropavlovsky, A. & Harris, R.P. (2010) Grazing by the copepods *Calanus helgolandicus* and *Acartia clausi* on the protozooplankton community at station L4 in the Western English Channel. *Journal of Plankton Research*, **32**, 709–724.

Finlay, K. & Roff, J.C. (2004) Radiotracer determination of the diet of calanoid copepod nauplii and copepodites in a temperate estuary. *ICES Journal of Marine Science*, **61**, 552.

Foissner, I. & Foissner, W. (1986) *Ciliomyces spectabilis*, nov. Gen., nov. Spec., a zoosporic fungus which parasitizes cysts of the ciliate Kahliella simplex. *Zeitschrift für Parasitenkunde*, **72**, 29–41.

Foissner, W. (1996) Ontogenesis in ciliated protozoa, with emphasis on stomatogenesis. In *Ciliates: Cells as Organisms* (K. Hausmann & P.C. Bradbury eds), pp. 95–177, G. Fischer, Stuttgart.

Foissner, W. (2005) The unusual, lepidosome-coated resting cyst of *Meseres corlissi* (Ciliophora: Oligotrichea): transmission electron microscopy and phylogeny. *Acta Protozoologica*, **44**, 217–230.

Foissner, W. & Foissner, I. (1984) First record of an ectoparasitic flagellate on ciliates: An ultrastructural investigation of the morphology and the mode of attachment of *Spiromonas gonderi* nov. spec. (Zoomastigophora, Spiromonadidae) invading the pelicle of ciliates of the genus *Colpoda* (Ciliophora, Colpodidae). *Protistologica*, **20**, 635–648.

Foissner, W. & Foissner, I. (1995) *Ciliasporidium platyphyrae* nov. gen., nov. spec. (Microspora incerta sedis), a parasite of Platyophrya terricola (Ciliophora, Colpodea). *European Journal of Protistology*, **31**, 248–259.

Foissner, W. & O'Donoghue, P.J. (1990) Morphology and infraciliature of some freshwater ciliates (Protozoa: Ciliophora) from Western and South Australia. *Invertebrate Taxonomy*, **3**, 661–696.

Foissner, W. & Wilbert, N. (1979) Morphologie, Infraciliatur und Ökologie der limnischen Tintinnina: *Tintinnidium fluviatile* Stein, *Tintinnidium pusillum* Entz, *Tintinnopsis cylindrata* Daday und *Codonella cratera* (Leidy) (Ciliophora, Polyhymenophora). *Journal of Protozoology*, **26**, 90–103.

Foissner, W., Blatterer, H., Berger, H. & Kohmann, F. (1991) Taxonomische und ökologische Revision der Ciliaten des Saprobiensystems – Band I: Cyrtophorida, Oligotrichida, Hypotrichia, Colpodea. *Informationsberichte des Bayerischen Landesamtes für Wasserwirtschaft*, **1/91**, 1–478.

Foissner, W., Berger, H. & Schaumburg, J. (1999) Identification and Ecology of Limnetic Plankton Ciliates. *Informationsberichte des Bayerischen Landesamtes für Wasserwirtschaft*, **3/99**, 1–793.

Foissner, W., Moon-van der Staay, S.Y., van der Staay, G.W.M. et al. (2004) Reconciling classical and molecular phylogenies in the stichotrichines (Ciliophora, Spirotrichea), including new sequences from some rare species. *European Journal of Protistology*, **40**, 265–281.

Foissner, W., Müller, H. & Weisse, T. (2005) The unusual, lepidosome-coated resting cyst of *Meseres corlissi* (Ciliophora: Oligotrichea): light and scanning electron microscopy, cytochemistry. *Acta Protozoologica*, **44**, 201–215.

Foissner, W., Pichler, M., Al-Rasheid, K.A.A.S. & Weisse, T. (2006) The unusual, lepidosome-coated resting cyst of *Meseres corlissi* (Ciliophora:Oligotrichea): Encystment and genesis and release of the lepidosomes. *Acta Protozoologica*, **45**, 323–338.

Foissner, W., Müller, H. & Agatha, S. (2007) A comparative fine structural and phylogenetic analysis of resting cysts in oligotrich and hypotrich Spirotrichea (Ciliophora). *European Journal of Protistology*, **43**, 295–314.

Foissner, W., Chao, A. & Katz, L.A. (2009) Diversity and geographic distribution of ciliates (Protista: Ciliophora). *Biodiversity and Conservation*, **8**, 111–129.

Foissner, W., Stoeck, T., Agatha, S. & Dunthorn, M. (2011) Intraclass evolution and classification of the Colpodea (Ciliophora). *Journal of Eukaryotic Microbiology*, **58**, 397–415.

Fokin, S.I., Schweikert, M., Görtz, H.-D. & Fujishima, M. (2003) Bacterial endocytobionts of Ciliophora. Diversity and some interactions with the host. *European Journal of Protistology*, **39**, 475–480.

Fokin, S.I., Di Giuseppe, G., Erra, F. & Dini, F. (2008) *Euplotespora binucleata* n. gen., n. sp (Protozoa: Microsporidia), a parasite infecting the hypotrichous ciliate *Euplotes woodruffi*, with observations on microsporidian infections in Ciliophora. *Journal of Eukaryotic Microbiology*, **55**, 214–228.

Fol, H. (1881) Contribution à la connaissance de la famille des Tintinnodea. *Archives des Sciences Physiques et Naturelles*, **5**, 5–24 + Plate 1.

Fol, H. (1883) Nouvelle contribution à la connaissance de la famille des Tintinnodea. *Archives des Sciences Physiques et Naturelles*, **9**, 554–578 + Plate 18.

Fol, H. (1884) Sur la famille des Tintinnodea. *Recueil Zoologique Suisse* **1**, 27–64.

*Fonda Umani, S., Honsell, G., Cabrini, M. & Milani, L. (1985) A tintinnid bloom in the Gulf of Trieste (Northern Adriatic Sea). *Oebalia*, **11**, 149–156.

*Fonda Umani, S., Beran, A., Parlato, S., Virgilio, D., Zollet, T., Olazabal, A.D., Lazzarini, B. & Cabrini, M. (2004) Noctiluca scintillans MaCartney in the Northern Adriatic Sea: long-term dynamics, relationships with temperature and eutrophication, and role in the food web. *Journal of Plankton Research*, **26**, 545–561.

*Fonda Umani, S., Monti, M., Cataletto, B. & Budillon, G. (2011) Tintinnid distributions in the Strait of Magellan (Chile). *Polar Biology*, **34**, 1285–1299.

Fontaneto, D. & Brodie, J. (2011) Why biogeography of microorganisms? In *Biogeography of Microscopic Organisms: Is Everything Small Everywhere?* (Fontaneto, D. ed.), pp. 3–10. Cambridge University Press, Cambridge.

*Fornshell, J.A. (1979) Microplankton patchiness in the northwest Atlantic Ocean. *Journal of Protozoology*, **26**, 270–272.

Fortier, L. & Harris, R.P. (1989) Optimal foraging and density dependent competition in marine fish larvae. *Marine Ecology Progress Series*, **51**, 19–33.

Fossum, P. & Ellertsen, B. (1994) Gut content analysis of first-feeding cod larvae (*Gadus morhua L.*) sampled at Lofoten, Norway, 1979–1986. ICES Marine Science Symposium, **198**, 430–437.

Foster, R.A., Carpenter, E.J. & Bergman, B. (2006) Unicellular cyanobionts in open ocean dinoflagellates, radiolarians, and tintinnids: ultrastructural characterization and immunolocalization of phycoerythrin and nitrogenase. *Journal of Phycology*, **42**, 453–463.

Fox, J.W., Nelson, W.A. & McCauley, E. (2010). Coexistence and the paradox of the plankton: quantifying selection from noisy data. *Ecology*, **91**, 1774–1786.

Fredrickson, K.A. & Strom, S.L. (2009) The algal osmolyte DMSP as a microzooplankton grazing deterrent in laboratory and field studies. *Journal of Plankton Research*, **31**, 135–152.

Frias-Lopez, J., Thompson, A., Waldbauer, J. & Chisholm, S.W. (2009) Use of stable isotope-labelled cells to identify active grazers of picocyanobacteria in ocean surface waters. *Environmental Microbiology*, **11**, 512–525.

Froneman, P.W. (2002) Trophic cascading in an oligotrophic temperate estuary, South Africa. *Journal of Plankton Research*, **24**, 807–816.

Froneman, P.W. & McQuaid, C.D. (1997) Preliminary Investigation of the Ecological Role of Microzooplankton in the Kariega Estuary, South Africa. *Estuarine, Coastal and Shelf Science*, **45**, 689–695.

Froneman, P.W. & Perissinotto, R. (1996a) Microzooplankton grazing and protozooplankton community structure in the South Atlantic and in the Atlantic sector of the Southern Ocean. *Deep-Sea Research I*, **43**, 703–721.

Froneman, P.W. & Perissinotto, R. (1996b) Structure and grazing of the microzooplankton communities of the Subtropical Convergence and a warm-core eddy in the Atlantic sector of the Southern Ocean. *Marine Ecology Progress Series*, **135**, 237–245.

Froneman, P.W., Pakhomov, E.A. & Meaton, V. (1998) Observations on the association between the diatom, *Fragilariopsis doliolus* Wallich, and the tintinnid, *Salpingella subconica* Kofoid. *South African Journal of Science*, **94**, 202–202.

Fuhrman, J.A. & Azam, F. (1980) Bacterioplankton secondary production estimates for coastal waters of British Columbia, Antarctica, and California. *Applied and Environmental Microbiology*, **39**, 1085–1095.

Fuhrman, J.A., Steele, J.A., Hewson, I., Schwalbach, M.S., Brown, M.V., Green, J.L. & Brown, J.H. (2008) A latitudinal gradient in planktonic marine bacteria. *Proceedings of the National Academy of Sciences of the United States of America*, **105**, 7774–7778.

Fukami, K., Watanabe, A., Fujita, S., Yamaoka, K. & Nishijima, T. (1999) Predation on naked protozoan microzooplankton by fish larvae. *Marine Ecology Progress Series*, **185**, 285–291.

Fukuyo, Y., Takano, H., Chihara, M., Matsuoka, K. (1990) *Red Tide Organisms in Japan, An Illustrated Taxonomic Guide*. Uchida Rokakuho, Tokyo.

Fulco, V.K. (2007) Harmful effects of the toxic dinoflagellate *Alexandrium tamarense* on the tintinnids *Favella taraikaensis* and *Eutintinnus* sp. *Journal of the Marine Biological Association of the United Kingdom*, **87**, 1085–1088.

Futuyma, D.J. (2009) *Evolution*, 2nd edn. Sinauer Associates Inc., Sunderland, Massachusetts.

*Gaarder, K.R. (1946) Tintinnoinea from the "Michael Sars" North Atlantic deep sea expedition 1910. *Reports of the Scientific Research of the "Michael Sars" North Atlantic Deep Sea Expedition* **2**, 1–37.

Gadomski, D.M. & Boehlert, G.W. (1984) Feeding ecology of pelagic larvae of English sole *Parophrys vetulus* and butter sole *Isopsetta isolepis* off the Oregon coast. *Marine Ecology Progress Series*, **20**, 1–12.

Gaines, G. & Elbrächter, M. (1987) Heterotrophic nutrition. In *The Biology of Dinoflagellates* (F.J.R. Taylor ed.), pp. 224–268, Blackwell Scientific Publishers, Oxford.

Gallegos, C.L. (1989) Microzooplankton grazing on phytoplankton in the Rhode River, Maryland: Nonlinear feeding kinetics. *Marine Ecology Progress Series*, **57**, 23–33.

Gao, S., Gong, J., Lynn, D., Lin, X. & Song, W. (2009) An updated phylogeny of oligotrich and choreotrich ciliates (Protozoa, Ciliophora, Spirotrichea) with representative taxa collected from Chinese coastal waters. *Systematics and Biodiversity*, **7**, 235–342.

*Gast, V. (1985) Bacteria as a food source for microzooplankton in the Schlei Fjord and Baltic Sea with special reference to ciliates. *Marine Ecology Progress Series*, **22**, 107–120.

Gaston, KJ. & He, F. (2011) Species occurrence and occupancy. In *Biological Diversity: frontiers in measurement and assessment* (Magurran, A.E. & McGill, B.J. eds), pp. 141–151, Oxford, Oxford University Press.

Gavrilova, N.A. (2010) Microzooplankton. In *The Introducers in the Biodiversity and Productivity of the Sea Azov and the Black Sea* (Matishov, G.G., Boltachyov, A.R., eds), Rostov-on-Don, pp. 63–39. SSC Ras Publishing.

Gavrilova, N. & Dolan, J.R. (2007) A note on species lists and ecosystem shifts: Black Sea tintinnids, ciliates of the microzooplankton. *Acta Protozoologica*, **46**, 279–288.

Gerritsen, J. & Strickler, J.R. (1977) Encounter probabilities and community structure in zooplankton: a mathematical model. *Journal of the Fisheries Research Board of Canada*, **34**, 73–82.

Gifford, D.J. (1985) Laboratory culture of marine planktonic oligotrichs (Ciliophora, Oligotrichida). *Marine Ecology Progress Series*, **23**, 257–267.

Gifford, D.J. (1991) The protozoan–metazoan trophic link in pelagic ecosystems. *Journal of Protozoology*, **38**, 81–86.

Gifford, D.J. & Dagg, M.J. (1988) Feeding of the estuarine copepod *Acartia tonsa* Dana: Carnivory vs. herbivory in natural microplankton assemblages. *Bulletin of Marine Science*, **43**, 458–468.

Gifford, D.J. & Dagg, M.J. (1991) The microzooplankton–mesozooplankton link: consumption of planktonic protozoa by the calanoid copepods *Acartia tonsa* Dana and *Neocalanus plumchrus* Murukawa. *Marine Microbial Food Webs*, **5**, 161–177.

Gilbert, J.J. (1994) Jumping behavior in the oligotrich ciliates *Strobilidium velox* and *Halteria grandinella*, and its significance as a defense against rotifer predators. *Microbial Ecology*, **27**, 189–200.

Gillies, C. & Hanson, E.D. (1963) A new species of *Leptomonas* parasitizing the macronucleus of *Paramecium trichium*. *Journal of Protozoology*, **10**, 467–473.

Gilmer, R.W. & Harbison, G.R. (1991) Diet of *Limacina helicina* (Gastropoda: Thecosomata) in Arctic waters in midsummer. *Marine Ecology Progress Series*, **77**, 125–134.

*Gilron, G.L. & Lynn, D.H. (1989a) Assuming a 50% cell occupancy of the lorica overestimates tintinnine ciliate biomass. *Marine Biology*, **103**, 413–416.

*Gilron, G.L. & Lynn, D.H. (1989b) Estimates of in situ population growth rates of four tintinnine ciliate species near Kingston Harbour, Jamaica. *Estuarine, Coastal and Shelf Science*, **29**, 1–10.

*Gilron, G.L., Lynn, D.H. & Smith, I.P. (1990) Taxonomic key and illustrated guide to the tintinnine ciliates of the Cays-Hellshire Region near Kingston Harbour, Jamaica. *Caribbean Journal of Science*, **26**, 152–162.

Gilron, G.L., Lynn, D.H. & Roff, J.C. (1991) The annual cycle of biomass and production of tintinnine ciliates in a tropical neritic region near Kingston, Jamaica. *Marine Microbial Food Webs*, **5**, 95–113.

Gismervik, I. (2005) Numerical and functional responses of choreo and oligotrich planktonic ciliates. *Aquatic Microbial Ecology*, **40**, 163–173.

Glazier, D.S. (2006) The 3/4-power law Is not universal: evolution of isometric, ontogenetic metabolic scaling inpelagic animals. *BioScience*, **56**, 325–332.

Glibert, P.M. (1982) Regional studies of daily, seasonal and size fraction variability in ammonium remineralization. *Marine Biology*, **70**, 209–222.

*Godhantaraman, N. & Uye, S. (2003) Geographical and seasonal variations in taxonomic composition, abundance and biomass of microzooplankton across a brackish-water lagoonal system of Japan. *Journal of Plankton Research*, **25**, 465–482.

*Godhantaraman, N. (1994) Species composition and abundance of tintinnids and copepods in the Pichavaram mangroves (South India). *Ciencias Marinas*, **20**, 371–391.

Godhantaraman, N. (2001) Seasonal variations in taxonomic composition, abundance and food web relationship of microzooplankton in estuarine and mangrove waters, Parangipettai region, southeast coast of India. *Indian Journal of Marine Sciences*, **30**, 151–160.

Godhantaraman, N. (2002) Seasonal variations in species comosition, abundance, biomass and estimated production rates of tintinnids at tropical estuarine and mangrove waters, Parangipettai, southeast coast of India. *Journal of Marine Systems*, **36**, 161–171.

Godhantaraman, N. & Uye, S. (2001) Geographical variations in abundance, biomass and trophodynamic role of microzooplankton across an inshore-offshore gradient in the Inland Sea of Japan and adjacent Pacific Ocean. *Plankton Biology and Ecology*, **48**, 19–27.

Goetz, E. & Ohman, M.D. (2010) Integrated molecular and morphological biogeography of the calanoid family Eucalanidae. *Deep-Sea Research II*, **57**, 2110–2129.

Gold, K. (1968) Some observations on the biology of *Tintinnopsis* sp. *Journal of Protozoology*, **15**, 193–194.

Gold, K. (1969a) The preservation of tintinnids. *Journal of Protozoology*, **16**, 126–128.

Gold, K. (1969b) Tintinnids: feeding experiments and lorica development. *Journal of Protozoology*, **16**, 507–509.

Gold, K. (1970) Cultivation of marine ciliates (Tintinida) and heterotrophic flagellates. *Helgolander Wissenschaftliche Meeresuntersuchungen*, **20**, 264–271.

Gold, K. (1971) Growth characteristics of the mass-reared tintinnid *Tintionnopsis beroidea*. *Marine Biology*, **8**, 105–108.

Gold, K. (1973) Methods for growing tintinnids in continous culture. *American Zoologist*, **13**, 203–208.

Gold, K. (1974) Studies on the size distribution of loricae and the growth status of Tintinnida in Eel Pond. *Biological Bulletin of the Marine Biological Laboratory, Woods Hole*, **147**, 477.

Gold, K. (1979) Scanning electron microscopy of *Tintinnopsis parva*: studies on particle accumulation and the striae. *Journal of Protozoology*, **26**, 415–419.

Gold, K. (1980) SEM studies on the lorica of various Tintinnina. *Scanning Electron Microscopy*, **3**, 537–541.

Gold, K. & Morales, E.A. (1974) Effects of temperature on 2 strains of *Tintinnopsis tubulosa*. *Journal of Protozoology*, **21**, 442.

Gold, K. & Morales, E.A. (1975a) Studies on Woods Hole tintinnids using scanning electron microscopy. *Biological Bulletin*, **149**, 427.

*Gold, K. & Morales, E.A. (1975b) Tintinnida of the New York Bight: Loricae of *Parafavella gigantea*, *P. parundentata*, and *Ptychocylis obtusa*. *Transactions of the American Microscopical Society*, **94**, 142–145.

Gold, K. & Morales, E.A. (1975c) Seasonal changes in lorica sizes and the species of Tintinnida in the New York Bight. *Journal of Protozoology*, **22**, 520–528.

Gold, K. & Morales, E.A. (1976a) Observations on the nature and significance of the particles used by Tintinnida during lorica-building. *Transactions of the American Microscopical Society*, **95**, 69–72.

*Gold, K. & Morales, E.A. (1976b) Studies on the sizes, shapes and the development of the lorica of agglutinated Tintinnida. *Biological Bulletin*, **150**, 377–392.

Gold, K. & Morales, E.A. (1976c) Studies on Tintinnida using scanning electron microscopy. *Transactions of the American Microscopical Society*, **95**, 707–711.

*Gold, K. & Morales, E.A. (1977) Studies on the Tintinnida of Enewetak Atoll. *Journal of Protozoology*, **24**, 580–587.

Gold, K. & Pollingher, U. (1971) Microgamete formation and the growth rate of *Tintinnopsis beroidea*. *Marine Biology*, **11**, 324–329.

Gomes, E.I., Gonçalves, C.P. & Pimenta, C.M.M. (2006) Os Tintinideos. In *Dinoflagelados e Tintiniideos da Regiao Central da Zona Economica Exclusiva Brasileira* (D.r. Tenenbaum, ed.), pp. 165–287. Museu Nacional: Rio de Janeiro.

*Gomez, F. (2007) On the consortium of the tintinnid *Eutintinnus* and the diatom *Chaetoceros* in the Pacific Ocean. *Marine Biology*, **151**, 1899–1906.

*Gomez, F. & Gorsky, G. (2003) Annual microplankton cycles in Villefranche Bay, Ligurian Sea, NW Mediterranean. *Journal of Plankton Research*, **25**, 323–339.

*Gong, J., Song, W. & Warren, A. (2005) Periphytic ciliate colonization: annual cycle and responses to environmental conditions. *Aquatic Microbial Ecology*, **39**, 156–170.

González, H.E., González, S.R. & Brummer, G.-J.A. (1994) Short-term sedimentation pattern of zooplankton faeces and microplankton at a permanent station in the Bjørnafjorden (Norway) during April–May 1992. *Marine Ecology Progress Series*, **105**, 31–45.

González, H.E., Hebbeln, D., Iriarte, J.L. & Marchant, M. (2004) Downward fluxes of faecal material and microplankton at 2300 m depth in the oceanic area off Coquimbo (30°S), Chile, during 1993–1995. *Deep-Sea Research II*, **51**, 2457–2474.

Gorsky, G., Ohman, M.D., Picheral, M., Gasparini, S., Stemmann, L., Romagnan, J.-B., Cawood, A., Pesant, S.T., Garcia-Comas, C. & Prejger, F. (2010) Digital zooplankton image analysis using the zooscan integrated system. *Journal of Plankton Research*, **32**, 285–303.

Görtz, H.D. (1988) Endocytobiosis. In *Paramecium* (H.D. Görtz ed.), pp. 393–405, Springer, Berlin.

Görtz, H.D. & Dieckmann, J. (1987) *Leptomonas ciliatorum* n. sp. (Kinetoplastida, Trypanosomatidae) in the macronucleus of a hypotrichous ciliate. *Journal of Protozoology*, **34**, 259–263.

Govoni, J.J. (2010) Feeding on protists and particulates by the leptocephali of the worm eels *Myrophis* spp. (Teleostei: Anguilliformes: Ophichthidae), and the potential energy contribution of large aloricate protozoa. *Scientia Marina*, **74**, 339–344.

Govoni, J.J. & Chester, A.J. (1990) Diet composition of larval *Leiostomus xanthurus* in and about the Mississippi River plume. *Journal of Plankton Research*, **12**, 819–830.

Govoni, J.J., Hoss, D.E. & Chester, A.J. (1983) Comparative feeding of three species of larval fishes in the Northern Gulf of Mexico: *Brevoortia patronus*, *Leiostomus xanthurus*, and *Micropogonias undulatus*. *Marine Ecology Progress Series*, **13**, 189–199.

Gowing, M.M. & Bentham, W.N. (1994) Feeding ecology of phaeodarian radiolarians at the VERTEX North Pacific time series site. *Journal of Plankton Research*, **16**, 707–719.

Gowing, M.M. & Wishner, K.F. (1986) Trophic relationships of deep-sea calanoid copepods from the benthic boundary layer of the Santa Catalina Basin, California. *Deep-Sea Research A*, **33**, 939–961.

Gowing, M.M. & Wishner, K.F. (1992) Feeding ecology of benthopelagic zooplankton on an eastern tropical Pacific seamount. *Marine Biology*, **112**, 451–467.

Graham, H.W. & Bronikovsky, N. (1944) *The Genus Ceratium in the Pacific and North Atlantic Oceans*. Carnegie Institute of Washington, Washington DC, Publication 565.

Graham, S.L. & Strom, S.L. (2010) Growth and grazing of microzooplankton in response to the harmful alga *Heterosigma akashiwo* in prey mixtures. *Aquatic Microbial Ecology*, **59**, 111–124.

Granhag, L., Moller, L.F. & Hansson, L.J. (2011) Size-specific clearance rates of the ctenophore *Mnemiopsis leidyi* based on in situ gut content analyses. *Journal of Plankton Research*, **33**, 1043–1052.

*Graziano, C. (1989) On the ecology of tintinnids (Ciliophora: Oligotrichida) in the North Irish Sea. *Estuarine, Coastal and Shelf Science*, **29**, 233–245.

Greuet, C., Gayol, P., Salvano, P. & Laval-Peuto, M. (1986) Preliminary report on the ultrastructural organization of the contractile appendix of *Tontonia appendiculariformis* (Ciliophora Oligotrichina). *Cell Motility and the Cytoskeleton*, **6**, 217–224.

Grover, J.J. (1998) Feeding habits of pelagic summer flounder, *Paralichthys dentatus*, larvae in oceanic and estuarine habitats. *Fishery Bulletin*, **96**, 248–257.

Grün, B. & Blau, J. (1996) Phylogenie, Systematik und Biostratigraphie der Calpionellidae Bonet, 1956: neue Daten aus dem Rosso Amonitico Superiore und dem Biancone (Oberjura/Unterkreide: Tithon-Valangin) von Ra Stua (Prov. Belluno, Italien). *Revue Paléobiologie*, **15**, 571–595.

Grün, B. & Blau, J. (1997) New aspects of calpinonellid biochronology: proposal for a revised calpionellid zonal and subzonal division. *Revue Paléobiologie*, **16**, 197–214.

Guillard, R.R.L. & Kilham, P. (1977) The ecology of marine planktonic diatoms. In *The Biology of Diatoms* (W. Werner ed.), pp. 372–469, University of California Press, Berkely.

Guillou, L., Viprey, M., Chambouvet, A., Welsh, R.M., Kirkham, A.R., Massana, R., Scanlan, D.J. & Worden, A.Z. (2008) Widespread occurrence and genetic diversity of marine parasitoids belonging to Syndiniales (Alveolata). *Environmental Microbiology*, **10**, 3349–3365.

Guindon, S., Dufayard, J.F., Lefort, V., Anisimova, M., Hordijk, W. & Gascuel, O. (2010) New algorithms and methods to estimate maximum-likelihood phylogenies: assessing the performance of PhyML3.0. *Systematic Biology*, **59**, 307–321.

Gunderson, J.H., John, S.A., Boman, W.C. & Coats, D.W. (2002) Multiple strains of the parasitic dinoflagellate *Amoebophrya* exist in Chesapeake Bay. *Journal of Eukaryotic Microbiology*, **49**, 469–474.

Gustafson, D.E., Stoecker, D.K., Johnson, M.D., Van Heukelem, W.F. & Sneider, K. (2000) Cryptophyte algae are robbed of their organelles by the marine ciliate *Mesodinium rubrum*. *Nature*, **405**, 1049–1052.

Gutiérrez, J.C., Martín-González, A. & Matsusaka, T. (1990) Towards a generalized model of encystment (cryptobiosis) in ciliates: a review and a hypothesis. *BioSystems*, **24**, 17–24.

Gutiérrez, J.C., Callejas, S., Boniquel, S., Benítez, L. & Martín-González, A. (2001) Ciliate cryptobiosis: a microbial strategy against environmental starvation. *International Microbiology*, **4**, 151–157.

Gutiérrez, J.C., Díaz, S., Ortega, R. & Martín-González, A. (2003) Ciliate resting cyst walls: a comparative review. *Recent Research Developments in Microbiology*, **7**, 361–379.

*Hada, Y. (1932a) Descriptions of two new neritic Tintinnoinea, *Tintinnopsis japonica* and *Tps. kofoidi* with a brief note on an unicellular organism parasitic on the latter. *Proceedings of the Imperial Academy of Japan* **8**, 209–212.

*Hada, Y. (1932b) Report of the biological survey of Mutsu Bay. 24. The pelagic Ciliata, suborder Tintinnoinea. *Science Reports of the Tohoku Imperial University*, **7**, 553–573.

*Hada, Y. (1932c) The Tintinnoinea from the Sea of Okhotsk and its neighborhood. *Journal of the Faculty of Science Hokkaido Imperial University*, **2**, 37–59.

*Hada, Y. (1937) The fauna of Akkeshi Bay IV. The pelagic Ciliata. *Journal of the Faculty of Science Hokkaido Imperial University*, **5**, 143–216.

Hada, Y. (1938) Studies on the Tintinnoinea from the Western Tropical Pacific. *Journal of the Faculty of Science Hokkaido Imperial University, Series 6, Zoology*, **6**, 87–190.

Hada, Y. (1964) New species of the Tintinnida found from the Inland Sea. *Bulletin Suzugamine Women's College, Natural Sciences*, **11**, 1–4.

*Hada, Y. (1970) The protozoan plankton of the Antarctic and Subantarctic Seas. *Japanese Antarctic Research Expedition, Scientific Reports, Series E*, **31**, 1–51.

*Hada, Y. (1972) Studies on plankton organisms in Akkeshi Bay, eastern district of Hokkaido, Japan. In *Biological Oceanography of the Northern North Pacific Ocean* (A.Y. Takenouti ed.), pp. 173–188, Idemitsu-shoten, Tokyo.

Haeckel, E. (1873) Ueber einige neue pelagische Infusorien. *Jenaische Zeitschrift für Naturwissenschaft*, **7** (year 1871), 561–568 + Plates 27, 28.

Haeckel, E. (1893) Plankton studies: a comparative investigation of the importance and constitution of the pelagic fauna and flora (translated by G.W. Field). *Report of the U.S. Commissioner of Fish and Fisheries, for 1889 to 1891*, pp 565–641.

Haeckel, E. (1904) *Kunstformen der Natur.* Verlag des Bibliographischen Instituts, Leipzig and Vienna.

Haley, S.T., Juhl, A.R., Keafer, B.A., Anderson, D.M. & Dyhrman, S.T. (2011) Detecting copepod grazing on low-concentration populations of *Alexandrium fundyense* using PCR identification of ingested prey. *Journal of Plankton Research*, **33**, 927–936.

Hall, M.S. & Katz, L.A. (2011) On the nature of species: insights from *Paramecium* and other ciliates. *Genetica*, **139**, 677–684.

Hall, N., Litaker, R., Fensin, E., Adolf, J., Bowers, H., Place, A. & Paerl, H. (2008) Environmental Factors Contributing to the Development and Demise of a Toxic Dinoflagellate (*Karlodinium veneficum*) Bloom in a Shallow, Eutrophic, Lagoonal Estuary. *Estuaries and Coasts* **31**, 402–418.

Hamm, C.E. & Smetacek, V. (2007). Armor: why, when, and how. In *Evolution of Primary Producers in the Sea.* (P.G. Falkowski and A.E. Knoll, A.E. eds.), pp. 311–332, Elsevier, Amsterdam.

Hammersmith, R.L. (1976) Differential cortical degradation in the two members of early conjugant pairs of *Oxytricha fallax*. *Journal of Experimental Zoology*, **196**, 45–69.

Hansen, B. Hansen, P.J., and Nielsen, T.G. (1991) Effects of large nongrazable particles on clearance and swimming behaviour of zooplankton. *Journal of Experimental Marine Biology and Ecology*, **152**, 257–269.

Hansen, B., Bjornsen, P.K. & Hansen, P.J. (1994) The size ratio between planktonic predators and their prey. *Limnology and Oceanography*, **39**, 395–403.

Hansen, B.W. & Jensen, F. (2000) Specific growth rates of protozooplankton in the marginal ice zone of the central Barents Sea during spring. *Journal of the Marine Biological Association of the United Kingdom*, **80**, 37–44.

Hansen, J., Bjornsen, P.K. & Hansen, B.W. (1997) Zooplankton grazing and growth: scaling within the 2–2,000 μm body size range. *Limnology and Oceanography*; **42**, 687–704.

*Hansen, P.J. (1989) The red tide dinoflagellate *Alexandrium tamarense*: effects on behaviour and growth of a tintinnid ciliate. *Marine Ecology Progress Series*, **53**, 105–116.

Hansen, P.J. (1995) Growth and grazing response of a ciliate feeding on the red tide dinoflagellate *Gyrodinium aureolum* in monoculture and in mixture with a non-toxic alga. *Marine Ecology Progress Series*, **121**, 65–72.

Hansen, P.J. (2011) The role of photosynthesis and food uptake for the growth of marine mixotrophic dinoflagellates. *Journal of Eukaryotic Microbiology*, **58**, 203–214.

Hansen, P.J., Cembella, A.D. & Moestrup, O. (1992) The marine dinoflagellate *Alexandrum Ostenfeldii*: paralytic shellfish toxin concentration, composition and toxicity to a tintinnid ciliate. *Journal of Phycology*, **28**, 597–603.

Hansen, V.K. & Andersen, K.P. (1962) Sampling the smaller zooplankton. *Rapports Proces-verbaux Réunion Conseil International Exploration Mer*, **153**, 39–47.

*Hara, S. & Tanoue, E. (1985) Protists along 150 E in the Southern Ocean: Its composition, stock, and distribution. *Transactions of the Tokyo University of Fisheries*, **6**, 99–115.

Harada, A., Ohtsuka, S. & Horiguchi, T. (2007) Species of the parasitic genus *Duboscquella* are members of the enigmatic marine alveolate Group I. *Protist*, **158**, 337–347.

Harding, G.C.H. (1974) The food of deep-sea copepods. *Journal of the Marine Biological Association of the United Kingdom*, **54**, 141–155.

Hardy, A.C. (1924) The herring in relation to its animate environment. Part 1. On food and feeding habits of the herring with special reference to the East Coast of England. *Fishery Investigations, Ministry of Agriculture, Fishery and Food, Series 2 Sea Fishery* **7**, 3.

Hardy, A.C. (1965) *The Open Sea: Its Natural History*, one volume edtion. Houghton Mifflin Company, Boston.

*Hargraves, P.E. (1981) Seasonal variations of tintinnids (Ciliophora, Oligotrichida) in Narragansett Bay, Rhode Island, U.S.A. *Journal of Plankton Research* **3**, 81–91.

Harman, R.A. (1972) The distribution of microbiogenic sediment near the mouth of the Columbia River. In *The Columbia River Estuary and Adjacent Ocean Waters* (A.T. Pruter, D.L. Alverson eds), pp. 265–278, University of Washington Press, Seattle.

Harrison, G.W. (1995) Comparing predator-prey models to Luckinbill's experiment with *Didinium* and *Paramecium*. *Ecology*, **76**, 357–374.

Harvey, E.L. and Menden-Deuer, S. (2011) Avoidance, movement, and mortality: The interactions between a protistan grazer and *Heterosigma akashiwo*, a harmful algal bloom species. *Limnology and Oceanography*, **56**, 371–378.

*Hattori, H., Yuki, K., Zaitsev, Y.P. & Motada, S. (1983) A preliminary observation on the neuston in Suruga Bay. *La Mer, Tokyo* **21**, 11–20.

Hausmann, K. (1978) Extrusive organelles in protists. *International Review of Cytology*, **52**, 197–276.

Hausmann, K. & Bradbury, P.C. (1996). *Ciliates: Cells as Organisms*. Gustav Fischer, Stuttgart.

Hausmann, K., Hülsmann, N. & Radek, R. (2003) *Protistology* 3rd Ed, E. Schweizerbart, Stuttgart.

Hebert, P.D.N., Cywinska, A., Ball, S.L. & deWaard, J.R. (2003a) Biological identifications through DNA barcodes. *Proceedings of the Royal Society, London, Series B*, **270**, 313–321.

Hebert, P.D.N., Ratnasingham, S. & deWaard, J.R. (2003b) Barcoding animal life: cytochrome c oxidase subunit 1 divergences among closely related species. *Proceedings of the Royal Society, London, Series B (Supplement)*, **270**, S96–S99.

Heckmann, K. (1983) Endosymbiosis of *Euplotes*. *International Review of Cytology*, **14** (Suppl.), 111–144.

Hedges, J.I. (2002) Why dissolved organics matter. In *Biogeochemistry of Marine Dissolved Organic Matter* (D.A. Hansell, C.E. Carlson eds), pp. 1–34, Academic Press, Amsterdam.

*Hedin, H. (1974) Tintinnids on the Swedish West Coast. *Zoon*, **2**, 123–133.

*Hedin, H. (1975a) On the ecology of tintinnids on the Swedish west coast. *Zoon*, **3**, 125–140.

*Hedin, H. (1975b) On the ultrastructure of *Favella ehrenbergii* (Claparède & Lachmann) and *Parafavella gigantea* (Brandt), Protozoa, Ciliata, Tintinnida. *Zoon*, **3**, 11–18.

*Hedin, H. (1976a) Examination of the tintinnid ciliate *Parafavella denticulata* (Ehrenberg) by scanning electron microscopy and the bodian protargol technique. *Acta Zoologica Fennica*, **57**, 113–118.

*Hedin, H. (1976b) Microtubules and microfilaments in the tintinnid ciliate *Ptychocylis minor* Jörgensen. *Zoon*, **4**, 3–10.

Hedin, H. (1976) On the Ecology and Cytology of Tintinnids. Ph.D. dissertation, Institute of Zoology, University of Upsalla, Sweden.

Hedley, R.H. (1960) The iron-containing shell of *Gromia oviformis* (Rhizopoda). *Quarterly Journal of Microscopical Science*, **101**, 279–293.

Hedley, R.H. (1963) Cement and iron in the arenaceous foraminifera. *Micropaleontology*, **9**, 433–441.

Hedley, R.H. & Rudall, K.M. (1974) Extracellular silk fibres in *Stannophyllum* (Rhizopodea: Protozoa). *Cell and Tissue Research*, **150**, 107–111.
*Heinbokel, J.F. (1977) Functional and Numerical Responses of Coastal Tintinnids: Implications for the Neritic Food Chain. Ph.D. Thesis, University of California, San Diego.
*Heinbokel, J.F. (1978a) Studies on the functional role of tintinnids in the southern California Bight. I. Grazing and growth rates in laboratory cultures. *Marine Biology*, **47**, 177–189.
*Heinbokel, J.F. (1978b) Studies on the functional role of tintinnids in the southern California Bight. II. Grazing rates of field populations. *Marine Biology*, **47**, 191–197.
*Heinbokel, J.F. (1987) Diel periodicities and rates of reproduction in natural populations of tintinnines in the oligotrophic waters off Hawaii, September 1982. *Marine Microbial Food Webs*, **2**, 1–14.
*Heinbokel, J.F. (1988) Reproductive rates and periodicities of oceanic tintinnine ciliates. *Marine Ecology Progress Series*, **47**, 239–248.
*Heinbokel, J.F. & Beers, J.R. (1979) Studies on the functional role of tintinnids in the southern California Bight. III. Grazing impact of natural assemblages. *Marine Biology*, **52**, 23–32.
*Heinbokel, J.F. & Coats, D.W. (1984) Reproductive dynamics of ciliates in the Antarctic ice-edge zone. *Antarctic Journal of the United States*, **19**, 111–113.
*Heinbokel, J.F. & Coats, D.W. (1986) Patterns of tintinnine abundance and reproduction near the edge of seasonal pack-ice in the Weddell Sea, November 1983. *Marine Ecology Progress Series*, **33**, 71–80.
Heinbokel, J.F., Coats, D.W., Henderson, K.W. & Tyler, M.A. (1988) Reproduction rates and secondary production of three species of the rotifer genus *Synchaeta* in the estuarine Potomac River. *Journal of Plankton Research*, **10**, 659–697.
Hemmersbach, R., Braun, M. (2006) Gravity-sensing and gravity-related signaling pathways in unicellular model systems of protists and plants. *Signal Transduction*, **6**, 432–442.
Hemmersbach-Krause, R. Briegleb, W. Hader, D.-P., Vogel, K., Klein, S. & Muslisch, M. (1994) Protozoa as model systems for the study of cellular responses to altered gravity conditions. *Advances Space Research*, **14**, 49–60.
Hennig, W. (1966) *Phylogenetic Systematics*. University of Illinois Press, Urbana, Chicago.
Hentschel, V.E. (1950) Die nahrung der heringslarven. *Helgolander Wissenschaftliche Meeresuntersuchungen*, **3**, 59–81.
Hermes, J.J. (1966) Tintinnids from the Silurian of the Betic Cordilleras, Spain. *Revue de Micropaleontologie*, **8**, 211–214.
*Hermosilla, J.G. (1972) Variacion estacional de los dinoflagelados y tintinidos en la Bahia de Concepcion, Chile. *Boln Soc. Biol. Concepcion*, **44**, 149–159.
*Hermosilla, J.G. (1973) Contribucion al conocimiento sistematico de los dinoflagelados y tintinidos del Archipelago de Juan Fernandez. *Boln Soc. Biol. Concepcion*, **46**, 11–36.
*Hermosilla, J.G. (1975) Contribucion al conocimiento de los dinoflagelados y tintinidos de Antarctica. I. Plancton colectado en Diciembre, 1969. *Gayana*, **34**, 3–55.
*Hernandez-Becerril, D.U. (1992) Observations on two closely related species, *Chaetoceros tetrastichon* and *C. dadayi* (Bacillariophyceae). *Nordic Journal of Botany*, **12**, 365–371.
*Hernroth, L. (1983) Marine pelagic rotifers and tintinnids – important trophic links in the spring plankton community of Gullmar Fjord, Sweden. *Journal of Plankton Research*, **5**, 835–846.
Hillebrand, H. (2004) Strength, slope and variability of marine latitudinal gradients. *Marine Ecology Progress Series*, **273**, 251–267.
*Hilliard, D.K. (1968). *Tintinnidium ephemeridum* n. sp. A new freshwater tintinnid from Alaska with notes on the algae attached to its test. *Hydrobiologia*, **31**, 385–401.
Hjort, J. (1914) Fluctuations in the Great Fisheries of Northern Europe. *Rapports et Procès-Verbaux des Réunions du Conseil International pour l'Exploration de la Mer*, **20**, 1–228.
Hobbie, J.E., Daley, R.J. & Jasper, S. (1977) Use of nuclepore filters for counting bacteria by fluorescence microscopy. *Applied and Environmental Microbiology*, **33**, 1225–1228.
Hofker, J. (1931a) Studien über Tintinnoidea. *Archiv für Protistenkunde*, **75**, 315–402.
Hofker, J. (1931b) Die Bildung der Tintinnengehäuse. *Tijdschrift der Nederlandsche Dierkundige Vereeniging, Serie 3*, **2**, 144–150.
Holder, M. & Lewis, P.O. (2003) Phylogeny estimation: traditional and Bayesian approaches. *Nature Reviews Genetics*, **4**, 275–284.
Holst, H., Zimmermann, H. Kausch, H. & Koste, W. (1998) Temporal and spatial dynamics of planktonic rotifers in the Elbe Estuary during spring. *Estuarine, Coastal and Shelf Science*, **47**, 261–274.
Holt, A.R., Gaston, K.J. & He, F.L. (2002) Occupancy–abundance relationships and spatial distribution: a review. *Basic and Applied Ecology*, **3**, 1–13.
Honjo, S., Okada, H. (1974) Community structure of coccolithophores in the photic layer of the mid-Pacific. *Micropaleontology*, **20**, 209–230.
*Hopkins, T.L. (1966) The plankton of the St. Andrew Bay system, Florida. *Publications of the Institute of Marine Science of the University of Texas* **11**, 12–64.
Hopkins, T.L. (1987) Midwater food web in McMurdo Sound, Ross Sea, Antarctica. *Marine Biology*, **96**, 93–106.
Hoppenrath, M., Yubuki, N., Bachvaroff, T.R. & Leander, B.S. (2010) Re-classification of *Pheopolykrikos hartmanii* as *Polykrikos* (Dinophyceae) based partly on the ultrastructure of complex extrusomes. *European Journal of Phycology*, **45**, 29–37.

Horsted, S.J., Nielsen, T.G., Riemann, B., Poc-Steen, J. & Bjornsen, P.K. (1988) Regulation of zooplankton by suspension-feeding bivalves and fish in estuarine enclosures. *Marine Ecology Progress Series*, **48**, 217–224.
Horwitz, P. & Wilcox, B.A. (2005) Parasites, ecosystems and sustainability: an ecological and complex systems perspective. *International Journal for Parasitology*, **35**, 725–732.
Houde, E.D. & Lovdal, J.A. (1984) Seasonality of occurrence, foods and food preferences of ichthyoplankton in Biscayne Bay, Florida. *Estuarine, Coastal and Shelf Science*, **18**, 403–419.
Huang, B., Xiang, W., Zeng, X., Chiang, K.-P., Tian, H., Hu, J., Lan, W. & Hong, H. (2011) Phytoplankton growth and microzooplankton grazing in a subtropical coastal upwelling system in the Taiwan Strait. *Continental Shelf Research*, **31**, S48–S56.
Hudson, P.J., Dobson, A.P. & Lafferty, K.D. (2006) Is a healthy ecosystem one that is rich in parasites? *Trends in Ecology and Evolution*, **21**, 381–385.
Huldtgren, T., Cunningham, J.A., Yin, C., Stampanoni, M., Marone, F., Donoghue, P.C.J. & Bengtson, S. (2011) Fossilized nuclei and germination structures identify Ediacaran "animal embryos" as encysting protists. *Science*, **334**, 1696–1699.
*Hunt, G.W. & Chien, S.M. (1983) Seasonal distribution, composition and abundance of the planktonic Ciliata and Testacea of Cayuga Lake. *Hydrobiologia*, **98**, 257–266.
Huntley, M., Sykes, P., Rohan, S. & Marin, V. (1986) Chemically-mediated rejection of dinoflagellate prey by the copepods *Calanus pacificus* and *Paracalanus parvus*: mechanism, occurrence and significance. *Marine Ecology Progress Series*, **28**, 105–120.
Hutchinson, G.E. (1961) The paradox of the plankton. *American Naturalist* **95**, 137–145.
Ichinomiya, M., Nakamachi, M. & Taniguchi, A. (2004) A practical method for enumerating cysts of ciliates in natural marine sediments. *Aquatic Microbial Ecology*, **37**, 305–310.
Ichinomiya, M., Nakamachi, M., Honda, M., Fukuchi, M. & Taniguchi, A. (2009) Role of heterotrophic dinoflagellates in the fate of diatoms released from fast ice in coastal water of Lützow-Holm Bay, East Antarctica. *Marine Ecology Progress Series*, **383**, 27–36.
ICZN [= International Commission on Zoological Nomenclature] (1999) *International Code of Zoological Nomenclature, Fourth Edition Adopted by the International Union of Biological Sciences*. The International Trust for Zoological Nomenclature, The Natural History Museum, London.
Imai, I. & Itakura, S. (1991) Densities of dormant cells of the red tide flagellate *Heterosigma akashiwo* (Raphidophyceae) in bottom sediments of northern Hiroshima Bay, Japan. *Bulletin of Japanese Society of Microbial Ecology*, **6**, 1–7.
Imai, I., Itoh, K. & Anraku, M. (1984) Extinction dilution method for enumeration of dormant cells of red tide organisms in marine sediments. *Bulletin of the Plankton Society of Japan*, **31**, 123–124.
Imai, I., Yamaguchi, M. & Watanabe, M. (1998) Ecophysiology, life cycle, and bloom dynamics of *Chattonella* in the Seto Inland Sea Japan. In *Physiological Ecology of Harmful Algal Blooms* (D.M. Anderson, A.D. Cembella, and G.M. Hallegraeff eds), pp. 95–112, Springer-Verlag, Berlin.
Irigoien, X., Flynn, K.J. & Harris, R.P. (2005) Phytoplankton blooms: a 'loophole' in microzooplankton grazing impact? *Journal of Plankton Research*, **27**, 313–321.
Ishikawa, A. & Taniguchi, A. (1994) The role of cysts on population dynamics of *Scrippsiella* spp. (Dinophyceae) in Onagawa Bay, northeast Japan. *Marine Biology*, **119**, 39–44.
Ishikawa, A. & Taniguchi, A. (1996) Contribution of benthic cysts to the population dynamics of *Scrippsiella* spp. (Dinophyceae) in Onagawa Bay, northeast Japan. *Marine Ecology Progress Series*, **140**, 169–178.
Itakura, S. & Yamaguchi, M. (2005) Morphological and physiological differences between the cysts of *Alexandrium catenella* and *A. tamarense* (Dinophyceae) in the Seto Inland Sea, Japan. *Plankton Biology and Ecology*, **52**, 85–91.
Itakura, S., Imai, I. & Itoh, K. (1991) Spatial distribution of cysts of the noxious red tide flagellates *Chattonella* (Raphidophyceae) in sediments of the eastern Seto Inland Sea, Japan. *Nippon Suisan Gakkaishi*, **57**, 1079–1088.
Itaoka, M. & Tamai, K. (1993) Effects of eutrophication on the structure of meiobenthic communities in Hiroshima Bay. *Benthos Research*, **45**, 19–28.
Jacobson, D.M. & Anderson, D.M. (1996) Widespread phagocytosis of ciliates and other protists by marine mixotrophic and heterotrophic thecate dinoflagellates. *Journal of Phycology*, **32**, 279–285.
Jakobsen, H.H. (2001) Escape response of planktonic protists to fluid mechanical signals. *Marine Ecology Progress Series*, **214**, 67–78.
Jakobsen, H.H. & Hansen, P.J. (1997) Prey size selection, grazing and growth response of the small heterotrophic dinoflagellate *Gymnodinium* sp. and the ciliate *Balanion comatum* – a comparative study. *Marine Ecology Progress Series*, **158**, 75–86.
Jakobsen, H.H. & Montagnes, D.J.S. (1999) A redescription of *Balanion comatum* Wulff, 1919 (Prorodontida, Ciliophora), with notes on its cultivation and behaviour. *Journal of Eukaryotic Microbiology*, **46**, 198–205.
Jakobsen, H.H., Hyatt, C. & Buskey, E.J. (2001) Growth and grazing on the 'Texas brown tide' alga *Aureoumbra lagunensis* by the tintinnid *Amphorides quadrilineata*. *Aquatic Microbial Ecology*, **23**, 245–252.
Jakobsen, H.H., Halvorsen, E., Hansen, B.W. & Visser, A.W. (2005) Effects of prey motility and concentration on feeding in *Acartia tonsa* and *Temora longicornis*: the importance of feeding modes. *Journal of Plankton Research*, **27**, 775–785.
*James, M.R. (1989) Role of zooplankton in the nitrogen cycle off the west coast of the South Island, New Zealand, winter

1987. *New Zealand Journal of Marine and Freshwater Research*, **23**, 507–518.

Janson, S., Gisselson, L.-Å., Salomon, P.S. & Granéli, E. (2000) Evidence for multiple species within the endoparasitic dinoflagellate *Amoebophrya ceratii* as based on 18S rRNA gene-sequence analysis. *Parasitology Research*, **86**, 929–933.

Jenkins, G.P. (1987) Comparative diets, prey selection, and predatory impact of co-occurring larvae of two flounder species. *Journal of Experimental Marine Biology and Ecology*, **110**, 147–170.

Jenkins, G.P. (1988) Micro- and fine-scale distribution of microplankton in the feeding environment of larval flounder. *Marine Ecology Progress Series*, **43**, 233–244.

*Jeong, H.J., Park, J.K., Kim, J.S., Kim, S.T., Yoon, J.E., Kim, S.K. & Park, Y.M. (2000) The outbreak of red tides in the coastal waters off Kohung, Chonnam, Korea. 3. The temporal and spatial variations in the heterotrophic dinoflagellates and ciliates in 1997. *The Sea, Journal of the Korean Society of Oceanography*, **5**, 37–46.

Jeong, H.J., Yoon, S.Y., Kim, J.S., Yoo, Y.D. & Seong, K.A. (2002) Growth and grazing rates of the prostomatid ciliate *Tiarina fusus* on red-tide and toxic algae. *Aquatic Microbial Ecology* **28**, 289–297.

Jeong, H.J., Yoo, Y.D., Kim, J.S., Seong, K.A., Kang, N.S. & Kim, T.H. (2010) Growth, feeding and ecological roles of the mixotrophic and heterotrophic dinoflagellates in marine planktonic food webs. *Ocean Science Journal*, **45**, 65–91.

Johannes, R.E. (1965) Influence of marine protozoa on nutrient regeneration. *Limnology and Oceanography*, **10**, 434–442.

*Johansen, P.L. (1976) A Study of Tintinnids and Other Feeding, Nitrogen Excretion and Reproduction Rates. Ph.D Thesis, Dalhousie University, Canada.

Johansson, M., Gorokhova, E. & Larsson, U. (2004) Annual variability in ciliate community structure, potential prey and predators in the open northern Baltic Sea proper. *Journal of Plankton Research*, **26**, 67–80.

Johnson, M.D., Rome, M. & Stoecker, D.K. (2003) Microzooplankton grazing on *Prorocentrum minimum* and *Karlodinium micrum* in Chesapeake Bay. *Limnology and Oceanography*, **48**, 238–248.

Johnson, M.D., Tengs, T., Oldach, D. & Stoecker, D.K. (2006) Sequestration, performance, and functional control of cryptophyte plastids in the ciliate *Myrionecta rubra* (Ciliophora). *Journal of Phycology*, **42**, 1235–1246.

Joint, I., Doney, S.C. & Karl, D.M. (2011) Will ocean acidification affect marine microbes? *ISME Journal*, **5**, 1–7.

Jonsson, P.R. (1986) Particle size selection, feeding rates and growth dynamics of marine planktonic oligotrichous ciliates (Ciliophora: Oligotrichina). *Marine Ecology Progress Series*, **33**, 265–277.

Jonsson, P.R. (1989) Vertical distribution of planktonic ciliates – an experimental analysis of swimming behaviour. *Marine Ecology Progress Series*, **52**, 39–53.

Jonsson, P.R. (1994) Tidal rhythm of cyst formation in the rock pool ciliate *Strombidium oculatum* Gruber (Ciliophora, Oligotrichida): a description of the functional biology and an analysis of the tidal synchronization of encystment. *Journal of Experimental Marine Biology and Ecology*, **175**, 77–103.

Jonsson, P.R., Johansson, M. & Pierce, R.W. (2004) Attachment to suspended particles may improve foraging and reduce predation risk for tintinnid ciliates. *Limnology and Oceanography*, **49**, 1907–1914.

Jörgensen, E. (1924) Mediterranean Tintinnids. Report on the Danish Oceanographical Expeditions 1908–10 2 J.3. (Biology), 1–110.

Jörgensen, E. (1927) Ciliata. Tintinnidae. In *Die Tierwelt der Nord- und Ostsee* 8, Teil II. c1 (G. Grimpe, E. Wagler eds), pp. 1–26. Akademische Verlagsgesellschaft, Leipzig.

Jyothibabu, R., Madhu, N.V., Jayalakshmi, K.V., Balachandran, K.K., Shiyas, C.A., Martin, G.D. & Nair, K.K.C. (2006) Impact of freshwater influx on microzooplankton mediated food web in a tropical estuary (Cochin backwaters – India). *Estuarine, Coastal and Shelf Science*, **69**, 505–518.

Kahl, A. (1932) Urtiere oder Protozoa I: Wimpertiere oder Ciliata (Infusoria) Eine Bearbeitung der freilebenden und ectocommensalen Infusorien der Erde, unter Ausschluß der marinen Tintinnidae. 3. Spirotricha. In *Die Tierwelt Deutschlands und der angrenzenden Meeresteile nach ihren Merkmalen und nach ihrer Lebensweise* 25 (ed. F. Dahl), pp. 399–650. G. Fischer, Jena.

Kaminski, L., Abu-Qarn, M., Guan, Z. et al. (2010) AglJ adds the first sugar of the N-linked pentasaccharide decorating the *Haloferax volcanii* S-layer glycoprotein. *Journal of Bacteriology*, **192**, 5572–5579.

*Kamiyama, T. (1994a) Effects of extracellular products from phytoplankton on the excystment of tintinnids from marine sediments. *Marine Ecology Progress Series*, **105**, 199–201.

*Kamiyama, T. (1994b) The impact of grazing by microzooplankton in northern Hiroshima Bay, the Seto Inland Sea, Japan. *Marine Biology*, **119**, 77–88.

Kamiyama, T. (1994c) Ecological studies on tintinnid ciliates in eutrophic embayment. Ph.D. Dissertation, Tokyo University, Tokyo.

Kamiyama, T. (1996) Determination of the abundance of viable tintinnid cysts in marine sediments in Hiroshima Bay, the Seto Inland Sea of Japan, using a modified MPN method. *Journal of Plankton Research*, **18**, 1253–1259.

*Kamiyama, T. (1997a) Effects of phytoplankton abundance on excystment of tintinnid ciliates from marine sediments. *Journal of Oceanography*, **53**, 299–302.

*Kamiyama, T. (1997b) Growth and grazing responses of tintinnid ciliates feeding on the toxic dinoflagellates *Heterocapsa circularisquama*. *Marine Biology*, **128**, 509–515.

Kamiyama, T. (2000) Application of a vital staining method to measure feeding rates of field ciliate assemblages on a harmful alga. *Marine Ecology Progress Series*, **197**, 299–303.

Kamiyama, T. (2011) Life cycles and ecological roles of planktonic ciliates. *Bulletin of the Plankton Society of Japan*, **58**, 81–86.

Kamiyama, T. (2011) Planktonic ciliates as a food source for the scyphozoan *Aurelia aurita* (s.l.): Feeding activity and assimilation of the polyp stage. *Journal of Experimental Marine Biology and Ecology*, **407**, 207–215.

*Kamiyama, T. & Aizawa, Y. (1987) Growth characteristics of two tintinnid ciliates, *Tintinnopsis beroidea* and *Amphorella quadrilineata*, in laboratory culture. *Bulletin of the Plankton Society of Japan*, **34**, 185–191.

*Kamiyama, T. & Aizawa, Y. (1990) Excystment of tintinnid ciliates from marine sediment. *Bulletin of the Plankton Society of Japan*, **36**, 137–139.

Kamiyama, T. & Aizawa, Y. (1992) Effects of temperature and light on tintinnid excystment from marine sediments. *Nippon Suisan Gakkaishi*, **58**, 877–884.

Kamiyama, T. & Arima, S. (1997) Lethal effect of the dinoflagellate *Heterocapsa circularisquama* upon the tintinnid ciliate *Favella taraikaensis*. *Marine Ecology Progress Series*, **160**, 27–33.

Kamiyama, T. & Arima, S. (2001) Feeding characteristics of two tintinnid ciliates species on phytoplankton including harmful species: effects on prey size on ingestion rates and selectivity. *Journal of Experimental Marine Biology and Ecology*, **257**, 281–296.

Kamiyama, T. & Matsuyama, Y. (2005) Temporal changes in the ciliate assemblage and consecutive estimates of their grazing effect during the course of a *Heterocapsa circularisquama* bloom. *Journal of Plankton Research*, **27**, 303–311.

Kamiyama, T. & Suzuki, T. (2006) Lack of accumulation of paralytic shellfish poisoning (PSP) toxins in the tintinnid ciliate *Favella taraikaensis* feeding on the toxic dinoflagellate *Alexandrium tamarense*. *Marine Ecology Progress Series*, **317**, 57–65.

*Kamiyama, T. & Tsujino, M. (1996) Seasonal variation in the species composition of tintinnid ciliates in Hiroshima Bay, the Seto Inland Sea of Japan. *Journal of Plankton Research*, **18**, 2313–2327.

*Kamiyama, T., Itakura, S. & Nagasaki, K. (1995) Effects of irradiance on excystment of tintinnids from marine sediments. *Journal of Oceanography*, **51**, 615–618.

Kamiyama, T., Takayama, H., Nishii, Y. & Uchida, T. (2001) Grazing impact of the field ciliate assemblage on a bloom of the toxic dinoflagellate *Heterocapsa circularisquama*. *Plankton Biology and Ecology*, **48**, 10–18.

Kamiyama, T., Tsujino, M., Matsuyama, Y. and Uchida, T. (2005) Growth and grazing rates of the tintinnid ciliate Favella taraikaensis on the toxic dinoflagellate Alexandrium tamarense. *Marine Biology*, **147**, 989–997.

Kamiyama, T., Suzuki, T. & Okumura, Y. (2006) Feeding of the tintinnid ciliate *Favella tarakaensis* on the toxic dinoflagellate *Alexandrium tamarense* and the fate of prey toxins. *African Journal of Marine Science*, **28**, 343–346.

Kanwar, K.C. (1960) Note on the specificity of mercuric bromophenol blue for the cytochemical detection of proteins. *Cellular and Molecular Life Sciences*, **16**, 355.

Karayanni, H., Christaki, U., Van Wambeke, F., Denis, M. & Moutin, T. (2005) Influence of ciliated protozoa and heterotrophic nanoflagellates on the fate of primary production in the northeast Atlantic Ocean. *Journal of Geophysical Research*, **110**, C07S15.

Katz, L.A., McManus, G.B., Snoeyenbos-West, O.L.O., Griffin, A., Pirog, K., Costas, B. & Foissner, W. (2005) Reframing the "Everything is everywhere" debate: evidence for high gene flow and diversity in ciliate morphospecies. *Aquatic Microbial Ecology*, **41**, 55–65.

Kauffman, T.A., Lindsay, J., Leithiser, R. (1981) Vertical distribution and food selection of larval atherinids. *Rapports et Proces-verbaux des Réunions. Conseil International pour l'Éxploration de la Mer*, **178**, 342–343.

Kaulich, B., Gianoncelli, A., Beran, A. et al. (2009) Low-energy X-ray fluorescence microscopy opening new opportunities for bio-related research. *Journal of the Royal Society Interface*, **6**, S641–S647.

Kemp, P.F., Sherr, B.F., Sherr, E.B. & Cole, J.J. (1993) *Handbook of Methods in Aquatic Microbial Ecology*. Lewis Publishers, London.

Kent, W. Saville (1880–1882) A Manual of the Infusoria: Including a Description of all known Flagellate, Ciliate, and Tentaculiferous Protozoa, British and Foreign, and an Account of the Organization and Affinities of the Sponges. David Bogue, London. p. i–x + 1–913 + Plates 1–51 (1880: p. i–x + 1–472 + Plates 1–23; 1881: 473–720 + Plates 24–40; 1882: 721–913 + Plates 41–51).

*Kim, D.-Y. & Kim, Y.-O. (1990) Tintinnina (Ciliophora: Oligotrichida) in the Marian Cove, King George Island. *Korean Journal of Polar Research*, **1**, 1–10.

Kim, D.Y., Countway, P.D., Gast, R.J. & Caron, D.A. (2011) Rapid shifts in the structure and composition of a protistan assemblage affect estimates of total protistan species richness. *Microbial Ecology*, **62**, 383–398.

Kim, J.S., Jeong, H.J., Strüeder-Kypke, M.C. et al. (2005) *Parastrombidinopsis shimi* n. gen., n. sp. (Ciliophora: Choreotrichia) from the coastal waters of Korea: morphology and small subunit ribosomal DNA sequence. *Journal of Eukaryotic Microbiology*, **52**, 514–522.

Kim, S. (2006) Patterns in host range for two strains of *Amoebophrya* (Dinophyta) infecting thecate dinoflagellates: *Amoebophrya* spp. ex *Alexandrium affine* and ex *Gonyaulax polygramma*. *Journal of Phycology*, **42**, 1170–1173.

*Kim, S.Y., Yang, E.J., Gong, J. & Choi, J.K. (2010) Redescription of *Favella ehrenbergii* (Claparède and Lachmann, 1858) Jörgensen, 1924 (Ciliophora: Choreotrichia), with phylogenetic analyses based on small subunit rRNA gene sequences. *Journal of Eukaryotic Microbiology*, **57**, 460–467.

Kim, Y.-O. (1995) Life Cycle Strategies of Choreotrich Ciliate Plankton in Onagawa Bay. Ph.D. Thesis, Tohoku University, Sendai, Japan.

Kim, Y.-O. & Han, M.S. (2000) Seasonal relationships between cyst germination and vegetative population of *Scrippsiella trochoidea* (Dinophyceae). *Marine Ecology Progress Series*, **204**, 111–118.

Kim, Y.-O. & Taniguchi, A. (1995) Excystment of the oligotrich ciliate *Strombidium conicum*. *Aquatic Microbial Ecology*, **9**, 149–156.

Kim, Y.-O. & Taniguchi, A. (1997) Seasonal variation of excystment pattern of the planktonic oligotrich ciliate *Strombidjum conicum*. *Marine Biology*, **128**, 207–212.

Kim, Y.-O., Ha, S. & Taniguchi, A. (2008) Morphology and in situ sedimentation of the cysts of a planktonic oligotrich ciliate, *Strombidium capitatum*. *Aquatic Microbial Ecology*, **53**, 173–179.

Kim, Y.-O., Park, M.H. & Han, M.S. (2002a) Role of cyst germination in the bloom initiation of *Alexandrium tamarense* (Dinophyceae) in Masan Bay, Korea. *Aquatic Microbial Ecology*, **29**, 279–286.

Kim, Y.-O., Suzuki, T. & Taniguchi, A. (2002b) A new species in the genus *Cyrtostrombidium* (Ciliophora, Oligotrichia, Oligotrichida): its morphology, seasonal cycle and resting stage. *Journal of Eukaryotic Microbiology*, **49**, 338–343.

Kim, Y.-O., Kim, S.Y., Lee, W.-J. & Choi, J.K. (2010) New observations on the choreotrich ciliate *Strombidinopsis acuminata* Fauré-Fremiet 1924, and comparison with *Strombidinopsis jeokjo* Jeong et al., 2004. *Journal of Eukaryotic Microbiology*, **57**, 48–54.

Kimmance, S., Atkinson, D. & Montagnes, D.J.S. (2006) Do temperature–food interactions matter? Responses of production and its components in the model heterotrophic flagellate *Oxyrrhis marina*. *Aquatic Microbial Ecology*, **42**, 63–73.

*Kimor, B. & Golandsky, B. (1977) Microplankton of the Gulf of Elat: aspects of seasonal and bathymetric distribution. *Marine Biology*, **42**, 55–67.

*Kimor, B. & Golandsky-Baras, B. (1981) Distribution and ecology of tintinnids in the Gulf of Elat (Aqaba) Red Sea. *Journal of Plankton Research*, **3**, 445–459.

*Kimor, B., Gordon, N. & Neori, A. (1992) Symbiotic associations among microplankton in oligotrophic marine environments, with special reference to the Gulf of Aqaba, Red Sea. *Journal of Plankton Research*, **14**, 1217–1231.

Kiørboe, T. (2008) *A Mechanistic Approach to Plankton Ecology*, Princeton University Press, Princeton, USA.

*Kivi, K., Kuosa, H. & Tanskanen, S. (1996) An experimental study on the role of crustacean and microprotozoan grazers in the planktonic food web. *Marine Ecology Progress Series*, **136**, 59–68.

Klaas, C. (1997) Microzooplankton distribution and their potential grazing impact in the Antarctic Circumpolar Current. *Deep-Sea Research* Part II, **44**, 375–393.

Kleppel, G.S. (1993) On the diets of calanoid copepods. *Marine Ecology Progres Series*, **99**, 183–195.

Knoll, A.H. (2004) *Life on a Young Planet: the first three billion years of evolution on Earth*. Princeton University Press, Princeton, USA.

Kofoid, C.A. (1897) On some important sources of error in the plankton method. *Science*, **6**, 829–832.

*Kofoid, C.A. (1905) Some new Tintinnidae from the plankton of the San Diego region. *University of California Publications in Zoology* **1**, 287–306.

Kofoid, C.A. (1930) Factors in the evolution of the pelagic ciliata, the Tintinnoinea. Contributions to marine biology; lectures and symposia given at the Hopkins marine station, December 20–21, 1929, at the midwinter meeting of the Western Society of Naturalists. Stanford University Press, Palo Alto, USA, pp. 1–39.

Kofoid, C.A. & Campbell, A.S. (1929) A conspectus of the marine and fresh-water Ciliata belonging to the suborder Tintinnoinea, with descriptions of new species principally from the Agassiz Expedition to the Eastern Tropical Pacific 1904–1905. *University of California Publications in Zoology*, **34**, 1–403.

*Kofoid, C.A. & Campbell, A.S. (1939) Reports on the scientific results of the expedition to the Eastern Tropical Pacific, in charge of Alexander Agassiz, by the U.S. Fish Commission Steamer "Albatross," from October, 1904, to March, 1905, Lieut.-Commander L.M. Garrett, U.S.N. commanding. 37. The Ciliata: The Tintinnoinea. *Bulletin of the Museum of Comparative Zoology, Harvard*, **84**, 1–473 + Plates 1–36.

*Kogan, M. (2005) Estudio de la Composición Especifíca, Abundancia y Distribución Espacial del Microzooplancton (Protozoos y Micrometazoos) en el Estuario del Río de la Plata (Argentina-Uruguay). *Tesis de Doctor en Ciencias Biologicas de la Universidad de Buenos Aires*, 1–174.

*Komarovsky, B. (1959) The Tintinnina of the Gulf of Eylath (Aqaba). *Bulletin of the Sea Fisheries Research Station, Israel* **21**, 1–40.

*Komarovsky, B. (1962) Tintinnina from the vicinity of the Straits of Tiran and Massawa Region. *Bulletin of the Sea Fisheries Research Station, Israel*, **30**, 48–56.

Kooistra, W.H.C.F., Sarno, D., Balzano, S., Gu, H., Andersen, R.A. & Zingone, A. (2008) Global diversity and biogeography of *Skeletonema* species (Bacillariophyta). *Protist*, **159**, 177–193.

Konovalova, G.V. (2007) Parasitic Peridineae (dinoflagellates) and Ellobiopsidae of coastal waters of the Sea of Japan. *Biologiya-Morya (Vladivostok)*, **33**, 167–175.

Kopylov, A.I. & Tumantseva, N.I. (1987) Analysis of the contents of tintinnid food vacuoles and evaluation of their contribution to the consumption of phytoplankton production off the Peru coast. *Oceanology*, **27**, 343–347.

Kormos, J. & Kormos, K. (1958) Die Zellteilungstypen der Protozoen. *Acta Biologica Hungarica*, **8**, 127–148.

*Krause, M. & Kattner, G. (1989) The influence of water exchange on zooplankton dynamics and species development in a south Norwegian fjord. *Journal of Plankton Research*, **11**, 85–103.

Kremp, A. & Anderson, D.M. (2000) Factors regulating germination of resting cysts of the spring bloom dinoflagellate

Scrippsiella hangoei from the northern Baltic Sea. *Journal of Plankton Research*, **22**, 1311–1327.

*Krishnamurthy, K. & Santhanam, R. (1978) Dictyocysta seshaiyai sp. nov. (Protozoa, Ciliatea, Tintinnida) from Porto Novo, India. *Archiv fur Protistenkunde*, **120**, 138–141.

*Krishnamurthy, K., Naidu, W.D. & Santhanam, R. (1979) Further studies on tintinnids (Protozoa: Ciliata). *Archiv fur Protistenkunde*, **122**, 171–180.

Krishnan, G. (1954) The epicuticle of an arachnid, *Palamneus swammerdami*. *Quarterly Journal of Microscopical Science*, **95**, 371–381.

Krock, B., Tillman, U., John, U. & Cembella, A.D. (2009) Characterization of azaspiracids in plankton size-fractions and isolation of an azaspiracid-producing dinoflagellate from the North Sea. *Harmful Algae*, **8**, 254–263.

Krsinc, F. (1980) Qualitative and quantitative investigtions of the tintinnids along the eastern coast of the Adriatic. *Acta Adriatica*, **21**, 19–104.

*Krsinic, F. (1982) On vertical distribution of tintinnines (Ciliata, Oligotrichida, Tintinnina) in the open waters of the South Adriatic. *Marine Biology*, **68**, 83–90.

*Krsinic, F. (1987a) On the ecology of tintinnines in the Bay of Mali Ston (Eastern Adriatic). *Estuarine, Coastal and Shelf Science*, **24**, 401–408.

*Krsinic, F. (1987b) Tintinnines (Ciliophora, Oligotrichida, Tintinnina) in eastern Adriatic bays. *Estuarine, Coastal and Shelf Science*, **24**, 527–538.

*Krsinic, F. (1988) The family Xystonellidae (Ciliophora, Tintinnina) in the Adriatic Sea. *Journal of Plankton Research*, **10**, 413–429.

Krsinic, F. (2010) *Tintinnids (Tintinnida, Choreotrichia, Ciliata) in the Adriatic Sea, Mediterranean. Part I Taxonomy, Part II Ecology*. Institute of Oceanography and Fisheries, Split, Croatia.

*Kruger, I. (1980) A checklist of southwest African marine phytoplankton with some phytogeographical relations. *Fisheries Bulletin, South Africa*, **13**, 31–53.

Kucera, F.P. (1992) Virus-like particles associated with the apostome ciliate *Hyalophysa chattoni*. *Diseases of Aquatic Organisms*, **12**, 151–153.

Kudela, R.M. (2010) Does horizontal mixing explain phytoplankon dynamics? *Proceedings of the National Academy of Sciences of the United States of America*, **107**, 18235–18236.

Kufferath, H. (1950) Recherches sur le plancton de la Mer flamande (mer du nord meridionale), I. Quelques flagellés, protistes et "caetera". *Bulletin of the Institut Royal des Sciences Naturelles de Belgique*, **26**, 1–43.

Kuris, A.M., Hechinger, R.F., Shaw, J.C., et al. (2008) Ecosystem energetic implications of parasite and free-living biomass in three estuaries. *Nature*, **454**, 515–518.

*Kuwahara, A., Kashiwai, M. & Sinoda, M. (1975) Effects of temperature and chlorinity on the microdistribution of *Favella tarakaensis* Hada (Ciliata) in Maizuru Bay. *Journal of the Oceanographic Society of Japan*, **31**, 109–114.

La Touche, R.W. & West, A.B. (1980) Observations on the food of *Antedon bifidia* (Echinodermata: Crinoidea). *Marine Biology*, **60**, 39–46.

Laackmann, H. (1906) Ungeschlechtliche und geschlechtliche Fortpflangzung der Tintinnen. *Zoologischer Anzeiger*, **30**, 440–443.

*Laackmann, H. (1907) Antarktische Tintinnen. *Zoologischer Anzeiger*, **31**, 235–239.

Laackmann, H. (1908) Ungeschlechtliche und geschlechtliche Fortpflanzung der Tintinnen. *Wissenschaftliche Meeresuntersuchungen, Abteilung Kiel*, **10**, 13–38 + Plates 1–3.

Laackmann, H. (1910) Die Tintinnodean der Deutschen Südpolar-expedition 1901–1903. *Deutsche Südpolar-Expedition. XI. Zoologie III*, **11**, 340–396.

*Lackey, J.B. (1936) Occurrence and distribution of marine protozoan species in the Woods Hole area. *Biological Bulletin*, **70**, 264–278.

*Lackey, J.B. & Balech, E. (1966) A new marine tintinnid. *Transactions of the American Microscopical Society*, **85**, 575–578.

*Lam-Hoai, T., Rougier, C. & Lasserre, G. (1997) Tintinnids and rotifers in a northern Mediterranean coastal lagoon. Structural diversity and function through biomass estimations. *Marine Ecology Progress Series*, **152**, 13–25.

Landry, M.R. & Calbet, A. (2004) Reality check on microbial food web interactions in dilution experiments: responses to the comments of Dolan and McKeon. *Ocean Science Discussions*, **1**, 65–76.

Landry, M.R. & Hassett, R.P. (1982) Estimating the grazing impact of marine microzooplankton. *Marine Biology*, **67**, 283–288.

Landry, M.R., Kirshtein, J. & and Constantinou, J. (1995) A Refined dilution technique for measuring the community grazing impact of microzooplankton, with experimental tests in the Central Equatorial Pacific. *Marine Ecology Progress Series*, **120**, 53–63.

Larson, R.J. (1987) Trophic ecology of planktonic gelatinous predators in Saanich Inlet, British Columbia: diets and prey selection. *Journal of Plankton Research*, **9**, 811–820.

Larson, R.J. (1991) Diet, prey selection and daily ration of *Stomolophus meleagris*, a filter-feeding Scyphomedusa from the NE Gulf of Mexico. *Estuarine, Coastal and Shelf Science*, **32**, 511–525.

Lasker, R. (1981) Factors contributing to variable recruitment of the northern anchovy (*Engraulis mordax*) in the California Current: contrasting years, 1975 through 1978. *Rapports et Procès-Verbaux des Réunions du Conseil International pour l'Exploration de la Mer*, **178**, 375–388.

Last, J.M. (1978a) The food of four species of pleuronectiform larvae in the eastern English Channel and southern North Sea. *Marine Biology*, **45**, 359–368.

Last, J.M. (1978b) The food of three species of gadoid larvae in the eastern English Channel and southern North Sea. *Marine Biology*, **48**, 377–386.

Lauckner, G. (1980) Diseases of protozoa. In *Diseases of Marine Animals* (O. Kinne ed.), Vol. 1, pp. 75–134, John Wiley & Sons, Chichester, UK.

Laval, M. (1971) Mise en évidence par la microscopie électronique d'un organite d'un type nouveau chez les ciliés Tintinnides. *Comptes Rendus Hebdomadaires des Seances de l Academie des Sciences, Paris, Série D*, **273**, 1383–1386.

Laval, M. (1972) Ultrastructure de *Petalotricha ampulla* (Fol). Comparaison avec d'autres tintinnides et avec les autres ordres de ciliés. *Protistologica*, **8**, 369–386.

Laval-Peuto, M. (1975) Cortex, périlemme et réticulum vésiculeux de *Cyttarocylis brandti* (cilié Tintinnide). Les ciliés a périlemme. *Protistologica*, **11**, 83–98.

Laval-Peuto, M. (1976a) Intervention des deux tomites du Tintinnide *Favella ehrenbergii* (Ciliata, Tintinnina) dans l'edification de sa lorica. *Comptes Rendus Hebdomadaires des Seances de l'Academie des Sciences Paris, Série D*, **287**, 2187–2190.

Laval-Peuto, M. (1976b) Ultrastructure de la ciliature et de l'infraciliature adorales de *Cyttarocylis brandti* (cilié). *Journal of Protozoology*, **23**, 13 A.

Laval-Peuto, M. (1977) Reconstruction d'une lorica de forme *Coxliella* par le trophonte nu de *Favella ehrenbergii* (Ciliata, Tintinnina). *Comptes Rendus Hebdomadaires des Seances de l'Academie des Sciences Paris, Série D*, **284**, 547–550.

Laval-Peuto, M. (1980) Ultrastructure of the lorica-wall of Tintinnina (Ciliata, Spirotricha). *Journal of Protozoology*, **27**, 83 A.

Laval-Peuto, M. (1981) Construction of the lorica in Ciliata Tintinnina. In vivo study of *Favella ehrenbergii*: variability of the phenotypes during the cycle, biology, statistics, biometry. *Protistologica*, **17**, 249–272.

Laval-Peuto, M. (1983) Sexual reproduction in *Favella ehrenbergii* (Ciliophora, Tintinnina). Taxonomical implications. *Protistologica*, **19**, 503–512.

Laval-Peuto, M. (1994) Classe des Oligotrichea Bütschli, 1887. Ordre des Tintinnida Kofoid et Campbell, 1929. In *Traité de Zoologie. Anatomie, systématique, biologie. II. Infusoires ciliés. 2. Systématique* (ed. P. de Puytorac), pp. 181–219. Masson, Paris.

Laval-Peuto, M. & Barria de Cao, M.S. (1987) Les capsules, extrusomes caractéristiques des Tintinnina (Ciliophora), permettent une classification evolutive des genres et des familles du sous-ordre. *Deuxième Réunion Scientifique du GRECO 88, Travaux du Comité Régional de la Méditerranée*, **8**, 53–59.

Laval-Peuto, M. & Brownlee, D.C. (1986) Identification and systematics of the Tintinnina (Ciliophora): evaluation and suggestion for improvement. *Annales de l'Institut Oceanographique, Paris*, **62**, 69–84.

*Laval-Peuto, M. & Rassoulzadegan, F. (1988) Autofluorescece of marine planktonic Oligotrichina and other ciliates. *Hydrobiologia*, **159**, 99–110.

Laval-Peuto, M., Gold, K. & Storm, E.R. (1979) The ultrastructure of *Tintinnopsis parva*. *Transactions of the American Microscopical Society*, **98**, 204–212.

Le Hégarat, G. & Remane, J. (1968) Tithonique supérieur et Berriasien de l'Ardèche et de l'Hérault. Correlation des Ammonites et des Calpionelles. *Géobios*, **1**, 7–70.

Leakey, R.J.G., Burkill, P.H. & Sleigh, M.A. (1993) Planktonic ciliates in Southampton Water: quantitative taxonomic studies. *Journal of the Marine Biological Association of the United Kingdom*, **73**, 579–594.

Leakey, R.J.G., Burkill, P.H. & Sleigh, M.A. (1994a) Ciliate Growth Rates from Plymouth Sound: Comparison of Direct and Indirect Estimates. *Journal of the Marine Biological Association of the United Kingdom*, **74**, 849–861.

*Leakey, R.J.G., Fenton, N. & Clarke, A. (1994b) The annual cycle of planktonic ciliates in nearshore waters at Signy Island, Antarctica. *Journal of Plankton Research*, **16**, 841–856.

Leakey, R.J.G., Burkill, P.H. & Sleigh, M.A. (1996) Planktonic ciliates in the northwestern Indian Ocean:their abundance and biomass in waters of contrasting productivity. *Journal of Plankton Research*, **18**, 1063–1071.

Leander, B.S., Clopton, R.E. & Keeling, P.J. (2003) Phylogeny of gregarines (Apicomplexa) as inferred from small-subunit rDNA and β-tubulin. *International Journal of Systematic and Evolutionary Microbiology*, **53**, 345–354.

Lebour, M.V. (1917) The microplankton of Plymouth Sound from the region beyond the breakwater. *Journal of the Marine Biological Association of the United Kingdom*, **11**, 133–182.

*Lebour, M.V. (1918) The food of post-larval fish. *Journal of the Marine Biological Association of the United Kingdom*, **11**, 433–469.

Lebour, M.V. (1919a) The food of plankton organisms. *Journal of the Marine Biological Association of the United Kingdom*, **12**, 644–677.

Lebour, M.V. (1919b) The food of young clupeoids. *Journal of the Marine Biological Association of the United Kingdom*, **12**, 458–467.

Lebour, M.V. (1919c) The food of young fish. *Journal of the Marine Biological Association of the United Kingdom*, **12**, 261–324.

Lebour, M.V. (1924) The food of young herring. *Journal of the Marine Biological Association of the United Kingdom*, **13**, 325–330.

Lecal, J. (1953) Sur deux nouveaux Tintinnides. *Archives de Zoologie Expérimentale et Générale*, **91**, 1–4.

Lecal, J. (1967) Le nannoplancton des côtes d'Israël. *Hydrobiologia*, **29**, 305–387.

Leegaard, C. (1915) Untersuchungen über einige Planktonciliaten des Meeres. *Nytt magazin for naturvidenskaberne*, **53**, 1–37.

*Leegaard, C. (1920) Microplankton from Finnish waters during the month of May 1912. *Acta Societatis Scientarium Fennic', Serie B*, **48**, 1–48.

Lehman, J.T. & Sandgren, C.D. (1985) Species-specific rates of growth and grazing loss among freshwater algae. *Limnology and Oceanography* **30**, 34–46.

Leising, A.W., Pierson, J.J., Halsband-Lenk, C., Horner, R. & Postel, J. (2005a) Copepod grazing during spring blooms:

Does *Calanus pacificus* avoid harmful diatoms? *Progress in Oceanography*, **67**, 384–405.

Leising, A.W., Pierson, J.J., Halsband-Lenk, C., Horner, R. & Postel, J. (2005b). Copepod grazing during spring blooms: Can *Pseudocalanus newmani* induce trophic cascades? *Progress in Oceanography*, **67**, 406–421.

Lefèvre, T., Lebarbenchon, C., Gauthier-Clerc, M., Missé, D., Poulin, R. & Thomas, F. (2008) The ecological significance of manipulative parasites. *Trends in Ecology and Evolution*, **24**, 41–48.

*Lepannen, J.-M. & Bruun, J.E. (1988) Cycling of organic matter during the vernal growth period in the open northern Baltic proper. IV. Ciliate and mesozooplankton species composition, biomass, food intake, respiration, and production. *Finnish Marine Research*, **255**, 55–78.

Lessard, E.J. (1991) The trophic role of heterotrophic dinoflagellates in diverse marine environments. *Marine Microbial Food Webs*, **5**, 49–58.

Lessard, E.J. & Swift, E. (1985) Species-specific grazing rates of heterotrophic dinoflagellates in oceanic waters, measured with a dual-label radioisotope technique. *Marine Biology*, **87**, 289–296.

Lessard, E.J. & Swift, E. (1986) Dinoflagellates from the North Atlantic classified as phototrophic or heterotrophic by epifluorescence microscopy. *Journal of Plankton Research*, **8**, 1209–1215.

*Levinsen, H., Nielsen, T.G. & Hansen, B.W. (1999) Plankton community structure and carbon cycling on the western coast of Greenland during the stratified summer situation. II. Heterotrophic dinoflagellates and ciliates. *Aquatic Microbial Ecology*, **16**, 217–232.

Levy, M.G., Litaker, R.W., Goldstein, R.J., Dykstra, M.J., Vandersea, M.W. & Noga, E.J. (2007) *Piscinoodinium*, a fish-ectoparasitic dinoflagellate, is a member of the class dinophyceae, subclass gymnodiniphycidae: Convergent evolution with Amyloodinium. *Journal of Parasitology*, **93**, 1006–1015.

Li, C.-W., Chen, J.-Y., Lipps, J.H., Gao, F., Chi, H.-M. & Wu, H.-J. (2007) Ciliated protozoans from the Precambrian Doushantuo Formation, Wengen, South China. *Geological Society, Special Publications*, **286**, 151–156.

Li, Y.-X. & Zhan, S.-X. (2006) New material of microfossils from Middle Proterozoic in Hubei Province. *Acta Palaeontologica Sinica*, **45**, 102–107.

Li, Y.-X., Zhang, S.-X. & Zhan, J. (2009) Mesoproterozoic calymmian tintinnids from central China. *Open Paleontology Journal*, **2**, 10–13.

*Li, Z., Yi, Z., Yang, J. et al. (2009) Phylogenetic investigation on five genera of tintinnid ciliates (Ciliophora, Choreotrichia), based on the small subunit ribosomal RNA gene sequences. *Progress in Natural Science*, **19**, 1097–1101.

*Lindley, J.A. (1975) Continuous plankton records: a plankton atlas of the North Atlantic and North Sea: Supplement 3 – Tintinnida (Protozoa, Ciliophora) in 1965. *Bulletin of Marine Ecology*, **8**, 201–213.

*Ling, H. (1965) The tintinnid *Parafavella gigantea* (Brandt) Kofoid & Campbell, **1929**, in the North Pacific Ocean. *Journal of Paleontology*, **39**, 721–723.

Ling, H.Y. (1992) *Tintinnids: A Taxon-vertical Distributional Study of Settling Assemblages from the Panama Basin*. Ocean Biocoenosis Series 4, Woods Hole Oceanographic Institution, Woods Hole, USA.

Linnaeus, C. (1758) *Systema naturae per regna tria naturae, secundum classes, ordines, genera, species, cum characteribus, differentiis, synonymis, locis. Tomus 1* (10th edn.). Laurentius Salvius, Holmiae.

Lipps, J.H. (2006) Major Features of Protistan Evolution: Controversies, Problems and a Few Answers. *Anuário do Instituto de Geociências – UFRJ*, **29**, 55–80.

Lipscomb, D. (1998) *Basics of Cladistic Analysis*. George Washington University, Washington, D.C., USA.

Lipscomb, D.L. (1992) Parsimony, homology and the analysis of multistate characters. *Cladistics*, **8**, 45–65.

Litaker, R.W., Tester, P.A., Colorni, A., Levy, M.G. & Noga, E.J. (1999) The phylogenetic relationship of *Pfiesteria piscicida, Cryptoperidiniopsoid* sp. *Amyloodinoum ocellatum* and a *Pfiesteria*-like dinoflagellate to other dinoflagellates and apicomplexans. *Journal of Phycology*, **35**, 1379–1389.

*Littlepage, J.L. (1968) Plankton investigations in McMurdo Sound. *Antarctic Journal of the United States*, **3**, 162–163.

Liu, H., Dagg, M.J. & Strom, S. (2005) Grazing by the calanoid copepod *Neocalanus cristatus* on the microbial food web in the coastal Gulf of Alaska. *Journal of Plankton Research*, **27**, 647–662.

Liu, W., Yi, Z., Lin, X. & Al-Rasheid, K.A.S. (2011) Morphologic and molecular data suggest that *Lynnella semiglobulosa* n. g., n. sp. represents a new family within the subclass Choreotrichia (Ciliophora, Spirotrichea). *Journal of Eukaryotic Microbiology*, **58**, 43–49.

Llopiz, J.K. & Cowen, R.K. (2009) Variability in the trophic role of coral reef fish larvae in the oceanic plankton. *Marine Ecology Progress Series*, **381**, 259–272.

Lobo-da-Cunha, A. & Azevedo, C. (1992) Virus-like particles in the fish parasite *Ichthyophthirius multifiliis* (Ciliophora). *Journal of Fish Diseases*, **14**, 273–277.

Loeblich, A.R. III (1976) Dinoflagellate evolution: speculation and evidence. *Journal of Protozoology*, **23**, 13–28.

Loeblich, A.R. III (1982) Dinophyceae. In *Synopsis and Classification of Living Organisms* (S.P. Parker ed.) Vol. 1, pp. 101–115, McGraw-Hill, New York.

Loeblich, A.R., Jr & Tappan, H. (1968) Annotated index to genera, subgenera and suprageneric taxa of the ciliate order Tintinnida. *Journal of Protozoology*, **15**, 185–192.

Logares, R., Rengefors, K., Kremp, A., Shalchian-Tabrizi, K., Boltovskoy, A., Tengs, T., Shurtleff, A. & Klaveness, D. (2007a) Phenotypically different microalgal morphospecies with identical ribosomal DNA: A case of rapid adaptive evolution? *Microbial Ecology*, **53**, 549–561.

Logares, R., Shalchian-Tabrizi, K., Boltovskoy, A. & Rengefors, K. (2007b) Extensive dinoflagellate phylogenies indicate infrequent marine-freshwater transitions. *Molecular Phylogenetics and Evolution*, **45**, 887–903.

Lohmann, H. (1901) Ueber das Fischen mit Netzen aus Müllergaze Nr 20 zu dem Zwecke quantitiven Untersuchgen des Auftriebs. Wissenschaftliche Meeresuntersuchungen, *Abteilung Kiel*, **5**, 46–66.

Lohmann, H. (1908) Untersuchungen zur Feststellung des vollständigen Gehaltes des Meeres an Plankton. *Wissenschaftliche Meeresuntersuchungen Kiel*, **10**, 131–370.

Lohmann, H. (1911) Über das Nannoplankton und die Zentrifugierung kleinster Wasserproben zur Gewinnung desselben in lebendem *Zustande*. *Internationale Revue der gesamten Hydrobiologie und Hydrographie*, **4**, 1–38.

Lonsdale, D.J., Caron, D.A., Dennett, M.R. & Schaffner, R. (2000) Predation by *Oithona* spp. on protozooplankton in the Ross Sea, Antarctica. *Deep-Sea Research II*, **47**, 3273–3283.

*Lubel, M.A.A. (1974) Distribucion de los tintinidos (Protozoa, Ciliatea) de la parte oeste, suroeste y sur del Golfo de Mexico. *Revista de la Sociedad Mexicana de Historia Natural*, **35**, 45–76.

Lugomela, C., Wallberg, P. & Nielsen, T.G. (2001) Plankton composition and cycling of carbon during the rainy season in a tropical coastal ecosystem, Zanzibar, Tanzania. *Journal of Plankton Research*, **23**, 1121–1136.

Lynn, D.H. (1981) The organization and evolution of microtubular organelles in ciliated protozoa. *Biological Reviews*, **56**, 243–292.

Lynn, D.H. (2008) The Ciliated Protozoa. Characterization, Classification and Guide to the Literature, 3rd edn. Springer, Berlin.

Lynn, D.H. & Montagnes, D.J.S. (1988) Taxonomic descriptions of some conspicuous species of strobilidiine ciliates (Ciliophora: Choreotrichida) from the Isles of Shoals, Gulf of Maine. *Journal of the Marine Biological Association of the United Kingdom*, **68**, 639–658.

Lynn, D.H. & Simpson, A.G.B. (2009) From the Editors: Describing new taxa of unicellular protists. *Journal of Eukaryotic Microbiology*, **56**, 403–405.

Lynn, D.H. & Small, E.B. (2002) Phylum Ciliophora Doflein, 1901. In An Illustrated Guide to the Protozoa. 2nd Edition. Organisms traditionally referred to as Protozoa, or newly discovered Groups (J.J. Lee, G.F. Leedale and P. Bradbury eds), pp. 371–656. Society of Protozoologists, year 2000. Allen Press, Lawrence.

Lynn, D.H. & Strüder-Kypke, M.C. (2006) Species of *Tetrahymena* identical by small subunit rRNA gene sequences are discriminated by mitochondrial cytochrome c oxidase I gene sequences. *Journal of Eukaryotic Microbiology*, **53**, 385–387.

Lynn, D.H., Montagnes, D.J.S., Dale, T., Gilron, G.L. & Strom, S.L. (1991) A reassessment of the genus *Strombidinopsis* (Ciliophora, Choreotrichida) with descriptions of four new planktonic species and remarks on its taxonomy and phylogeny. *Journal of the Marine Biological Association of the United Kingdom*, **71**, 597–612.

Macpherson, E. (2002) Large-scale species-richness gradients in the Atantic Ocean. *Proceedings of the Royal Society of London Series B*, **269**, 1715–1720.

Malone, T.C. (1980) Algal size. In *The Physiological Ecology of Phytoplankton* (I. Morris ed.), pp 433–463, University of California Press, Berkeley.

*Mamaeva, N.V. (1982) Planktonic infusorians of the central part of the Indian Ocean. *Soviet Journal of Marine Biology*, **8**, 254–258.

*Mamaeva, N.V. (1984) Planktonic infusorians in the Bering Sea. *Soviet Journal of Marine Biology*, **9**, 180–185.

Maneiro, I., Frangopulos, M., Guisande, C., Fernandez, M., Reguera, B. & Riveiro, I. (2000) Zooplankton as a potential vector of diarrhetic shellfish poisoning toxins through the food web. *Marine Ecology Progress Series*, **201**, 155–163.

Margalef, R. (1957) Fitoplancton de las costas de Blanes (Gerona) de agosto de 1952 a junio de 1956. *Investigacion Pesquera*, **8**, 89–96.

Margalef, R. (1982) Some thoughts on the dynamics of populations of ciliates. *Annales Institut Océanographique Paris* **58** (S), 15–18.

Margalef, R. & Durán, M. (1953) Microplancton de Vigo, de octubre de 1951 a septiembre de 1952. *Publicaciones del Instituto de Biología Aplicada*, **13**, 5–78 + Table 1.

*Margalef, R., Estrada, M. (1987) Synoptic distribution of summer microplankton (algae and protozoa) across the principal front in the western Mediterranean. *Investigacion Pesquera*, Barcelona, **51**, 121–140.

*Margalef, R. & Morales, E. (1960) Fitoplancton de las costas de Blanes (Gerona), de julio de 1956 a junio de 1959. *Investigacion Pesquera, Barcelona*, **16**, 3–31.

Marret, F. & Zonneveld, K.A.F. (2003) Atlas of modern organic-walled dinoflagellate cyst distribution. *Review of Paleobotany and Palynology*, **125**, 1–200.

*Marron-Aguilar, M.A. & Lopez-Ochoterena, E. (1969) Protozarios ciliados de Mexico XVI. Sistematica de algunas especies del orden Tintinnida Kofoid y Campbell, de la Laguna de Terminos, Campeche. *Revista de la Sociedad Mexicana de Historia Natural*, **30**, 43–64.

Marshall, S. (1924) The food of *Calanus finmarchicus* during 1923. *Journal of the Marine Biological Association of the United Kingdom*, **13**, 473–479.

*Marshall, S.M. (1934) The Silicoflagellata and Tintinnoinea. *Great Barrier Reef Expedition 1928–29, Scientific Reports, British Museum of Natural History, London*, **4** (no. 15), 623–664.

Martin, A.J. & Montagnes, D.J.S. (1993) Winter ciliates in a British Columbian fjord: six new species and an analysis of ciliate putative prey. *Journal of Eukaryotic Microbiology*, **40**, 535–549.

Martín-González, A., Wierzchos, J., Gutiérrez, J.C., Alonso, J. & Ascaso, C. (2008) Morphological stasis of protists in lower Cretaceous amber. *Protist*, **159**, 251–257.

*Massuti Alzamora, M. (1929) Contribucion al estudio de los infusorios de la Bahia de Palma Mallorca. Nota segunda. *Publicaciones del Instituto Español de Oceanografia*, **32**, 1–11.

*Massuti Alzamora, M. (1930) El plancton de la Bahia de Palma de Mallorca en 1929. *Publicaciones del Instituto Español de Oceanografia*, **43**, 1–28.

*Massuti Alzamora, M. (1933) Contribucion al estudio de los infusorios (Nota tercera) Nuevos tintinidos de la Bahia de Palma de Mallorca. *Publicaciones del Instituto Español de Oceanografia*, **2**, 1–14.

Matsuoka, K. & Fukuyo, Y. (2000) *Technical Guide for Modern Dinoflagellate Cyst Study*. WESTPAC-HAB Asian Natural Environmental Science Center, Tokyo, p. 29.

Matsusaka, T. (2006) On the resting cyst of ciliates. *Japanese Journal of Protozoology*, **39**, 205–216.

Mauchline, J. (1980) The biology of mysids and euphausiids. *Advances in Marine Biology*, **18**, 3–681.

Mauchline, J. (1998) The biology of calanoid copepods. *Advances in Marine Biology*, **33**, 1–710.

Mauchline, J. & Fisher, L.R. (1969) The biology of euphausiids. *Advances in Marine Biology*, **7**, 1–421.

Mayr, E. & Bock, W.J. (2002) Classifications and other ordering systems. *Journal of Zoological Systematics and Evolutionary Research*, **40**, 169–194.

McAllister, C.D., Parsons, T.R. & Strickland, J.D.H. (1960) Primary productivity and fertility at Station "P" in the North-East Pacific Ocean. *Journal du Conseil International pour l'Exploration de la Mer*, **25**, 240–259.

McKinley, M.P., Bolton, D.C. & Prusiner, S.B. (1983) A protease-resistant protein is a structural component of the Scrapie prion. *Cell*, **35**, 57–62.

McManus, G.B. & Ederington-Cantrell, M.C. (1992) Phytoplankton pigments and growth rates, and microzooplankton grazing in a large temperate estuary. *Marine Ecology Progress Series*, **87**, 77–85.

McManus, G.B. & Katz, L.A. (2009) Molecular and morphological methods for identifying plankton: what makes a successful marriage? *Journal of Plankton Research*, **31**, 1119–1129.

McManus, G.B. & Okubo, A. (1991) On the use of surrogate food particles to measure protistan ingestion. *Limnology and Oceanography*, **36**, 613–617.

McManus, G.B., Costas, B.A., Dam, H.G., Lopes, R.M., Gaeta, S.A., Susini, S. & and Rosetta, C. (2007) Microzooplankton grazing of phytoplankton in a tropical upwelling region. *Hydrobiologia*, **575**, 69–81.

McQuoid, M.R. & Hobson, L.A.(1996) Diatom resting stages. *Journal of Phycology*, **32**, 889–902.

Meckler, A.N., Schubert, C.J., Hochuli, P.A., Plessen, B., Birgel, D., Flower, B.P., Hinrichs, K.-U. & Haug, G.H. (2008) Glacial to Holocene terrigenous organic matter input to sediments from Orca Basin, Gulf of Mexico – A combined optical and biomarker approach. *Earth and Plantary Science Letters*, **272**, 251–263.

Menden-Deuer, S. & Lessard, E.J. (2000). Carbon to volume relationships for dinoflagellates, diatoms, and other protist plankton. *Limnology and Oceanography*, **45**, 569–579.

Merkle, H. (1909) Untersuchung an Tintinnodeen der Ost- und Nordsee. *Wissenschaftliche Meeresuntersuchungen, Abteilung Kiel*, **11**, 139–186 + Plates 2, 3.

Meunier, A. (1910) *Campagne arctique de 1907. Microplankton des mers de Barents et de Kara*. Bulens, Bruxelles, Belgium.

Meunier, A. (1919) Microplankton de la mer Flamande: Les Tintinnides et caetera. *Mémoires du Musée Royal d'Histoire Naturelle de Belgique*, **8**, 1–59 + Plates 22, 23.

*Middlebrook, K., Emerson, C.W., Roff, J.C. & Lynn, D.H. (1987) Distribution and abundance of tintinnids in the Quoddy Region of the Bay of Fundy. *Canadian Journal of Zoology*, **65**, 594–601.

Miller, C.B. (2004) *Biological Oceanography*. Blackwell Publishing, Oxford, UK.

Miller, J.J., Delwiche, C.F. & Coats, D.W. (2012). Ultrastructure of *Amoebophrya* sp. and its changes during the course of infection. *Protist*, in press, DOI:10.1016/j.protis.2011.11.007.

Mills, E.L. (1989) *Biological Oceanography: an Early History, 1870–1960*. Cornell University Press, Ithaca, USA.

Minter, E.J.A., Fenton, A., Cooper, J.C. & Montagnes, D.J.S. (2011) Prey-dependent mortality rate: a critical parameter in microbial models. *Microbial Ecology*, **62**, 155–161.

Modigh, M. (2001) Seasonal variations of photosynthetic ciliates at a Mediterranean coastal site. *Aquatic Microbial Ecology*, **23**, 163–175.

Modigh, M. & Castalado, S. (2002) Variability and persistence in tintinid assemblages at a Mediterranean coastal site. *Aquatic Microbial Ecology*, **28**, 299–311.

Moestrup, O., Garcia-Cuetos, L., Hansen, P.J. & Fenchel, T. (2012) Studies on the ciliate *Mesodinium* I. Ultrastructure and description of *M. chamaeleon* sp. nov., a benthic marine species with green or red chloroplasts. *Journal of Eukaryotic Microbiology* **59**, in press.

Montagnes, D.J.S. (1996) Growth responses of planktonic ciliates in the genera *Strobilidium* and *Strombidium*. *Marine Ecology Progress Series*, **130**, 241–254.

Montagnes, D.J.S. & Berges, J.A. (2004) Determining parameters of the numerical response. *Microbial Ecology*, **48**, 139–144.

Montagnes, D.J.S. & Lessard, E.J. (1999) Population dynamics of the marine planktonic ciliate *Strombidinopsis multiauris*: its potential to control phytoplankton blooms. *Aquatic Microbial Ecology*, **20**, 167–181.

Montagnes, D.J.S. & Lynn, D.H. (1987) A quantitative protargol stain (QPS) for ciliates: Method description and test of its quantitative nature. *Marine Microbial Food Webs*, **2**, 83–93.

Montagnes, D.J.S. & Lynn, D.H. (1991) Taxonomy of Choreotrichs, the major marine ciliates, with emphasis on the aloricate forms. *Marine Microbial Food Webs*, **5**, 59–74.

Montagnes, D.J.S., Lynn, D.H., Stoecker, D.K. & Small, E.B. (1988) Taxonomic descriptions of one new species and redescription of four species in the family Strombidiidae (Ciliophora, Oligotrichida). *Journal of Protozoology*, **35**, 189–197.

Montagnes, D.J.S., Berges, J.A., Harrison, P.J. & Taylor, F.J.R. (1994) Estimating carbon, nitrogen, protein, and chlorophyll a from cell volume in marine phytoplankton. *Limnology and Oceanography*, **39**, 1044–1060.

Montagnes, D.J.S., Berger, J.D. & Taylor, F.J.R. (1996) Growth rate of the marine planktonic ciliate *Strombidinopsis cheshiri* Snyder and Ohman as a function of food concentration and interclonal variability. *Journal of Experimental Marine Biology and Ecology*, **206**, 121–132.

Montagnes, D.J.S., Wilson, D., Brooks, S.J., Lowe, C. & Campey, M. (2002a) Cyclical behaviour of the tide-pool ciliate *Strombidium oculatum*. *Aquatic Microbial Ecology*, **28**, 55–68.

Montagnes, D.J.S., Lowe, C.D., Poulton, A. & Jonsson, P.R. (2002b) Redescription of *Strombidium oculatum* Gruber 1884 (Ciliophora, Oligotrichia). *Journal of Eukaryotic Microbiology*, **49**, 329–337.

Montagnes, D.J.S., Kimmance, S.A. & Atkinson, D. (2003) Using Q10: Can growth rates increase linearly with temperature? *Aquatic Micobial Ecology*, **32**, 307–313.

Montagnes, D.J.S., Barbosa, A.B., Boenigk, J., Davidson, K., Jürgens, K., Macek, M., Parry, J., Roberts, E.C. & Simek, K. (2008a) Selective feeding behaviour of free-living protists: avenues for continued study. *Aquatic Microbial Ecology*, **53**, 83–98.

Montagnes, D.J.S., Chambouvet, A., Guillou, L. & Fenton, A. (2008b) Responsibility of microzooplankton and parasite pressure be responsible for the demise of toxic dinoflagellate blooms. *Aquatic Microbial Ecology*, **53**, 211–225.

Montagnes, D.J.S., Dower, J.F. & Figueiredo, G.M. (2010) The protozooplankton–ichthyoplankton trophic link: an overlooked aspect of aquatic food webs. *Journal of Eukaryotic Microbiology*, **57**, 223–228.

Montagnes, D.J.S., Lowe, C.D., Roberts, E.C., Breckels, M.N., Boakes, D.E., Davidson, K., Keeling, P.J., Slamovits, C.H., Steinke, M., Yang, Z. & Watts, P.C. (2011) An introduction to the special issue: *Oxyrrhis marina*, a model organism? *Journal of Plankton Research*, **33**, 549–554.

Montagnes, D.J.S., Roberts, E., Lukes, J. & Lowe, C. (2012) The rise of model protozoa. *Trends in Microbiology*, **20**, 184–191.

Montani, S., Ichimi, K., Meksumpun, S. & Okaichi, T. (1995) The effects of dissolved oxygen and sulfide on germination of the cysts of some different phytoflagellates. In *Harmful Marine Algal Blooms* (P. Lassus, G. Arzul, E. Erard, P. Gentien and C. Marcaillou eds), pp. 627–632, Lavoisier, Paris.

*Monti, M. & Umani, S.F. (1995) Tintinnids in Terra Nova Bay – Ross Sea during two austral summers (1987/88 and 1989/90). *Acta Protozoologica*, **34**, 193–201.

Montresor, M., Lovejoy, C., Orsini, L. & Procaccini, G. (2003) Bipolar distribution of the cyst-forming dinoflagellate *Polarella glacialis*. *Polar Biology*, **26**, 186–194.

Moon-van der Staay, S., De Wachter, R. & Vaulot, D. (2001) Oceanic 18S rDNA sequences from picoplankton reveal unsuspected eukaryotic diversity. *Nature*, **409**, 607–610.

Moraczewski, J. (1970) Etudes radioautographiques sur l'amibe *Arcella*. *Septième Congrès International de Microscopie Électronique, Grenoble*, 405–406.

Moraczewski, J. (1971a) La composition chimique de la coque d'*Arcella discoides* Ehrbg. *Acta Protozoologica*, **8**, 407–421 + Plate 1.

Moraczewski, J. (1971b) Structure et formation de la coque d'*Arcella*. *Acta Protozoologica*, **8**, 423–437 + Plates 1–5.

Mori, K., Yamamoto, K., Teruya, K., et al. (2007) Endoparasitic dinoflagellate of the genus *Ichthyodinium* infecting fertilized eggs and hatched larvae observed in the seed production of leopard coral grouper *Plectropomus leopardus*. *Fish Pathology*, **42**, 49–57.

Morote, E., Olivar, M.P., Pankhurst, P.M., Villate, F. & Uriate, I. (2008a) Trophic ecology of bullet tuna *Auxis rochei* larvae and ontogeny of feeding-related organs. *Marine Ecology Progress Series*, **353**, 243–254.

Morote, E., Olivar, M.P., Villate, F. & Uriate, I. (2008b) Diet of round sardinella, *Sardinella aurita*, larvae in relation to plankton availability in the NW Mediterranean. *Journal of Plankton Research*, **30**, 807–816.

Morote, E., Olivar, M.P., Villate, F. & Uriate, I. (2010) Comparison of anchovy (*Engraulis encrasicolus*) and sardine (*Sardina pilchardus*) larvae feeding in the Northwest Medierranean: influence of prey availability and ontogeny. *ICES Jornal of Marine Science*, **67**, 879–908.

Müller, H. (1996) Encystment of the freshwater ciliate *Pelagostrombidium fallax* (Ciliophora, Oligotrichida) in laboratory culture. *Aquatic Microbial Ecology*, **11**, 289–295.

Müller, H. (2002) Laboratory study of the life cycle of a freshwater strombidiid ciliate. *Aquatic Microbial Ecology*, **29**, 189–197.

Müller, H. & Wünsch, C. (1999) Seasonal dynamics of cyst formation of pelagic strombidiid ciliates in a deep prealpine lake. *Aquatic Microbial Ecology*, **17**, 37–47.

Müller, H., Stadler, P. & Weisse, T. (2002) Seasonal dynamics of cyst formation of strombidiid ciliates in alpine Lake Mondsee, Austria. *Aquatic Microbial Ecology*, **29**, 181–188.

Müller, O.F. (1776) *Zoologiae Danicae prodromus, seu animalium Daniae et Norvegiae indigenarum characteres, nomina, et synonyma imprimis popularium*. Hallageriis, Havniae.

Müller, O.F. (1779) *Zoologia danica seu animalium Daniae et Norvegiae rariorum ac minus notorum descriptiones et historia 1*. Weygandinis, Havniae et Lipsiae.

Munn, C.B. (2006) Viruses as pathogens of marine organisms from bacteria to whales. *Journal of the Marine Biological Association of the United Kingdom*, **86**, 453–467.

*Murdoch, R.C. (1990) Diet of hoki larvae (*Macruronus novaezelandiae*) off Westland, New Zealand. *New Zealand Journal of Marine and Freshwater Research*, **24**, 519–527.

Murray, J.W. & Taylor, F.J.R. (1965) Early calpionellids from the Upper Devonian of western Canada, with a note on pyrite inclusions: *Bulletin Canadian Petroleum Geology*, **13**, 327–334.

Murray, S., Jørgensen, M.F., Ho, S.Y.W., Patterson, D.J. & Jermlin, L.S. (2005) Improving the analysis of dinoflagellate phylogeny based on rDNA. *Protist*, **156**, 269–286.

Nagano, N. and Uye, S. (2002) Seasonal variations in abundance, biomass, in situ growth rate and production of tintinnid ciliates in Kure Port, the Inland Sea of Japan. *Bulletin of the Society of Sea Water Science*, **56**, 142–149.

Nagano, N., Iwatsuki, Y., Kamiyama, T. & Nakata, H. (2000) Effects of marine ciliates on survivability of the first-feeding larval surgeonfish, *Paracanthurus hepatus*: laboratory rearing experiments. *Hydrobiologia*, **432**, 149–157.

Nagano, N., Iwatsuki, Y., Okazaki, Y. & Nakata, H. (2001) Feeding strategy of Japanese sand lance larvae in relation to ciliated protozoa in the vicinity of a thermohaline front. *Journal of Oceanography*, **57**, 155–163.

*Naidu, W.D. (1983) Tintinnida (Protozoa: Ciliata) – a vital link in the estuarine food web. *Mahasagar – Bulletin of the National Institute of Oceanograph*, **16**, 403–407.

*Naidu, W.D. & Krishnamurthy, K. (1985) Biogeographical distribution of tintinnids (Protozoa: Ciliata) from Porto Novo waters. *Mahasagar – Bulletin of the National Institute of Oceanography*, **18**, 417–423.

Nakagawa, Y., Endo, Y. & Taki, K. (2001) Diet of *Euphausia pacifica* Hansen in Sanriku waters off northeastern Japan. *Plankton Biology Ecology*, **48**, 68–77.

Nakagawa, Y., Ota, T., Endo, Y., Taki, K. & Sugisaki, H. (2004). Importance of ciliates as prey of the euphausiid *Euphausia pacifica* in the NW North Pacific. *Marine Ecology Progress Series*, **271**, 261–266.

Nakamachi, M. & Iwasaki, N. (1998) List of tintinnids (Protozoa: Ciliata) in Uranouchi Inlet, Kochi, Japan. *Bulletin of Marine Sciences and Fisheries, Kochi University*, **18**, 65–76.

Nakamura, T. & Matsusaka, T. (1991) Effects of cyst age on excystment of the ciliate *Histriculus cavicola* (Kahl, 1935) (Ciliophora: Stichotrichia). *European Journal of Protistology*, **27**, 375–380.

Nakamura, Y. & Hirata, A. (2006) Plankton community structure and trophic interactions in a shallow and eutrophic estuarine system, Ariake Sound, Japan. *Aquatic Microbial Ecology*, **44**, 45–57.

Nakane, T., Nakaka, K., Bouman H. & Platt, T. (2008) Environmental control of short-term variation in the plankton community of inner Tokyo Bay, Japan. *Estuarine, Coastal and Shelf Science*, **78**, 796–810.

*Nakashima, J. & Kimoto, K. (1987) Comparison of the two counting methods for microzooplankton. *Bulletin of the Seikai Regional Fishery Research Laboratory*, **64**, 77–82.

Nazarov, B.B. (1973) Radiolaria from Lower Cambrian horizons. *Trudi Instituta Geologii (Akademia Nauk USSR)*, **49**, 5–13.

Nejstgaard, J.C., Naustvoll, L.J. & Sazhin, A. (2001) Correcting for underestimation of microzooplankton grazing in bottle incubation experiments with mesozooplankton. *Marine Ecology Progress Series*, **221**, 59–75.

Nejstgaard, J.C., Frischer, M.E., Simonelli, P., Troedsson, C., Brkel, M., Adiyaman, F., Sazhin, A.F. & Artigas, L.F. (2008) Quantitative PCR to estimate copepod feeding. *Marine Biology*, **153**, 565–577.

Neuer, S. & Cowles, T.J. (1995) Comparative size-specific grazing rates in field populations of ciliates and dinoflagellates. *Marine Ecology Progress Series*, **125**, 259–267.

*Neumann-Leitao, S., Paranaguá, M.N. & Valentin, J.L. (1992) The planktonic rotifers of the estuarine lagunar complex of Suape (Pernambuco, Brazil). *Hydrobiologia*, **232**, 133–143.

Nichols, R. (2001) Gene trees and species trees are not the same. *Trends in Ecology and Evolution*, **16**, 358–364.

Nie, D. (1934) Notes on Tintinnoinea from the Bay of Amoy. *Annual Report of the Marine Biology Association of China*, **3**, 71–80.

Nielsen, T.G. & Kiorboe, T. (1994) Regulation of zooplankton biomass and production in a temperate, coastal ecosystem. 2. Ciliates. *Limnology and Oceanography*, **39**, 508–519.

Nishibe, Y., Kobari, T. & Ota, T. (2010) Feeding by the cyclopoid copepod *Oithona similis* on the microplankton assemblage in the Oyashio region during spring. *Plankton Benthos Research*, **5**, 74–78.

Nomura, H., Ishimaru, T. & Murano, M. (1992) Microzooplankton assemblage and its seasonal variation in Tokyo Bay. *La Mer, Tokyo* **30**, 57–72.

Noren, F., Moestrup, Ø. & Rehnstam-Holm, A.S. (1999) *Parvilucifera infectans* gen.et sp. nov. (Perkinsozoa phylum nov.): a parasitic flagellate capable of killing toxic algae. *European Journal of Protistology*, **35**, 233–254.

*Ogilvie, H.S. (1923) Microplankton from the south coast of Ireland. *Rapports et Process-Verbaux Réunion de la Conseil International pour Exploration de la Mer*, **29**, 30–71.

*Ogilvie, H.S. (1938) The food of post-larval Haddock with reference to the annual fluctuations in the Haddock broods. *Rapports et Process-Verbaux Réunion de la Conseil International pour Exploration de la Mer*, **107**, 57–66.

Ohman, M.D. & Runge, J.A. (1994) Sustained fecundity when phytoplankton resources are in short supply: omnivory by *Calanus finmarchicus* in the Gulf of St. Lawrence. *Limnology and Oceanography*, **39**, 21–36.

Ohman, M.D., Theilacker, G.H. & Kaupp, S.E. (1991) Immunochemical detection of predation on ciliate protists by larvae of the northern anchovy (*Engraulis mordax*). *Biological Bulletin*, **181**, 500–504.

Olesen, N.J., Purcell, J.E. & Stoecker, D.K. (1996) Feeding and growth by ephyrae of scyphomedusae *Chrysaora quinquecirrha*. *Marine Ecology Progress Series*, **137**, 149–159.

Olson, M.B. & Strom, S.L. (2002) Phytoplankton growth, microzooplankton herbivory and community structure in the southeast Bering Sea: insight into the formation and temporal persistence of an *Emiliania huxleyi* bloom. *Deep-Sea Research II*, **49**, 5969–5990.

Olson, M.B., Lessard, E.J., Wong, C.W.J. & Bernhardt, M.J. (2006) Copepod feeding selectivity on microplankton, including the toxigenic diatoms *Pseudo-nitzschia* spp., in the coastal Pacific Northwest. *Marine Ecology Progress Series*, **326**, 207–220.

Omori, M. (1974) The biology of pelagic shrimps in the ocean. *Advances in Marine Biology*, **12**, 233–324.

*Osorio-Tafall, B.F. (1941) Tintinnidos nuevos o poco conocidos del plancton neritico de Mexico. *Revista de la Sociedad Mexicana de Historia Natural*, **2**, 147–173.

*Ostenfeld, C.H. & Schmidt, J. (1902) Plankton fra det Rode Hav og Adenbugten. *Videnskabelige Meddelelser den naturhistoriske Forening i Kjobenhavn for 1901*, 141–182.

Ota, T. & Taniguchi, A. (2003) Conjugation in the marine aloricate oligotrich *Pelagostrobilidium* (Ciliophora: Oligotrichia). *European Journal of Protistology*, **39**, 149–160.

Pace, M.L., Findlay, S.E. & Fischer, D. (1998) Effects of an invasive bivalve on the zooplankton community of the Hudson River. *Freshwater Biology*, **39**, 103–116.

Pace, M.L., Glasser, E.J. & Pomeroy, L.R. (1984) A simulation analysis of continental shelf food webs. *Marine Biology*, **82**, 47–63.

Paiva, T.S., Borges, B.N., Harada, M.L. & Silva-Neto, I.D. (2009) Comparative phylogenetic study of *Stichotrichia* (Alveolata: Ciliophora: Spirotrichea) based on 18S-rDNA sequences. *Genetics and Molecular Research*, **8**, 223–246.

Paivinen, J., Grappuo, A., Kaitala, V., Komonen, A., Kotiaho, J.S., Saarinen, K. & Wahlberg, N. (2005) Negative density–distribution relationship in butterflies. *BMC Biology*, **3**, 5.

Paranjape, M.A. (1968) The egg mass and veligers of *Limacina helicina*. *Veliger*, **10**, 322–326.

*Paranjape, M.A. (1980) Occurrence and significance of resting cysts in a hyaline tintinnid, *Helicostomella subulata* (Ehre.) Jorgensen. *Journal of Experimental Marine Biology and Ecology*, **48**, 23–33.

*Paranjape, M.A. (1987a) Grazing by microzooplankton in the eastern Canadian arctic in summer 1983. *Marine Ecology Progress Series*, **40**, 239–246.

*Paranjape, M.A. (1987b) The seasonal cycles and vertical distribution of tintinnines in Bedford Basin, Nova Scotia, Canada. *Canadian Journal of Zoology*, **65**, 41–48.

*Paranjape, M.A., Conover, R.J., Harding, G.C. & Prouse, N.J. (1985) Micro- and macrozooplankton on the Nova Scotian Shelf in the prespring bloom period: A comparison of their potential resource utilization. *Canadian Journal of Fisheries and Aquatic Science*, **42**, 1484–1492.

*Paredes, J.F. (1969–1970) Subsidios para o conhecimento do plancton marinho de Cabo Verde. II – Tintinideos, Foraminiferos e Radiolarios. *Mems Inst. Invest. cient. Motamb.*, **10**, 109–123.

Parfrey, L.W., Lahr, D.J.G., Knoll, A.H. & Katz, L.A. (2011) Estimating the timing of early eukarotic diversification with mutigene molecular clocks. *Proceedings of the National Academy of Sciences of the United States of America*, **108**, 13624–13629.

Park, G.S. & Marshall, H.G. (2000) The trophic contributions of rotifers in tidal freshwater and estuarine habitats. *Estuarine, Coastal and Shelf Science*, **51**, 729–742.

Park, M.G., Yih, W. & Coats, D.W. (2004) Parasites and phytoplankton, with special emphasis on dinoflagellate infections. *Journal of Eukaryotic Microbiology*, **51**, 145–155.

Parrow, M.W. & Burkholder, J.M. (2003) Reproduction and sexuality in *Pfiesteria shumwayae* (Dinophyceae). *Journal of Phycology*, **39**, 697–711.

Passow, U. & Peinert, R. (1993) The role of plankton in particle flux: two case studies from the northeast Atlantic. *Deep-Sea Research II*, **40**, 573–585.

Paulmier, G. (1997) *Tintinnides (Ciliophora, Oligotrichida, Tintinnina) de l'Atlantique Boréal, de l'Océan Indien et quelques mers adjactentes: Méditerranée, Mer Caraibe, Mer Rouge. Inventaire et distribution. Observations basée sur les loricas.* Rapport IFREMER DRV/RH/97-17, Brest, France.

Pearcy, W.G. (1962) Ecology of an estuarine population of winter flounder, *Pseudopleuronectes americanus* (Walbaum). *Bulletin of the Bingham Oceanographic Collection*, **18**, 1–78.

Pearre, S. Jr (1973) Vertical migration and feeding in *Sagitta elegans* Verrill. *Ecology*, **54**, 300–314.

Pearre, S. Jr (1981) Feeding by chaetognatha – energy balance and importance of various components of the diet of *Sagitta elegans*. *Marine Ecology Progress Series*, **5**, 46–54.

Pedersen, M.F. & Hansen, P.J. (2003) Effects of high pH on the growth and survival of six marine heterotrophic protists. *Marine Ecology Progress Series*, **260**, 33–41.

Perez-Ruzafa, A., Gilabert, J., Gutierrez, J.M., Fernandez, A.I., Marcos, C. & Sabah, S. (2002) Evidence of a planktonic food web response to changes in nutrient input dynamics in the Mar Menor coastal lagoon, Spain. *Hydrobiologia*, **475/476**, 359–369.

Persson, A. & Smith, B.C. (2009) Grazing on a natural assemblage of ciliate and dinoflagellate cysts by the eastern oyster *Crasostrea virginica*. *Aquatic Biology*, **6**, 227–233.

Petran, A. (1958) Contributii la cunoasterea speciilor de Tintinnoidee din dreptul litoralului Rominesc al Marrii Negre. *Hidrobiolgia*, **1**, 75–88.

Petz, W. & Foissner, W. (1992) Morphology and morphogenesis of *Strobilidium caudatum* (Fromentel), *Meseres corlissi* n. sp., *Halteria grandinella* (Müller), and *Strombidium rehwaldi* n. sp., and a proposed phylogenetic system for oligotrich ciliates (Protozoa, Ciliophora). *Journal of Protozoology* **39**, 159–176.

Petz, W. & Foissner, W. (1993) Morphogenesis in some freshwater tintinnids (Ciliophora, Oligotrichida). *European Journal of Protistology*, **29**, 106–120.

Petz, W., Song, W. & Wilbert, N. (1995) Taxonomy and ecology of the ciliate fauna (Protozoa, Ciliophora) in the endopagial and pelagial of the Weddell Sea, Antarctica. *Stapfia*, **40**, 1–223.

Peulvé, S., De Leeuw, J.W., Sicre, M.-A., Baas, M. & Saliot, A. (1996) Characterization of macromolecular organic matter in sediment traps from the northwestern Mediterranean Sea. *Geochimica et Cosmochimica Acta*, **60**, 1239–1259.

Pfiester, L.A. & Anderson, D.M. (1987) Dinoflagellate life cycles and their environmental control. In *The Biology of Dinoflagellates* (Taylor, F.J.R. ed.), pp. 611–648, Blackwell, Oxford.

Pierce, R.W. & Turner, J.T. (1992) Ecology of planktonic ciliates in marine food webs. *Reviews in Aquatic Sciences*, **6**, 139–181.

Pierce, R.W. & Turner, J.T. (1993) Global biogeography of marine tintinnids. *Marine Ecology Progress Series*, **94**, 11–26.

*Pierce, R.W. & Turner, J.T. (1994) Plankton studies in Buzzards Bay, Massachusetts, USA. IV. Tintinnids, 1987 to 1988. *Marine Ecology Progress Series*, **112**, 235–240.

Pierce, R.W., Carlton, J.T., Carlton, D.A. & Geller, J.B. (1997) Ballast water as a vector for tintinnid transport. *Marine Ecology Progress Series*, **149**, 295–297.

Pierce, S., Kossoy, V., Valenti, R. & Smetana, D.G. (1968) Cytochemical studies on the test of *Allogromia laticollare*. *Micropaleontology*, **14**, 242–246.

Pinheiro, M.D.O., Power, M.E., Butler, B.J., Dayeh, V.R., Slawson, R., Lee, L.E.J., Lynn, D.H. & Bols, N.C. (2007) Use of *Tetrahymena thermophila* to study the role of protozoa in inactivation of viruses in water. *Applied and Environmental Microbiology*, **73**, 643–649.

Pinheiro, M.D.O., Power, M.E., Butler, B.J., Dayeh, V.R., Slawson, R., Lee, L.E.J., Lynn, D.H. & Bols, N.C. (2008) Inactivation of the bacteriophage MS2 by the ciliated protozoan, *Tetrahymena thermophila*. *Water Quality Research Journal Canada*, **43**, 69–76.

Pitta, P. & Giannakourou, A. (2000) Planktonic ciliates in the oligotrophic Eastern Mediterranean: vertical, spatial distribution and mixotrophy. *Marine Ecology Progress Series*, **194**, 269–282.

Pitta, P., Giannakourou, A. & Christaki, U. (2001) Planktonic ciliates in the oligotrophic Mediterranean Sea: longitudinal trends of standing stocks, distributions and analysis of food vacuole contents. *Aquatic Microbial Ecology*, **24**, 297–311.

Platt, T. (1985) Structure of the marine ecosystem: its allometric basis. *Canadian Bulletin of Fisheries and Aquatic Science*, **213**, 55–64.

Pomeroy, L.R. (1974) The ocean's food web: a changing paradigm. *Bioscience*, **24**, 499–504.

Pomeroy, L.R. & Wiebe, W.J. (1988) Energetics of microbial food webs. *Hydrobiologia*, **159**, 7–18.

Pop, G. (1976) Tithonian-Valanginian calpionellid zones from Cuba. *Dari de Seama ale Sedintelor, Institutul de Geologie si Geofizica (3. Paleontologie)*, **62**, 237–266.

Porter, S. (2011) The rise of predators. *Geology*, **39**, 607–608.

*Posta, A. (1963) Relation entre l'evolution de quelques tintinnides de la rade de Villefranche et la température de l'eau. *Cahiers de Biologie Marine*, **4**, 201–210.

*Prasad, R.R. (1956) Further studies on the plankton of the inshore waters off Mandapam. *Indian Journal of Fisheries*, **3**, 1–42.

Price, A.M. & Pospelova, V. (2011) High-resolution sediment trap study of organic-walled dinoflagellate cyst production and biogenic silica flux in Saanich Inlet (BC, Canada). *Marine Micropaleontology*, **80**, 18–43.

Probyn, T.A. (1987) Ammonium regeneration by microplankton in an upwelling environment. *Marine Ecology Progress Series*, **37**, 64.

Purcell. E.M. (1977) Life at low Reynolds number. *American Journal of Physics*, **45**, 3–11.

Puri, H.S. (1953) Taxonomic Comment on: "Ostracoda from Wells in North Carolina Part I: Cenozoic Ostracoda" by F. M. Swain. *Journal of Paleontology*, **27**, 750–752.

Quinlan, E., Jett, C. & Phlips, E. (2009) Microzooplankton grazing and the control of phytoplankton biomass in the Suwannee River estuary, USA. *Hydrobiologia*, **632**, 127–137.

Raikov, I.B. (1982) *The Protozoan Nucleus, Morphology and Evolution*. Cell Biology Monographs 9. Springer Verlag, Vienna, Austria.

*Rampi, L. (1948) I tintinnoidi delle acque di San Remo. *Bolletin di Pesca, di Piscicultura e di Idrobiologia*, **3**, 50–56.

*Rampi, L. (1950) I tintinnoidi delle acque di Monaco raccolti dall'Eider nell'anno 1913. *Bulletin de L'institut Océanographique, Monaco*, **965**, 1–8.

Rampi, L. & Zattera, A. (1982) Chiave per la determinazione dei Tintinnidi mediterranei. *Comitato nazionale per la ricerca e per lo sviluppo dell'energia nucleare e delle energie alternative, ENEA-RT/BIO, (82)* **28**, 1–104.

Rassoulzadegan, F. (1975) Ecologie et Relations Trophiiques du Microzooplancton dans un Ecosystème Nértique. Thèse Doctorale, Océanographie Biologique, Université Pierre et Marie Curies (Paris VI), Paris, France.

Rassoulzadegan, F. (1977) Evolution annuelle des cilies pelagiques en mediterranee nord-occidentale cilies oligotriches "non tintinnides" (oligotrichina). *Annales Institut Océanographique, Paris*, **53**, 125–134.

Rassoulzadegan, F. (1978) Dimensions et taux d'ingestion des particules consommées par un tintinnide: *Favella ehrenbergii* (Clap. & Lach.) Jörg., cilié pelagique marine. *Annales de l'Institute Océanographique, Paris*, **54**, 17–24.

Rassouzadegan, F. (1979) Annual evolution of pelagic ciliates in the North-western Mediterranean sea, 2: Oligotrichous ciliates. Tintinnids (Tintinnidae). *Investigacion Pesquera*, **43**, 417–448.

Rassoulzadegan, F. (1980) Granulometric analysis of the particles used by a tintinnid *Stenosemella ventricosa* (Clap. & Lachm.) Jorg., during lorica building. *Protistologica*, **16**, 507–510.

Rassoulzadegan, F. & Etienne, M. (1981) Grazing rate of the tintinnid *Stenosemella ventricosa* (Clap. & Lach.) Jörg. on the spectrum of naturally occurring particulate matter from Mediterranean neritic area. *Limnology and Oceanography*, **26**, 258–270.

*Rassoulzadegan, F., Laval-Peuto, M. & Sheldon, R.W. (1988) Partitioning of the food ration of marine ciliates between pico- and nanoplankton. *Hydrobiologia*, **159**, 75–88.

Raybaud, V., Tunin-Ley, A., Ritchie, M.E., Dolan, J.R. (2009) Similar patterns of community organization characterize distinct groups of different trophic levels in the plankton of the NW Mediterranean Sea. *Biogeosciences*, **6**, 431–438.

Reaugh, M.L., Roman, M.R. & Stoecker, D.K. (2007) Changes in plankton community structure and function in response to variable freshwater flow in two tributaries of the Chesapeake Bay. *Estuaries and Coasts*, **30**, 403–417.

Reck, E.M. (1988) Lorica-splitting by the Tintinnina. *Naturwissenschaften*, **75**, 45–47.

Redden, A.M., Sanderson, B.G. & Rissik, D. (2002.) Extending the analysis of the dilution method to obtain the phytoplankton concentration at which microzooplankton grazing becomes saturated. *Marine Ecology Progress Series*, **226**, 27–33.

Reid, P.C. (1987) Mass encystment of a planktonic oligotrich ciliate. *Marine Biology*, **95**, 221–230.

*Reid, P.C. & John, A.W.G. (1978) Tintinnid cysts. *Journal of the Marine Biological Association of the United Kingdom*, **58**, 551–557 + Plate 1.

Reid, P.C. & John, A.W.G. (1983) Resting cysts in the ciliate class Polyhymenophorea: phylogenetic implications. *Journal of Protozoology*, **30**, 710–713.

Remane, J. (1963) Les Calpionelles dans les couches de passage jurassique-crétacé de la fosse vocontienne. *Travaux du Laboratoire de Géologie de la Faculté des Sciences de la Université de Grenoble*, **39**, 25–82.

Remane, J. (1964) Untersuchungen zur Systematik und Stratigraphie der Calpionellen in den Jura-Kreide-Grenzchichten des Vocontischen Troges. *Palaeontographica A*, **123**, 1–57.

Remane, J. (1971) Les Calpionelles protozoaires planktoniques des mers mesogéennes de l'époque secondaire. *Annales Guebhard Neuchatel*, **47**, 370–393.

Remane, J. (1985) Calpionellids. *Plankton Statigraphy* (Bolli, H.M., Saunders, J.B., Perch-Nielsen, K. eds), pp. 555–572, Cambridge University Press, Cambridge.

Remane, J. (1998) Calpionellids. *Introduction to Marine Micropaleontology* 2nd Ed. (Haq, B.U., Boersma, A. eds), pp. 161–170. Elsevier, Amsterdam.

Remane, J., Bakalova-Ivanova, D., Borza, K., Knauer, J., Nagy, I., Pop, G. & Tardi-Filacz, E. (1986) Agreement of the subdivision of the standard Calpionellid-Zones defined at the second Planktonic Conference, Roma 1970. *Acta Geologica Hungarica*, **29**, 5–13.

*Revelante, N. & Gilmartin, M. (1987) Seasonal cycle of the ciliated protozoan and micrometazoan biomass in a Gulf of Maine estuary. *Estuarine, Coastal and Shelf Science*, **25**, 581–598.

*Revelante, N. & Gilmartin, M. (1990) Vertical water column resource partitioning by a ciliated protozoan population under stratified conditions in the northern Adriatic. *Journal of Plankton Research*, **12**, 89–107.

Richards, T.A. & Bass, D. (2005) Molecular screening of free-living microbial eukaryotes: diversity and distribution using a met-analysis. *Current Opinion in Microbiology*, **8**, 240–252.

Rigby, S. & Milsom, C.Y. (2000) Origins, evolution and diversification of zooplankton. *Annual Review of Ecology and Systematics*, **31**, 293–313.

Ris, H. & Kubai, D.F. (1974) An unusual mitotic mechanism in the parasitic protozoan *Syndinium* sp. *The Journal of Cell Biology*, **60**, 702–720.

Roberts, E.C., Legrand, C., Steinke, M. & Wootton, E.C. (2011) Mechanisms underlying chemical interactions between predatory planktonic protists and their prey. *Journal of Plankton Research*, **33**, 833–841.

*Robertson, J.R. (1983) Predation by estuarine zooplankton on tintinnid ciliates. *Estuarine, Coastal and Shelf Science*, **16**, 27–36.

*Rogers, G.F., Roff, J.C. & Lynn, D.H. (1981) Tintinnids of Chesterfield Inlet, Northwest Territories. *Canadian Journal of Zoology*, **59**, 2360–2364.

Roja de Mendiola, B. (1974) Food of the larval anchoveta *Engraulis ringens* J. In *The Early Life History of Fish* (J.H.S. Blaxter ed.), pp. 277–285, Springer-Verlag, Berlin.

Rollwagen Bollens, G.C.R. & Penry, D. (2003) Feeding dynamics of *Acartia* spp. copepods in a large, temperate estuary (San Francisco Bay, CA). *Marine Ecology Progress Series*, **257**, 139–158.

Roman, M.R. & Rublee, P.A. (1981) A method to determine in situ zooplankton grazing rates on natural particle assemblages. *Marine Biology* **65**, 303–309.

Rombouts, I., Beaugrand, G., Ibanez, F., Gasparini, S., Chiba, S. & Legendre, L. (2009) Global latitudinal variations in marine copepod diversity and environmental factors. *Proceedings of the Royal Society B*, **267**, 293–299.

Rombouts, I., Beaugrand, G., Ibanez, F., Gasparini, S., Chiba, S. & Legendre, L. (2010) A multivariate approach to large-scale variation in marine planktonic copepod diversity and its environmental correlates. *Limnology and Oceanography*, **55**, 2219–2229.

Roncaglia, L. (2004) Palynofacies analysis and organic-walled dinoflagellate cysts as indicators of palaeo-hydrographic changes: an example from Holocene sediments in

Skalafjord, Faroe Islands. *Marine Micropaleontology*, **50**, 21–42.

Rosati, G. & Modeo, L. (2003) Extrusomes in ciliates: diversification, distribution, and phylogenetic implications. *Journal of Eukaryotic Microbiology*, **50**, 383–402.

Rosetta, C.H. & McManus, G.B. (2003) Feeding by ciliates on two harmful algal bloom species, *Prymnesium parvum* and *Prorocentrum minimum*. *Harmful Algae*, **2**, 109–126.

Rougier, C., Pourriot, R., Lam-Hoai, T. & Guiral, D. (2005) Ecological patterns of the rotifer communities in the Kaw River estuary (French Guiana). *Estuarine, Coastal and Shelf Science*, **63**, 83–91.

*Roxas, H.A. (1941) Marine protozoa of the Philippines. *Philippine Journal of Science*, **74**, 91–139.

Rublee, P.A. & Gallegos, C.L. (1989) Use of fluorescently labelled algae (FLA) to estimate microzooplankton grazing. *Marine Ecology Progress Series*, **51**, 221–227.

Rudstam, L.G., Danielsson, K., Hansson, S. & Johnsson, S. (1989) Diel vertical migration and feeding patterns of *Mysis-mixta* (Crustacea, Mysidacea) in the Baltic Sea. *Marine Biology*, **101**, 43–52.

Rüst, D. (1885) Beiträge zur Kenntniss der fossilen Radiolerien aus Gesteinen de Jura. *Palaeontographica*, **11**, 273–321.

Rutherford, S., D'Hondt, S. & Prell, W. (1999) Environmental controls on the geographic distribution of zooplankton diversity. *Nature*, **400**, 749–753.

Rutman, J. & Fishelson, L. (1969) Food composition and feeding behavior of shallow-water crinoids at Eilat (Red Sea). *Marine Biology*, **3**, 46–57.

Ryazanova, T.V., Eliselkina, M.G., Kukhlevsky, A.D. & Kharlamenko, V.I. (2010) *Hematodinium* sp. infection of red *Paralithodes camtschaticus* and blue *Paralithodes platypus* king crabs from the Sea of Okhotsk, Russia. *Journal of Invertebrate Pathology*, **105**, 329–334.

Ryther, J.H. (1969) Photosynthesis and fish production in the sea. *Science*, **166**, 72–76.

Saccà, A., Guglielmo, L. & Bruni, V. (2008) Vertical and temporal microbial community patterns in a meromictic coastal lake influenced by the Straits of Messina upwelling system. *Hydrobiologia*, **600**, 89–104.

Saccà, A., Stüder-Kype, M.C. & Lynn, D.H. (2012) Redescription of *Rhizodomus tagatzi* (Ciliophora: Spirotrichea: Tintinnida), Based on morphology, ciliary pattern and small subunit ribosomal RNA. gene sequence. *Journal of Eukaryotic Microbiology*, **59**, 218–231.

Sakka Hlaili, A., Grami, B., Hadj Mabrouk, H., Gosselin, M. & Hamel, D. (2007) Phytoplankton growth and microzooplankton grazing rates in a restricted Mediterranean lagoon (Bizerte Lagoon, Tunisia). *Marine Biology*, **151**, 767–783.

Saldarriaga, J.F., Taylor, F.J.R., Keeling, P.J. & Cavalier-Smith, T. (2001) Dinoflagellate nuclear SSU rRNA phylogeny suggests multiple plastid losses and replacements. *Journal Molecular Evolution*, **53**, 204–213.

Saldarriaga, J.F., McEwan, M.L., Fast, N.M., Taylor, F.J.R. & Keeling, P.J. (2003) Multiple protein phylogenies show that *Oxyrrhis marina* and *Perkinsus marinus* are early branches of the dinoflagellate lineage. *International Journal of Systematic and Evolutionary Microbiology*, **53**, 355–365.

Salomon, P.S., Janson, S. & Granéli, E. (2003) Multiple species of the dinophagous dinoflagellate genus *Amoebophrya* infect the same host species. *Environmental Microbiology*, **5**, 1046–1052.

Salvano, P. (1974) Analyse microspectrographique dans l'ultraviolet du macronoyau de *Strombidium sulcatum* Clap. et L., 1859 (Cilié Hypotriche), pendant l'intercinèse. *Comptes Rendus Hebdomadaires des Seances de l'Academie des Sciences Paris, Série D*, **278**, 2445–2447.

Salvano, P. (1975) Comparaison du fonctionnement des bandes de réorganisation d'*Euplotes crassus* (Dujardin) et de *Strombidium sulcatum* Claparède et Linné après analyse microspectrographique en UV. *Journal of Protozoology*, **22**, 230–232.

Sanchez, N., González, H.E. & Iriate, J.L. (2011) Trophic interactions of pelagic crustaceans in Comau Fjord (Chile): their role in the food web structure. *Journal of Plankton Research*, **33**, 1212–1229.

*Sanders, R.W. (1987) Tintinnids and other microzooplankton – seasonal distributions and relationships to resources and hydrography in a Maine estuary. *Journal of Plankton Research*, **9**, 65–77.

Sanders, R.W. & Wickham, S.A. (1993) Planktonic protozoa and metazoa: predation, food quality and population control. *Marine Microbial Food Webs*, **7**, 197–223.

Santoferrara, L.F. & Alder, V. (2009) Abundance trends and ecology of planktonic ciliates of the south-western Atlantic (35-63o S): a comparison between neritic and oceanic environments. *Journal of Plankton Research*, **31**, 837–851.

*Santoferrara, L.F. & Alder, V.A. (2009a) Morphological variability, spatial distribution and abundance of *Helicostomella* species (Ciliophora: Tintinnina) in relation to environmental factors (Argentine shelf; 40–55°S). *Scientia Marina*, **73**, 701–716.

Santoferrara, L.F. & Alder, V.A. (2012) Abundance and diversity of tintinnids planktonic ciliates) under contrasting levels of productivity in the Argentine Shelf and Drake Passage. *Journal of Sea Research*, **71**, 25–30.

Santoferrara, L.F., Gómez, M.I. & Alder, V.A. (2011) Bathymetric, latitudinal and vertical distribution of protozooplankton in a cold-temperate shelf (southern Patagonian watres) during winter. *Journal of Plankton Research*, **33**, 457–468.

Santoferrara, L.F., McManus, G.B. & Alder, V.A. (2012a) Utility of genetic markers and morphology for species discrimination within the order Tintinnida (Ciliophora, Spirotrichea). *Protist*, **163**, in press.

Santoferrara L., McManus G. & Alder V. (2012b) Phylogeny of the order Tintinnida (Ciliophora, Spirotrichea) inferred from small and large subunit rRNA genes. *Journal of Eukaryotic Microbiology*, **59**, in press.

Sarnelle, O. (1997) *Daphnia* effects on microzooplankton: comparisons of enclosure and whole lake responses. *Ecology*, **78**, 913–928.

*Sassi, R. & Melo, G.D.N. (1986) Tintinnina (Protozoa – Ciliophora – Oligotrichida) from the First Brazilian Expedition to the Antarctic. *Anais Acad. Bras. Cienc.*, **58**, 63–84.

*Sassi, R. & Melo, G.N. (1989) Hyaline tintinnina (Protozoa – Ciliophora – Oligotrichida) from northeast Brazilian coastal reefs. *Boletim del Instituto Oceanografica, San Paulo* **37**, 59–74.

Sato, N.E., Hernandez, D. & Vinas, M.D. (2010) Feeding habits of *Noctiluca scintillans* in coastal waters off Buenos Aires Province, Argentina. *Latin American Journal of Aquatic Research*, **38**, 403–412.

Saunders, G.W., Hill, D.R.A., Sexton, J.P. & Andersen, R.A. (1997). Small-subunit ribosomal RNA sequences from selected dinoflagellates: testing classical evolutionary hypotheses with molecular systematic methods. In *Origins of Algae and their Plastids* (D. Bhattacharya ed.), pp. 237–259, Springer, Vienna.

Saville Kent, W. (1880–1882) *A Manual of the Infusoria: Including a description of all known Flagellate, Ciliate, and Tentaculiferous Protozoa, British and Foreign, and an Account of the Organization and Affinities of the Sponges.* David Bogue, London.

Schemske, D.W., Mittelbach,G.G., Cornell, H.V., Sobel, J.M. & Roy, K. (2009) Is there a latitudinal gradient in the importance of biotic interactions? *Annual Review of Ecology, Evolution and Systematics*, **40**, 245–269.

Schmidt, A.R., von Eynatten, H. & Wagreich, M. (2001) The Mesozoic amber of Schliersee (southern Germany) is Cretaceous in age. *Cretaceous Research*, **22**, 423–428.

*Schmidt, J. (1902) Some Tintinnodea from the Gulf of Siam. *Videnskabelige Meddelelser den naturhistoriske Forening i Kjobenhavn for 1901*, 183–190.

Schmidt-Nielsen, K. (1990) *Animal Physiology: adaptation and environment*, 4th ed. Cambridge University Press, Cambridge, UK.

Schmitt, P.D. (1986) Feeding by larvae of *Hypoatherina tropicalis* (Pisces: Atherinidae) and its relation to prey availability in One Tree Lagoon, Great Barrier Reef, Australia. *Environmental Biology of Fishes*, **16**, 79–94.

*Schönberger, M. (1994) Planktonic ciliated protozoa of Neusiedler See (Austria / Hungary) – a comparison between the turbid open lake and a reedless brown-water pond. *Marine Microbial Food Webs*, **8**, 251–263.

Schönborn, W., Dörfelt, H., Foissner, W., Krientiz, L. & Schäfer, U. (1999) A fossilized microcenosis in Triassic amber. *Journal of Eukaryotic Microbiology*, **46**, 571–584.

Schönfeld, C. (1959). Über das parasitische Verhalten einer Astasia-Art in *Stentor coerulus*. *Archiv für Protistenkunde*, **104**, 261–264.

Schopf, J.W. (2001) *Cradle of Life: the discovery of Earth's earliest fossils*. Princeton University Press, Princeton, USA.

Schrank, F.P. (1803) *Fauna Boica: durchgedachte Geschichte der in Baiern einheimschen und zahmen.* Band 3b: Arbtheilung 2.

Schuster, B. & Sleytr, U.B. (2000) S-layer-supported lipid membranes. *Reviews in Molecular Biotechnology*, **74**, 233–254.

*Schwamborn, R., Ekau, W., Silva, A.P., Schwamborn, S.H.L., Silva, T.A., Neumann-Leita, S. & Saint-Paul, U. (2006) Ingestion of large centric diatoms, mangrove detritus, and zooplankton by zoeae of *Aratus pisonii* (Crustacea: Brachyura: Grapsidae). *Hydrobiologia*, **560**, 1–13.

Schweyer, A. (1909) Zur Kenntnis des Tintinnodeenweichkörpers, nebst einleitenden Worten über die Hulsenstruktur und die Hülsenbildung. *Archiv für Protistenkunde*, **18**, 134–189 + Plates 10, 11.

Schweyer, A.W. (1905) Ueber den Bau und die Vermehrung der Tintinnoidae (inf. cil.). *Travaux de la Société Impériale des Naturalistes de St. Pétersbourg*, **35**, 1–5.

*Scott, D.B., Schell, T., Rochon, A. & Blasco, S. (2008) Benthic foraminifera in the surface sediments of the Beaufort Shelf and slope, Beaufort Sea, Canada: Applications and implications for past sea-ice conditions. *Journal of Marine Systems*, **74**, 840–863.

Sellner, K.G. (1981) Primary productivity and the flux of dissolved organic matter in several marine environments. *Marine Biology*, **65**, 101–112.

Sellner, K.G., Brownlee, D.C., Bundy, M.H., Brownlee, S.G. & Braun, K.R. (1993) Zooplankton grazing in a Potomac River cyanobacteria bloom. *Estuaries and Coasts*, **16**, 859–872.

Seuthe, L., Rokkan Iversen, K. & Narcy, F. (2011) Microbial processes in a high-latitude fjord (Kongsfjorden, Svalbard): II. Ciliates and dinoflagellates. *Polar Biology*, **34**, 751–766.

Shaheen, P.A., Manderson, J.P. & Fahay, M.P. (2004) Stage-specific spatial and temporal variability in the diets of larval winter flounder (*Pseudopleuronectes americanus*) in a North-eastern U.S. estuarine nursery. *Estuaries*, **27**, 958–965.

Shalchian-Tabrizi, K., Minge, M., Cavalier-Smith, T., Nedreklepp, J.M., Klaveness, D. & Jakobsen, K.S. (2006) Combined heat shock protein 90 and ribosomal RNA sequence phylogeny supports multiple replacements of dinoflagellate plastids. *Journal of Eukaryotic Microbiology*, **53**, 217–224.

Sharaf, G.M. (1995) On Some Tintinnids (Protozoa: Ciliata) From The Strait Of Hormoz And The United Arab Emirates Waters. *Qatar University Science Journal*, **15**, 389–398.

Sharp, J.H. (1977) Excretion of organic matter by marine phytoplankton: do healthy cells do it? *Limnology and Oceanography*, **22**, 381–399.

Sherman, K. & Honey, K.A. (1971) Seasonal variations in the food of larval herring in coastal waters of central Maine. *Rapports et Proces Verbaux Réunion Consiel Internationale pou l'Exploration de la Mer*, **160**, 121–124.

Sherr, E. & Sherr, B. (1988). Role of microbes in pelagic food webs: a revised concept. *Limnology and Oceanography*, **33**, 1225–1227.

Sherr, E. & Sherr, B. (2008a) Understanding roles of microbes in marine pelagic food webs: a brief history. In *Microbial Ecology of the Oceans*, 2nd Ed (D.L. Kirchman ed.), Wiley-Blackwell, Hoboken, NJ, USA.

Sherr, E.B. & Sherr, B.F. (2008b) Heterotrophic dinoflagellates: a significant component of microzooplankton biomass and major grazers of diatoms in the sea. *Marine Ecology Progress Series*, **352**, 187–197.

Sherr, E.B., Sherr, B.F., Fallon, R.D. & Newell, S.Y. (1986) Small aloricate ciliates as a major component of the marine heterotrophic nanoplankton. *Limnology and Oceanography*, **31**, 177–183.

Shimeta, J., Jumars, P.A. & Lessard, E.J. (1995) Influences of turbulence on suspension feeding by planktonic protozoa; experiments in laminar shear fields. *Limnology and Oceanography*, **40**, 845–859.

Shoresh, N., Hegreness, M. & Kishony, R. (2008) Evolution exacerbates the paradox of the plankton. *Proceedings of the National Academy of Sciences of the United States of America*, **105**, 12365–12369.

Sieburth, J. McN., Smetacek, V. & Lenz, J. (1978) Pelagic ecosystem structure: heterotrophic compartments of the plankton and their relationship to plankton size fractions. *Limnology and Oceanography*, **23**, 1256–1263.

Silberman, J.D., Collins, A.G., Gershwin, L.A., Johnson, P.J. & Roger, A.J. (2004) Ellobiopsids of the genus *Thallassomyces* are alveolates. *Journal of Eukaryotic Microbiology*, **5**, 246–252.

*Silva, E.S. (1950) Les tintinnides de la Baie de Cascais (Portugal). *Bulletin de L'institut Oceanographique, Monaco*, **979**, 1–28.

*Silva, E.S. (1953) Estudios de plancton na Lagoa de Obidos. II. Tintinnoinea. *Rev. Fac. Cienc. Lisboa, Serie 2* **2**, 97–116.

Simonelli, P., Troedsson, C., Nejstgaard, J.C., Zech, K., Larsen, J.B. & Frischer, M.E. (2009) Evaluation of DNA extraction and handling procedures for PCR-based copepod feeding studies. *Journal of Plankton Research*, **31**, 1465–1474.

Simpson, A.G.B. & Patterson, D.J. (1996) Ultrastructure and identification of the predatory flagellate *Colpodella pugnax* Cienkowski (Apicomplexa) with a description of *Colpodella turpis* n. sp. and a review of the genus. *Systematic Parasitology*, **33**, 187–198.

Sipura, J., Lores, E. & Snyder, R.A. (2003) Effect of copepods on estuarine microbial plankton in short-term microcosms. *Aquatic Microbial Ecology*, **33**, 181–190.

*Sitran, R., Bergamasco, A., Decembrini, E. & Guglielmo, L. (2007) Temporal succession of tintinnids in the northern Ionian Sea, Central Mediterranean. *Journal of Plankton Research*, **29**, 495–508.

*Sitran, R., Bergamasco, A., Decembrini, E. & Guglielmo, L. (2009) Microzooplankton (tintinnid ciliates) diversity: coastal community structure and driving mechanisms in the southern Tyrrhenian Sea (Western Mediterranean). *Journal of Plankton Research*, **31**, 153–170.

Sjoqvist, O. & Lindholm, T.J. (2011) Natural co-occurrence of *Dinophysis acuminata* (Dinoflagellata) and *Mesodinium rubrum* (Ciliophora) in thin layers in a coastal inlet. *Journal of Eukaryotic Microbiology*, **58**, 365–372.

Skovgaard, A. & Legrand, C. (2005) Observation of live specimens of *Pseudotontonia cornuta* (Ciliophora: Oligotrichida) reveals new distinctive characters. *Journal of the Marine Biological Association of the United Kingdom*, **85**, 783–786.

Skovgaard, A., Massana, R., Balague, V. & Saiz, E. (2005) Phylogenetic position of the copepod-infesting parasite *Syndinium turbo* (Dinoflagellata, Syndinea). *Protist*, **156**, 413–423.

Skovgaard, A., Massana, R. & Saiz, E. (2007) Parasitic species of the genus *Blastodinium* (Blastodiniphyceae) are peridinioid dinoflagellates. *Journal of Phycology*, **43**, 553–560.

Skovgaard, A., Meneses, I. & Angélico, M.M. (2009) Identifying the lethal fish egg parasite *Ichthyodinium chabelardi* as a member of Marine Alveolate Group I. *Environmental Microbiology*, **11**, 2030–2041.

*Skryabin, V.A. & Al-Yamani, F.Y. (2006) New Species of Genera *Tintinnopsis*, *Metacylis* and *Eutintinnus* (Ciliophora, Spirotrichea, Tintinnida) from the Kuwaiti Waters of the Arabian Gulf. *Russian Journal of Marine Biology*, **32**, 302–307.

*Skryabin, V.A. & Al-Yamani, F.Y. (2007) Composition, distribution and seasonal changes of tintinnids in the northwestern Arabian Gulf around the Bubiyan Island. *International Journal of Oceans and Oceanography*, **2**, 99–106.

Slapeta, J., Lopez-Garcia, P., Moreira, D. 2006. Global dispersal and ancient cryptic spieces in the smallest marine eukaryotes. *Molecular Biology and Evolution*, **23**, 23–29.

Sleytr, U.B., Sára, M., Pum, D. & Schuster, B. (2001) Characterization and use of crystalline bacterial cell surface layers. *Progress in Surface Science*, **68**, 231–278.

Small, E.B. & Lynn, D.H. (1985) Phylum Ciliophora Doflein, 1901. In *An Illustrated Guide to the Protozoa* (J.J. Lee, S.H. Hutner & E.C. Bovee eds), pp. 393–575, Society of Protozoologists, Allen Press, Lawrence, Kansas.

Small, H.J., Neil, D.M., Taylor, A.C., Atkinson, R.J. & Coombs, G.H. (2006) Molecular detection of *Hematodinium* spp. in Norway lobster *Nephrops norvegicus* and other crustaceans. *Diseases of Aquatic Organisms*, **69**, 185–195.

Small, H.J., Shields, J.D., Moss, J.A. & Reece, K.S. (2007) Conservation in the first internal transcribed spacer region (ITS1) in *Hematodinium* species infecting crustacean hosts found in the UK and Newfoundland. *Diseases of Aquatic Organisms*, **75**, 251–258.

Smalley, G.W. & Coats, D.W. (2002). Ecology of the red-tide dinoflagellate *Ceratium furca*: Distribution, mixotrophy, and grazing impact on ciliate populations of Chesapeake Bay. *Journal of Eukaryotic Microbiology*, **49**, 63–73.

Smalley, G.W., Coats, D.W. & Adam, E.J. (1999) A new method using fluorescent microspheres to determine grazing on ciliates by the mixotrophic dinoflagellate *Ceratium furca*. *Aquatic Microbial Ecology*, **17**, 167–179.

Smalley, G.W., Coats, D.W. & Stoecker, D.K. (2003) Feeding in the mixotrophic dinoflagellate *Ceratium furca* is influenced by intracellular nutrient concentrations. *Marine Ecology Progress Series*, **262**, 137–151.

*Smayda, T.J. (1958) Phytoplankton studies around Jan Mayen Island. March – April, 1955. *Nytt Magasin for Botanikk*, **6**, 75–96.

*Smayda, T.J. (1966) A quantitative analysis of the phytoplankton of the Gulf of Panama. *Bulletin of the Inter-American Tropical Tuna Commission*, **11**, 353–612.

*Smetacek, V. (1981) The annual cycle of protozooplankton in the Kiel Bight. *Marine Biology*, **63**, 1–11.

Smetacek, V. (1985) The annual cycle of Kiel Bight plankton: a long-term analysis. *Estuaries and Coasts*, **8**, 145–157.

Smetacek, V. (2001) A watery arms race. *Nature*, **411**, 745.

*Sniezek, J.H., Capriulo, G.M., Small, E.B. & Russo, A. (1991) *Nolaclusilis hudsonicus* n. sp. (Nolaclusiliidae n. fam.) a bilaterally symmetrical tintinnine ciliate from the lower Hudson River estuary. *Journal of Protozoology*, **38**, 589–594.

*Snoeyenbos-West, O.L.O., Salcedo, T., McManus, G.B. & Katz, L.A. (2002) Insights into the diversity of choreotrich and oligotrich ciliates (Class: Spirotrichea) based on genealogical analyses of multiple loci. *International Journal of Systematic and Evolutionary Microbiology*, **52**, 1901–1913.

*Snyder, R.A. & Brownlee, D.C. (1991) *Nolaclusilis bicornis* n. g., n. sp. (Tintinnina: Tintinnidiidae): a tintinnine ciliate with novel lorica and cell morphology from the Chesapeake Bay estuary. *Journal of Protozoology*, **38**, 583–589.

Sokolova, Y.Y. & Gerassimova, Z.P. (1984) The ultrastructure of the ciliates *Parafavella denticulata* Ehrenberg, 1840. *Tsitologiya*, **26**, 1237–1245 + Plates 1–4.

Sokolova, Y.Y., Seravin, L.N. & Gerassimova, Z.P. (1986) Membrane structures and their supposed involvement in the digestive events with the ciliate *Parafavella denticulata* Ehrenberg, 1890 (Tintinnina). *Tsitologiya*, **28**, 220–224.

Song, W. (1993) Studies on the cortical morphogenesis during cell division in *Halteria grandinella* (Müller, 1773) (Ciliophora, Oligotrichida). *Chinese Journal of Oceanology and Limnology*, **11**, 122–129.

Song, W. & Wilbert, N. (1989) Taxonomische Untersuchungen an Aufwuchsciliaten (Protozoa, Ciliophora) im Poppelsdorfer Weiher, Bonn. *Lauterbornia* **3**, 1–221.

Song, W.B. & Wilbert, N. (2002) Faunistic studies on marine ciliates from the Antarctic benthic area, including descriptions of one epizoic form, 6 new species and, 2 new genera (Protozoa: Ciliophora). *Acta Protozoologica*, **41**, 23–61.

*Sorokin, Y.I. (1977) The heterotrophic phase of plankton succession in the Japan Sea. *Marine Biology*, **41**, 107–117.

Sorokin, Y.I. & Kogelschatz, J.E. (1979) Analysis of heterotrophic microplankton in an upwelling area. *Hydrobiologia*, **66**, 195–208.

Sorokin, Y.I., Kopylov, A.I. & Mamaeva, M.V. (1985) Abundance and dynamics of microplankton in the central tropical Indian Ocean. *Marine Ecology Progress Series*, **24**, 27–41.

Sorokin, Y.I., Sorokin, P.I. & Zakuskina, O.Y. (1998) Microplankton and its functional activity in zones of shallow hydrotherms in the Western Pacific. *Journal of Plankton Research*, **20**, 1015–1031.

Southward, A.J. (1955) Observations on the ciliary currents of the jelly-fish *Aurelia aurita* L. *Journal of the Marine Biological Association of the United Kingdom*, **34**, 201–216.

*Souto, S. (1970a) Tintinnidos de la costa Atlantica entre los 31 degree y 35 degree de latitude sur (Uruguay y sur de Brasil) (Protozoa, Ciliata). *Physis, Buenos Aires*, **30**, 187–208.

*Souto, S. (1970b) Tintinnidos de la costa Brasilera colectados pro el Walther Herwig (Protozoa, Ciliata). *Physis, Buenos Aires*, **30**, 209–224.

*Souto, S. (1972) Tintinnidos subantarticos del Mar Argentino (Protozoa, Ciliata). *Physis, Buenos Aires*, **31**, 451–462.

*Souto, S. (1973) Contribucion al conocimiento de los Tintinnidos de agua dulce de la republica Argentina. I. Rio de la Plata y del Parana. *Physis, Buenos Aires*, **32**, 249–254.

*Souto, S. (1974) Tintinnidos del Rio de la Plata y su zona de influencia. *Physis, Buenos Aires*, **33**, 201–205.

Souto, S. (1979) Titinnidos del Atlantico sudoccidental y Antartica Argentina (Protozoa), Cilato). Estudio Taxonomico, Distribucional y Ecologica. Tesis Diss, Univ. Buenos Aires, Argentina.

Spector, D.L. (1984). Dinoflagellate nuclei. In *Dinoflagellates* (D.L. Spector ed.), pp. 107–147, Academic Press, London.

Spittler, P. (1973) Feeding experiments with tintinnids. *Oikos*, **15**, 128–132.

*Srinivasan, A., Santhanam, R. & Jegatheesan, G. (1988) Biomass & seasonal distribution of planktonic tintinnids of Pullavazhi Estuary, southeast coast of India. *Indian Journal of Marine Sciences*, **17**, 131–133.

Steele, T.W. & McLennan, A.M. (1996) Infection of *Tetrahymena pyriformis* by *Legionella longbeachae* and other *Legionella* species found in potting mixes. *Applied and Environmenal Microbology*, **62**, 1081–1083.

Steidinger, K.A. (1975) Basic factors influencing red tides. In *Proceedings of the 1st International Conference on Toxic Dinoflagellate Blooms* (V.R. LoCicero ed.), pp. 153–162, Massachusetts Science and Technology Foundation, Wakefield, USA.

Steinberg, D.K. (1995) Diet of copepods (*Scopalatum vorax*) associated with mesopelagic detritus (giant larvacean houses) in Monterey Bay, California. *Marine Biology*, **122**, 571–584.

Stepanjants, S.D., Cortese, G., Kruglikova, S.B. & Bjorklund, K.R. (2006) A review of bipolarity concepts: history and examples from radiolaria and medusozoa (Cnidaria). *Marine Biology Research*, **2**, 200–241.

Stepien, C.A. (1986) Life history and larval development of the giant kelpfish, *Heterostichus rostratus* Girard, 1854. *Fishery Bulletin*, **84**, 809–826.

Stepien W.P., Jr. (1976) Feeding of laboratory-reared larvae of the sea bream *Archosargus rhomboidalis* (Sparidae) *Marine Biology* **38**, 1–16.

Stibor, H., Vadstein, O., Diehl, S., Gelzleichter, A., Hansen, T., Hantzsche, F. & Katechakis, A. (2004) Copepods act as a switch between alternative trophic cascades in marine pelagic food webs. *Ecology Letters*, **7**, 321–328.

Stickney, H.L., Hood, R.R. & Stoecker, D.K. (2000) The impact of mixotrophy on planktonic marine ecosystems. *Ecological Modeling*, **125**, 203–230.

Stoecker, D.K. (1984) Particle production by planktonic ciliates. *Limnology and Oceanography*, **29**, 930–940.

Stoecker, D.K. (1988) Are marine planktonic ciliates suspension-feeders? *Journal of Protozoology*, **35**, 252–255.

Stoecker, D.K. (1999) Mixotrophy among dinoflagellates. *Journal of Eukaryotic Microbiology*, **46**, 397–401.

Stoecker, D.K. & Egloff, D.A. (1987) Predation by *Acartia tonsa* Dana on planktonic ciliates and rotifers. *Journal Experimental Marine Biology and Ecology*, **110**, 53–68.

Stoecker, D.K. & Evans, G.T. (1985) Effects of protozoan herbivory and carnivory in a microplankton food web. *Marine Ecology Progress Series*, **25**, 159–167.

Stoecker, D.K. & Govoni, J.J. (1984) Food selection by young larval Gulf menhaden (*Brevoortia patronus*). *Marine Biology*, **80**, 299–306.

Stoecker, D.K. & Guillard, R.R.L. (1982) Effects of temperature and light on the feeding rate of *Favella* sp. (ciliated Protozoa, Suborder Tintinnina). *Annales Institut Oceanographique, Paris*, **58** (Supplement), 309–318.

Stoecker, D.K. & & McDowell Capuzzo, J. (1990) Predation on protozoa: its importance to zooplankton. *Journal of Plankton Research*, **12**, 891–908.

Stoecker, D.K. & Sanders, N.K. (1985). Differential grazing by *Acartia tonsa* on a dinoflagellate and a tintinnid. *Journal of Plankton Research* **7**, 85–100.

*Stoecker, D.K., Guillard, R.R.L. & Kavee, R.M. (1981) Selective predation by *Favella ehrenbergii* (Tintinnia) on and among dinoflagellates. *Biological Bulletin*, **160**, 136–145.

*Stoecker, D.K., Davis, L.H. & Provan, A. (1983) Growth of *Favella* sp. (Ciliata: Tintinnina) and other microzooplankters in cages incubated in situ and comparison to growth in vitro. *Marine Biology*, **75**, 293–302.

Stoecker, D.K., Davis, H.L. & Anderson, D.M. (1984) Fine scale spatial correlations between planktonic ciliates and dinoflagellates. *Journal of Plankton Research*, **6**, 829–842.

Stoecker, D.K., Sunda, W.G. & Davis, L.H. (1986) Effects of copper and zinc on two planktonic ciliates. *Marine Biology*, **92**, 21–29.

Stoecker, D.K., Cucci, T.L., Hulburt, E.M. & Yentsch, C.M. (1986) Selective feeding by *Balanion* sp. (Ciliata: Balanionidae) on phytoplankton that best support its growth. *Journal of Experimental Marine Biology and Ecology* **95**, 113–130.

Stoecker, D.K., Michaels, A.E. & Davis, L.H. (1987a) Grazing by the jellyfish, *Aurelia aurita*, on microzooplankton. *Journal of Plankton Research*, **9**, 901–915.

Stoecker, D.K., Verity, P.G., Michaels, A.E. & Davis, L.H. (1987b) Grazing by larval and post-larval ctenophores on microzooplankton. *Journal of Plankton Research*, **9**, 667–683.

Stoecker, D.K., Taniguchi, A. & Michaels, A.E. (1989) Abundance of autotrophic, mixotrophic and heterotrophic planktonic ciliates in shelf and slope waters. *Marine Ecology Progress Series*, **50**, 241–254.

Stoecker, D.K., Gallager, S.M., Langdon, C.J. & Davis, L.H. (1995) Particle capture by *Favella* sp. (Ciliata, Tintinnina). *Journal of Plankton Research*, **17**, 1105–1124.

Stoecker, D.K., Gustafson, D.E. & Verity, P.G. (1996) Micro- and mesoprotozooplankton at 140° W in the equatorial Pacific: heterotrophs and mixotrophs. *Aquatic Microbial Ecology*, **10**, 273–282.

*Stoecker, D.K., Stevens, K. & Gustafson, D.E. (2000) Grazing on *Pfiesteria piscicida* by microzooplankton. *Aquatic Microbial Ecology*, **22**, 261–270.

Stoecker, D.K., Parrow, M.W., Burkholder, J.M. & Glasgow, H.B. Jr (2002) Grazing by microzooplankton on *Pfiesteria piscicida* cultures with different histories of toxicity. *Aquatic Microbial Ecology*, **28**, 79–85.

Stoecker, D.K., Johnson, M.D., de Vargas, C. & Not, F. (2009) Acquired phototrophy in aquatic protists. *Aquatic Microbial Ecology*, **57**, 279–310.

Straile, D. (1997) Gross growth efficiencies of protozoan and metazoan zooplankton and their dependence on food concentration, predator–prey weight ratio, and taxonomic group. *Limnology and Oceanography*, **42**, 1375–1385.

Strathmann, R.R. & Grünbaum, D. (2006) Good eaters, poor swimmers: compromises in larval form. *Integrative and Comparative Biology*, **46**, 312–322.

Strelkow, A.A. & Wirketis, M.A. (1950) New planktonic infusoria (suborder Tintinnoinea) from Peter the Great Bay. *Doklady Akademii Nauk S.S.S.R.*, **74**, 389–391.

Strom, S.L. (2001) Light-aided digestion, grazing and growth in herbivorous protists. *Aquatic Microbial Ecology*, **23**, 253–261.

Strom, S.L. & Loukos, H. (1998) Selective feeding by protozoa: model and experimental behaviors and their consequences for population stability. *Journal of Plankton Research*, **20**, 831–846.

Strom, S.L. & Welschmeyer, N.A. (1991) Pigment-specific rates of phytoplankton growth and microzooplankton grazing in the open subarctic Pacific Ocean. *Limnology and Oceanography*, **36**, 50–63.

Strom, S.L., Morello, T.A. & Bright, K.J. (1998) Protozoan size influences algal pigment degradation during grazing. *Marine Ecology Progress Series*, **164**, 189–197.

Strom, S.L., Wolfe, G.V. & Bright, K.J. (2007) Responses of marine planktonic protists to amino acids: feeding inhibition and swimming behavior in the ciliate Favella sp. *Aquatic Microbial Ecology*, **47**, 107–121.

Strüder-Kypke, M.C. & Lynn, D.H. (2003) Sequence analyses of the small subunit rRNA gene confirm the paraphyly of oligotrich ciliates sensu lato and support the mono-

phyly of the subclasses Oligotrichia and Choreotrichia (Ciliophora, Spirotrichea). *Journal of Zoology, London* **260**, 87–97.
Strüder-Kypke, M.C. & Lynn, D.H. (2008) Morphological versus molecular data – phylogeny of tintinnid ciliates (Ciliophora, Choreotrichia) inferred from small subunit rRNA gene sequences. *Denisia*, **23**, 417–424.
Strüder-Kypke, M.C. & Lynn, D.H. (2010) Comparative analysis of the mitochondrial cytochrome c oxidase subunit I (COI) gene in ciliates (Alveolata, Ciliophora) and evaluation of its suitability as a biodiversity marker. *Systematics and Biodiversity*, **8**, 131–148.
Strüder-Kypke, M.C. & Montagnes, D.J.S. (2002) Development of web-based guides to planktonic protists. *Aquatic Microbial Ecology*, **27**, 203–207.
Sullivan, L.J. (2010) Gut evacution of larval *Mnemiopsis leidyi* A. Agassiz (Ctenophora, Lobata). *Journal of Plankton Research*, **32**, 69–74.
Sullivan, L.J. & Gifford, D.J. (2007) Growth and feeding rates of the newly hatched larval ctenophore *Mnemiopsis leidyi* A. Agassiz (Ctenophora, Lobata). *Journal of Plankton Research*, **29**, 949–965.
Sundt-Hansen, L.E., Olsen, Y., Stibor, H., Heldal, M. & Vadstein, O. (2006) Trophic cascades mediated by copepods, not nutrient supply rate, determine the development of picocyanobacteria. *Aquatic Microbial Ecology*, **45**, 207–218.
Suzuki, N., Kurihara, T. & Matsuoka, A. (2009) Sporogenesis of an extracellular cell chain from the spheroidal radiolarian host *Haliommilla capillaceum* (Haeckel), Polycystina, Protista. *Marine Micropaleontology*, **72**, 157–164.
Suzuki, T. (2000) Role of planktonic ciliates in partitioning of carbon flux in the ocean. *Oceanography in Japan*, **9**, 27–41.
Suzuki, T. & Taniguchi, A. (1995) Sinking rate of loricae of some common tintinnid ciliates. *Fisheries Oceanography*, **4**, 257–263.
Swanberg, N.R. & Anderson, O.R. (1981) *Collozoum caudatum* sp. nov.: A giant colonial radiolarian from equatorial and Gulf Steam waters. *Deep-Sea Research*, **28**, 1033–1047.
Swanberg, N.R. & Anderson, O.R. (1985) The nutrition of radiolarians: Trophic activity of some solitary Spumellaria. *Limnology and Oceanography*, **30**, 646–652.
Swanberg, N.R. & Caron, D.A. (1991) Patterns of sarcodine feeding in epipelagic oceanic plankton. *Journal of Plankton Research*, **13**, 287–312.
Swanberg, N.R., Harbison G.R. (1980) The ecology of *Collozoum longiforme*. sp. nov. a new colonial radiolarian from the equatorial Atlantic Ocean. *Deep-Sea Research*, **27**, 715–731.
Swanberg, N.R. & Bennett, P., Lindsey, J.L. & Anderson, O.R. (1986) A comparative study of predation in two Caribbean Radiolarian populations. *Marine Microbial Food Webs*, **1**, 105–118.
Swofford, D.L. (2002) PAUP*. Phylogenetic analysis using parsimony (* and other methods). Version 4.0b10. Sinauer Associates, Sunderland, Massachusetts.
Sykes, P.F. & Huntley, M.E. (1987) Acute physiological reactions of *Calanus pacificus* to selected dinoflagellates: direct observations. *Marine Biology*, **94**, 19–24.
Symonds, M.R.E. & Johnson, C.N. (2006) Range size–abundance relationships in Australian Passerines. *Global Ecology and Biogeography*, **15**, 143–152.
Tackx, M., Irigoien, X., Daro, N., Castel, J., Zhu, L., Zhang, X. & Nijs, J. (1995) Copepod feeding in the Westerschelde and the Gironde. *Hydrobiologia*, **311**, 71–83.
*Taguchi, S. (1976) Microzooplankton and seston in Akkeshi Bay, Japan. *Hydrobiologia*, **50**, 195–204.
*Takahashi, K. & Ling, H.Y. (1984) Particle selectivity of pelagic tintinnid agglutination. *Marine Micropaleontology*, **9**, 87–92.
*Takahashi, M. & Hoskins, K.D. (1978) Winter condition of marine plankton populations in Saanich Inlet, B.C., Canada. II. Micro-zooplankton. *Journal of Experimental Marine Biology and Ecology*, **32**, 27–37.
Tamigneaux, E., Mingelbier, M., Klein, B. & Legendre, L. (1997) Grazing by protists and seasonal changes in the size structure of protozooplankton and phytoplankton in a temperate nearshore environment (western Gulf of St. Lawrence, Canada). *Marine Ecology Progress Series*, **146**, 231–247.
Tamura, M., Katz, L.A. & McManus, G.B. (2011) Distribution and diversity of oligotrich and choreotrich ciliates across an environmental gradient in a large temperate estuary. *Aquatic Microbial Ecology*, **64**, 51–67.
*Taniguchi, A. (1977) Biomass and size composition of copepod nauplii and tintinnids in the Philippine Sea and the Celebes Sea, summer 1972. *Bulletin of the Plankton Society of Japan*, **24**, 1–10.
*Taniguchi, A. (1977) Distribution of microzooplankton in the Philippine Sea and the Celebes Sea in Summer, 1972. *Journal of the Oceanographic Society of Japan*, **33**, 82–89.
*Taniguchi, A. (1983) Microzooplankton distribution along a transverse section crossing a marked oceanographic front. *La Mer, Tokyo*, **21**, 95–101.
*Taniguchi, A. (1984) Microzooplankton biomass in the Arctic and Subarctic Pacific Ocean in the Summer. *Memoirs of the National Institute of Polar Research, Special Issue* **32**, 63–76.
*Taniguchi, A. & Hada, Y. (1981) *Eutintinnus haslae* n. sp., a new ciliated protozoa from the tropical Pacific and Indian Oceans. *La Mer, Tokyo*, **19**, 18–22.
*Taniguchi, A. & Kawakami, R. (1983) Growth rates of ciliate *Eutintinnus lususundae* and *Favella taraikaensis* observed in laboratory culture experiments. *Bulletin of the Plankton Society of Japan*, **30**, 33–40.
Taniguchi, A. & Kawakami, R. (1985) Feeding activity of a tintinnid ciliate *Favella taraikaensis* and its variability observed in laboratory cultures. *Marine Microbial Food Webs*, **1**, 17–34.
*Taniguchi, A. & Takeda, Y. (1988) Feeding rate and behavior of the tintinnid ciliate *Favella taraikaensis* observed with a

high speed VTR system. *Marine Microbial Food Webs*, **3**, 21–34.

Tappan, H. (1993) Tintiinnids. In *Fossil Prokaryotes and Protists* (J.H. Lipps ed.), pp. 285–303, Blackwell Scientific Publications, Boston, USA.

Tappan, H. & Loeblich, A.R. Jr (1968) Lorica composition of modern and fossil Tintinnida (ciliate Protozoa), systematicsm geological distribution, and some new tertiary taxa. *Journal of Paleontology*, **42**, 1378–1394.

Taylor, F.J.R. (1982) Symbiosis in marine microplankton. *Annales de l'Institut Océanographique de Paris*, **58** (Supplement), 61–90.

Taylor, F.J.R. (1987) Ecology of dinoflagellates: General and marine ecosystems. In *The Biology of Dinoflagellates* (F.J.R. Taylor ed.), pp. 399–501. Blackwell Scientific Publications, London.

*Thompson, G.A. & Alder, V.A. (2005) Patterns in tintinnid species composition and abundance in relation to hydrological conditions of the southwestern Atlantic during austral spring. *Aquatic Microbial Ecology*, **40**, 85–101.

*Thompson, G.A., Alder, V.A., Boltovskoy, D. & Brandini, F.P. (1999) Abundance and biogeography of tintinnids (Ciliophora) and associated microzooplankton in the Southwestern Atlantic Ocean. *Journal of Plankton Research*, **21**, 1265–1298.

*Thompson, G.A., Alder, V.A. & Boltovskoy, D. (2001) Tintinnids (Ciliophora) and other net microzooplankton (>30 um) in southwestern Atlantic shelf break waters. *P.S.Z.N.I. Marine Ecology*, **22**, 343–355.

Tiselius, P. (1989) Contribution of aloricate ciliates to the diet of *Acartia clausi* and *Centropages hamatus* in coastal waters. *Marine Ecology Progress Series*, **56**, 49–56.

*Tiselius, P. (1992) Behavior of *Acartia tonsa* in patchy environments. *Limnology and Oceanography*, **37**, 1640–1651.

*Toklu-Aliçh, B., Balkis, N., Toklu, A.S. & Höbek, A. (2010) First Record of *Amphorellopsis tetragona* (Protozoa: Ciliophora: Tintinnina) from the Sea of Marmara. *IUFS Journal of Biology*, **69**, 41–44.

*Travers, A. & Travers, M. (1971) Catalogue des tintinnides (Cilies Oligotriches) recoltes dans le Golfe de Marseille de 1962 a 1964. *Tethys*, **2**, 639–646.

*Travers, M. (1973) Le microplancton du Golfe de Marseille: Variations de la composition systematique et de la densité des populations. *Tethys*, **5**, 31–53.

Troedsson, C., Simonelli, P., Nagele, V., Nejstgaard, J.C. & Frischer, M.E. (2009) Quantification of copepod gut content by differential length amplification quantitative PCR (dla-qPCR). *Marine Biology*, **156**, 253–259.

Tsuda, A., Furuya, K. & Nemoto, T. (1989) Feeding of micro- and macrozooplankton at the subsurface chlorophyll maximum in the subtropical North Pacific. *Journal of Experimental Marine Biology and Ecology*, **132**, 41–52.

Tsujino, M. & Uchida, T. (2004) Fate of resting cysts of *Alexandrium* spp. ingested by *Perinereis nuntia* (Polychaeta) and *Theola fragilis* (Mollusca). *Journal of Experimental Marine Biology and Ecology*, **303**, 1–10.

*Tumantseva, N.I. (1983) Ecology of tintinnids in the equatorial Pacific and coastal Peruvian region. *Oceanology*, **23**, 244–248.

*Tumantseva, N.I. (1989) Role of mass species of infusoria in the consumption of phytoplankton in Antarctic and Subantarctic waters of the Pacific Ocean. *Oceanology*, **29**, 96–99.

Turchin, P. (2003) *Complex Population Dynamics: a Theoretical/Empirical Synthesis*. Princeton University Press, Princeton, USA.

Turner, J.T. (1981) Latitudinal patterns of calanoid and cyclopoid diversity in estuarine waters of eastern North America. *Journal of Biogeography*, **8**, 369–382.

Turner, J.T. (1984a) The feeding ecology of some zooplankters that are important prey items of larval fish. *NOAA Technical Report NMFS* **7**, U.S. Department of Commerce, National Oceanic and Atmospheric Adminstration, National Marine Fisheries Service.

Turner, J.T. (1984b) Zooplankton feeding ecology: contents of fecal pellets of the copepods *Eucalanus pileatus* and *Paracalanus quasimodo* from continental shelf waters of the Gulf of Mexico. *Marine Ecology Progress Series*, **15**, 27–46.

*Turner, J.T. & Anderson, D.M. (1983) Zooplankton grazing during dinoflagellate blooms in a Cape Cod embayment, with observations of predation upon tintinnids by copepods. *Pubblicazioni della Stazione Zoologica di Napoli I: Marine Ecology*, **4**, 359–374.

*Turner, J.T. & Tester, P.A. (1992) Zooplankton feeding ecology: bacterivory by metazoan microzooplankton. *Journal of Experimental Marine Biology and Ecology*, **160**, 149–167.

Turner, J.T., Doucette, G.J., Powell, C.L., Kulis, D.M., Keafer, B.A. & Anderson, D.M. (2000) Accumulation of red tide toxins in larger size fractions of zooplankton assemblages from Massachusetts Bay, USA. *Marine Ecology Progress Series*, **202**, 95–107.

Turner, J.T., Doucette, G.J., Keafer, B.A. & Anderson, D.M. (2005) Trophic accumulation of PSP toxins in zooplankton during *Alexandrium fundyense* blooms in Casco Bay, Gulf of Maine, April–June 1998. II. Zooplankton abundance and size-fractionated community composition. *Deep-Sea Research II*, **52**, 2784–2800.

Turriff, N., Runge, J.A. & Cembella, A.D. (1995) Toxin accumulation and feeding behaviour of the planktonic copepod Calanus finmarchicus exposed to the red-tide dinoflagellate *Alexandrium excavatum*. *Marine Biology*, **123**, 55–64.

Uchida, T. Matsuyama, Y., Yamaguchi, M. & Honjo, T. (1996) The life cycle of *Gyrodinium instriatum* (Dinophyceae) in culture. *Phycological Research*, **44**, 119–123.

Uchida, T., Kamiyama, T. & Matsuyama, Y. (1997) Predation by a photosynthetic dinoflagellate *Gyrodinium instriatum* on loricated ciliates. *Journal of Plankton Research*, **19**, 603–608.

Urban, J.L., McKenzie, C.H. & Deibel, D. (1992) Seasonal differences in content of *Oikopleura vanoeffeni* and *Calanus finmarchicus* faecal pellets: illustrations of zooplankton food web shifts in coastal Newfoundland waters. *Marine Ecology Progress Series*, **84**, 255–264.

*Uribe, E., Castillo, J. (1982) Tintinnidos indicadores de masas agua. *Investnes Mar., Valparaįso* **10**, 15–34.

*Urrutxurtu, I. (2004) Seasonal succession of tintinnids in the Nervión River estuary, Basque Country, Spain. *Journal of Plankton Research*, **26**, 307–314.

Van Breemen, P.J. (1905) Plankton van Noordzee en Zuiderzee. PhD Thesis, University of Amsterdam, Netherlands.

Van Der Meeren, T. & Naess, T. (1993) How does cod (*Gadus morhua*) cope with variability in feeding conditions during early larval stages. *Marine Biology*, **116**, 637–647.

van Hengstum, P.J. & Scott, D.B. (2011) Ecology of foraminifera and habitat variability in an underwater cave; distinguishing anchialine versus submarine cave environments. *Journal of Foraminiferal Research*, **41**, 201–229.

Vanhöffen, E. (1897) Die Fauna und Flora Grönlands. In *Grönland-Expedition der Gesellschaft für Erdkunde zu Berlin 1891–1893*, Vol. 2 (E. von Drygalski ed.).

Vannini, C., Petroni, G., Schena, A., Verni, F. & Rosati, G. (2003) Well-established mutualistic associations between ciliates and prokaryotes might be more widespread and diversified than so far supposed. *European Journal of Protistology*, **39**, 481–485.

Van Wagtendonk, W.J. (1955) Encystment and excystment of protozoa. In *Biochemistry and Physiology of Protozoa*. (S.H. Hunter and A. Lwoff eds), Vol. II, pp. 85–90, Academic Press, New York, USA.

Vargas, C.A. & Gonzalez, H.E. (2004) Plankton community structure and carbon cycling in a coastal upwelling system. I. Bacteria, microprotozoans and phytoplankton in the diet of copepods and appendicularians. *Aquatic Microbial Ecology*, **34**, 151–164.

Vargas, C.A. & Martínez, R.A. (2009) Grazing impact of natural populations of ciliates and dinoflagellates in a river-influenced continental shelf. *Aquatic Microbial Ecology*, **56**, 93–108.

*Verity, P.G. (1984) The Physiology and Ecology of Tintinnids in Narragansett Bay, Rhode Island. Ph.D. Dissertation, University of Rhode Island, Kingston, USA.

Verity, P.G. (1985) Grazing, respiration, excretion, and growth rates of tintinnids. *Limnology and Oceanography*, **30**, 1268–1282.

*Verity, P.G. (1986a) Growth rates of natural tintinnid populations in Narragansett Bay. *Marine Ecology Progress Series*, **29**, 117–126.

Verity, P.G. (1986b) Grazing of phototrophic nanoplankton by microzooplankton in Narragansett Bay. *Marine Ecology Progress Series*, **29**, 105–115.

*Verity, P.G. (1987) Abundance, community composition, size distribution, and production rates of tintinnids in Narragansett Bay, Rhode Island. *Estuarine, Coastal and Shelf Science*, **24**, 671–690.

Verity, P.G. (1988) Chemosensory behaviour in marine ciliates. *Bulletin of Marine Science*, **43**, 772–782.

Verity, P.G. (1991a) Measurement and simulation of prey uptake by marine planktonic ciliates fed plastidic and aplastidic nanoplankton. *Limnology and Oceanography*, **36**, 729–750.

Verity, P.G. (1991b) Feeding in planktonic protozoans: evidence for non-random acquisition of prey. *Journal of Protozoology*, **38**, 69–76.

Verity, P.G. & Langdon, C. (1984) Relationships between lorica volume, carbon, nitrogen, and ATP content of tintinnids in Narragansett Bay. *Journal of Plankton Research*, **6**, 859–868.

Verity, P.G. & Stoecker, D.K. (1982) Effects of *Olisthodiscus luteus* on the Growth and Abundance of Tintinnids. *Marine Biology*, **72**, 79–87.

Verity, P.G. & Villareal, T.A. (1986). The relative food value of diatoms, dinoflagellates, flagellates and cyanobacteria for tintinnid cultures. *Archiv fur Protistenkunde*, **131**, 71–84.

Verity, P.G., Robertson, C.Y., Tronzo, C.R., Andrews, M.G., Nelson, J.R. & Sieracki, M.E. (1992) Relationships between cell volume and the carbon and nitrogen content of marine photosynthetic nanoplankton. *Limnology and Oceanography*, **37**, 1434–1446.

Verity, P.G., Stoecker, D.K., Sieracki, M.E., Burkhill, P.H., Edwards, E.A. & Tronzo, C.R. (1993) Abundance, biomass and distribution of heterotrophic dinoflagellates during the North Atlantic spring bloom. *Deep-Sea Research*, **40**, 227–244.

Verity, P.G., Paffenhofer, G.A., Wallace, D., Sherr, E. & Sherr, B. (1996) Composition and biomass of plankton in spring on the Cape Hatteras shelf, with implications for carbon flux. *Continental Shelf Research*, **16**, 1087–1116.

Verni, F. & Rosati, G. (2011) Resting cysts: a survival strategy in Protozoa Ciliophora. *Italian Journal of Zoology*, **78**, 134–145.

Verni, F., Rosati, G. & Nobili, R. (1977) Infection mechanism of eta killing particles in sensitive cells of the ciliate *Euplotes crassus*. *Transactions of the American Microscopical Society*, **96**, 363–369.

Verweij, G.L., Esselink, K., Fockens, K. & Koeman, R.P.T. (2005) Biomonitoring van microzooplankton in de Nederlandse zoute watern 204. Rapport 2005-024, Koeman en Bijkerk Ecologisch Onderzoek en Advies, Haren, Netherlands.

*Vitello, P. (1964) Contribution a l'etude des tintinnides de la Baie d'Alger. *Pelagos*, **2**, 5–41.

Vogel, S. (1996) *Life in Moving Fluids: the Physical Biology of Flow*, 2nd Ed. Princeton University Press, Princeton, USA.

von Herbing, I.H. & Gallager, S.M. (2000) Foraging behavior in early Atlantic cod larvae (*Gadus morhua*) feeding on a protozoan (*Balanion* sp.) and a copepod (*Pseudodiatomus* sp). *Marine Biology*, **136**, 591–602.

Voss, H.-J. (1989) Vergleichende morphogenetische Untersuchung von 13 Arten der Gattung *Euplotes* (Ciliophora, Hypotrichida). *Archiv für Protistenkunde*, **137**, 331–344.

Waggoner, B.J. (1994) An aquatic microfossil assemblage from Cenomanian amber of France. *Lethaia*, **27**, 77–84.

*Wailes, G.H. (1925) Tintinnidae from the Strait of Georgia. B.C. *British Columbia Contributions to Canadian Biology, New Series* **2**, 533–539.

*Wang, C.C. (1936) Notes on Tintinnoinea from the Gulf of Pe-Hai. *Sinensia, Shanghai*, **8**, 353–370.

*Wang, C.C. & Nie, D. (1932) A survey of the marine protozoa of Amoy. *Contributions of the Biological Laboratory, Science Society of China Zoological Series*, **8**, 283–385.

Wasik, A. (1998) Antarctic tintinnids: their ecology, morphology, ultrastructure and polymorphism. *Acta Protozoologica*, **37**, 5–15.

*Wasik, A. & Mikołajczyk, E. (1990) Tintinnids near pack-ice between South Shetland and the South Orkney Islands (26 Dec. 1988–18 Jan. 1989). *Acta Protozoologica*, **29**, 229–244.

Wasik, A. & Mikołajczyk, E. (1992) The morphology and ultrastructure of the Antarctic ciliate, *Cymatocylis convallaria* (Tintinnina). *Acta Protozoologica*, **31**, 233–239.

*Wasik, A. & Mikołajczyk, E. (1994a) Annual cycle of tintinnids in Admiralty Bay with an emphasis on seasonal variability in *Cymatocylis affinis/convallaria* lorica morphology. *Journal of Plankton Research*, **16**, 1–8.

Wasik, A. & Mikołajczyk, E. (1994b) Infraciliature of *Cymatocylis affinis/convallaria* (Tintinnina). *Acta Protozoologica*, **33**, 79–85.

Wasik, A., Mikołajczyk, E. & Ligowski, R. (1996) Agglutinated loricae of some Baltic and Antarctic Tintinnina species (Ciliophora). *Journal of Plankton Research*, **18**, 1931–1940.

Wasik, A., Mikolajczyk, E. & Gołębiowska, M. (1997a) Morphology and microstructure of selected Tintinnina loricae. *Acta Protozoologica*, **36**, 31–38.

Wasik, A., Mikołajczyk, E., Gołębiowska, M. & Sikora, J. (1997b) X-ray analysis and cytochemical staining of some tintinnid loricae. *Acta Protozoologica*, **36**, 153–155.

Watanabe, M.M., Watanabe, M., Fukuyo, Y. (1982) Encystment and excystment of red tide flagellates. I. *Scrippsiella trochoidea*. *Research Report from the National Institute for Environrnental Studies*, **30**, 27–42.

Watras, C.J., Garcon, V.C., Olson, R.J., Chisholm, S.W. & Anderson, D.M. (1985) The effect of zooplankton grazing on estuarine blooms of the toxic dinoflagellate *Gonyaulax tamarensis*. *Journal of Plankton Research*, **7**, 891–908.

Watson, W. & Davis, R.L. Jr (1989) Larval fish diets in shallow coastal waters off San Onofre, California. *Fishery Bulletin*, **87**, 569–591.

White, A.W. (1981) Marine zooplankton can accumulate and retain dinoflagellate toxins and cause fish kills. *Limnology and Oceanography*, **26**, 103–109.

Wickham, S.A. (1995) Cyclops predation on ciliates: species-specific differences and functional responses. *Journal of Plankton Research*, **17**, 1633–1646.

Wille, J.J. Jr, Weidner, E. & Steffens, W.L. (1981) Intranuclear parasitism of the ciliate *Euplotes* by a trypanosomatid flagellate. *Journal of Protozoology*, **28**, 223–227.

Williams, R., McCall, H., Pierce, R.W. & Turner, J.T. (1994) Speciation of the tintinnid genus *Cymatocylis* by morphometric analysis of the loricae. *Marine Ecology Progress Series*, **107**, 263–272.

Williamson, M. (1997) Marine biodiversity in its global context. In *Marine Biodiversity: Patterns and Processes* (R.F.G. Ormond, J.D. Gage, J.D. and M.V. Angel eds), pp. 1–17, Cambridge University Press, Cambridge, UK.

Wilson, C.B. (1942) The copepods of the plankton gathered during the last cruise of the Carnegie. Carnegie Institution of Washington, Publication 536, Washington DC, USA.

*Winter, A., Stockwell, D. & Hargraves, P.E. (1986) Tintinnid agglutination of coccoliths: a selective or random process? *Marine Micropaleontology*, **10**, 375–379.

Wittenrich, M.L., Rhody, N.R., Turingan, R.G. & Main, K.L. (2009) Coupling osteological development of the feeding apparatus with feeding performance in common snook, *Centropomus undecimalis*, larvae: identifying morphological constraints to feeding. *Aquaculture*, **294**, 221–227.

Won, M.-Z. & Below, R. (1999) Cambrian Radiolaria from the Georgina basin, Queensland, Australia. *Micropaleontology* **45**, 325–363.

Woodd-Walker, R.S., Ward, P. & Clarke, A. (2002) Large-scale patterns in diversity and community structure of surface water copepods from the Atlantic Ocean. *Marine Ecology Progress Series*, **236**, 189–203.

Wright, A.-D.G. & Lynn, D.H. (1997) Maximum ages of ciliate lineages estimated using a small subunit rRNA molecular clock: crown eukaryotes date back to the paleoprotoerozoic. *Archive fur Protistenkunde*, **148**, 329–341.

Xu, D., Song, W., Warren, A., Roberts, D. & Hu, X. (2007) Redescriptions of two marine planktonic ciliates from China, *Parastrombidium faurei* (Kahl, 1932) Maeda, 1986 and *Strombidium capitatum* (Leegaard, 1915) Kahl, 1932 (Ciliophora, Oligotrichea). *European Journal of Protistology*, **43**, 27–35.

Xu, D., Sun, P., Shin, M.K. & Kim, Y.O. (2012) Species boundaries in tintinnid ciliates: a case study – morphometric variability, molecular characterisation, and temporal distribution of *Helicostomella* species (Ciliophora, Tintinnina). *Journal of Eukaryotic Microbioplogy*, **59**, in press.

Xu, K., Choi, J.K., Lei, Y. & Yang, E.J. (2011) Marine ciliate community in relation to eutrophication of coastal waters in the Yellow Sea. *Chinese Journal of Oceanology and Limnology*, **29**, 118–127.

Yamaguchi, M., Itakura, S., Imai, I. & Ishida, Y. (1995) A rapid and precise technique for enumeration of resting cysts of *Alexandrium* spp. (Dinophyceae) in natural sediments. *Phycologia*, **34**, 207–214.

Yamaguchi, M., S. Itakura, K. Nagasaki & Y. Kotani (2002). Distribution and abundance of resting cysts of the toxic *Alexandrium* spp. in sediments of the western Seto Inland Sea, Japan. *Fisheries Science*, **68**, 1012–1019.

*Yamamoto, T., Taniguchi, A. & Nishizawa, S. (1981) Microplankton distribution at an oceanic front formed in the Sanriku Waters off Northeast Japan. *Bulletin of the Plankton Society of Japan*, **28**, 111–120.

Yang, E.J., Choi, J.K. & Hyun, J.-H. (2008) Seasonal variation in the community and size structure of nano- and microzooplankton in Gyeonggi Bay, Yellow Sea. *Estuarine, Coastal and Shelf Science*, **77**, 320–330.

Yang, E.J., Ju, S-J., Choi, J.-K. (2010) Feeding activity of the copepod *Acartia hongi* on phytoplankton and micro-zooplankton in Gyeonggi Bay, Yellow Sea. *Estuarine, Coastal and Shelf Science*, **88**, 292–301.

Yoo, K.-I. & Kim, Y.-O. (1990) Taxonomical studies on tintinnids (Protozoa: Ciliata) in Korean coastal waters 2. Yongil Bay. *Bulletin of the Korean Fisheries Society*, **6**, 87–122.

*Yoo, K.-I. & Lee, J.-B. (1987) On the trophic correlation between tintinnids and dinoflagellates in Masan Bay, Korea. *Bulletin of the Korean Fisheries Society*, **20**, 230–236.

Yoo, K.I., Kim, Y.-O. & Kim, D.-Y. (1988) Taxonomical studies on tintinnids (Protozoa: Ciliata) in Korean coastal waters. 1. Chinhae Bay. *Bulletin of the Korean Fisheries Society*, **4**, 67–90.

Young, J.R. & Geisen, M. (2002) Xenospheres – associations of coccoliths resembling coccospheres. *Journal of Nannoplankton Research*, **24**, 27–34.

Yuasa, K., Kamaishi, T., Koh-ichiro, M., Hutapea, J.H., Permana, G.N. & Nakazawa, A. (2007) Infection by a protozoan endoparasite of the genus *Ichthyodinium* in embryos and yolk-sac larvae of yellowfin tuna *Thunnus albacares*. *Fish Pathology*, **42**, 59–66.

Zarauz, L., Irigoien, X. & Fernandes, J.A. (2009) Changes in the plankton size structure and composition, during the generation of a phytoplankton bloom, in the central Cantabrian sea. *Journal of Plankton Research*, **31**, 193–207.

*Zeitzschel, B. (1982) Zoogeography of pelagic marine protozoa. *Annales de L'institute Oceanographique, Paris* **58** (S), 91–116.

Zeldis, J., Robinson, K., Ross, A. & Hayden, B. (2004) First observations of predation by New Zealand greenshell mussels (*Perna canaliculus*) on zooplankton. *Journal of Experimental Marine Biology and Ecology*, **311**, 287–299.

*Zhang, C., Zhang, W., Xiao, T., Lü, R., Sun, S. & Song, W. (2008) Meso-scale spatial distribution of large tintinnids in early summer in southern Yellow Sea. *Chinese Journal of Oceanology and Limnology*, **26**, 81–90.

*Zhang, C., Zhang, W., Xiao, T., Lü, R., Sun, S. & Song, W. (2009) Wintertime meso-scale horizontal distribution of large tintinnids in the southern Yellow Sea. *Chinese Journal of Oceanology and Limnology*, **27**, 31–37.

Zhang, Q., Yi, Z., Xu, D., Al-Rasheid, K.A.S., Gong, J. & Song, W. (2010) Molecular phylogeny of oligotrich genera *Omegastrombidium* and *Novistrombidium* (Protozoa, Ciliophora) for the systematical relationships within Family Strombidiidae. *Chinese Journal of Oceanology and Limnology*, **28**, 769–777.

Zhang, W. & Wang, R. (2000) Summertime ciliate and copepod Nauplii distributions and micro-zooplankton herbivorous activity in the Laizhou Bay, Bohai Sea, China. *Estuarine, Coastal and Shelf Science*, **51**, 103–114.

*Zhang, W. & Wang, R. (2002) Short term dynamics of ciliate abundance in the Bohai Sea (China). *Chinese Journal of Oceanology and Limnology*, **20**, 135–141.

ORGANISM INDEX

The Biology and Ecology of Tintinnid Ciliates: Models for Marine Plankton, First Edition. Edited by John R. Dolan, David J.S. Montagnes, Sabine Agatha, D. Wayne Coats, and Diane K. Stoecker.